AF601887

Simplicial Methods for Higher Categories

Algebra and Applications

Volume 26

Algebra and Applications aims to publish well-written and carefully refereed monographs with up-to-date expositions of research in all fields of algebra, including its classical impact on commutative and noncommutative algebraic and differential geometry, K-theory and algebraic topology, and further applications in related domains, such as number theory, homotopy and (co)homology theory through to discrete mathematics and mathematical physics.

Particular emphasis will be put on state-of-the-art topics such as rings of differential operators, Lie algebras and super-algebras, group rings and algebras, Kac-Moody theory, arithmetic algebraic geometry, Hopf algebras and quantum groups, as well as their applications within mathematics and beyond. Books dedicated to computational aspects of these topics will also be welcome.

More information about this series at http://www.springer.com/series/6253

Simona Paoli

Simplicial Methods for Higher Categories

Segal-type Models of Weak n-Categories

Simona Paoli
Department of Mathematics
University of Leicester
Leicester, UK

ISSN 1572-5553 ISSN 2192-2950 (electronic)
Algebra and Applications
ISBN 978-3-030-05673-5 ISBN 978-3-030-05674-2 (eBook)
https://doi.org/10.1007/978-3-030-05674-2

Mathematics Subject Classification (2010): 18-XX, 55P-XX

This Springer imprint is published by the registered company Springer Nature Switzerland AG.
The registered company address is: Gewerbestrasse 11, 6330 Cham, Switzerland

To my parents

Preface

The theory of higher categories is a very active area of research which has penetrated diverse fields of science.

Historically the subject was motivated by questions in algebraic topology and mathematical physics, two areas where the most important applications are currently found. Algebraic geometry also makes use of higher categorical notions. More recently higher categories have found their way into logic and computer science, and are also starting to appear in algebra and representation theory. Higher categories can sometimes be used as a common language to describe complex phenomena occurring in these areas.

A plethora of different approaches to higher categories have been developed over the years. Each one represents certain relevant aspects of the abstract notion being modelled, often with a view to supporting a particular ecosystem of applications. At this stage no single approach suits all such contexts, and indeed one might doubt the viability of a universally applicable model. It appears instead that the most prudent approach is to continue the development of all of these important strands, relating them where necessary by explicit comparisons.

The purpose of this monograph is to introduce a new approach to working with higher categories: this is based on a simple higher categorical structure consisting of iterated internal categories (also called n-fold categories) as well as on a new paradigm to weaker higher categorical structures, which is the idea of weak globularity.

We show that our new model, called weakly globular n-fold categories, is suitably equivalent to a model of higher categories that has been studied in great depth, the one introduced by Tamsamani [126] and further studied by Simpson [119].

We achieve this comparison by developing a larger context of 'Segal-type models of weak n-categories', based on multi-simplicial structures, of which both the Tamsamani model and weakly globular n-fold categories are special cases.

The use of simplicial structures to capture higher coherence phenomena has a long history in algebraic topology, starting with the work of Graeme Segal [117], and then the study of categories enriched in simplicial sets by Dwyer and Kan

[48–50], Dwyer et al. [52] and others. More recently, simplicial techniques have underpinned the development of so-called (∞, n)-categories, and several models have been developed and studied by Bergner [23], Bergner and Rezk [27], Barwick and Kan [9], Lurie [89], Joyal [74], Rezk [109, 110] and others.

Simplicial models of (∞, ∞)-categories have been developed by Verity [137], building upon insights from the study of simplicial nerves of strict n-categories initiated by Street [123].

In this work we concentrate on higher structures in the 'truncated' case, where there are higher morphisms only in dimensions 0 up to n. This is intimately connected to the Postnikov tower in algebraic topology. In fact, the algebraic modelling of the building blocks of spaces, the n-types, which are Postnikov sections of spaces, is related to models of weak n-categories via the so-called 'homotopy hypothesis': a good model of weak n-categories should give an algebraic model of n-types in the weak n-groupoid case. We show that our models do satisfy the homotopy hypothesis.

There are long-standing open questions about weak n-categories, both within category theory and in its applications to homotopy theory: for instance the comparison between the simplicial and higher operadic models of higher categories and the algebraic description of the k-invariants of spaces and simplicial categories.

The present work provides a platform where these and other open questions can be studied, as we outline in the last chapter. These questions however go beyond the scope of this work, whose goal is to lay the foundations of this theory.

The potential of our model to tackle these open questions comes from one of the main novelties of our approach: the use of an entirely rigid structure, namely a subcategory of n-fold categories, to model weak n-categories.

The terminology 'rigid structure' refers to the fact that n-fold categories, being iterated internal categories, have associative and unital compositions in n different simplicial directions. In this sense, n-fold categories are a strict higher categorical structure, though they are not the same as strict n-categories, since the higher morphisms in dimensions 0 to up n do not form just a set. In our model, the higher morphisms in dimension k (for each $0 \leq k \leq n-2$) have themselves an $(n-1+k)$-fold categorical structure of a special type which is suitably equivalent to a discrete structure (that is, a set): we call this property the 'weak globularity condition'.

n-Fold structures were used in homotopy theory by Loday [85] for the modelling of connected $(n+1)$-types via cat^n-groups. The idea of weak globularity was first introduced by the author in [102] in an internal setting for the category of cat^n-groups: weakly globular cat^n-groups were shown in [102] to be algebraic models of connected $(n+1)$-types; weak globularity was extended and further studied by Blanc and the author in [29] in the context of general n-types, for which an analogue of Loday's model was not available.

However, none of these works captured the general categorical case. This case necessitates many novel ideas and techniques, such as the use of pseudo-functors to model higher structures and the construction of a rigidification functor from the Tamsamani model to weakly globular n-fold categories.

This work uses a blend of techniques from category theory and simplicial homotopy theory, reviewed in Part I. We therefore hope it will be accessible both to category theorists and to algebraic topologists.

This book is organized into four parts:

Part I Higher Categories: Introduction and Background.
This Part aims to provide the reader with a guide to the rest of the book. It contains a broad introduction to higher categories, some historical development of the notion of weak globularity and a non-technical overview of the main ideas and results we shall encounter. It also covers the main techniques that will be used from category theory and simplicial homotopy theory.

Part II The Three Segal-type Models and Segalic Pseudo-Functors.
In this Part we introduce the three Segal-type models, that is, the categories $\mathsf{Ta}^n_{\mathsf{wg}}$ of weakly globular Tamsamani n-categories and its subcategories Ta^n (Tamsamani n-categories) and $\mathsf{Cat}^n_{\mathsf{wg}}$ (weakly globular n-fold categories). We establish the relation between the category $\mathsf{Cat}^n_{\mathsf{wg}}$ and a class of pseudo-functors which we call Segalic pseudo-functors.

Part III Rigidification of Weakly Globular Tamsamani n-Categories.
The main goal of this Part is the construction of the rigidification functor from weakly globular Tamsamani n-categories to weakly globular n-fold categories.

Part IV Weakly Globular n-Fold Categories as a Model of Weak n-Categories.
This Part contains the construction of the discretization functor from weakly globular n-fold categories to Tamsamani n-categories, and the final results: the equivalence after localization of $\mathsf{Cat}^n_{\mathsf{wg}}$ and Ta^n, exhibiting $\mathsf{Cat}^n_{\mathsf{wg}}$ as a model of weak n-categories, and the proof of the homotopy hypothesis. The last chapter of this Part contains an outline of further directions of applications and of open questions arising from this work.

Leicester, UK Simona Paoli

Acknowledgements

This work was supported by the EU International Reintegration Grant HOMALGHIGH No 256341, of which I was principal investigator. It also received financial support from the University of Leicester, where I have worked since 2011 and which supported my study leave in 2015; from the Centre of Australian Category Theory at Macquarie University, which hosted me during August–December 2015; and from the University of Chicago, where I visited in 2016.

This work was presented in several talks, where I could gather invaluable feedback from colleagues. In particular I thank the Centre of Australian Category Theory for the opportunity to give a long series of talks at the Australian Category Seminar, and I am particularly indebted for useful feedback from its members, especially Michael Batanin, Steve Lack, Richard Garner, Ross Street, Mark Weber, and Dominic Verity.

I would also like to thank Peter May for giving me the chance to give a series of talks at the University of Chicago and for many useful comments.

I am grateful for the opportunity to present this material at international conferences and at seminars and for the discussions with experts in the field that followed. I thank in particular Joachim Kock, Martin Hyland, and Dorette Pronk for interesting discussions and feedback. I also thank David Blanc, Frank Neumann, Alexander Kurz and Dominic Verity for reading the introductory part of the manuscript and for their helpful comments, and Nicola Gambino for some bibliographical suggestions.

I thank my colleagues in the Department of Mathematics of the University of Leicester for their ongoing support and encouragement.

I express my deep gratitude to the four reviewers of this book, for their careful reading and numerous helpful comments and suggestions.

I finally thank my parents for their invaluable moral support throughout my mathematical life.

ACKNOWLEDGMENTS

Contents

Part I Higher Categories: Introduction and Background

1 An Introduction to Higher Categories 3
1.1 Motivation and Context 3
1.2 Different Types of Higher Structures 6
1.2.1 ω-Categories 6
1.2.2 Truncated Higher Categories 7
1.2.3 Strict Versus Weak n-Categories 7
1.2.4 n-Fold Categories 8
1.2.5 n-Fold Structures Versus Strict and Weak n-Categories 10
1.3 The Homotopy Hypothesis 11
1.3.1 Homotopy Types and Their Algebraic Models 12
1.3.2 Modelling Homotopy Types with n-Fold Structures 14

2 Multi-Simplicial Techniques 17
2.1 Multi-Simplicial Objects and Segal Maps 18
2.1.1 Simplicial Objects and Their Segal Maps 18
2.1.2 Multi-Simplicial Objects 20
2.2 Multi-Simplicial Sets 23
2.2.1 The Functors $p^{(r)}$ and $q^{(r)}$ 24
2.2.2 Closure Properties 27
2.3 n-Fold Internal Categories 27
2.4 Multi-Nerve Functors 29
2.5 n-Fold Categories 34
2.6 A Multi-Simplicial Description of Strict n-Categories 39
2.7 The Functor Décalage 46

3 An Introduction to the Three Segal-Type Models 49
3.1 Geometric Versus Higher Categorical Equivalences 50
3.2 Multi-Simplicial Structures as an Environment for Higher Categories 51

3.3 The Idea of Weak Globularity 53
3.4 The Three Segal-Type Models 54
3.4.1 Notational Conventions 56
3.4.2 Common Features of the Three Segal-Type Models 57
3.4.3 Main Results 60
3.4.4 Organization of This Work 65
3.4.5 Informal Discussions 66

4 Techniques from 2-Category Theory 71
4.1 Some Functors on Cat 72
4.2 Pseudo-Functors and Their Strictification 78
4.2.1 Adjunctions and Equivalences in 2-Categories 79
4.2.2 The Notion of Pseudo-Functor 79
4.2.3 Pseudo T-Algebras 81
4.2.4 Strictification of Pseudo-Functors 82
4.3 Transport of Structure 83

Part II The Three Segal-Type Models and Segalic Pseudo-Functors

5 Homotopically Discrete n-Fold Categories 91
5.1 The Definition of Homotopically Discrete n-Fold Categories 92
5.1.1 The Idea of a Homotopically Discrete n-Fold Category 92
5.1.2 The Formal Definition of $\mathsf{Cat}^n_{\mathsf{hd}}$ 93
5.1.3 Homotopically Discrete n-Fold Categories As Internal Equivalence Relations 95
5.2 Properties of Homotopically Discrete n-Fold Categories 99
5.2.1 Closure Properties of $\mathsf{Cat}^n_{\mathsf{hd}}$ 100
5.2.2 n-Equivalences in $\mathsf{Cat}^n_{\mathsf{hd}}$ 102
5.3 Homotopically Discrete n-Fold Categories and 0-Types 105

6 The Definition of the Three Segal-Type Models 107
6.1 Weakly Globular Tamsamani n-Categories 108
6.1.1 The Idea of Weakly Globular Tamsamani n-Categories 108
6.1.2 Closure Properties 109
6.1.3 The Formal Definition of the Category $\mathsf{Ta}^n_{\mathsf{wg}}$ 113
6.2 Tamsamani n-Categories 117
6.3 Weakly Globular n-Fold Categories 118
6.3.1 The Idea of Weakly Globular n-Fold Categories 118
6.3.2 The Formal Definition of the Category $\mathsf{Cat}^n_{\mathsf{wg}}$ 119

7 Properties of the Segal-Type Models 127
7.1 Properties of Weakly Globular Tamsamani n-Categories 128
7.1.1 Properties of n-Equivalences 128
7.1.2 The Functor $q^{(n-1)}$ 135
7.1.3 Pullback Constructions Using $q^{(n-1)}$ 138

7.2 Properties of Weakly Globular n-Fold Categories 144
7.2.1 Weakly Globular n-Fold Categories and n-Equivalences 145
7.2.2 A Criterion for an n-Fold Category to Be Weakly Globular 146
7.2.3 A Geometric Interpretation.................................. 154

8 Pseudo-Functors Modelling Higher Structures 163
8.1 The Definition of a Segalic Pseudo-Functor 164
8.1.1 Notational Conventions for Segalic Pseudo-Functors 164
8.1.2 The Idea of a Segalic Pseudo-Functor 165
8.1.3 The Formal Definition of a Segalic Pseudo-Functor...... 166
8.2 Strictification of Segalic Pseudo-Functors 169

Part III Rigidification of Weakly Globular Tamsamani n-Categories

9 Approximating Weakly Globular Tamsamani n-Categories by Simpler Ones 181
9.1 The Category LTa^n_{wg} 182
9.1.1 The Idea of the Category LTa^n_{wg} 182
9.1.2 The Formal Definition of the Category LTa^n_{wg} 183
9.1.3 Properties of the Category LTa^n_{wg} 184
9.1.4 Cat^n_{wg} and the Category LTa^n_{wg} 186
9.2 Approximating Ta^n_{wg} by LTa^n_{wg} 188
9.2.1 The Main Steps in Approximating Ta^n_{wg} by LTa^n_{wg} 188
9.2.2 Approximating Ta^n_{wg} by LTa^n_{wg}: The Formal Proofs 190

10 Rigidifying Weakly Globular Tamsamani n-Categories................ 201
10.1 From LTa^n_{wg} to Pseudo-Functors.................................... 203
10.1.1 The Idea of the Functor Tr_n 203
10.1.2 The Formal Construction of the Functor Tr_n 205
10.2 Rigidifying Weakly Globular Tamsamani n-Categories........... 214
10.2.1 The Rigidification Functor Q_n: Main Steps............... 214
10.2.2 The Rigidification Functor: The Formal Proof........... 215

Part IV Weakly Globular n-Fold Categories as a Model of Weak n-Categories

11 Functoriality of Homotopically Discrete Objects 223
11.1 A Construction on Cat^n_{wg} 225
11.1.1 The Idea of the Construction $X(f_0)$ 226
11.2 Weakly Globular n-Fold Categories and Functoriality of Homotopically Discrete Objects.................................... 239
11.2.1 The Idea of the Functors V_n and F_n 239
11.2.2 The Functors V_n and F_n 242

11.3 The Category FCat^n_{wg} 257
11.3.1 The Idea of the Category FCat^n_{wg} 257
11.3.2 The Formal Definition of the Category FCat^n_{wg} 259
11.3.3 The Idea of the Functor G_n 263
11.3.4 The Functor G_n: The Formal Proof 265

12 Weakly Globular n-Fold Categories as a Model of Weak n-Categories 275
12.1 From FCat^n_{wg} to Tamsamani n-Categories 277
12.1.1 The Idea of the Functor D_n 277
12.1.2 The Functor D_n: Definition and Properties 279
12.2 The Discretization Functor and the Comparison Result 285
12.2.1 The Idea of the Functor $Disc_n$ 286
12.2.2 The Comparison Result 287
12.3 Groupoidal Weakly Globular n-Fold Categories 298
12.4 An Alternative Fundamental Functor 303
12.4.1 The Functor $\mathscr{H}_n$ 304
12.4.2 Some Examples 308

13 Conclusions and Further Directions 315
13.1 Algebraic Description of Postnikov Systems 316
13.2 Model Comparisons 317
13.3 Intermediate Levels of Weakness 318
13.4 Model Structures 319
13.5 A Weakly Globular Approach to (∞, n)-Categories 319

A Proof of Lemma 10.1.4 321

References 335

Index 341

List of Symbols

$[\mathscr{A}, \mathscr{B}]$	Category of functors from $\mathscr{A}$ to $\mathscr{B}$ and natural transformations, Sect. 1.2.4
$A[f]$	Internal equivalence relation corresponding to $f : A \to B$, Definition 5.1.8
$\widetilde{B}$	Functor $\mathsf{GCat}^{\mathsf{n}}_{\mathsf{wg}}/\!\sim\ \to$ Ho(n-types), Remark 12.4.5
B	Classifying space. Definition 5.3.1, Sects. 12.3, 12.4
$\mathsf{Cat}^{\mathsf{n}}_{\mathsf{hd}}$	Category of homotopically discrete n-fold categories, Definition 5.1.2
$\mathsf{Cat}\,\mathscr{C}$	Category of internal categories in $\mathscr{C}$, Definition 2.3.1
$\mathsf{Cat}^{\mathsf{n}}(\mathscr{C})$	Category of n-fold internal categories in $\mathscr{C}$, Sect. 2.3
$\mathsf{Cat}^{\mathsf{n}}$	Category of n-fold categories, Sect. 2.5
$\mathsf{Cat}^{\mathsf{n}}_{\mathsf{wg}}$	Category of weakly globular n-fold categories, Definition 6.3.3
Cube(n, t)	Set of (n, t)-hypercubes, Definition 7.2.18
Dec, Dec$'$	Functors décalage, Sect. 2.7
Δ	Simplicial category, Sect. 2.1.2
$\Delta^{n^{op}}$	Product of n copies of $\Delta^{^{op}}$, Sect. 2.1.2
$[\Delta^{^{op}}, \mathscr{C}]$	Category of simplicial objects in $\mathscr{C}$, Sect. 2.1.1
$[\Delta^{n^{op}}, \mathscr{C}]$	Category of n-fold simplicial objects in $\mathscr{C}$, Sect. 2.1.2
$[\Delta^{n^{op}}, \mathsf{Set}]$	Category of n-fold simplicial sets, Sect. 2.2
$d^{(n)}$	Functor $\mathsf{Cat}^{n-1}(\mathscr{C}) \to \mathsf{Cat}^{\mathsf{n}}(\mathscr{C})$, Definition 2.4.11
d_n	Functor $[\Delta^{n-1^{op}}, \mathscr{C}] \to [\Delta^{n^{op}}, \mathscr{C}]$, Definition 2.4.11
$Diag_n$	Multi-diagonal functor, Definition 5.3.1
$Disc_n$	Discretization functor, Definition 12.2.1
D_n	Functor $\mathsf{FCat}^{\mathsf{n}}_{\mathsf{wg}} \to \mathsf{Ta}^{\mathsf{n}}$, Proposition 12.1.4
F_n	Functor $\mathsf{Cat}^{\mathsf{n}}_{\mathsf{wg}} \to \mathsf{Cat}^{\mathsf{n}}_{\mathsf{wg}}$, Proposition 11.2.5
$\overline{F}$	Functor $[\mathscr{I}, \mathscr{C}] \to [\mathscr{I}, \mathscr{D}]$ for $F : \mathscr{C} \to \mathscr{D}$, Definition 2.1.1
$f_n(X)$	Map $F_n X \to X$, Proposition 11.2.5
$\mathsf{FCat}^{\mathsf{n}}_{\mathsf{wg}}$	Definition 11.3.1

$\mathsf{GCat}^{\mathsf{n}}_{\mathsf{wg}}$	Category of groupoidal weakly globular n-fold categories, Definition 12.3.6
$g_n(X)$	Map $G_n X \to X$, Theorem 11.3.6
G_n	Functor $\mathsf{Cat}^{\mathsf{n}}_{\mathsf{wg}} \to \mathsf{FCat}^{\mathsf{n}}_{\mathsf{wg}}$, Theorem 11.3.6
$\mathscr{G}_n$	Fundamental groupoidal weakly globular n-fold category functor, Sect. 12.4, Eq. (12.26)
$\mathsf{Gpd}\,\mathscr{C}$	Category of internal groupoids in $\mathscr{C}$, Definition 2.3.3
Gpd^n	Category of n-fold groupoids, Sect. 2.5
$\mathsf{Gpd}^{\mathsf{n}}_{\mathsf{wg}}$	Category of weakly globular n-fold groupoids, Sect. 12.4
$\mathsf{GSeg}_{\mathsf{n}}$	Groupoidal Segal-type model, Sect. 3.4.3
$\mathsf{GTa}^{\mathsf{n}}$	Category of groupoidal Tamsamani n-categories, Definition 12.3.6
$\mathsf{GTa}^{\mathsf{n}}_{\mathsf{wg}}$	Category of groupoidal weakly globular Tamsamani n-categories, Definition 12.3.1
$\mathscr{H}_n$	Alternative fundamental groupoidal weakly globular n-fold category functor, Definition 12.4.2
$h_n(X)$	Map $V_n(X) \to F_n(X)$, $X \in \mathsf{Cat}^{\mathsf{n}}_{\mathsf{hd}}$, Proof of Proposition 11.2.3
$\underline{k}(1,i)$, $\underline{k}(0,i)$	Remark 8.1.3
Ho(n-types)	Homotopy category of n-types, Sect. 1.3
J_n	Functor n-$\mathsf{Cat} \to [\Delta^{{n-1}^{op}}, \mathsf{Cat}]$, Definition 2.6.3
$\mathsf{LTa}^{\mathsf{n}}_{\mathsf{wg}}$	Definition 9.1.1
n-Cat	Category of strict n-categories, Definition 2.6.1
n-Gpd	Category of strict n-groupoids, Sect. 2.6
N	Nerve functor $N : \mathsf{Cat}\,\mathscr{C} \to [\Delta^{op}, \mathscr{C}]$, Sect. 2.4
$N^{(k)}$	Nerve functor in the kth direction, Definition 2.4.8
$N_{(n)}$	Multinerve functor, Definition 2.4.3
$\mathrm{Or}_{(2)}$	Functor $[\Delta^{op}, \mathsf{Set}] \to [\Delta^{2^{op}}, \mathsf{Set}]$, Fig. 12.4
$\mathrm{Or}_{(3)}$	Functor $[\Delta^{op}, \mathsf{Set}] \to [\Delta^{3^{op}}, \mathsf{Set}]$, Fig. 12.6
$\mathrm{Or}_{(n)}$	Functor $[\Delta^{op}, \mathsf{Set}] \to [\Delta^{n^{op}}, \mathsf{Set}]$, Sect. 12.4
or_n	Ordinal sum $\Delta^n \to \Delta$, Sect. 12.4
p	Functor $[\Delta^{n^{op}}, \mathsf{Set}] \to \mathsf{Set}$, Notational Convention 2.2.4
p_n	Functor $[\Delta^{n^{op}}, \mathsf{Set}] \to \mathsf{Set}$, Definition 2.2.3
$p^{(r)}$	Functor $\mathsf{Seg}_n \to \mathsf{Seg}_r$, Definition 6.3.3, Definition 6.1.8, Definition 5.1.2
$p^{(n)}$	Functor $\mathsf{SegPs}[\Delta^{n^{op}}, \mathsf{Cat}] \to \mathsf{Cat}^{\mathsf{n}}_{\mathsf{wg}}$, Definition 8.1.2
P_n	Functor $P_n : \mathsf{Ta}^{\mathsf{n}}_{\mathsf{wg}} \to \mathsf{LTa}^{\mathsf{n}}_{\mathsf{wg}}$, Proof of Theorem 10.2.1
$\mathscr{P}$	Homotopy category functor $[\Delta^{op}, \mathsf{Set}] \to \mathsf{Cat}$, Definition 2.2.2
$\mathscr{P}_n$	Left adjoint to the n-fold nerve, Sect. 12.4
$\mathsf{Ps}[\Delta^{n^{op}}, \mathsf{Cat}]$	Category of pseudo-functors, Sect. 4.2
q	Connected components functor, Notational Convention 2.2.4
q_n	Functor $[\Delta^{n^{op}}, \mathsf{Set}] \to \mathsf{Set}$, Definition 2.2.3
$q^{(r)}$	Functor $\mathsf{Seg}_n \to \mathsf{Seg}_r$, Proposition 7.1.7, Corollary 7.1.8

Q_n	Rigidification functor, Theorem 10.2.1
R_0	Functor $\mathsf{FCat}^n_{\mathsf{wg}} \to [\Delta^{op}, \mathsf{FCat}^{n-1}_{\mathsf{wg}}]$, Definition 12.1.1
$\mathscr{R}_n$	Composite $\mathscr{P}_n\,\mathrm{Or}_{(n)}$, Sect. 12.4.2
$\mathscr{S}$	Singular functor $\mathsf{Top} \to [\Delta^{op}, \mathsf{Set}]$, Sect. 12.4
Seg_n	Segal-type model, Sect. 3.4.2
$\mathsf{SegPs}[\Delta^{n^{op}}, \mathsf{Cat}]$	Category of Segalic pseudo-functors, Definition 8.1.2
$s_n(X)$	Map $Q_n X \to X$, Theorem 10.2.1
St	Strictification functor, Sect. 4.2
Ta^n	Category of Tamsamani n-categories, Definition 6.2.1
$\mathsf{Ta}^n_{\mathsf{wg}}$	Category of weakly globular Tamsamani n-categories, Definition 6.1.8
$t_n(X)$	Pseudo-natural transformation $Tr_n X \to X$, Theorem 10.1.1
$\mathscr{T}_n$	Fundamental Tamsamani n-groupoid functor, Theorem 12.3.11
Tr_n	Functor $\mathsf{LTa}^n_{\mathsf{wg}} \to \mathsf{SegPs}[\Delta^{n-1^{op}}, \mathsf{Cat}]$, Theorem 10.1.1
u_X	Map $\mathrm{Dec}\, X \to X$, Sect. 2.7
$v_n(X)$	Map $V_n X \to X$, Proposition 11.2.3
$v_k^{\{r\}}$	Map $X_k^{\{r\}} \to X_1^{\{r\}} \times_{p^{(n-2)}X_0^{\{r\}}} \overset{k}{\cdots} \times_{p^{(n-2)}X_0^{\{r\}}} X_1^{\{r\}}$, Sect. 9.1.2
V_n	Functor $\mathsf{Cat}^n_{\mathsf{hd}} \to \mathsf{Cat}^n_{\mathsf{hd}}$, Proposition 11.2.3
$V(X)$	Map $X(f_0) \to X$, Lemma 11.1.1, Proposition 11.1.5
$w_n(X)$	Map $P_n X \to X$, Proof of Theorem 10.2.1
$X(a,b)$	Hom-$(n-1)$-category of $X \in \mathsf{Seg}_n$, Notation 5.2.4, Definition 6.3.3, Definition 6.1.8
$X(f_0)$	Construction on $X \in \mathsf{Cat}\,\mathscr{C}$ and $f_0 : X'_0 \to X_0$, Lemma 11.1.1
X^d	Discretization of $X \in \mathsf{Cat}^n_{\mathsf{hd}}$, Definition 5.1.2
$X^\alpha, X^{\{r\}}$	Definition 2.1.5, Notational Convention 2.5.4
$\gamma_{(n)}$	Discretization map $X \to X^d$ for $X \in \mathsf{Cat}^n_{\mathsf{hd}}$, Definition 5.1.4
$\gamma^{(r)}$	Map $X \to q^{(r)}X$, Lemma 2.2.7, Remark 7.1.9, Remark 7.2.13
μ_k	kth Segal map, Definition 2.1.2
$\hat{\mu}_k$	kth induced Segal map, Definition 2.1.3
$\widetilde{\xi}_i$	Map $\mathsf{Cat}^n(\mathscr{C}) \to \mathsf{Cat}\,(\mathsf{Cat}^{n-1}(\mathscr{C}))$, Proposition 2.4.6
ξ_i	Map $[\Delta^{n^{op}}, \mathscr{C}] \to [\Delta^{op}, [\Delta^{n-1^{op}}, \mathscr{C}]]$, Lemma 2.1.7
Σ_n	Symmetric group, Definition 2.1.5
$\hat{\pi}_1^{(i)}$	Fundamental groupoid functor in direction i, Theorem 12.4.3
$\sim^n$	n-equivalences in Seg_n, Sect. 3.4.2

List of Figures

Fig. 2.1	Corner of a bisimplicial object X	23
Fig. 2.2	Corner of the double nerve of a double category X	37
Fig. 2.3	Geometric picture of the corner of the double nerve of a double category	38
Fig. 2.4	Corner of the 3-fold nerve of a 3-fold category X	39
Fig. 2.5	Geometric picture of the corner of the 3-fold nerve of a 3-fold category	40
Fig. 2.6	Corner of the double nerve of a strict 2-category X	43
Fig. 2.7	Geometric picture of the corner of the double nerve of a strict 2-category	44
Fig. 2.8	Corner of the 3-fold nerve of a strict 3-category X	45
Fig. 2.9	Geometric picture of the corner of the 3-fold nerve of a strict 3-category	46
Fig. 3.1	Diagram of connections between the topics in the list of informal discussions	69
Fig. 3.2	Summary of overall organization and main results	70
Fig. 4.1	The three Segal-type models and Segalic pseudo-functors	89
Fig. 6.1	Corner of the double nerve of a weakly globular double category X	122
Fig. 6.2	Geometric picture of the corner of the double nerve of a weakly globular double category	122
Fig. 6.3	Corner of the multinerve of a weakly globular 3-fold category X	124
Fig. 6.4	Geometric picture of the corner of the multinerve of a weakly globular 3-fold category	124
Fig. 7.1	$(r, 1)$-staircase of $(3, 1)$-hypercubes in $X \in \mathsf{Cat}^3_{\mathsf{wg}}$	160
Fig. 7.2	Lifting condition for the staircase in Fig. 7.1	160
Fig. 7.3	A $(3, 1)$-staircase of $(3, 2)$-hypercubes in $X \in \mathsf{Cat}^3_{\mathsf{wg}}$ with orientation $(1, 1, 0)$	161

Fig. 7.4 Lifting condition for the staircase in Fig. 7.3 161

Fig. 8.1 Picture of the corner of $X \in \mathsf{SegPs}[\Delta^{2^{op}}, \mathsf{Cat}]$ 168
Fig. 8.2 Picture of the corner of $p^{(2)}X$, for $X \in \mathsf{SegPs}[\Delta^{2^{op}}, \mathsf{Cat}]$ 169
Fig. 8.3 The construction of the rigidification functor Q_n 179

Fig. 10.1 Construction of the discretization functor 220
Fig. 10.2 $\mathsf{Cat}^{\mathsf{n}}_{\mathsf{wg}}$ as a model of weak n-categories 221

Fig. 12.1 Corner of $X \in \mathsf{FCat}^3_{\mathsf{wg}} \subset [\Delta^{2^{op}}, \mathsf{Cat}]$ 284
Fig. 12.2 Corner of $D_3X \in [\Delta^{2^{op}}, \mathsf{Cat}]$ for $X \in \mathsf{FCat}^3_{\mathsf{wg}}$ 284
Fig. 12.3 Corner of $\mathscr{L}_\bullet Y$ 309
Fig. 12.4 Corner of $\mathrm{Or}_{(2)}\, Y$ 310
Fig. 12.5 Corner of the double nerve of $\mathscr{H}_2 X$ for $Y = \mathscr{S}X$ 311
Fig. 12.6 Corner of $\mathrm{Or}_{(3)}\, Y$ 312
Fig. 12.7 Corner of the 3-fold nerve of $\mathscr{H}_3 X$, with $Y = \mathscr{S}X$ 312

Part I
Higher Categories: Introduction and Background

This Part aims to provide the reader with a guide to the rest of the book. Chapter 1 contains an introduction to higher categories while Chap. 3 consists of an introduction to the three Segal-type models studied herein: the Tamsamani model and the two new models we introduce here, called weakly globular n-fold categories and weakly globular Tamsamani n-categories.

The techniques used throughout draw from two sources: simplicial homotopy theory and category theory. This Part reviews these techniques, more precisely we cover multi-simplicial techniques in Chap. 2, and the 2-categorical background in Chap. 4.

The reader who wishes to obtain an overall picture of this work may want at first to read Chap. 3 straight after Chap. 1: however, the material in Chap. 2 establishes the notation that some of Chap. 3 refers to and that is used in the rest of the book and it is therefore necessary for a more detailed reading.

Our overview of higher categories in Chap. 1 aims to highlight the open questions that led to our approach to weak higher categories and also provides some context behind their development. We do not aim to give a comprehensive detailed survey of different models of weak higher categories, and we have provided several bibliographical references where further information can be found.

After giving some motivation and context, we describe in Chap. 1 the main classes of higher structures and highlight their differences: strict versus weak higher structures, truncated versus non-truncated, and the class of n-fold categories, which is central to this work.

We then concentrate in Sect. 1.3 on one of the most important connections between higher category theory and homotopy theory, which is the algebraic modelling of the building blocks of spaces, the n-types. Once again, rather than giving a detailed comprehensive survey of all the different models, we point the reader towards bibliographical references and then concentrate on a model of path-connected $(n+1)$-types via n-fold structures due to Loday, called cat^n-groups, which developed independently of models of weak higher categories. The question of how cat^n-groups compare to models of weak higher groupoids is what first led the author to introduce the notion of weak globularity, in an internal context inside the category of groups.

We begin Chap. 2 by recalling in Sect. 2.1 the central notion of Segal maps for (multi-simplicial) objects and fix our notation for multi-simplicial objects. In the case of multi-simplicial sets, we introduce notational conventions and special functors used throughout the book. We then review in Sect. 2.3 internal n-fold categories and in Sect. 2.4 their multi-nerves. In Sect. 2.5 we specialize to n-fold categories and we introduce a notational convention related to their multi-nerves.

We use these multi-simplicial notions to give in Sect. 2.6 a multi-simplicial description of strict n-categories. This description is very important when building a geometric intuition around the Segal-type models we will encounter. For this reason, we illustrate in detail some low-dimensional cases and draw corresponding pictures. In Sect. 2.7 we recall the functor décalage, which is used in Chaps. 11 and 12.

In Chap. 3 we explain the idea of weak globularity in the general categorical context. Our aim in this chapter is to convey some of the ideas behind our constructions and to give a summary of the main results. In particular, in Sect. 3.4.2 we give an account of the main common features of the three Segal-type models we shall consider. Section 3.4.4 ends with a description of the overall organization of the rest of the book, and with some diagrammatic summaries in Figs. 3.1 and 3.2.

We begin Chap. 4 by recalling in Sect. 4.1 two important functors from Cat to Set, the isomorphism classes of objects and the connected components functors, which play a crucial role in the Segal-type models.

The last part of Chap. 4 reviews the notion of pseudo-functors and their strictification, as well as a standard technique to produce pseudo-functors, which is an instance of 'transport of structure along an adjunction' in the sense of [78]. These techniques play a crucial role in this work, in relation to the new notion of Segalic pseudo-functors studied in Chaps. 8 and 9.

The material in Chaps. 2 and 4 is essentially known, but we have presented it in a way that is best suited for the remaining material, emphasizing the multi-simplicial description of strict n-categories and n-fold categories: the latter is not always spelled out at this level of detail in the literature, and it is crucial for building an intuition around our Segal-type models.

Suitable references for Chaps. 2 and 4 include [32, 63, 78, 81, 91, 94], and [106].

Chapter 1
An Introduction to Higher Categories

Abstract In this chapter we give a non-technical introduction to higher categories. We describe some of the contexts that inspired and motivated their development, explaining the idea of higher categories, and the different classes of higher structures. We discuss one of the most important occurrences of higher categories in algebraic topology, which is the algebraic modelling of homotopy types; in particular, we give an account of the use of internal n-fold structures in modelling path-connected $(n+1)$-types. This provides an historical development of the notion of weak globularity, which is central to this work.

1.1 Motivation and Context

The language of categories and functors permeates modern mathematics. In a category we have objects, morphisms, compositions of morphisms and identity morphisms for each object, such that compositions are associative and unital. When each morphism is invertible we obtain a groupoid. A one-object groupoid is the familiar notion of a group, while a one-object category is a monoid.

The maps between categories are the functors: These associate objects to objects and arrows to arrows in a way that is compatible with the composition and with the identities.

Many familiar mathematical structures form a category: for instance vector spaces and linear maps, topological spaces and continuous maps, and so on.

The idea of a higher category was prompted by several inputs. First of all, there are many natural examples of higher structures. An important one is the 2-category Cat of small categories. This structure comprises 0-dimensional data which are the categories (the objects), 1-dimensional data which are the functors (1-morphisms between the objects), and 2-dimensional data which are the natural transformations between functors (2-morphisms between 1-morphisms).

Functors can be composed: given functors $F : \mathscr{A} \to \mathscr{B}$ and $G : \mathscr{B} \to \mathscr{C}$ the composite functor $G \circ F : \mathscr{A} \to \mathscr{C}$ associates to each object $a \in \mathscr{A}$ the object $G(F(a)) \in ob\,(\mathscr{C})$ and to each morphism $f \in \mathscr{A}$ the morphism $G(F(f)) \in mor\,(\mathscr{C})$. Natural transformations can also be composed, but in two different ways.

S. Paoli, *Simplicial Methods for Higher Categories*, Algebra and Applications 26,
https://doi.org/10.1007/978-3-030-05674-2_1

Given functors $F, G, H : \mathscr{A} \to \mathscr{B}$ between categories $\mathscr{A}$ and $\mathscr{B}$, and natural transformations $\alpha : F \Rightarrow G$ and $\beta : G \Rightarrow H$, we can form a 'vertical' composite natural transformation $\beta \circ_v \alpha : G \to H$ with components given by the composites

$$(\beta \circ_v \alpha)_a = \beta_a \alpha_a : F(a) \to H(a)$$

for each object $a \in \mathscr{A}$. Given functors $F, F' : \mathscr{A} \to \mathscr{B}$ between categories $\mathscr{A}$ and $\mathscr{B}$, functors $G, G' : \mathscr{B} \to \mathscr{C}$ between categories $\mathscr{B}$ and $\mathscr{C}$ and natural transformations $\alpha : F \Rightarrow F'$ and $\beta : G \Rightarrow G'$, we can form the 'horizontal' composite natural transformation $\beta \circ_h \alpha : GF \to G'F'$ with components given by the composites

$$(\beta \circ_h \alpha)_a = (\beta_{F'(a)})\,(G\alpha_a) : GF(a) \to G'F'(a)$$

for each object $a \in \mathscr{A}$. We can build a geometric picture from these data by associating points to categories (the objects), arrows to functors (the 1-morphisms) and globes to natural transformations (the 2-morphisms), with the two different compositions of natural transformations pictured as vertical and horizontal compositions of globes, as illustrated below:

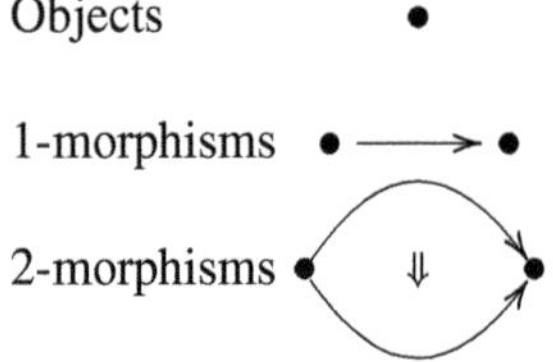

Vertical and horizontal compositions

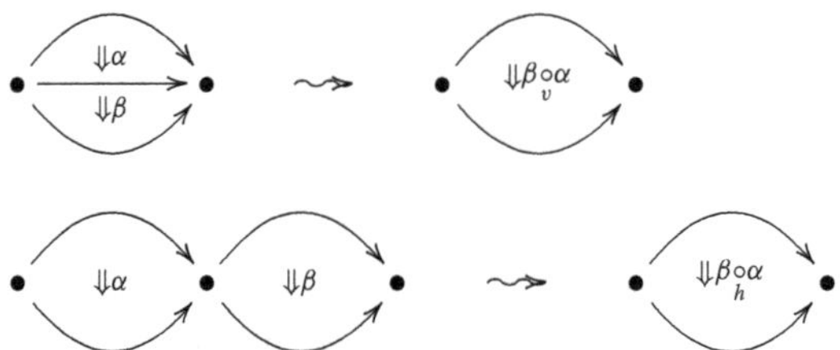

Every 2-category comprises data as above, and we also call the objects '0-cells', the 1-morphisms '1-cells', the 2-morphisms '2-cells'. So we see that every k-cell (for $k = 1, 2$) has a $(k - 1)$-cell as its source and target. The highest dimension of cells in this example is 2, and is called the 'dimension' of the higher structure.

The idea of a higher category in dimension greater than 2 is to have higher cells of globular shape, with each k-dimensional cell having a $(k - 1)$-cell as its source and target. When cells are present only in dimensions 0 up to n we say that the higher structure is n-dimensional. So for instance when $n = 3$ we can think of a 3-cell as a 'globe between globes', that is, a sphere.

Many higher categorical ideas have their root in the notion of homotopy coherence in algebraic topology. The latter developed along several directions, and has been used in diverse applications. One was the study of loop spaces, leading in an algebraic setting to the notions of H_∞ and E_∞ spaces: the works of Boardman and Vogt [31], May [96], Stasheff [121], Segal [116], and Sugawara [124] are relevant here. Operads also developed as a way to encode higher homotopy coherences, see for instance the works of May [95], Loday [86], and Markl et al. [93].

The abstraction and 'categorification' of these models of homotopy coherence led to several combinatorial approaches to higher categorical structures, in particular the Segal-type model of Simpson [119], Tamsamani [126] and the higher operadic models of Batanin [12], also studied by Leinster [84], Batanin et al. [17], Batanin et al. [16], and others.

Another way to encode higher order homotopical information is via the notion of model category pioneered by Quillen [108], and that of simplicial categories and their localizations, which was studied by Dwyer and Kan [48–50] and Dwyer et al. [51, 52].

Quillen model categories remain a key tool in algebraic topology and are increasingly tackled from a categorical perspective, as shown for instance in the works of Garner [61] and Riehl [111].

Simplicial categories are one of the models of a class of higher structures called $(\infty, 1)$-categories, which have become a central object of study in modern homotopy theory, encoding the idea of a 'homotopy theory of homotopy theories'. Several other models have also been developed, leading to a variety of applications, as further explained in the next section.

Algebraic structures in $(\infty, 1)$-categories led to the notion of ∞-operads: See the work of Lurie [90], as well as the dendroidal sets model of Moerdijk and Weiss [99], further studied by Heuts et al. [70], Cisinski and Moerdijk [42, 43], and the complete Segal operads of Barwick [7], also studied in [41]. A further extension of these approaches was developed by Hackney et al. [69].

Another motivating force for the development of higher category theory coming from algebraic topology was the algebraic modelling of the Postnikov systems of spaces. This is quite central to this work, as we explain in further detail in Sect. 1.3.

Mathematical physics has also inspired many developments in higher categories, in the pursuit of models for TQFT and higher cobordism categories. Several conjectures in this direction were formulated by Baez and Dolan [3], the cobordism hypothesis was recently tackled by Lurie [88], higher categories and low-dimensional TQFTs were investigated by Schommer-Pries [114], Schommer-Pries and Christopher [115], and an (∞, n)-category of bordisms was studied by Calanque and Scheimbauer [37]. See also [25].

Algebraic geometry also has seen the use of higher categorical ideas in the pursuit of the notion of higher and derived stacks as well as higher non-abelian cohomology: see for instance the works of Simpson [119], Hirschowitz and Simpson [71], Pridham [107], Moerdijk and Töen [98], Toën [129, 130], and Toën and Vezzosi [131, 132].

More recently higher categories entered logic and computer science in the area of homotopy type theory, see for instance the book of the Univalent Foundations Project [133], and the works of Voevodsky [139], Kapulkin [76], Lumsdaine [87], Awodey and Warren [2], and van den Berg and Garner [134]. Other recent applications of categorical structures are in the areas of quantum computing, see for instance the works of Coecke and Kissinger [44], Vicary [138]. Higher categories have also given rise to interesting software implementations, see the works of Bar et al. [6], Bar and Vicary [5].

1.2 Different Types of Higher Structures

The behaviour of compositions of cells in a higher category determines two main classes: strict and weak higher categories. As further explained in Sect. 1.2.3 below, in the strict case compositions are associative and unital; in the weak case, they are associative and unital only up to coherent isomorphisms.

For each of these classes, there are higher categories which admit cells in every dimension (ω-categories), and those that have cells in dimensions only 0 up to n (truncated n-categories). A further class of higher structures central to this work is that of n-fold structures. Below we give a description of these different types of higher structures and some of their relationships.

1.2.1 ω-Categories

ω-Categories have been studied extensively in relation to applications to homotopy theory, mathematical physics and algebraic geometry, giving rise to several models of (∞, n)-categories; intuitively, the latter are weak higher categories admitting cells in all dimensions and with weakly invertible arrows in dimension higher than n.

There are several models of $(\infty, 1)$-categories, all of which are Quillen equivalent: quasicategories, introduced by Boardman and Vogt [31] under the name of 'weak Kan complexes' and much developed by Joyal [74] and Lurie [89]; simplicial categories, introduced by Dwyer and Kan [48, 50], more recently studied with a model category approach by Bergner [23]; complete Segal spaces, studied by Rezk [109]; and relative categories, studied by Barwick and Kan [9]. The survey paper of Bergner [24] gives a description of these different models and their Quillen equivalences, see also her monograph [26]. More recently quasi-categories have been studied using techniques of 2-category theory and monad theory by Riehl and Verity [112, 113], and a model of $(\infty, 1)$-categories in terms of internal categories in simplicial sets was studied by Horel [72].

Models of (∞, n)-categories for $n > 1$ have been studied by Ara [1], Barwick and Kan [10], Bergner [25, 27], Rezk [110], and Lurie [88] and played an important role in Lurie's approach to the cobordism hypothesis, see also [37]. An axiomatic approach to (∞, n)-categories was developed by Barwick and Schommer-Pries [11].

The most general kind of weak ω-category possible would admit cells at all dimensions without stipulating that all cells should be weakly invertible above some finite dimension. Verity developed the theory of complicial sets [135, 137], to model these (∞, ∞)-categories as an adaptation of the theory of strict complicial sets [136], which he developed to prove the Street–Roberts conjecture [123] on the characterization of nerves of strict n-categories.

1.2.2 Truncated Higher Categories

In this work we concentrate on 'truncated' higher categories, with cells only in dimensions 0 up to n. These structures relate to one of the original motivations for the development of higher categories, namely the algebraic modelling of the Postnikov systems of spaces, whose sections are the n-types, that is spaces with trivial homotopy groups in dimension higher than n.

The largely open problem of understanding algebraic invariants such as the higher homotopy and cohomology operations leads us inexorably to the question of unravelling the combinatorics of Postnikov systems of spaces and simplicial categories. This is connected with achieving a useful combinatorial description of the k-invariants of spaces, another open problem of some significant merit. The work of Baues [19] provides a low dimensional and stable exemplar in this direction, by demonstrating the utility of this approach to computations of some differentials in the Adams spectral sequence.

Applications of the truncated case to E_n-structures and to Hochschild cohomology were developed in the context of the higher operadic model of Batanin [13, 14], see also the work of Tamarkin on the Deligne conjecture [15, 125]. Simpson in [119] envisages the use of Tamsamani n-categories for applications to algebraic geometry, in the theory of higher stacks and of non-abelian cohomology.

As we discuss in Sect. 13.1, we envisage that our new model of weak n-categories will lead to a better understanding of the k-invariants of spaces and of simplicial categories.

We thus see that the applications of weak n-categories, though slightly different in nature than those of (∞, n)-categories, are important and worth pursuing.

1.2.3 Strict Versus Weak n-Categories

In a strict higher category, compositions of cells are associative and unital, and there is a simple way to describe strict n-categories via iterated enrichment. Although simple to define, strict n-categories are insufficient for many applications and the wider class of weak n-categories is needed. For instance, strict n-groupoids do not model n-types in dimension $n > 2$ (see [118] for a counterexample in dimension $n = 3$).

In a weak n-category, higher cells compose in a way that is associative and unital only up to an invertible cell in the next dimension, and these associativity and unit isomorphisms are suitably compatible or coherent.

In dimensions $n = 2$ and $n = 3$ the idea of a weak n-category is embodied in the classical notions of bicategory due to Bénabou [21] and tricategory due to Gordon et al. [64], and more recently studied by Gurski [68], Garner and Gurski [62] and others. In these structures, explicit diagrams encode the coherence axioms for the associativity and unit isomorphisms.

Capturing the coherence axioms explicitly in dimension $n > 3$ seems intractable. Instead, various combinatorial machines have emerged to automate the process of defining weak n-categories [83]: in these approaches the coherence data for the higher associativity are not given explicitly but they are automatically encoded in the combinatorics defining the models.

Different types of combinatorics have been used, including multi-simplicial structures as in Simpson [119], Tamsamani [126], higher operads as in Batanin [12], Leinster [84], Batanin and Weber [16], Batanin et al. [17] and Trimble [40], opetopes as in Baez and Dolan [4] and as in Cheng [39] and several others. The comparison between these different approaches is still largely an open problem.

1.2.4 n-Fold Categories

There is a third class of higher structures besides strict n-categories and weak n-categories which is central to this work: the class of n-fold categories. n-Fold categories were introduced by Ehresmann and Ehresmann [54–56]. There is an extensive literature for the case $n = 2$ (when they are called double categories), developed among others by Dawson et al., see for instance [46], and Grandis and Paré [65, 66]. Model structures on double categories were developed in joint work by Fiore et al. [60] while Fiore et al. built a model structure on n-fold categories [59] generalizing Thomason's model structure on categories [128]. A recent application of n-fold categories to algebraic geometry can be found in [140].

The definition of an n-fold category is elementary, and is based on the notion of an internal category (see Sect. 2.3 for more details). To understand the latter, remember that the data for a small category can be presented by a diagram

$$X_1 \times_{X_0} X_1 \xrightarrow{m} X_1 \underset{\xleftarrow[s]{}}{\overset{\xrightarrow{d_0}}{\xrightarrow{d_1}}} X_0$$

where X_0 is the set of objects, X_1 the set of arrows, the maps d_0, d_1 are the source and target maps, s is the identity map and m is the composition. These maps satisfy the axioms of a category, giving associativity of composition and identity laws. Such a diagram and axioms make sense in any category $\mathscr{C}$ with pullbacks and this defines the notion of an internal category in $\mathscr{C}$.

Let Δ be the category of non-empty finite ordinals and morphisms the non-decreasing maps between them. This category is the basis for the combinatorial model of topological spaces called simplicial sets, which are functors from Δ^{op} to the category of sets. Simplicial sets are ubiquitous in algebraic topology.

Functors from the product of n copies of Δ^{op} (which we denote by $\Delta^{n^{op}}$) to Set are called multi-simplicial sets and are also prominent in algebraic topology, more specifically in simplicial homotopy theory. Given categories $\mathscr{A}$ and $\mathscr{B}$, we denote by $[\mathscr{A}, \mathscr{B}]$ the category whose objects are the functors from $\mathscr{A}$ to $\mathscr{B}$ and whose morphisms are the natural transformations. Thus the category of n-fold simplicial sets is denoted by $[\Delta^{n^{op}}, \mathsf{Set}]$.

The relation between categories and simplicial sets comes from the nerve functor

$$N : \mathsf{Cat} \to [\Delta^{^{op}}, \mathsf{Set}] \ .$$

Given a category X, $(NX)[0]$ is the set of objects of X while for $k \geq 1$ $(NX)[k]$ consists of the set of composable sequences of arrows in X of length k. Conversely, a characterization of those simplicial sets that are nerves of categories can be given in terms of the so-called Segal condition on a simplicial set.

Similarly, for any category $\mathscr{C}$ with pullbacks there is a nerve functor

$$N : \mathsf{Cat}\,(\mathscr{C}) \to [\Delta^{^{op}}, \mathscr{C}]$$

whose essential image can be characterized in terms of Segal maps.

The definition of internal categories can be iterated: if $\mathscr{C}$ is the category Cat, we can consider internal categories in Cat, then repeating this until the n-th iteration affords the category $\mathsf{Cat}^{\mathsf{n}}$ of n-fold categories. By iterating the nerve construction one obtains a full and faithful nerve functor

$$N_{(n)} : \mathsf{Cat}^{\mathsf{n}} \to [\Delta^{n^{op}}, \mathsf{Set}]$$

and a characterization of the essential image of this functor can be given in terms of (iterated) Segal conditions (see Proposition 2.4.5 for more details).

Thus we can think of n-fold categories as structures whose elements carry an intrinsic n-cube geometry and which may be composed along any one of the n axes of that geometry.

For instance, when $n = 2$, we can visualize a double category as having objects, 1-morphisms in two different directions (horizontal and vertical), both composable in the respective directions, and squares which can be composed both horizontally and vertically, as in the picture below:

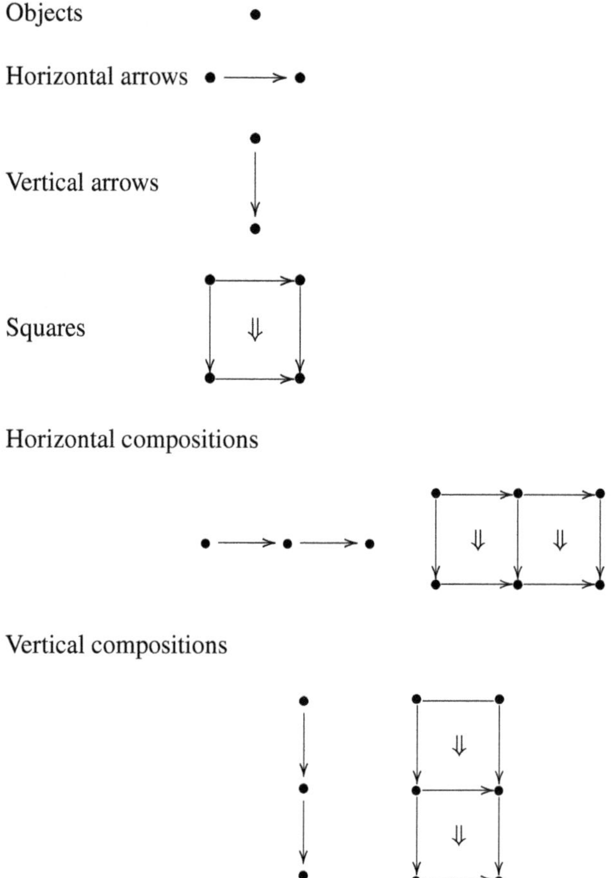

These compositions are all associative and unital, hence these structures are, in this sense, 'strict'. However, they are much wider than strict n-categories because, unlike in strict n-categories, these n directions are completely symmetric and we cannot identify in an n-fold category any 'sets of k-cells' for $k = 0, \cdots, n$.

1.2.5 n-Fold Structures Versus Strict and Weak n-Categories

An important application of n-fold structures appeared in the context of homotopy theory, in the modelling of connected $(n + 1)$-types using n-fold categories internal to the category of groups [85], as we are going to explain in Sect. 1.3.2 below.

This leads to the question of how n-fold categories relate to strict and weak n-categories. The first question has an easy answer, and as we will illustrate in detail in Sect. 2.6, there is a full and faithful embedding

$$n\text{-}\mathsf{Cat} \hookrightarrow \mathsf{Cat}^{\mathsf{n}}\ .$$

For instance, a strict 2-category is a double category in which the category of objects and vertical arrows is discrete. Consequently the vertical sides of the squares in the picture on page 10 are identities which we may contract in our diagrams, thereby depicting those squares as globes:

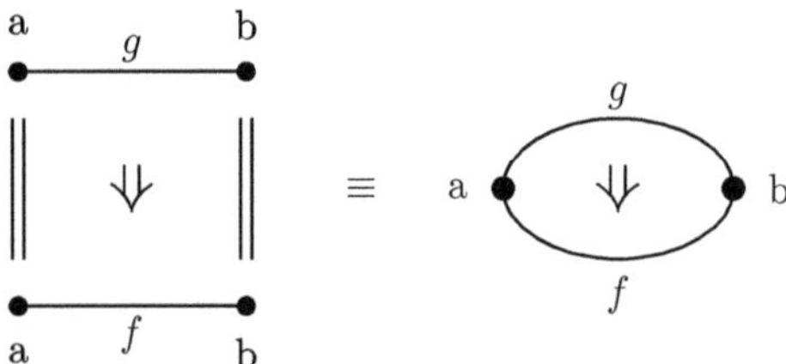

Thus the picture on page 10 for a double category reduces to the picture on page 4 for a strict 2-category.

The second question, on the relationship between n-fold categories and weak n-categories, is in general much harder to answer. Indeed much of the work presented here addresses itself directly to answering the following motivating question of this kind:

Can we identify a suitable subcategory of n-fold categories which gives a model of weak n-categories?

We positively answer this question with the introduction of the category of weakly globular n-fold categories and the proof of a suitable equivalence to the Tamsamani model of weak n-categories. Our model is based on a new paradigm to weaken higher categorical structures, which is the notion of weak globularity. We explain the idea of weak globularity in the general higher categorical setting in Chap. 3, while in what follows we give an account of its first appearance in the context of the algebraic modelling of homotopy types.

1.3 The Homotopy Hypothesis

As already pointed out, one of the most important connections between homotopy theory and higher category theory is the 'homotopy hypothesis': any good model of weak n-categories should give an algebraic model of n-types in the weak n-groupoid case.

In this section we first give a summary of the notions of algebraic models of n-types, and we then concentrate on the use of n-fold structures to give such models in the path-connected case. The latter was the first context to see the development of the notion of weak globularity, in the work by Paoli [102]. In Chap. 3 we will introduce the idea of weak globularity in a more general categorical context.

1.3.1 Homotopy Types and Their Algebraic Models

A topological space whose homotopy groups vanish in dimension higher than n is called an n-type. The n-types are building blocks of spaces thanks to a classical construction in algebraic topology, namely the Postnikov decomposition [63, 94, 105, 120]. More precisely, the Postnikov system of a space consists of its n-type constituents together with some cohomological invariants called k-invariants. One of the fundamental theorems in algebraic topology is that the Postnikov system of a space determines its homotopy type [63].

Postnikov systems exist for more complex structures than spaces, for instance for categories enriched in simplicial sets, also called simplicial categories, which are models of $(\infty, 1)$-categories. Their Postnikov sections consist of categories enriched in (simplicial) n-types, while the k-invariants are cohomology classes in the Dwyer–Kan–Smith cohomology [51].

A fundamental question in algebraic topology is the search for algebraic models of n-types. Such a model consists of a category $\mathcal{G}_n$, built only from combinatorial and categorical data, together with a pair of functors

$$\Pi_n : n\text{-types} \to \mathcal{G}_n \qquad B : \mathcal{G}_n \to n\text{-types}$$

inducing an equivalence of categories

$$\mathcal{G}_n/{\sim^n} \simeq \text{Ho}(n\text{-types}) . \tag{1.1}$$

Here $\mathcal{G}_n/\sim^n$ is the localization of $\mathcal{G}_n$ with respect to some algebraically defined weak equivalences, and Ho(n-types) is the homotopy category of n-types. The equivalence of categories (1.1) means that the right-hand side, which has a purely topological input, is described by the left-hand side in an entirely algebraic or categorical way. Hence we call $\mathcal{G}_n$ an 'algebraic model of n-types'.

Since the Postnikov sections of simplicial categories are categories enriched in (simplicial) n-types, an algebraic model of the latter via a product-preserving functor Π_n gives rise to an algebraic model for the Postnikov sections of simplicial categories, namely consisting of categories enriched in $\mathcal{G}_n$. It is crucial for this application to work with a model of general n-types, not merely path-connected ones, since the mapping spaces of simplicial categories are general simplicial sets, not path-connected ones.

The simplest case is $n = 1$. In this case, $\mathcal{G}_1$ is the category of groupoids. The functor B is the classifying space functor, which can be obtained by taking the geometric realization of the nerve of the groupoid; equivalence of groupoids is equivalence of categories. The functor Π_1 is the classical fundamental groupoid functor [63]. The latter associates to a space the groupoid whose objects are the points of the space and whose morphisms are the homotopy classes of paths.

The extension to the case $n > 1$ amounts to finding higher order analogues of the fundamental groupoid. It was noted by Grothendieck in 'Pursuing Stacks' [67] that this question would naturally lead to some type of higher groupoidal structure. For a given space, there are notions not only of 'homotopy between paths' but also of 'homotopies between homotopies' and then 'homotopies between homotopies between homotopies' and so on. Intuitively we expect the 'fundamental n-groupoid' $\Pi_n(X)$ of a space X to have these higher homotopies as higher cells, with only the n-cells requiring dividing out by the homotopies one level higher.

Converting this intuition into rigorously defined models is far from trivial and has been the object of much study both by algebraic topologists and category theorists.

The category n-Gpd of strict n-groupoids is insufficient for the purpose of modelling n-types. For instance, in [118] there is a counterexample showing that one cannot use strict 3-groupoids to model 3-types of spaces with non-trivial Whitehead products. Thus a more complex notion of higher groupoid is needed.

A good notion of weak n-category needs to satisfy the homotopy hypothesis: namely, it should provide an algebraic model of n-types when the cells in the structure are weakly invertible, that is, in the weak n-groupoid case. For several models of higher categories this hypothesis has been shown to hold.

Some algebraic models of n-types arose as the higher groupoidal case of models of weak n-categories, such as the Tamsamani [126] and the Batanin [12] models. Other algebraic models of n-types developed independently within algebraic topology; these models were build algebraically and combinatorially and were shown to satisfy the equivalence of categories (1.1) but they were not built as the higher groupoid version of a model of weak n-category. Examples of these models include the cat^n-groups of Loday [85], the hypercrossed complexes of Carrasco and Cegarra [38], and the crossed n-cubes of Porter [104] and of Ellis and Steiner [57]. In low dimension other examples are the crossed modules in groups introduced by MacLane and Whitehead [92], the double groupoids with connections of Brown and Spencer [35], the quadratic modules of Baues [18], and the crossed modules of length 2 of Conduché [45].

Of particular relevance for this work is Loday's model of connected $(n+1)$-types, which is based on n-fold categories internal to the category of groups. This model was developed independently in an algebraic topological context and gave rise to interesting topological applications, such as a higher order version of the Van Kampen theorem [34].

A crucial question, which lies at the very origin of this work, is the following:

Is Loday's model of connected $(n+1)$-types, based on n-fold categories internal to groups, the higher groupoid version of some general model of weak higher category?

In order to answer this question one must first understand how the n-fold model of Loday relates to some other model of higher categories that is known to satisfy the homotopy hypothesis, such as the Tamsamani model.

This question is far from trivial. As we explain in the next section, the key to this comparison is the notion of weak globularity, introduced in the context of cat^n-groups by Paoli in [102].

1.3.2 Modelling Homotopy Types with n-Fold Structures

One of the earliest appearances of higher categorical structures in algebraic topology is due to MacLane and Whitehead [92], who used crossed modules in groups to model connected 2-types. It is well known that the category of crossed modules in groups is equivalent to the category of internal categories in groups. The latter is the same as the category of strict 2-groupoids with one object, also called strict 2-groups. Thus the Maclane–Whitehead model amounts to modelling of connected 2-types with a strict 2-dimensional categorical structure.

The modelling of homotopy types becomes easier to handle in the path connected case, since it allows us to work within the entirely algebraic context of the category of groups. The Kan loop group functor from the category of based topological spaces to simplicial groups [63] is the reason why the internalisation in groups works in the path-connected case.

As mentioned in the previous section, several models of connected $(n+1)$-types were developed independently of the pursuit of higher categorical models, even though they all exhibit some type of higher structure.

Among these, cat^n-groups are very appealing from a categorical perspective as their underlying higher categorical structure is particularly simple: it is easy to prove (see for instance [102] for details) that they are equivalent to the category $\mathsf{Cat}^n(\mathsf{Gp})$ of n-fold categories internal to groups.

At the same time, models of higher categories were being developed, and in particular the Tamsamani model for which a proof of the homotopy hypothesis was given in [126]: thus Tamsamani $(n+1)$-groupoids model $(n+1)$-types. A suitable subcategory of Tamsamani $(n+1)$-groupoids can be identified with model connected $(n+1)$-types.

This naturally leads to the question of finding an explicit comparison between the Tamsamani model and the cat^n-groups model for the path-connected case. Such a comparison is highly non-trivial because the two higher categorical structures have a fundamental difference: the Tamsamani model has sets of cells in dimensions 0 up to n, together with compositions coming from a multi-simplicial structure. The higher cells have a globular shape as in the case of strict n-groupoids, but the compositions are no longer associative and unital.

In the cat^n-groups model, however, there is no immediate way to identify sets of higher cells. The n-fold structure is symmetric in all the n different simplicial directions, and the only higher cells which we can identify in the structure have a (hyper-)cubical shape. Hence the question:

How can we connect the cubical model of cat^n*-groups to a globular model like the Tamsamani model of connected* $(n+1)$*-types?*

The notion of weak globularity, introduced by Paoli [102] for the category of cat^n-groups, is the key to answering this question.

We showed in [102] that the category $\mathsf{Cat}^n(\mathsf{Gp})$ of n-fold categories in groups can be replaced by the category $\mathsf{Cat}^n(\mathsf{Gp})_{wg}$ of weakly globular cat^n-groups (which is also called in [102] the category of 'special cat^n-groups') without loss of homotopical information. That is, weakly globular cat^n-groups model connected $(n+1)$-types.

The idea of weakly globular cat^n-groups is to impose additional conditions on cat^n-groups to break the symmetry of the n simplicial directions and identify 'substructures' in a cat^n-group from which one can recover the sets of higher cells. To identify which substructures in a cat^n-group should play this role we consider the full and faithful embedding

$$n\text{-}\mathsf{Cat}\,(\mathsf{Gp}) \hookrightarrow \mathsf{Cat}^n(\mathsf{Gp})$$

of strict n-categories internal to groups to n-fold categories internal to groups, which is formally analogous to the embedding of strict n-categories in n-fold categories mentioned in Sect. 1.2.

It is not difficult to see (see Sect. 2.6 for more details) that a strict n-category in groups amounts to an n-fold category in groups in which certain substructures (which are cat^k-groups, for $0 < k \leq n-1$) are discrete. By discrete here we mean they are groups, seen as discrete cat^k-groups in which all structure maps are identities. The corresponding underlying sets are the sets of higher cells in a strict n-category internal to groups. We call this property the *globularity condition* since it is the condition that gives rise to the globular shape of higher cells in the structure.

In a weakly globular cat^n-group these substructures are no longer required to be discrete, but 'homotopically discrete' in a precise sense that allows iteration (these are called in [102] 'strongly contractible cat^n-groups'). Each of these substructures has a higher groupoidal structure on its own which is however not a general k-fold structure (for the respective dimension k) like in a general cat^n-group, but is equivalent to a discrete one. The sets underlying the latter correspond to the 'sets of higher cells'.

Further, we show in [102] that there is a comparison functor

$$\mathsf{Cat}^n(\mathsf{Gp})_{wg} \to \mathsf{GTa}^{n+1}$$

which preserves the homotopy type. Here GTa^{n+1} denotes groupoidal Tamsamani $(n+1)$-groupoids. This functor is obtained by discretizing the homotopically discrete substructures in a weakly globular cat^n-group, so as to recover the globularity condition. Full details can be found in [102].

Besides the comparison with the Tamsamani model, another advantage of weakly globular cat^n-groups over general cat^n-groups to model $(n+1)$-types is that the latter do not have an algebraic version of the Postnikov functor from cat^n-groups to cat^{n-1}-groups: in other words this functor can be produced only by passing to classifying spaces, using the topological Postnikov functor and then applying the fundamental cat^{n-1}-group functor. Instead, weakly globular cat^n-groups come equipped with a very simple and entirely algebraically defined functor

$$p^{(n-1)} : \mathsf{Cat}^{\mathsf{n}}(\mathsf{Gp})_{wg} \to \mathsf{Cat}^{\mathsf{n-1}}(\mathsf{Gp})_{wg}$$

exactly corresponding to the Postnikov truncation; that is, such that for each $X \in \mathsf{Cat}^{\mathsf{n}}(\mathsf{Gp})_{wg}$ (whose classifying space BX is therefore a connected $(n+1)$-type) there is a map $BX \to Bp^{(n-1)}X$ inducing isomorphisms of homotopy groups in dimensions 0 up to n.

In [29] Blanc and Paoli developed the notion of weakly globular n-fold groupoids in view of using n-fold structures for modelling general n-types. This case is considerably more complex than the path-connected case. In particular, several features of weakly globular cat^n-groups that could be deduced from their definition now need to become part of the definition of a weakly globular n-fold groupoid.

In [29] we build a functor from spaces to weakly globular n-fold groupoids as well as from weakly globular n-fold groupoids to n-types, but we do not exhibit a proof that these functors induce an equivalence of categories after localization, that is, that weakly globular n-fold groupoids are an algebraic model of n-types.

More precisely, in [29] we show that the functor from the localization of weakly globular n-fold groupoids to the homotopy category of n-types is essentially surjective on objects, but we do not give a proof of the fully faithfulness needed to realize an equivalence of categories. We do realize an equivalence of categories in [29] by enlarging the category of weakly globular n-fold groupoids, but at the price that the larger category is no longer an n-fold structure.

The realization of the modelling of general n-types via weakly globular n-fold structures remained open until the present work, where it is one of our main results. For this purpose we introduce the category of *groupoidal weakly globular n-fold categories*, which strictly contains the weakly globular n-fold groupoids of [29]. Further, we show that the functor from spaces to the latter given in [29] can still be used as a convenient model: such a model is very explicit and it is independent of other models of n-types.

Chapter 2
Multi-Simplicial Techniques

Abstract In this chapter we review the most important techniques used in this work, which are multi-simplicial techniques. Although known in the literature, we present them in a way that is most suitable for the development of the Segal-type models. We discuss multi-simplicial objects and their Segal maps, and we also introduce the higher categorical notions of n-fold category and strict n-category, as well as their multi-nerves. We provide a multi-simplicial description of these structures. This description is important to build the intuition around the Segal-type models. We also introduce notational conventions that will be used throughout this work.

This chapter reviews (multi)-simplicial techniques. Multi-simplicial objects are functors from $\Delta^{n^{op}}$ to a category $\mathscr{C}$ (which we will assume to have finite limits). When $\mathscr{C}$ is the category Set, multi-simplicial sets are combinatorial models of spaces and are used in homotopy theory, see for instance [63].

As outlined in Chap. 1, the connection between higher categorical structures and multi-simplicial structures arises via the notions of nerves (and their iterated versions, called multi-nerves) and of Segal maps. We review these in this chapter, where we also spell out in detail the multi-simplicial description of strict n-categories and of n-fold categories. In the cases $n = 2$ and $n = 3$ these descriptions admit a geometric interpretation which is easy to visualize (see Examples 2.5.8 and 2.6.9). This geometric visualization is very helpful in building one's intuition around the modifications needed to define weak models of higher categories using multi-simplicial structures.

We also discuss in this chapter the functors

$$p, q : [\Delta^{n^{op}}, \mathsf{Set}] \to \mathsf{Set}$$

and the corresponding functors $p^{(r)}, q^{(r)} : [\Delta^{n^{op}}, \mathsf{Set}] \to [\Delta^{r^{op}}, \mathsf{Set}]$ for $0 \leq r \leq n-1$. These functors are crucial to this work.

S. Paoli, *Simplicial Methods for Higher Categories*, Algebra and Applications 26,
https://doi.org/10.1007/978-3-030-05674-2_2

2.1 Multi-Simplicial Objects and Segal Maps

We start by reviewing the notion of multi-simplicial objects and their associated Segal maps. These structures play a crucial role throughout this work since, as explained in the next chapter (see Sect. 3.2), multi-simplicial objects form a natural environment for the definition of higher categorical structures.

Recall from Sect. 1.2.4 that given categories $\mathscr{A}$ and $\mathscr{B}$, $[\mathscr{A}, \mathscr{B}]$ denotes the category of functors from $\mathscr{A}$ to $\mathscr{B}$ and natural transformations. The following notation will be used throughout this work.

Definition 2.1.1 Let $F : \mathscr{C} \to \mathscr{D}$ be a functor, $\mathscr{I}$ a small category. Denote by

$$\overline{F} : [\mathscr{I}, \mathscr{C}] \to [\mathscr{I}, \mathscr{D}]$$

the functor given, for all $i \in \mathscr{I}$, by

$$(\overline{F}X)_i = F(X(i)).$$

2.1.1 Simplicial Objects and Their Segal Maps

Let Δ be the simplicial category. Its objects are finite ordered sets

$$[n] = \{0 < 1 < \cdots < n\}$$

for integers $n \geq 0$ and its morphisms are non-decreasing monotone functions. If $\mathscr{C}$ is any category, a simplicial object X in $\mathscr{C}$ is a contravariant functor from Δ^{op} to $\mathscr{C}$, that is, $X : \Delta^{op} \to \mathscr{C}$. We write X_n for $X([n])$. A simplicial map $F : X \to Y$ between simplicial objects in $\mathscr{C}$ is a natural transformation. We denote by $[\Delta^{op}, \mathscr{C}]$ the category of simplicial objects and simplicial maps.

It is well known (see for instance [94]) that to give a simplicial object X in $\mathscr{C}$ is the same as to give a sequence $X_0, X_1, X_2, \ldots$ of objects of $\mathscr{C}$ together with face operators $\partial_i : X_n \to X_{n-1}$ and degeneracy operators $\sigma_i : X_n \to X_{n+1}$ ($i = 0, \ldots, n$) which satisfy the following simplicial identities:

$$\begin{aligned}
\partial_i \partial_j &= \partial_{j-1} \partial_i && \text{if } i < j \\
\sigma_i \sigma_j &= \sigma_{j+1} \sigma_i && \text{if } i \leq j \\
\partial_i \sigma_j &= \begin{cases} \sigma_{j-1} \partial_i, & \text{if } i < j \\ \mathrm{Id}, & \text{if } i = j \text{ or } i = j+1 \\ \sigma_j \partial_{i-1}, & \text{if } i > j+1. \end{cases}
\end{aligned} \tag{2.1}$$

Under this correspondence $\partial_i = X(\varepsilon_i)$ and $\sigma_i = X(\eta_i)$, where

$$\varepsilon_i : [n-1] \to [n] \qquad\qquad \eta_i : [n+1] \to [n]$$
$$\varepsilon_i(j) = \begin{cases} j, & \text{if } j < i \\ j+1, & \text{if } j \geq i \end{cases} \qquad\qquad \eta_i(j) = \begin{cases} j, & \text{if } j \leq i \\ j-1, & \text{if } j > i. \end{cases}$$

If $X \in [\Delta^{^{op}}, \mathscr{C}]$ is such that $X_n = X_m$ for all $n, m \in \Delta^{op}$ and all face and degeneracy maps are identities, we say the simplicial object X is discrete.

Definition 2.1.2 Let $X \in [\Delta^{^{op}}, \mathscr{C}]$ be a simplicial object in a category $\mathscr{C}$ with pullbacks. For each $1 \leq j \leq k$ and $k \geq 2$, let $\nu_j : X_k \to X_1$ be induced by the map $[1] \to [k]$ in Δ sending 0 to $j-1$ and 1 to j. Then the following diagram commutes:

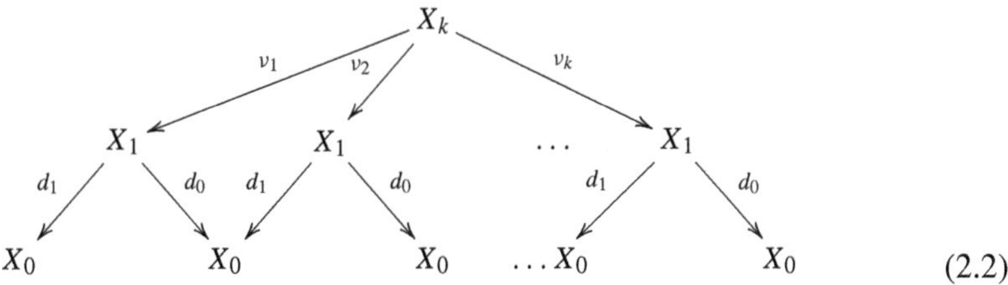

(2.2)

If $X_1 \times_{X_0} \overset{k}{\cdots} \times_{X_0} X_1$ denotes the limit of the lower part of the diagram (2.2), the *k-th Segal map of* X is the unique map

$$\mu_k : X_k \to X_1 \times_{X_0} \overset{k}{\cdots} \times_{X_0} X_1$$

such that $\mathrm{pr}_j\ \mu_k = \nu_j$ where pr_j is the jth projection.

Definition 2.1.3 Let $X \in [\Delta^{^{op}}, \mathscr{C}]$ and suppose that there is a map

$$\gamma : X_0 \to Y$$

in $\mathscr{C}$ such that the limit of the diagram

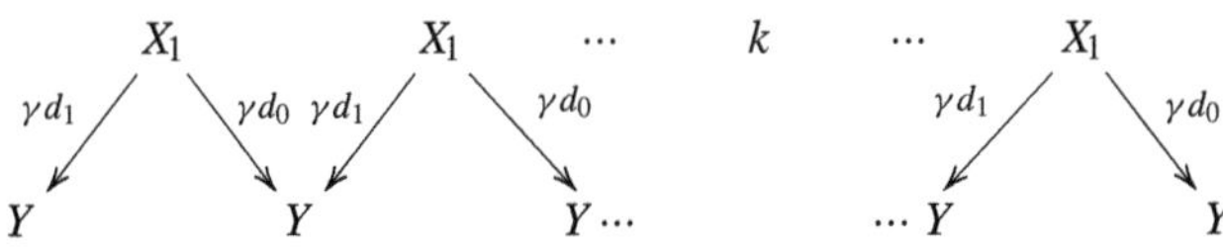

exists; denote the latter by $X_1 \times_Y \overset{k}{\cdots} \times_Y X_1$. Then the following diagram commutes, where ν_j is as in Definition 2.1.2, and $k \geq 2$

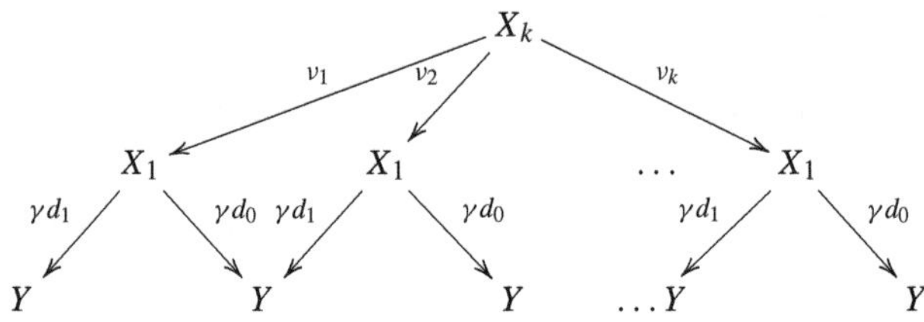

The *k-th induced Segal map of* X is the unique map

$$\hat{\mu}_k : X_k \rightarrow X_1 \times_Y \overset{k}{\cdots} \times_Y X_1$$

such that $\mathrm{pr}_j\, \hat{\mu}_k = \nu_j$ where pr_j is the jth projection. If $Y = X_0$ and γ is the identity, the induced Segal map coincides with the Segal map of Definition 2.1.2.

2.1.2 Multi-Simplicial Objects

Let $\Delta^{n^{op}}$ denote the product of n copies of Δ^{op}, that is

$$\Delta^{n^{op}} = (\Delta^{\mathrm{op}})^n = \Delta^{op} \times \overset{n}{\cdots} \times \Delta^{op}.$$

Given a category $\mathscr{C}$, $[\Delta^{n^{op}}, \mathscr{C}]$ is called the category of n-fold simplicial objects in $\mathscr{C}$ (or multi-simplicial objects for short).

Notation 2.1.4 If $X \in [\Delta^{n^{op}}, \mathscr{C}]$ and $\underline{k} = ([k_1], \dots, [k_n]) \in \Delta^{n^{op}}$, we shall use several alternative notations as follows

$$X([k_1], \dots, [k_n]) = X(k_1, \dots, k_n) = X_{k_1,\dots,k_n} = X_{\underline{k}}.$$

Definition 2.1.5 Let Σ_n be the symmetric group and $\alpha \in \Sigma_n$. Given $\underline{k} = (k_1, \dots, k_n) \in \Delta^{n^{op}}$, let $\alpha(\underline{k}) = (k_{\alpha(1)}, \dots, k_{\alpha(n)})$. Define a map

$$(\text{-})^{\alpha} : [\Delta^{n^{op}}, \mathscr{C}] \rightarrow [\Delta^{n^{op}}, \mathscr{C}]$$

by $(X^{\alpha})_{\underline{k}} = X_{\alpha(\underline{k})}$ for each $\underline{k} \in \Delta^{n^{op}}$ and $X \in [\Delta^{n^{op}}, \mathscr{C}]$.

Note that, if $\alpha, \beta \in \Sigma_n$ and $X \in [\Delta^{n^{op}}, \mathscr{C}]$, for each $\underline{k} \in \Delta^{n^{op}}$

$$X^{\alpha\circ\beta}_{\underline{k}} = X_{\alpha\beta(\underline{k})} = X^{\alpha}_{\beta(\underline{k})} = (X^{\alpha})^{\beta}_{\underline{k}}\,,$$

so that $X^{\alpha\circ\beta} = (X^{\alpha})^{\beta}$. In particular, if $\{r\} \in \Sigma_n$ denotes the r-cycle $(1, 2, \dots, r)$ (for $1 \leq r \leq n$) we have, for each $(k_1, \dots, k_n) \in \Delta^{n^{op}}$

$$X^{\{r\}}_{k_1\dots k_n} = \begin{cases} X_{k_2k_3\dots k_r k_1 k_{r+1}\dots k_n}, & \text{if} \quad 1 \leq r < n \\ X_{k_2k_3\dots k_{n-1}k_1} & \text{if} \quad r = n\ . \end{cases} \tag{2.3}$$

The functor sending X to $X^{\{r\}}$ can be thought of as 'bringing the rth index of X to the fore'. Note that

$$X^{\{1\}} = X\ . \tag{2.4}$$

Remark 2.1.6 For each $X \in [\Delta^{n^{op}}, \mathscr{C}]$, $(k_1, \dots, k_{n-1}) \in \Delta^{n-1^{op}}$, $s \in \Delta^{^{op}}$

$$\begin{aligned}(X^{\{r\}}_s)_{k_1}(k_2, \dots, k_{n-1}) &= X^{\{r\}}_s(k_1, \dots, k_{n-1}) \\ &= X_{k_1\dots k_r\, s\, k_{r+1}\dots k_{n-1}} = (X_{k_1})^{\{r-1\}}_s(k_2, \dots, k_{n-1}),\end{aligned}$$

therefore

$$(X^{\{r\}}_s)_{k_1} = (X_{k_1})^{\{r-1\}}_s\ .$$

The following is an elementary fact:

Lemma 2.1.7 *Every n-simplicial object in $\mathscr{C}$ can be regarded as a simplicial object in $[\Delta^{n-1^{op}}, \mathscr{C}]$ in n possible ways.*

Proof For each $1 \leq i \leq n$ define the isomorphism

$$\xi_i : [\Delta^{n^{op}}, \mathscr{C}] \to [\Delta^{^{op}}, [\Delta^{n-1^{op}}, \mathscr{C}]]$$

by $(\xi_i X)_{k_1} = X^{\{i\}}_{k_1}$ for each $X \in [\Delta^{n^{op}}, \mathscr{C}]$, $k_1 \in \Delta^{op}$. Thus, for each $(k_2, \dots, k_n) \in \Delta^{n-1^{op}}$

$$(\xi_i X)_{k_1}(k_2, \dots, k_n) = \begin{cases} X_{k_2\dots k_i\, k_1\, k_{i+1}\dots k_n}, & 1 \leq i < n \\ X_{k_2\dots k_{n-1}\, k_1}, & i = n\ . \end{cases}$$

Note that, since as observed in (2.4), $X^{\{1\}} = X$, we have

$$(\xi_1 X)_k = X_k\ . \tag{2.5}$$

□

Remark 2.1.8 From Lemma 2.1.7, given an n-fold simplicial object $X \in [\Delta^{n^{op}}, \mathscr{C}]$, for each simplicial direction i there are Segal maps for $\xi_i X \in [\Delta^{^{op}}, [\Delta^{n-1^{op}}, \mathscr{C}]]$, with $1 \leq i \leq n$. For each $r \geq 2$ these are maps in $[\Delta^{n-1^{op}}, \mathscr{C}]$

$$X^{\{i\}}_r \to X^{\{i\}}_1 \times_{X^{\{i\}}_0} \overset{r}{\cdots} \times_{X^{\{i\}}_0} X^{\{i\}}_1,$$

where $(\xi_i X)_r = X_r^{\{i\}}$ is as in the proof of Lemma 2.1.7. By (2.4) when $i = 1$ the Segal maps are simply denoted by

$$X_r \to X_1 \times_{X_0} \overset{r}{\cdots} \times_{X_0} X_1 .$$

Remark 2.1.9 Note that for each $1 \leq i \leq n$ the following diagram commutes

$$\begin{array}{ccc} [\Delta^{n^{op}}, \mathcal{C}] & \xrightarrow{(\text{-})^{\{i\}}} & [\Delta^{n^{op}}, \mathcal{C}] \\ & \searrow^{\xi_i} & \downarrow{\xi_1} \\ & & [\Delta^{op}, [\Delta^{n-1^{op}}, \mathcal{C}]] \end{array}$$

In fact, for each $X \in [\Delta^{n^{op}}, \mathcal{C}]$ and $k \in \Delta^{op}$, by (2.4)

$$(\xi_i X)_k = (X^{\{i\}})_k = (\xi_1 X^{\{i\}})_k .$$

Since this holds for each $k \in \Delta^{op}$, $\xi_i X = \xi_1 X^{\{i\}}$.

Example 2.1.10 In the case $n = 2$, $X \in [\Delta^{2^{op}}, \mathcal{C}]$ is called a bisimplicial object in $\mathcal{C}$. It is equivalent to a bigraded sequence of objects X_{st} of $\mathcal{C}$ $(s, t \geq 0)$ together with horizontal face and degeneracy maps

$$\partial_i^h : X_{st} \to X_{s-1,t} \qquad \sigma_i^h : X_{st} \to X_{s+1,t}$$

as well as vertical face and degeneracy maps

$$\partial_i^v : X_{st} \to X_{s,t-1} \qquad \sigma_i^v : X_{st} \to X_{s,t+1} .$$

These face and degeneracy maps must satisfy the simplicial identities (horizontally and vertically) and every horizontal map must commute with every vertical map. We call these equalities bisimplicial identities. In Fig. 2.1 we draw a pictorial representation of the corner of a bisimplicial object X, where direction 1 is horizontal and direction 2 is vertical.

We see from the picture that a bisimplicial object is a simplicial object in simplicial objects in two directions, vertical and horizontal.

For each $r \geq 0$, $X_r^{\{1\}} \in [\Delta^{op}, \mathcal{C}]$ is the vertical simplicial object with

$$(X_r^{\{1\}})_k = X_{rk}$$

and $X_r^{\{2\}} \in [\Delta^{op}, \mathcal{C}]$ is the horizontal simplicial object with

$$(X_r^{\{2\}})_k = X_{kr} .$$

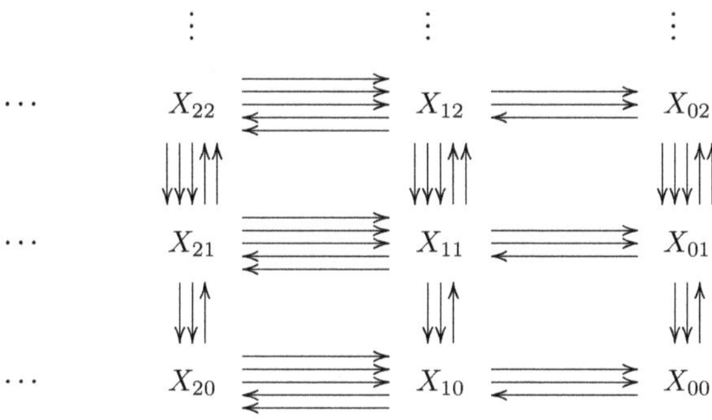

Fig. 2.1 Corner of a bisimplicial object X

Then for each $r \geq 2$ there are horizontal Segal maps in $[\Delta^{^{op}}, \mathscr{C}]$

$$X_r^{\{1\}} = X_1^{\{1\}} \times_{X_0^{\{1\}}} \overset{r}{\cdots} \times_{X_0^{\{1\}}} X_1^{\{1\}}$$

and vertical Segal maps in $[\Delta^{^{op}}, \mathscr{C}]$

$$X_r^{\{2\}} = X_1^{\{2\}} \times_{X_0^{\{2\}}} \overset{r}{\cdots} \times_{X_0^{\{2\}}} X_1^{\{2\}} .$$

2.2 Multi-Simplicial Sets

When $\mathscr{C} = \mathsf{Set}$ the category $[\Delta^{n^{op}}, \mathsf{Set}]$ is the category of n-fold simplicial sets. This category plays a special role in this work, since our Segal-type models are all full subcategories of $[\Delta^{n^{op}}, \mathsf{Set}]$. We also discuss in this section some closure properties of subcategories of multi-simplicial sets, which will be satisfied by our Segal-type models.

In order to efficiently handle the relationship between Segal-type models of the same type for different dimensions, we introduce a special notational convention as follows.

Let $d : \mathsf{Set} \to [\Delta^{^{op}}, \mathsf{Set}]$ be the discrete simplicial set functor. Clearly d is fully faithful, and therefore so is the induced functor

$$\bar{d} : [\Delta^{{n-1}^{op}}, \mathsf{Set}] \to [\Delta^{{n-1}^{op}}, [\Delta^{^{op}}, \mathsf{Set}]]$$

with $(\bar{d}X)_{\underline{k}} = dX_{\underline{k}}$ for all $\underline{k} \in \Delta^{{n-1}^{op}}$ (see Definition 2.1.1). There is an isomorphism

$$[\Delta^{{n-1}^{op}}, [\Delta^{^{op}}, \mathsf{Set}]] \cong [\Delta^{n^{op}}, \mathsf{Set}]$$

associating to $Y \in [\Delta^{n-1^{op}}, [\Delta^{^{op}}, \mathsf{Set}]]$ the n-fold simplicial set taking $(k_1, \dots, k_n) \in \Delta^{n^{op}}$ to $(Y_{k_1 \dots k_{n-1}})_{k_n}$. Hence the composite functor

$$[\Delta^{n-1^{op}}, \mathsf{Set}] \xrightarrow{\bar{d}} [\Delta^{n-1^{op}}, [\Delta^{^{op}}, \mathsf{Set}]] \cong [\Delta^{n^{op}}, \mathsf{Set}] \tag{2.6}$$

is fully faithful. This property justifies the following

Notational Convention 2.2.1 We identify $[\Delta^{n-1^{op}}, \mathsf{Set}]$ with its essential image in $[\Delta^{n^{op}}, \mathsf{Set}]$ under the functor (2.6). In other words, we identify $(n-1)$-fold simplicial sets with those n-fold simplicial sets X for which the simplicial set $X_{k_1 \dots k_{n-1}}$ is discrete for all $(k_1 \dots k_{n-1}) \in \Delta^{n-1^{op}}$.

2.2.1 The Functors $p^{(r)}$ and $q^{(r)}$

Let Cat be the category of small categories. There is a fully faithful nerve functor $N : \mathsf{Cat} \to [\Delta^{^{op}}, \mathsf{Set}]$ with a left adjoint $\mathcal{P}$ (the homotopy category construction).

Definition 2.2.2 Let $q : [\Delta^{^{op}}, \mathsf{Set}] \to \mathsf{Set}$ be the connected components functor. Let $p : [\Delta^{^{op}}, \mathsf{Set}] \to \mathsf{Set}$ be obtained by applying $\mathcal{P}$ to $X \in [\Delta^{^{op}}, \mathsf{Set}]$ and then taking the set of isomorphism classes of the category $\mathcal{P}X$.

We can extend the functors p and q as follows.

Definition 2.2.3 Define inductively

$$p_n, q_n : [\Delta^{n^{op}}, \mathsf{Set}] \to \mathsf{Set}$$

by setting $p_1 = p$, $q_1 = q$ and given p_{n-1}, q_{n-1} let p_n and q_n be the composites

$$\begin{aligned} &[\Delta^{n^{op}}, \mathsf{Set}] \cong [\Delta^{n-1^{op}}, [\Delta^{^{op}}, \mathsf{Set}]] \xrightarrow{\bar{p}} [\Delta^{n-1^{op}}, \mathsf{Set}] \xrightarrow{p_{n-1}} \mathsf{Set}\,, \\ &[\Delta^{n^{op}}, \mathsf{Set}] \cong [\Delta^{n-1^{op}}, [\Delta^{^{op}}, \mathsf{Set}]] \xrightarrow{\bar{q}} [\Delta^{n-1^{op}}, \mathsf{Set}] \xrightarrow{q_{n-1}} \mathsf{Set}\,, \end{aligned} \tag{2.7}$$

where the isomorphism on the left-hand side of (2.7) associates to $X \in [\Delta^{n^{op}}, \mathsf{Set}]$ the object of $[\Delta^{n-1^{op}}, [\Delta^{^{op}}, \mathsf{Set}]]$ taking $(k_1 \dots k_{n-1}) \in \Delta^{n-1^{op}}$ to the simplicial set $X_{k_1 \dots k_{n-1}}$.

Notational Convention 2.2.4 Under the embedding

$$[\Delta^{n-1^{op}}, \mathsf{Set}] \hookrightarrow [\Delta^{n^{op}}, \mathsf{Set}]$$

of Notational Convention 2.2.1 we see from Definition 2.2.3 that the following diagram commutes:

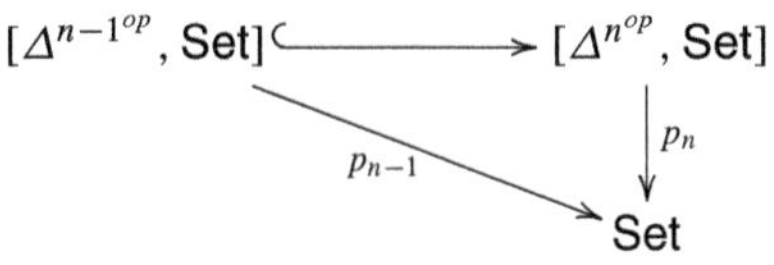

$$\begin{array}{ccc} [\Delta^{n-1^{op}}, \mathsf{Set}] & \hookrightarrow & [\Delta^{n^{op}}, \mathsf{Set}] \\ & \searrow^{p_{n-1}} & \downarrow p_n \\ & & \mathsf{Set} \end{array}$$

and similarly for q_n. Consequently, no ambiguity can arise by dropping subscripts in the definition of p_n, q_n and simply writing

$$p, q : [\Delta^{n^{op}}, \mathsf{Set}] \to \mathsf{Set}\ . \tag{2.8}$$

We now extend the definition of p and q as follows.

Definition 2.2.5 Let $p, q : [\Delta^{n^{op}}, \mathsf{Set}] \to \mathsf{Set}$ be as in (2.8). For each $0 \le r \le n-1$ define

$$p^{(r)}, q^{(r)} : [\Delta^{n^{op}}, \mathsf{Set}] \to [\Delta^{r^{op}}, \mathsf{Set}]$$

by $p^{(0)} = p, q^{(0)} = q$ and for $1 \le r \le n-1$, $(k_1 \dots k_r) \in \Delta^{r^{op}}$ and $X \in [\Delta^{n^{op}}, \mathsf{Set}]$

$$\begin{aligned} (p^{(r)}X)_{k_1 \dots k_r} &= p(X_{k_1 \dots k_r})\ , \\ (q^{(r)}X)_{k_1 \dots k_r} &= q(X_{k_1 \dots k_r})\ . \end{aligned}$$

Remark 2.2.6 Using the Notational Convention 2.2.1 we have a commuting diagram for each $0 \le r \le n-1$

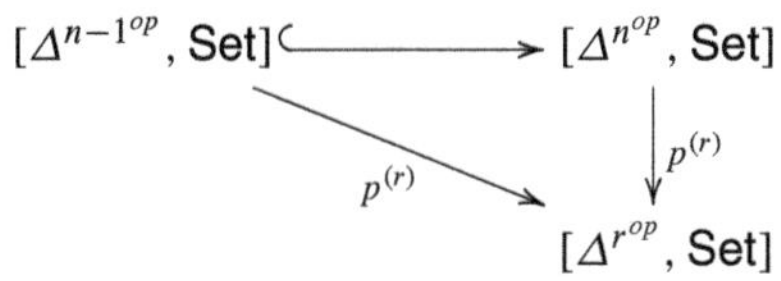

$$\begin{array}{ccc} [\Delta^{n-1^{op}}, \mathsf{Set}] & \hookrightarrow & [\Delta^{n^{op}}, \mathsf{Set}] \\ & \searrow^{p^{(r)}} & \downarrow p^{(r)} \\ & & [\Delta^{r^{op}}, \mathsf{Set}] \end{array}$$

In the definition of $p^{(r)}$ we can therefore drop explicit mention of the source dimension n.

The following lemma establishes some elementary properties of the functors $p^{(r)}$ and $q^{(r)}$.

Lemma 2.2.7 *For each $0 \le r \le n-1$, let $p^{(r)}$, $q^{(r)}$ be as in Definition 2.2.5. Then*

a) For each $s \in \Delta^{op}$,

$$(p^{(r)}X)_s = p^{(r-1)}X_s, \qquad (q^{(r)}X)_s = q^{(r-1)}X_s\ .$$

b) *For each* $X \in [\Delta^{n^{op}}, \mathsf{Set}]$ *there is a map, natural in* X, $\gamma_{(n)} : X \to qX$.
c) *For each* $X \in [\Delta^{n^{op}}, \mathsf{Set}]$ *there is a map, natural in* X, $\gamma_X^{(r)} : X \to q^{(r)}X$.
d) *For each* $1 \leq i \leq r$ *we have*

$$p^{(r)}(X^{\{i\}}) = (p^{(r)}X)^{\{i\}}, \qquad q^{(r)}(X^{\{i\}}) = (q^{(r)}X)^{\{i\}},$$

where we regard $\{i\}$ *as an element of* Σ_n *fixing* $i+1, \dots, n$.
e) *For each* $0 \leq r \leq n-1$ *the functor* $p^{(r)} : [\Delta^{n^{op}}, \mathsf{Set}] \to [\Delta^{r^{op}}, \mathsf{Set}]$ *factors as*

$$[\Delta^{n^{op}}, \mathsf{Set}] \xrightarrow{p^{(n-1)}} [\Delta^{n-1^{op}}, \mathsf{Set}] \xrightarrow{p^{(n-2)}} [\Delta^{n-2^{op}}, \mathsf{Set}] \cdots \xrightarrow{p^{(r)}} [\Delta^{r^{op}}, \mathsf{Set}]$$

Proof

a) For each $(k_1, \dots, k_{r-1}) \in \Delta^{r-1^{op}}$, we have

$$(p^{(r)}X)_s(k_1, \dots, k_{r-1}) = (p^{(r)}X)_{s\,k_1 \dots k_{r-1}} = pX_{s\,k_1 \dots k_{r-1}} = (p^{(r-1)}X_s)_{k_1 \dots k_{r-1}} \,.$$

Since this holds for each $(k_1, \dots, k_{r-1}) \in \Delta^{r-1^{op}}$ it follows that $(p^{(r)}X)_s = p^{(r-1)}X_s$. The proof for $q^{(r)}X$ is similar.
b) By induction on n. It is clear for $n = 1$. Assume the lemma holds for $(n-1)$. Then we have a map $X \to \bar{q}X$ given levelwise by $X_{\underline{k}} \to qX_{\underline{k}}$ for each $\underline{k} \in \Delta^{n^{op}}$ and by induction we have a map $\gamma_{(n-1)} : \bar{q}X \to qX$. By composition this gives $\gamma_{(n)}$.
c) For each $(k_1, \dots, k_r) \in \Delta^{r^{op}}$ by b) there are maps

$$X_{k_1 \dots k_r} \to q(X_{k_1 \dots k_r}) = (q^{(r)}X)_{k_1 \dots k_r}$$

natural in X and therefore a map $\gamma_X^{(r)} : X \to q^{(r)}X$ natural in X.
d) For each $\underline{k} \in \Delta^{r^{op}}$ we have

$$(p^{(r)}(X^{\{i\}}))_{\underline{k}} = p(X_{\underline{k}}^{\{i\}}) = (p^{(r)}X)_{\underline{k}}^{\{i\}} \,.$$

Since this holds for each $\underline{k} \in \Delta^{r^{op}}$ we conclude that $p^{(r)}(X^{\{i\}}) = (p^{(r)}X)^{\{i\}}$. The proof for $q^{(r)}$ is similar.
e) By induction on n. It is clear when $n = 1$. Suppose, inductively, that the lemma holds for $(n-1)$. By the inductive hypothesis applied to X_s (for $s \in \Delta^{op}$) and by a) we have

$$(p^{(r)} \dots p^{(n-2)} p^{(n-1)} X)_s = p^{(r-1)} \dots p^{(n-3)} p^{(n-2)} X_s = p^{(r-1)} X_s = (p^{(r)}X)_s \,.$$

Therefore $p^{(r)} \dots p^{(n-2)} p^{(n-1)} X = p^{(r)}X$. □

2.2.2 Closure Properties

The Segal-type models in this work are full subcategories of $[\Delta^{n^{op}}, \mathsf{Set}]$ satisfying a number of closure properties, which we discuss below for future reference.

Let $\mathscr{C} \subset [\Delta^{n^{op}}, \mathsf{Set}]$ be a full subcategory of n-fold simplicial sets containing the terminal object. We consider the following closure properties of $\mathscr{C}$:

C1 Repletion under isomorphisms; that is, if $A \cong B$ in $[\Delta^{n^{op}}, \mathsf{Set}]$ and $A \in \mathscr{C}$ then $B \in \mathscr{C}$.
C2 Closure under finite products.
C3 Closure under small coproducts.
C4 If the small coproduct $\coprod_i A_i$ is in $\mathscr{C}$, then each $A_i \in \mathscr{C}$.

Lemma 2.2.8 *Let $\mathscr{C}$ be a full subcategory of $[\Delta^{n^{op}}, \mathsf{Set}]$ satisfying the closure properties* ***C1–C4****. Then:*

C5 Every discrete n-fold simplicial set is an object of $\mathscr{C}$.
C6 If X is a discrete n-fold simplicial set and $f : E \to X$ is a morphism in $\mathscr{C}$ then for each $x \in X$ the fiber E_x of f at x is in $\mathscr{C}$.
C7 If $A \xrightarrow{f} X \xleftarrow{g} B$ is a diagram in $\mathscr{C}$ with X discrete, then $A \times_X B \in \mathscr{C}$.

Proof

C5 Every discrete n-fold simplicial set is a small coproduct of the terminal object, thus it is in $\mathscr{C}$ by **C3**.
C6 We have $E = \coprod_{x \in X} E_x$. Thus, since $E \in \mathscr{C}$, each $E_x \in \mathscr{C}$ by **C4**.
C7 We have

$$A \times_X B = \coprod_{x \in X} A_x \times B_x\ . \tag{2.9}$$

By **C6**, $A_x \in \mathscr{C}$ and $B_x \in \mathscr{C}$, so by **C2**, $A_x \times B_x \in \mathscr{C}$ and by **C3** and (2.9) we conclude that $A \times_X B \in \mathscr{C}$.

□

2.3 n-Fold Internal Categories

Definition 2.3.1 Let $\mathscr{C}$ be a category with finite limits. An internal category X in $\mathscr{C}$ is a diagram in $\mathscr{C}$

$$X_1 \times_{X_0} X_1 \xrightarrow{\ m\ } X_1 \underset{s}{\overset{d_0,\ d_1}{\rightrightarrows\!\!\leftarrow}} X_0 \tag{2.10}$$

where the maps d_0, d_1, s, m satisfy the axioms:

(1) $d_0 s = d_1 s = \mathrm{Id}_{X_0}$.
(2) $d_1 p_2 = d_1 m$, $d_0 p_1 = d_0 m$, where $p_i : X_1 \times_{X_0} X_1 \to X_1$, $i = 1, 2$ are the two projections.
(3) $m \begin{pmatrix} \mathrm{Id}_{X_1} \\ s\, d_0 \end{pmatrix} = \mathrm{Id}_{X_1} = \begin{pmatrix} s\, d_1 \\ \mathrm{Id}_{X_1} \end{pmatrix}$.
(4) $m\,(\mathrm{Id}_{X_1} \times_{X_0} m) = m\,(m \times_{X_0} \mathrm{Id}_{X_1})$.

X_0 is called the 'object of objects', X_1 the 'object of arrows', d_0 and d_1 are called respectively the 'source' and 'target' maps, s is the 'identity map' and m is the 'composition map'. When $\mathscr{C} = \mathsf{Set}$ we obtain the axioms of a small category and $X_1 \times_{X_0} X_1$ is the set of composable arrows.

Definition 2.3.2 If X and Y are internal categories in $\mathscr{C}$, an internal functor $F : X \to Y$ is a diagram in $\mathscr{C}$

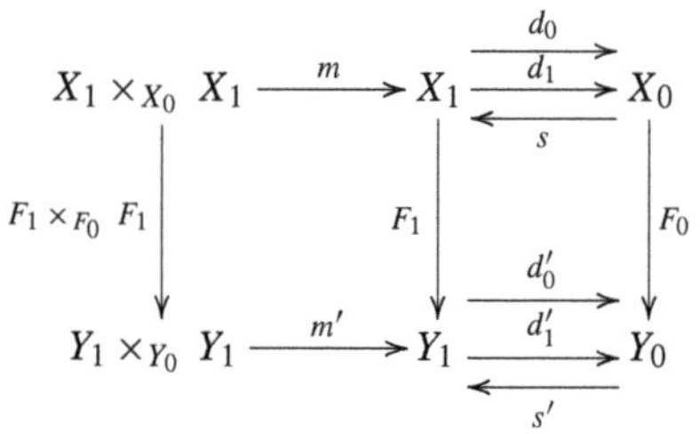

$$\begin{array}{ccccc}
X_1 \times_{X_0} X_1 & \xrightarrow{m} & X_1 & \overset{d_0,\ d_1}{\underset{s}{\rightrightarrows\!\!\leftarrow}} & X_0 \\
\downarrow{\scriptstyle F_1 \times_{F_0} F_1} & & \downarrow{\scriptstyle F_1} & & \downarrow{\scriptstyle F_0} \\
Y_1 \times_{Y_0} Y_1 & \xrightarrow{m'} & Y_1 & \overset{d'_0,\ d'_1}{\underset{s'}{\rightrightarrows\!\!\leftarrow}} & Y_0
\end{array}$$

where the corresponding maps commute, that is, they satisfy

(1) $d'_0\, F_1 = F_0\, d_0, \qquad d'_1\, F_1 = F_0\, d_1$.
(2) $F_1\, s = s'\, F_0$.
(3) $F_1\, m = m'\,(F_1 \times_{F_0} F_1)$.

We denote by $\mathsf{Cat}\,\mathscr{C}$ the category of internal categories and internal functors. When $\mathscr{C} = \mathsf{Set}$, this is the category Cat of small categories.

Definition 2.3.3 The category $\mathsf{Gpd}\,\mathscr{C}$ of internal groupoids in $\mathscr{C}$ is the full subcategory of $\mathsf{Cat}\,\mathscr{C}$ whose objects X are such that there is a map

$$i : X_1 \to X_1$$

such that

$$m(\mathrm{Id}_{X_1}, i) = s d_1, \qquad m(i, \mathrm{Id}_{X_1}) = s d_0\ .$$

The map i gives the inverse to each (internal) arrow. When $\mathscr{C} = \mathsf{Set}$, $\mathsf{Gpd}\mathscr{C}$ is the category Gpd of small groupoids.

Definition 2.3.4 The category $\mathsf{Cat}^n(\mathscr{C})$ of n-fold categories in $\mathscr{C}$ is defined inductively by iterating n times the internal category construction. That is, $\mathsf{Cat}^1(\mathscr{C}) = \mathsf{Cat}\,\mathscr{C}$ and, for $n > 1$,

$$\mathsf{Cat}^n(\mathscr{C}) = \mathsf{Cat}\,(\mathsf{Cat}^{n-1}(\mathscr{C})).$$

Definition 2.3.5 The category $\mathsf{Gpd}^n(\mathscr{C})$ of n-fold groupoids in $\mathscr{C}$ is defined inductively by iterating n times the internal groupoid construction. That is, $\mathsf{Gpd}^1(\mathscr{C}) = \mathsf{Gpd}\mathscr{C}$ and, for $n > 1$,

$$\mathsf{Gpd}^n(\mathscr{C}) = \mathsf{Gpd}(\mathsf{Gpd}^{n-1}(\mathscr{C})).$$

2.4 Multi-Nerve Functors

There is a nerve functor

$$N : \mathsf{Cat}\,\mathscr{C} \to [\Delta^{op}, \mathscr{C}]$$

such that, for $X \in \mathsf{Cat}\,\mathscr{C}$

$$(NX)_k = \begin{cases} X_0, & k = 0; \\ X_1, & k = 1; \\ X_1 \times_{X_0} \overset{k}{\cdots} \times_{X_0} X_1, & k > 1. \end{cases}$$

The following fact is well known (see for instance [117]):

Proposition 2.4.1 *A simplicial object in $\mathscr{C}$ is the nerve of an internal category in $\mathscr{C}$ if and only if all the Segal maps are isomorphisms.*

Remark 2.4.2 If $\mathscr{C}$ admits some class of limits, then these are inherited levelwise by $[\Delta^{op}, \mathscr{C}]$. Suppose that $D : \mathscr{I} \to [\Delta^{n^{op}}, \mathscr{C}]$ admits such a levelwise limit $\mathscr{L} = \lim_i D_i$ and each of its vertices satisfies the Segal condition; that is, for each $k \geq 2$

$$(D_i)_k \cong (D_i)_1 \times_{(D_i)_0} \overset{k}{\cdots} \times_{(D_i)_0} (D_i)_1 \,.$$

We claim that

$$\mathscr{L}_k \cong \mathscr{L}_1 \times_{\mathscr{L}_0} \overset{k}{\cdots} \times_{\mathscr{L}_0} \mathscr{L}_1 \,. \tag{2.11}$$

In fact, since limits in $[\Delta^{op}, \mathscr{C}]$ are computed pointwise, $\mathscr{L}_k = (\lim_i D_i)_k = \lim_i (D_i)_k$ for all $k \geq 0$. Since the pullbacks in the Segal condition commute with all limits that exist in $\mathscr{C}$, we obtain

$$\mathscr{L}_k = (\lim_i D_i)_k = \lim_i (D_i)_k = \lim_i \{(D_i)_1 \times_{(D_i)_0} \overset{k}{\cdots} \times_{(D_i)_0} (D_i)_1\}$$

$$= \lim_i (D_i)_1 \times_{\lim_i (D_i)_0} \overset{k}{\cdots} \times_{\lim_i (D_i)_0} \lim_i (D_i)_1 = \mathscr{L}_1 \times_{\mathscr{L}_0} \overset{k}{\cdots} \times_{\mathscr{L}_0} \mathscr{L}_1 \,,$$

proving (2.11).

Definition 2.4.3 By iterating the nerve construction, we obtain the multinerve functor

$$N_{(n)} : \mathsf{Cat}^{\mathsf{n}}(\mathscr{C}) \to [\Delta^{n^{op}}, \mathscr{C}] \,.$$

More precisely, $N_{(n)}$ is defined recursively as

$$N_{(1)} = N : \mathsf{Cat}\,\mathscr{C} \to [\Delta^{^{op}}, \mathscr{C}] \,.$$

Given $N_{(n-1)} : \mathsf{Cat}^{\mathsf{n-1}}(\mathscr{C}) \to [\Delta^{n-1^{op}}, \mathscr{C}]$, let $N_{(n)}$ be the composite

$$N_{(n)} : \mathsf{Cat}^{\mathsf{n}}(\mathscr{C}) \xrightarrow{N_{(1)}} [\Delta^{^{op}}, \mathsf{Cat}^{\mathsf{n-1}}(\mathscr{C})] \xrightarrow{\overline{N}_{(n-1)}} [\Delta^{^{op}}, [\Delta^{n-1^{op}}, \mathscr{C}]] \xrightarrow{\xi_1^{-1}} [\Delta^{n^{op}}, \mathscr{C}] \,,$$

where $\overline{N}_{(n-1)}$ is as in Definition 2.1.1 and ξ_1 is as in the proof of Lemma 2.1.7.

Definition 2.4.4 An internal n-fold category $X \in \mathsf{Cat}^{\mathsf{n}}(\mathscr{C})$ is discrete when $N_{(n)}X$ is a constant functor.

Thus in a discrete internal n-fold category all structure maps are identities.

We next give a characterization of those multi-simplicial objects $X \in [\Delta^{n^{op}}, \mathscr{C}]$ which are in the essential image of the multinerve functor $N_{(n)} : \mathsf{Cat}^{\mathsf{n}}(\mathscr{C}) \to [\Delta^{n^{op}}, \mathscr{C}]$. We are going to show that they are those multi-simplicial objects whose Segal maps in all directions are isomorphisms, as illustrated in the following Proposition.

Proposition 2.4.5 *$X \in [\Delta^{n^{op}}, \mathscr{C}]$ is in the essential image of the multinerve functor $N_{(n)}$ of Definition 2.4.3 if and only if, for each $1 \le r \le n$, $X^{\{r\}}$ satisfies the Segal condition that for all $s \ge 2$ there is an isomorphism in $[\Delta^{n-1^{op}}, \mathscr{C}]$*

$$X_s^{\{r\}} \cong X_1^{\{r\}} \times_{X_0^{\{r\}}} \overset{s}{\cdots} \times_{X_0^{\{r\}}} X_1^{\{r\}} . \tag{2.12}$$

Proof By induction on n. For $n = 1$ it is Proposition 2.4.1. Suppose the lemma holds for $(n-1)$. Let $X \in [\Delta^{n^{op}}, \mathscr{C}]$ satisfy (2.12). Then for each $k_1 \in \Delta^{op}$

$$(X_s^{\{r\}})_{k_1} \cong (X_1^{\{r\}})_{k_1} \times_{(X_0^{\{r\}})_{k_1}} \overset{s}{\cdots} \times_{(X_0^{\{r\}})_{k_1}} (X_1^{\{r\}})_{k_1} \,, \tag{2.13}$$

which, by Remark 2.1.6 is the same as

$$(X_{k_1})_s^{\{r-1\}} \cong (X_{k_1})_1^{\{r-1\}} \times_{(X_{k_1})_0^{\{r-1\}}} \overset{s}{\cdots} \times_{(X_{k_1})_0^{\{r-1\}}} (X_{k_1})_1^{\{r-1\}}. \tag{2.14}$$

This shows that $X_{k_1} \in [\Delta^{n-1^{op}}, \mathscr{C}]$ satisfies the inductive hypothesis and therefore X_{k_1} is in the essential image of N_{n-1} for all $k_1 \geq 0$. Taking $r = 1$ in (2.12) we obtain $X_s \cong X_1 \times_{X_0} \overset{s}{\cdots} \times_{X_0} X_1$. We conclude by Proposition 2.4.1 that X is in the essential image of $N_{(n)}$.

Conversely, let $X = N_{(n)} Y$ with $Y \in \mathsf{Cat}^{\mathsf{n}}(\mathscr{C})$. Since $N_{(n)} = \xi_1^{-1} \overline{N}_{(n-1)} N_1$ by Proposition 2.4.1 we obtain

$$\begin{aligned} X_s &= (N_{(n)}Y)_s \cong N_{(n-1)}Y_s \\ &\cong N_{(n-1)}Y_1 \times_{N_{(n-1)}Y_0} \overset{s}{\cdots} \times_{N_{(n-1)}Y_0} N_{(n-1)}Y_1 = X_1 \times_{X_0} \overset{s}{\cdots} \times_{X_0} X_1 \,, \end{aligned}$$

proving (2.12) when $r = 1$. For $1 < r \leq n$, applying the inductive hypothesis to X_{k_1} (which is in the essential image of $N_{(n-1)}$) we obtain (2.14), which by Remark 2.1.6 is the same as (2.13). Since this holds for each $k_1 \in \Delta^{op}$, (2.12) follows. □

We now deduce that every object of $\mathsf{Cat}^{\mathsf{n}}(\mathscr{C})$ can be considered as an internal category in $\mathsf{Cat}^{\mathsf{n-1}}(\mathscr{C})$ in n possible ways, corresponding to the n simplicial directions of the multinerve.

Proposition 2.4.6 *For each $1 \leq i \leq n$ there is an isomorphism*

$$\widetilde{\xi}_i : \mathsf{Cat}^{\mathsf{n}}(\mathscr{C}) \to \mathsf{Cat}\,(\mathsf{Cat}^{\mathsf{n-1}}(\mathscr{C}))$$

making the following diagram commute:

$$\begin{array}{ccc} \mathsf{Cat}^{\mathsf{n}}(\mathscr{C}) & \xrightarrow{N_{(n)}} & [\Delta^{n^{op}}, \mathscr{C}] \\ \downarrow{\scriptstyle \widetilde{\xi}_i} & & \downarrow{\scriptstyle \xi_i} \\ \mathsf{Cat}\,(\mathsf{Cat}^{\mathsf{n-1}}(\mathscr{C})) & \xrightarrow[\overline{N}_{(n-1)}N_1]{} & [\Delta^{op}, [\Delta^{n-1^{op}}, \mathscr{C}]] \end{array} \tag{2.15}$$

where ξ_i is as in the proof of Lemma 2.1.7.

Proof We claim that, for each $Y \in \mathsf{Cat}^{\mathsf{n}}(\mathscr{C})$, $(\xi_i N_{(n)} Y)_k$ is the multinerve of an object $Z_{ik} \in \mathsf{Cat}^{\mathsf{n-1}}(\mathscr{C})$ and

$$(\xi_i N_{(n)} Y)_k = N_{(n-1)} Z_{ik} \,.$$

In fact, by Proposition 2.4.5, $N_{(n)}Y$ has all the Segal maps isomorphisms, hence the same holds for $(\xi_i N_{(n)} Y)_k$ so that, by Proposition 2.4.5 again, the claim follows.

Consider the simplicial object

$$\xi_i N_{(n)} Y = \overline{N}_{(n-1)} Z_i \in [\Delta^{op}, [\Delta^{n-1^{op}}, \mathscr{C}]].$$

By Proposition 2.4.5 its Segal maps are isomorphisms. Hence Z_i is the nerve of an internal category $\widetilde{\xi}_i Y \in \mathsf{Cat}\,(\mathsf{Cat}^{n-1}(\mathscr{C}))$, that is,

$$\overline{N}_{(n-1)} N_1 \widetilde{\xi}_i Y = \xi_i N_{(n)} Y \ .$$

This defines $\widetilde{\xi}_i$ and shows that (2.15) commutes.

Corollary 2.4.7 *For each $1 \leq i \leq n$, the following diagram commutes:*

$$\begin{array}{ccc} \mathsf{Cat}^n(\mathscr{C}) & \xrightarrow{N_{(n)}} & [\Delta^{n^{op}}, \mathscr{C}] \\ {\scriptstyle \widetilde{\xi}_1^{-1}\widetilde{\xi}_i} \downarrow & & \downarrow {\scriptstyle (\text{-})^{\{i\}}} \\ \mathsf{Cat}^n(\mathscr{C}) & \xrightarrow[N_{(n)}]{} & [\Delta^{n^{op}}, \mathscr{C}] \end{array}$$

Proof By Proposition 2.4.6, $\xi_1 N_{(n)} = \overline{N}_{(n-1)} N_1 \widetilde{\xi}_1$, therefore $N_{(n)} \widetilde{\xi}_1^{-1} = \xi_1^{-1} \overline{N}_{(n-1)} N_1$. Since, by Remark 2.1.9, $\xi_1^{-1} \xi_i = (\text{-})^{\{i\}}$ it follows that $N_{(n)} \widetilde{\xi}_1^{-1} \widetilde{\xi}_i = \xi_1^{-1} \overline{N}_{(n-1)} N_1 \widetilde{\xi}_i = \xi_1^{-1} \xi_i N_{(n)} = (\text{-})^{\{i\}} N_{(n)}$. □

Definition 2.4.8 Let $1 \leq i \leq n$. The nerve functor in the ith direction is defined as the composite

$$N^{(i)} : \mathsf{Cat}^n(\mathscr{C}) \xrightarrow{\widetilde{\xi}_i} \mathsf{Cat}\,(\mathsf{Cat}^{n-1}(\mathscr{C})) \xrightarrow{N} [\Delta^{op}, \mathsf{Cat}^{n-1}(\mathscr{C})] \ ,$$

where $\widetilde{\xi}_i$ is as in Proposition 2.4.6.

Note that

$$N_{(n)} = N^{(n)} \ldots N^{(2)} N^{(1)}. \tag{2.16}$$

Lemma 2.4.9 *For each $1 \leq i \leq n$ there is a commuting diagram*

$$\begin{array}{ccccc} \mathsf{Cat}^n(\mathscr{C}) & & \xrightarrow{\quad N^{(i)} \quad} & & [\Delta^{op}, \mathsf{Cat}^{n-1}(\mathscr{C})] \\ {\scriptstyle N_{(n)}} \downarrow & & & & \downarrow {\scriptstyle \overline{N}_{(n-1)}} \\ [\Delta^{n^{op}}, \mathscr{C}] & \xrightarrow[(\text{-})^{\{i\}}]{} & [\Delta^{n^{op}}, \mathscr{C}] & \xrightarrow[\xi_1]{} & [\Delta^{op}, [\Delta^{n-1^{op}}, \mathscr{C}]] \end{array}$$

Proof By Definition 2.4.8, Proposition 2.4.6 and Remark 2.1.9 we have

$$\overline{N}_{(n-1)} N^{(i)} = \overline{N}_{(n-1)} N \widetilde{\xi}_i = \xi_i N_{(n)} = \xi_1 (\text{-})^{\{i\}} N_{(n)} \ .$$

□

Remark 2.4.10 Let $ob : \mathsf{Cat}\mathcal{C} \to \mathcal{C}$ be the object of objects functor. The left adjoint to ob is the discrete internal category functor

$$d : \mathcal{C} \to \mathsf{Cat}\mathcal{C}$$

associating to an object $X \in \mathcal{C}$ the discrete internal category on X. By Definition 2.4.4, the nerve of the discrete internal category on X is the constant simplicial object on X.

By Proposition 2.4.6 we then have

$$\mathsf{Cat}^{\mathsf{n}}(\mathcal{C}) \overset{\xi_n}{\cong} \mathsf{Cat}\,(\mathsf{Cat}^{\mathsf{n-1}}(\mathcal{C})) \underset{d}{\overset{ob}{\rightleftarrows}} \mathsf{Cat}^{\mathsf{n-1}}(\mathcal{C})\,.$$

Definition 2.4.11 We define $d^{(1)} = d : \mathcal{C} \to \mathsf{Cat}\,\mathcal{C}$ and for $n > 1$,

$$d^{(n)} = \xi_n^{-1} \circ d : \mathsf{Cat}^{\mathsf{n-1}}(\mathcal{C}) \overset{d}{\to} \mathsf{Cat}\,(\mathsf{Cat}^{\mathsf{n-1}}(\mathcal{C})) \xrightarrow{\xi_n^{-1}} \mathsf{Cat}^{\mathsf{n}}(\mathcal{C})\,.$$

Thus $d^{(n)}$ is the discrete inclusion of $\mathsf{Cat}^{\mathsf{n-1}}(\mathcal{C})$ into $\mathsf{Cat}^{\mathsf{n}}(\mathcal{C})$ in the nth direction.

Let $d_n : [\Delta^{n-1^{op}}, \mathcal{C}] \to [\Delta^{n^{op}}, \mathcal{C}]$ be defined by

$$(d_n Y)_{k_1 \dots k_n} = Y_{k_1 \dots k_n} \tag{2.17}$$

for all $Y \in [\Delta^{n-1^{op}}, \mathcal{C}]$ and $(k_1 \dots k_n) \in \Delta^{n^{op}}$.

Lemma 2.4.12 *Let $d^{(n)}$ be as in Definition 2.4.11 and d_n as in* (2.17). *Then the following diagram commutes:*

$$\begin{array}{ccc} \mathsf{Cat}^{\mathsf{n-1}}(\mathcal{C}) & \xrightarrow{N_{(n-1)}} & [\Delta^{n-1^{op}}, \mathcal{C}] \\ {\scriptstyle d^{(n)}}\downarrow & & \downarrow{\scriptstyle d_n} \\ \mathsf{Cat}^{\mathsf{n}}(\mathcal{C}) & \xrightarrow[N_{(n)}]{} & [\Delta^{n^{op}}, \mathcal{C}] \end{array} \tag{2.18}$$

Proof We note that d_n is the composite

$$[\Delta^{n-1^{op}}, \mathcal{C}] \overset{d}{\to} [\Delta^{op}, [\Delta^{n-1^{op}}, \mathcal{C}]] \xrightarrow{\xi_n^{-1}} [\Delta^{n^{op}}, \mathcal{C}]\,,$$

where d is the constant simplicial object functor. In fact, for each $Y \in [\Delta^{n-1^{op}}, \mathcal{C}]$ and $(k_1 \dots k_n) \in \Delta^{n^{op}}$,

$$(\xi_n^{-1} dY)(k_1 \dots k_n) = (dY)_{k_n}(k_1 \dots k_{n-1}) = Y(k_1 \dots k_n) = (d_n Y)(k_1 \dots k_n)\,,$$

therefore the commutativity of (2.18) is equivalent to the commutativity of the outer rectangle in the diagram:

$$\begin{array}{ccc} \mathsf{Cat}^{n-1}(\mathscr{C}) & \xrightarrow{N_{(n-1)}} & [\Delta^{n-1^{op}}, \mathscr{C}] \\ \Big\downarrow{\scriptstyle d} & & \Big\downarrow{\scriptstyle d} \\ \mathsf{Cat}(\mathsf{Cat}^{n-1}(\mathscr{C})) & \xrightarrow[\overline{N}_{(n-1)}N_1]{} & [\Delta^{op}, [\Delta^{n-1^{op}}, \mathscr{C}]] \\ \Big\downarrow{\scriptstyle \tilde{\xi}_n^{-1}} & & \Big\downarrow{\scriptstyle \xi_n^{-1}} \\ \mathsf{Cat}^n(\mathscr{C}) & \xrightarrow[N_{(n)}]{} & [\Delta^{n^{op}}, \mathscr{C}] \end{array} \tag{2.19}$$

The top diagram in (2.19) commutes by the definition of d and the bottom diagram in (2.19) commutes by Proposition 2.4.6. Hence the outer rectangle in (2.19) commutes, as required.

2.5 n-Fold Categories

When $\mathscr{C} = \mathsf{Set}$, $\mathsf{Cat}^n(\mathsf{Set})$ is simply denoted by Cat^n and called the category of n-fold categories (double categories when $n = 2$), and $\mathsf{Gpd}^n(\mathsf{Set})$ is denoted by Gpd^n and called the category of n-fold groupoids. For $n \geq 1$ the multinerve functor

$$N_{(n)} : \mathsf{Cat}^n \to [\Delta^{n^{op}}, \mathsf{Set}]$$

is fully faithful. This property justifies the following notational convention, which will be adopted throughout this book to streamline notation.

Notational Convention 2.5.1 We shall identify Cat^n with the subcategory $N_{(n)}\mathsf{Cat}^n$ of $[\Delta^{n^{op}}, \mathsf{Set}]$ which is the essential image of the functor $N_{(n)}$.

If we identify $[\Delta^{n-1^{op}}, [\Delta^{op}, \mathsf{Set}]]$ with $[\Delta^{n^{op}}, \mathsf{Set}]$ under the isomorphism associating to $Y \in [\Delta^{n-1^{op}}, [\Delta^{op}, \mathsf{Set}]]$ the n-fold simplicial set taking $(k_1, \dots, k_n) \in \Delta^{n^{op}}$ to $(Y_{k_1 \dots k_{n-1}})_{k_n}$, we obtain the following

Notational Convention 2.5.2 We identify $[\Delta^{n-1^{op}}, \mathsf{Cat}]$ with the full subcategory of $[\Delta^{n^{op}}, \mathsf{Set}]$ which is the essential image of the functor

$$[\Delta^{n-1^{op}}, \mathsf{Cat}] \xrightarrow{\overline{N}} [\Delta^{n-1^{op}}, [\Delta^{op}, \mathsf{Set}]] \xrightarrow{\cong} [\Delta^{n^{op}}, \mathsf{Set}]\,.$$

That is, objects of $[\Delta^{n-1^{op}}, \mathsf{Cat}]$ are n-fold simplicial sets X such that for all $(k_1, \dots, k_{n-1}) \in \Delta^{n-1^{op}}$, the simplicial set $X_{k_1 \dots k_{n-1}}$ is the nerve of a category.

Remark 2.5.3 Using the fact (see (2.16)) that $N_{(n)} = N^{(n)}N^{(n-1)}\dots N^{(1)}$ we see that $N^{(n)}$ is the composite

$$\mathsf{Cat}^{\mathsf{n}} \xhookrightarrow{N^{(n-1)}\dots N^{(1)}} [\Delta^{n-1^{op}}, \mathsf{Cat}] \xrightarrow{\overline{N}} [\Delta^{n-1^{op}}, [\Delta^{op}, \mathsf{Set}]] \cong [\Delta^{n^{op}}, \mathsf{Set}]\,,$$

where the last isomorphism is as in Notational Convention 2.5.2.

With these conventions, the diagram of Corollary 2.4.7 becomes, for each $1 \le i \le n$,

$$\begin{array}{ccc} \mathsf{Cat}^{\mathsf{n}} & \hookrightarrow & [\Delta^{n^{op}}, \mathsf{Set}] \\ \downarrow{\scriptstyle \widetilde{\xi}_1^{-1}\widetilde{\xi}_i} & & \downarrow{\scriptstyle (\text{-})^{\{i\}}} \\ \mathsf{Cat}^{\mathsf{n}} & \hookrightarrow & [\Delta^{n^{op}}, \mathsf{Set}] \end{array}$$

Thus, given $X \in \mathsf{Cat}^{\mathsf{n}}$, $\widetilde{\xi}_1^{-1}\widetilde{\xi}_i X$ can be identified with $X^{\{i\}}$, and we obtain

Notational Convention 2.5.4 If $X \in \mathsf{Cat}^{\mathsf{n}}$, then $X^{\{i\}}$ is the n-fold category X viewed as an internal category in $\mathsf{Cat}^{\mathsf{n-1}}$ along direction i, and we denote $X^{\{1\}}$ by X.

Similarly, the diagram of Lemma 2.4.9 becomes

$$\begin{array}{ccc} \mathsf{Cat}^{\mathsf{n}} & \xrightarrow{N^{(i)}} & [\Delta^{op}, \mathsf{Cat}^{\mathsf{n-1}}] \\ \downarrow & & \downarrow \\ [\Delta^{n^{op}}, \mathsf{Set}] & \xrightarrow[\xi_1(\text{-})^{\{i\}}]{} & [\Delta^{op}, [\Delta^{n-1^{op}}, \mathsf{Set}]] \end{array}$$

Thus given $X \in \mathsf{Cat}^{\mathsf{n}}$ we can identify $N^{(i)}X$ with $\xi_1 X^{\{i\}}$ where $(\xi_1 X^{\{i\}})_k = X_k^{\{i\}}$ for all $k \in \Delta^{op}$. In other words, the simplicial object

$$\cdots X_1^{\{i\}} \times_{X_0^{\{i\}}} X_1^{\{i\}} \substack{\longrightarrow\\ \longrightarrow\\ \longrightarrow} X_1^{\{i\}} \substack{\longrightarrow\\ \longleftarrow\\ \longrightarrow} X_0^{\{i\}}$$

gives the nerve of X in the ith direction.

Finally, the diagram of Lemma 2.4.12 becomes

$$\begin{array}{ccc} \mathsf{Cat}^{\mathsf{n-1}} & \hookrightarrow & [\Delta^{n-1^{op}}, \mathsf{Set}] \\ \downarrow{\scriptstyle d^{(n)}} & & \downarrow \\ \mathsf{Cat}^{\mathsf{n}} & \hookrightarrow & [\Delta^{n^{op}}, \mathsf{Set}] \end{array}$$

We can therefore identify $d^{(n)}$ with the inclusion

$$\mathsf{Cat^{n-1}} \hookrightarrow \mathsf{Cat^n}\,.$$

That is, we obtain

Notational Convention 2.5.5 We identify $(n-1)$-fold categories with n-fold categories X such that, for all $(k_1 \dots k_{n-1}) \in \Delta^{n-1^{op}}$, the category $X_{k_1\dots k_{n-1}}$ is discrete.

Since, as observed before, (see (2.4) and (2.5)) $X^{\{1\}} = X$ and $(\xi_1 X)_k = X_k$ for each $X \in \mathsf{Cat^n}$, $k \in \Delta^{op}$, using the Notational Convention 2.5.2, the commutative diagram of Lemma 2.4.9 when $i = 1$ and $\mathscr{C} = \mathsf{Set}$ becomes

$$\begin{array}{ccc} \mathsf{Cat^n} & \xrightarrow{N^{(1)}} & [\Delta^{op}, \mathsf{Cat^{n-1}}] \\ \big\downarrow & & \big\downarrow \\ [\Delta^{n^{op}}, \mathsf{Set}] & \xrightarrow[\xi_1]{} & [\Delta^{op}, [\Delta^{n-1^{op}}, \mathsf{Set}]] \end{array}$$

Thus, for each $X \in \mathsf{Cat^n}$ and $k \in \Delta^{op}$,

$$(N^{(1)}X)_k = (\xi_1 X)_k = X_k\,. \tag{2.20}$$

It also follows that X, viewed as an object of $[\Delta^{op}, \mathsf{Cat^{n-1}}]$ along direction 1 has the form

$$\cdots X_1 \times_{X_0} X_1 \substack{\longrightarrow\\[-1em]\longrightarrow\\[-1em]\longrightarrow} X_1 \substack{\longrightarrow\\[-1em]\longleftarrow\\[-1em]\longrightarrow} X_0$$

Finally, we will adopt a common terminology for levelwise concepts. Suppose that P_r is a property that applies to r-fold simplicial sets and assume that X is an n-fold simplicial set with $n \geq r$. Then we say that X has the property P_r levelwise if for all sequences of indices $k_1, \dots, k_{n-r}$ the r-fold simplicial set $X_{k_1\dots k_{n-r}}$ has the property P_r. So under the previous notational conventions we obtain

Notational Convention 2.5.6

1) An object of $[\Delta^{n^{op}}, \mathsf{Set}]$ is in the subcategory $[\Delta^{(n-r)}, \mathsf{Cat^r}]$ if and only if it is a levelwise r-fold category.
2) An object of $[\Delta^{n^{op}}, \mathsf{Set}]$ is in the subcategory $[\Delta^{n-r^{op}}, \mathsf{Set}]$ if and only if it is levelwise r-fold discrete.
3) Let $\mathsf{Cat_{hd}}$ be the category of equivalence relations. An object of $[\Delta^{n^{op}}, \mathsf{Set}]$ is in the subcategory $[\Delta^{n-1^{op}}, \mathsf{Cat_{hd}}]$ if and only if it is a levelwise equivalence relation.

We next consider some low-dimensional examples of n-fold categories in which the multinerves can be visualized geometrically.

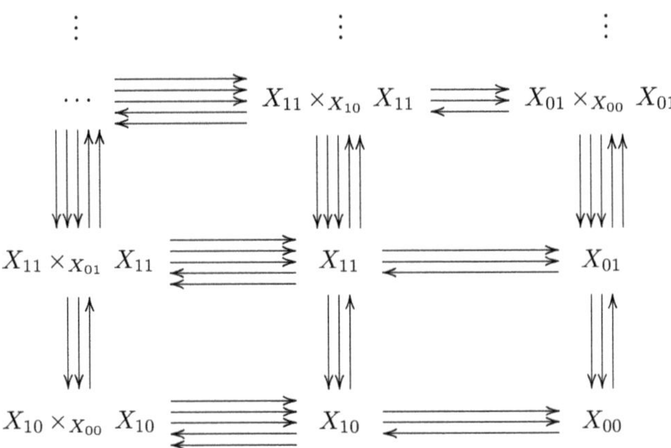

Fig. 2.2 Corner of the double nerve of a double category X

Example 2.5.7 (Double Nerves of Double Categories)
Consider the double nerve functor

$$N_{(2)} : \mathsf{Cat}(\mathsf{Cat}) \to [\Delta^{2^{op}}, \mathsf{Set}]$$

from double categories to bisimplicial sets. Given $X \in \mathsf{Cat}(\mathsf{Cat})$ we can visualize the corner of its double nerve as in Fig. 2.2 on page 37.

We see that all the Segal maps in both the vertical and horizontal directions are isomorphisms. It is possible to give a geometric interpretation to this double nerve by thinking of X_{00} as sets of points, X_{01} as sets of vertical arrows, X_{10} as sets of horizontal arrows, and X_{11} as sets of squares. Then horizontal and vertical arrows compose (respectively horizontally and vertically) while squares compose horizontally via

$$X_{11} \times_{X_{01}} X_{11} \to X_{11}$$

and vertically via

$$X_{11} \times_{X_{10}} X_{11} \to X_{11}\ .$$

This geometric data, and the relative axioms, are an alternative way to define double categories: see for instance [53] for more details.

Although this presentation is useful for the study of some aspects of double categories, for this work the presentation in terms of internal categories in Cat and the one in terms of their double nerves is sufficient. Figure 2.3 on page 38 is a geometric picture of the corner of the double nerve of a double category.

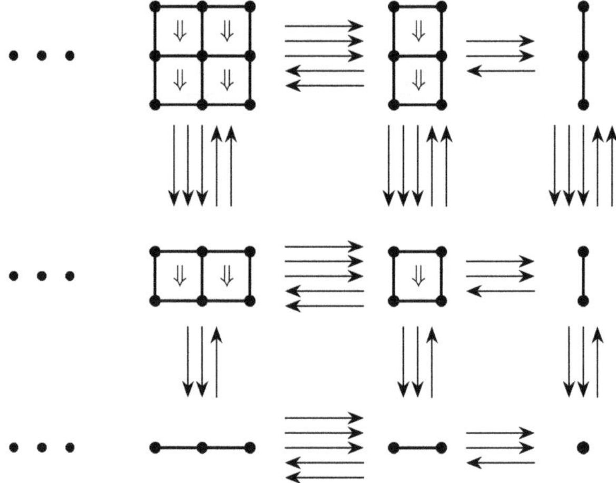

Fig. 2.3 Geometric picture of the corner of the double nerve of a double category

Example 2.5.8 (3-Fold Nerves of 3-Fold Categories) Consider the 3-fold nerve functor

$$N_{(3)} : \mathsf{Cat}^3 \to [\Delta^{3^{op}}, \mathsf{Set}] \ .$$

Given $X \in \mathsf{Cat}^3$, a picture of the corner of $N_{(3)}X$ is given in Fig. 2.4 on page 39, where direction 1 is horizontal and direction 3 is vertical. We have omitted drawing the degeneracy operators for simplicity.

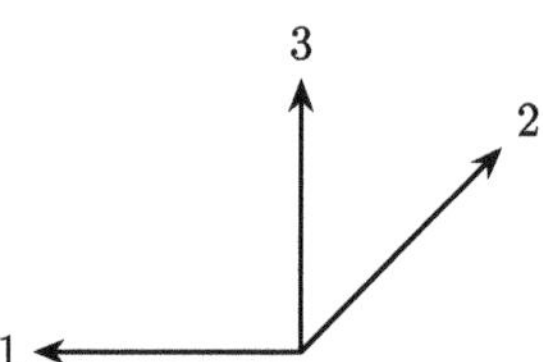

The isomorphisms above Fig. 2.4 correspond to the Segal condition.

We can obtain a geometric visualisation of Fig. 2.4 by setting

X_{000} = set of objects;
X_{010} = set of arrows in direction 2;
X_{100} = set of arrows in direction 1;
X_{001} = set of arrows in direction 3;
X_{110} = set of squares in directions 1,2;
X_{011} = set of squares in directions 2,3;
X_{101} = set of squares in directions 1,3;
X_{111} = set of cubes.

In the following picture, for all $i, j, k \in \Delta^{op}$
$X_{2jk} \cong X_{1jk} \times_{X_{0jk}} X_{1jk}$, $X_{i2k} \cong X_{i1k} \times_{X_{i0k}} X_{i1k}$, $X_{ij2} \cong X_{ij1} \times_{X_{ij0}} X_{ij1}$.

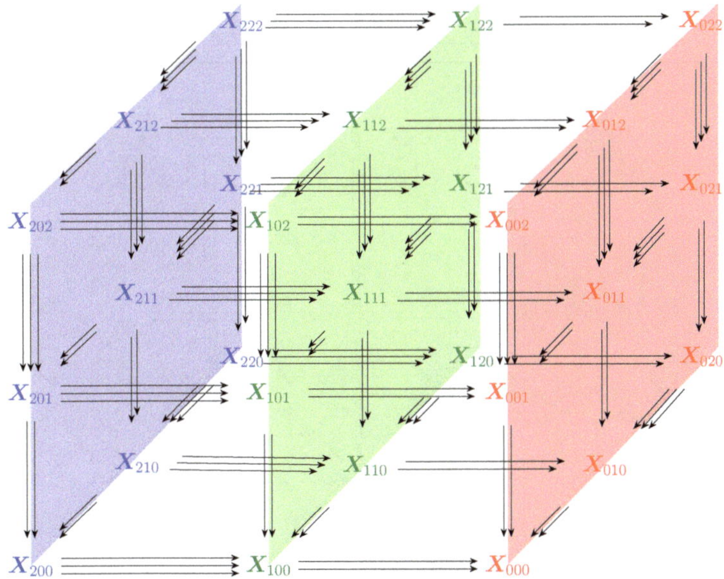

Fig. 2.4 Corner of the 3-fold nerve of a 3-fold category X

Arrows in direction i can be composed in direction i (for $i = 1, 2, 3$). Squares in direction i, j ($i, j = 1, 2, 3$) can be composed in direction i and in direction j. Cubes can be composed in all three directions. A geometric picture can be found in Fig. 2.5 on page 40.

2.6 A Multi-Simplicial Description of Strict n-Categories

In this section we establish the analogue of Proposition 2.4.5 for strict n-categories. We recall the definition of a strict n-category, which is given by iterated enrichment. We refer to [77] for background on enriched categories.

Definition 2.6.1 The category $n\text{-}\mathsf{Cat}$ of strict n-categories is defined by induction of n. For $n = 1$, $1\text{-}\mathsf{Cat} = \mathsf{Cat}$. Given $(n-1)\text{-}\mathsf{Cat}$, let

$$n\text{-}\mathsf{Cat} = ((n-1)\text{-}\mathsf{Cat})\text{-}\mathsf{Cat}\ ,$$

where the enrichment is with respect to the cartesian monoidal structure.

Unravelling this inductive definition we see that strict n-categories have sets of cells in dimensions 0 up to n, and these compose in a way which is associative and unital. Below we describe these properties using the multi-simplicial language.

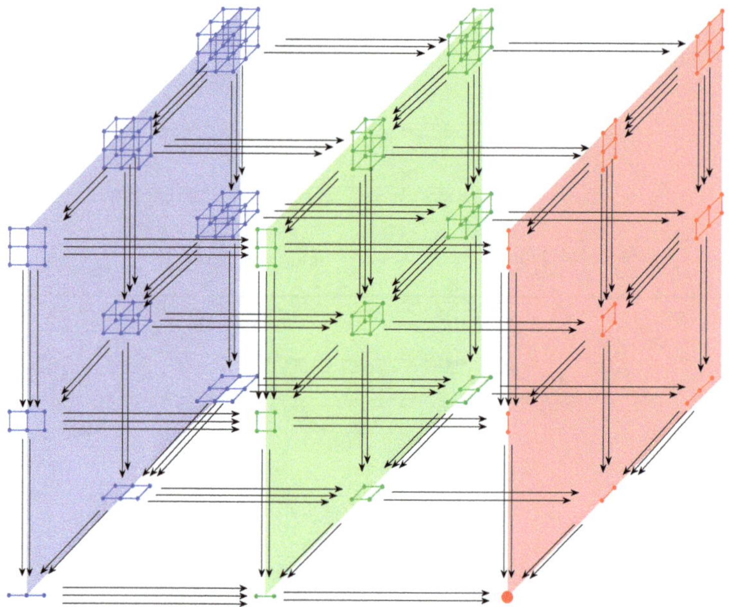

Fig. 2.5 Geometric picture of the corner of the 3-fold nerve of a 3-fold category

A strict n-category in which all cells are invertible is called a strict n-groupoid. We denote by n-Gpd the category of strict n-groupoids, where 1-$\mathsf{Gpd} = \mathsf{Gpd}$.

We prove a characterization of the image of the functors

$$J_n : n\text{-}\mathsf{Cat} \to [\Delta^{n-1^{op}}, \mathsf{Cat}\,] \quad \text{and} \quad N_{(n)} : n\text{-}\mathsf{Cat} \to [\Delta^{n^{op}}, \mathsf{Set}]\,.$$

The resulting multi-simplicial description of strict n-categories is helpful to build one's intuition about the weakening of the structure achieved with the Segal-type models of this work, which are also full subcategories of $[\Delta^{n-1^{op}}, \mathsf{Cat}\,]$.

To establish this characterization, we use the fact, proved in the Appendix of [55], that if $\mathcal{V}$ is a category satisfying mild conditions, categories enriched in $\mathcal{V}$ are internal categories in $\mathcal{V}$ whose object of objects is discrete, that is, it is the coproduct of copies of the terminal object. As observed in [55], these conditions are satisfied when $\mathcal{V} = \mathsf{Cat}^{\mathsf{n}}$ and $\mathcal{V} = n$-Cat.

Remark 2.6.2 Consider the composite

$$N : \mathcal{V}\text{-}\mathsf{Cat} \hookrightarrow \mathsf{Cat}\,\mathcal{V} \to [\Delta^{^{op}}, \mathcal{V}]\,,$$

where $\mathcal{V}$ is as in [55]. It follows by Ehresmann and Ehresmann [55] and by Proposition 2.4.1 that $X \in [\Delta^{^{op}}, \mathcal{V}]$ is in the essential image of N if and only if the Segal maps of X are isomorphisms and X_0 is discrete.

When $\mathcal{V} = (n-1)\text{-}\mathsf{Cat}$, we obtain the functor

$$N : n\text{-}\mathsf{Cat} = ((n-1)\text{-}\mathsf{Cat})\text{-}\mathsf{Cat} \to [\Delta^{^{op}}, (n-1)\text{-}\mathsf{Cat}]\,.$$

Iterating this construction we obtain the multinerve

$$J_n : n\text{-}\mathsf{Cat} \to [\Delta^{n-1^{op}}, \mathsf{Cat}]\,. \tag{2.21}$$

More precisely

Definition 2.6.3 J_n is defined recursively as

$$J_2 : 2\text{-}\mathsf{Cat} \hookrightarrow \mathsf{Cat}(\mathsf{Cat}) \xrightarrow{N} [\Delta^{^{op}}, \mathsf{Cat}]\,.$$

Given $J_{n-1} : (n-1)\text{-}\mathsf{Cat} \to [\Delta^{n-2^{op}}, \mathsf{Cat}]$, let J_n be the composite

$$J_n : n\text{-}\mathsf{Cat} = ((n-1)\text{-}\mathsf{Cat})\text{-}\mathsf{Cat} \hookrightarrow \mathsf{Cat}((n-1)\text{-}\mathsf{Cat}) \xrightarrow{N} [\Delta^{^{op}}, (n-1)\text{-}\mathsf{Cat}]$$
$$\xrightarrow{\overline{J}_{n-1}} [\Delta^{^{op}}, [\Delta^{n-2^{op}}, \mathsf{Cat}]] \overset{\xi_1^{-1}}{\cong} [\Delta^{n-1^{op}}, \mathsf{Cat}]\,,$$

where ξ_1 is as in Lemma 2.1.7.

We next give a characterization of the essential image of J_n.

Proposition 2.6.4 *Let $X \in [\Delta^{n-1^{op}}, \mathsf{Cat}]$. Then X is in the essential image of the functor J_n as in Definition 2.6.3 if and only if:*

a) The Segal maps in all directions are isomorphisms.
b) $X_0 \in [\Delta^{n-2^{op}}, \mathsf{Cat}]$ and $X_{\underset{1\ldots10}{r}} \in [\Delta^{n-r-2^{op}}, \mathsf{Cat}]$ are constant functors taking values in a discrete category for all $1 \le r \le n-2$.

Proof By induction on n. For $n = 2$ it follows by Remark 2.6.2 taking $\mathcal{V} = \mathsf{Cat}$. Suppose, inductively, that the lemma holds for $(n-1)$ and let $X = J_n Y$ with $Y \in n\text{-}\mathsf{Cat}$. By construction, J_n is the composite:

$$n\text{-}\mathsf{Cat} \hookrightarrow [\Delta^{^{op}}, (n-1)\text{-}\mathsf{Cat}] \xrightarrow{\overline{J}_{n-1}} [\Delta^{^{op}}, [\Delta^{n-2^{op}}, \mathsf{Cat}]] \cong [\Delta^{n-1^{op}}, \mathsf{Cat}]\,.$$

By Remark 2.6.2 when $\mathcal{V} = (n-1)\text{-}\mathsf{Cat}$, $(NY)_0$ is a constant functor taking values in a discrete category and the Segal maps of NY are isomorphisms. Hence $X_0 = (J_n Y)_0 = J_{n-1}(NY)_0$ is constant with values in a discrete category and the Segal maps of X in direction 1 are isomorphisms.

The Segal maps of $J_n X$ in direction $k > 1$ are levelwise the Segal maps of $X_t = J_{n-1}(NY)_t$ for each $t \ge 0$ and these are isomorphisms by the inductive hypothesis applied to $(NY)_t \in (n-1)\text{-}\mathsf{Cat}$. So in conclusion the Segal maps in all directions are isomorphisms, which is a).

Further $X_1 = J_{n-1}(NY)_1$ and since $(NY)_1 \in (n-1)\text{-}\mathsf{Cat}$, by the inductive hypothesis X_{10} and $X_{\underset{1\ldots10}{s}}$ are constant with values in a discrete category for all

$1 \leq s \leq n-3$. So in conclusion $X_{\underset{1...10}{r}}$ is constant with values in a discrete category for all $1 \leq r \leq n-2$, proving that condition b) holds.

Conversely, let $X \in [\Delta^{n-1^{op}}, \mathsf{Cat}]$ satisfy a) and b). By Proposition 2.4.5 it follows from condition a) that $X = J_n Y$ for $Y \in \mathsf{Cat}^n$. Further, for each $t \geq 0$, $(NY)_t \in \mathsf{Cat}^{n-1}$ and $J_{n-1}(NY)_t$ satisfies a) and b), so by the inductive hypothesis $(NY)_t \in (n-1)\text{-}\mathsf{Cat}$. Also, since by b) X_0 is constant with values in a discrete category, $(NY)_0$ is discrete.

In conclusion, $NY \in [\Delta^{op}, (n-1)\text{-}\mathsf{Cat}]$ is such that $(NY)_0$ is discrete and the Segal maps are isomorphisms. From Remark 2.6.2 it follows that $Y \in n\text{-}\mathsf{Cat}$. □

By taking nerves of categories dimensionwise in $[\Delta^{n-1^{op}}, \mathsf{Cat}]$ we obtain the multinerve functor:

$$N_{(n)} : n\text{-}\mathsf{Cat} \xrightarrow{J_{n-1}} [\Delta^{n-1^{op}}, \mathsf{Cat}] \hookrightarrow [\Delta^{n^{op}}, \mathsf{Set}],$$

where the second inclusion map is the one in Notational convention 2.5.2. Using Proposition 2.6.4 we immediately deduce the following characterization of the image of $N_{(n)}$, which affords a multi-simplicial description of strict n-categories.

Corollary 2.6.5 *Let $X \in [\Delta^{n^{op}}, \mathsf{Set}]$. Then X is in the image of the functor*

$$N_{(n)} : n\text{-}\mathsf{Cat} \to [\Delta^{n^{op}}, \mathsf{Set}]$$

if and only if

a) The Segal maps of X in all directions are isomorphisms.
b) $X_0 \in [\Delta^{n-1^{op}}, \mathsf{Set}]$ and $X_{\underset{1...10}{r}} \in [\Delta^{n-r-1^{op}}, \mathsf{Set}]$ are constant functors for all $1 \leq r \leq n-2$.

Proof It follows immediately from Proposition 2.6.4, Proposition 2.4.1 and the fact that the nerve of a discrete category is a constant simplicial set. □

Remark 2.6.6 This follows immediately from Proposition 2.4.5 and Corollary 2.6.5 that n-Cat is the full subcategory of Cat^n whose objects X satisfy condition a) and there are commuting diagrams

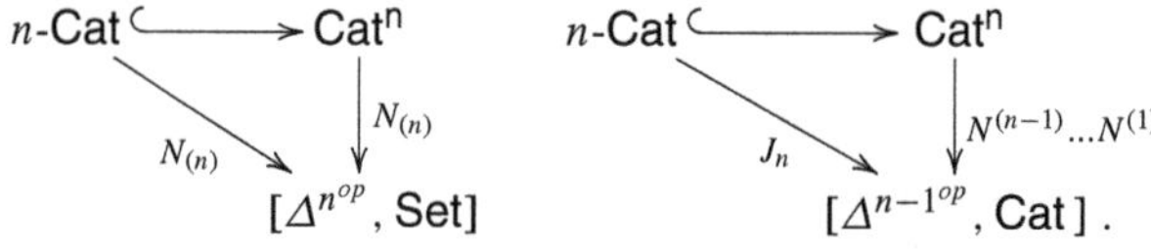

Remark 2.6.7 Condition b) in Proposition 2.6.4 can be replaced by:

b') $X_0 \in [\Delta^{n-2^{op}}, \mathsf{Cat}]$ and $X_{k_1\dots k_r\,0} \in [\Delta^{n-r-2^{op}}, \mathsf{Cat}]$ are constant functors taking values in a discrete category for all $1 \leq r \leq n-2$ and all $(k_1, \dots, k_r) \in \Delta^{r^{op}}$.

In fact, condition b') clearly implies b); further, we claim that conditions a) and b) together imply b'). We prove this claim by induction on n. When $n = 3$, for each $k_1 \geq 2$, by condition a), $X_{k_1} \cong X_1 \times_{X_0} \overset{k_1}{\cdots} \times_{X_0} X_1$. Since, by b), X_1 and X_0 are discrete, so is X_{k_1}. Suppose the claim holds for $(n-1)$. For each $k_1 \geq 2$,

$$X_{k_1\dots k_r\,0} = X_{1\,k_2\dots k_r\,0} \times_{X_{0\,k_2\dots k_r\,0}} \overset{k_1}{\cdots} \times_{X_{0\,k_2\dots k_r\,0}} X_{1\,k_2\dots k_r\,0}\,. \tag{2.22}$$

Note that $X_0, X_1 \in [\Delta^{n-2^{op}}, \mathsf{Cat}]$ satisfy conditions a) and b). Thus by the inductive hypothesis they satisfy b'), so that $X_{1\,k_2\dots k_r\,0}$ and $X_{0\,k_2\dots k_r\,0}$ are discrete. By (2.22) it follows that $X_{k_1\dots k_r\,0}$ is also discrete, which is b').

Condition b') in Remark 2.6.6 is called the *globularity condition*, since it gives rise to the globular shape of the higher cells in a strict n-category. Below are some low-dimensional examples.

Example 2.6.8 (Double Nerves of Strict 2-Categories) The double nerve functor

$$N_{(2)} : 2\text{-}\mathsf{Cat} \to [\Delta^{2^{op}}, \mathsf{Set}]$$

associates to $X \in 2\text{-}\mathsf{Cat}$ the bisimplicial set $N_{(2)}X$ such that the Segal maps in both horizontal and vertical directions are isomorphisms and $(N_{(2)}X)_0$ is a constant functor.

Denoting $(N_{(2)}X)_{ij}$ by X_{ij}, we can visualize the corner of its double nerve as in Fig. 2.6 on page 43.

A geometric picture can be obtained by thinking of X_{00} as sets of objects or 0-cells, X_{10} as sets of 1-cells and X_{20} as sets of 2-cells, see Fig. 2.7 on page 44.

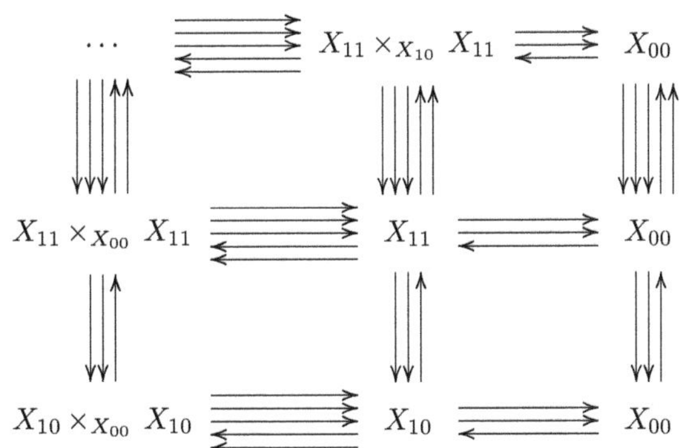

Fig. 2.6 Corner of the double nerve of a strict 2-category X

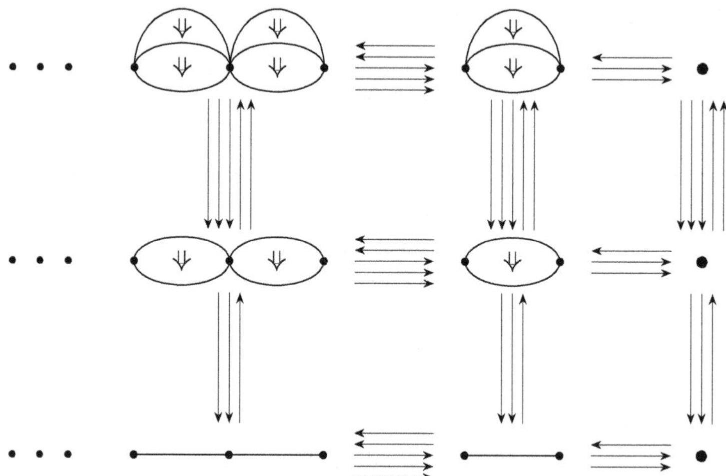

Fig. 2.7 Geometric picture of the corner of the double nerve of a strict 2-category

This figure should be compared with Fig. 2.3 in Example 2.5.7: the globularity condition that X_0 is a constant simplicial set means that the vertical sides of the squares in Fig. 2.3 are identities and thus can be represented as globes under the identification

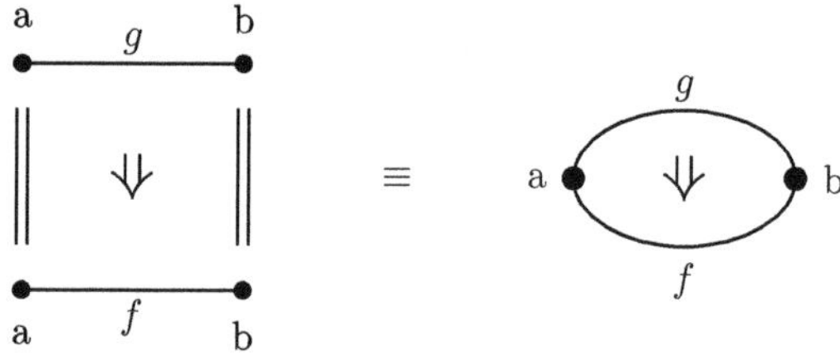

Example 2.6.9 (3-Fold Nerves of Strict 3-Categories) The 3-fold nerve functor

$$N_{(3)} : \mathsf{3\text{-}Cat} \to [\Delta^{3^{op}}, \mathsf{Set}]$$

associates to $Y \in \mathsf{3\text{-}Cat}$ the 3-fold simplicial set $N_{(3)}Y = X$ such that

a) All Segal maps in directions 1,2,3 are isomorphisms.
b) X_0 and X_{10} are constant functors.

The corner of X can be visualized as in Fig. 2.8 on page 45, where for simplicity we omitted drawing the face operators. The isomorphisms above Fig. 2.8 correspond to the Segal condition.

In the following picture, for all $i, j, k \in \Delta^{op}$
$X_{2jk} \cong X_{1jk} \times_{X_{0jk}} X_{1jk}$, $X_{i2k} \cong X_{i1k} \times_{X_{i0k}} X_{i1k}$, $X_{ij2} \cong X_{ij1} \times_{X_{ij0}} X_{ij1}$.

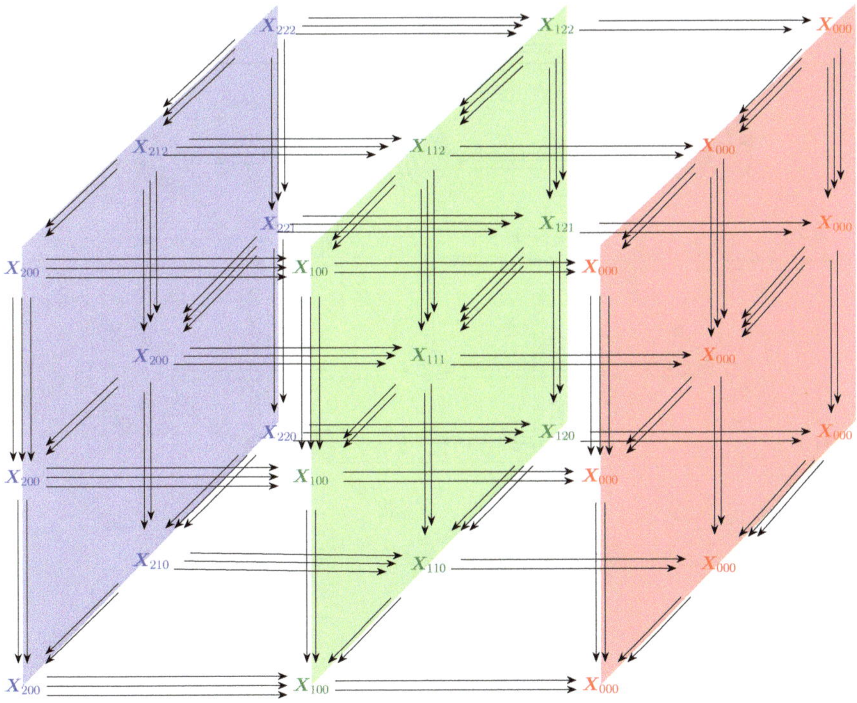

Fig. 2.8 Corner of the 3-fold nerve of a strict 3-category X

We obtain a geometric visualization of Fig. 2.8 by setting

X_{000} = set of points;
X_{100} = set of arrows in direction 1;
X_{010} = set of arrows in direction 2;
X_{001} = set of arrows in direction 3;
X_{110} = set of globes in directions 1,2;
X_{011} = set of globes in directions 2,3;
X_{101} = set of globes in directions 1,3;
X_{111} = set of spheres.

The points are the 0-cells. The arrows (1-cells) in direction i can be composed in direction i $(i = 1, 2, 3)$. The globes (2-cells) in direction i, j $(i, j = 1, \ldots, 3)$ can be composed in direction i and in direction j. The spheres (3-cells) can be composed in all three directions. A geometric picture can be found in Fig. 2.9 on page 46.

This figure should be compared with Fig. 2.5 in Example 2.5.8: the globularity condition identifies the squares with globes and the cubes with spheres.

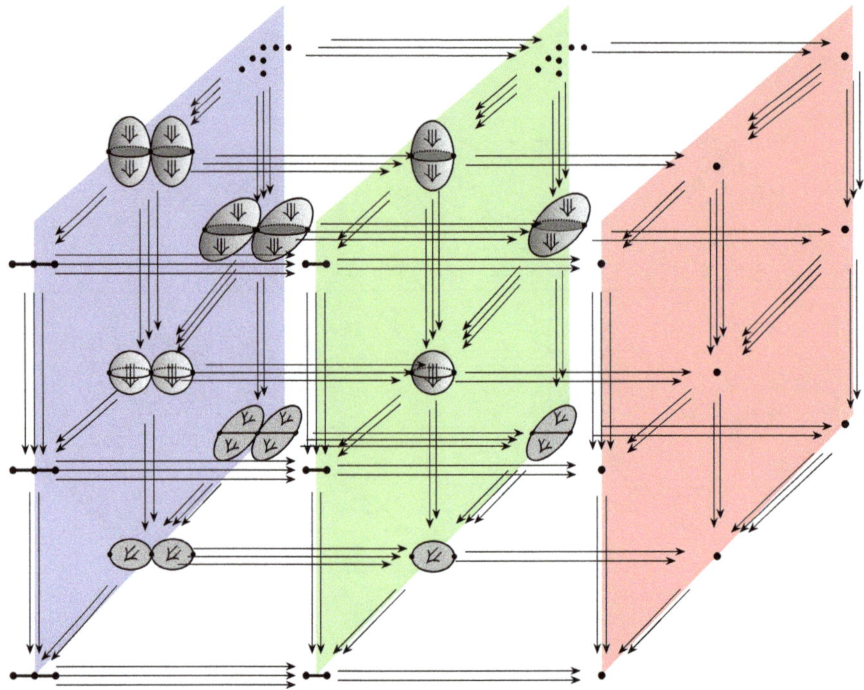

Fig. 2.9 Geometric picture of the corner of the 3-fold nerve of a strict 3-category

2.7 The Functor Décalage

Recall from Duskin [47] the décalage comonad

$$\mathrm{Dec} : [\Delta^{op}, \mathsf{Set}] \to [\Delta^{op}, \mathsf{Set}]$$

on simplicial sets. Given $X \in [\Delta^{op}, \mathsf{Set}]$, $\mathrm{Dec}\,X \in [\Delta^{op}, \mathsf{Set}]$ has

$$(\mathrm{Dec}\,X)_n = X_{n+1} \qquad n \geq 0$$

with face and degeneracy given by

$$d_i : (\mathrm{Dec}\,X)_n = X_{n+1} \to (\mathrm{Dec}\,X)_{n-1} = X_n, \qquad 1 \leq i \leq n\ ,$$
$$s_i : (\mathrm{Dec}\,X)_n = X_{n+1} \to (\mathrm{Dec}\,X)_{n+1} = X_{n+2}, \qquad 0 \leq i \leq n\ .$$

That is, the last face and degeneracy operators are omitted in each dimension. The omitted last face map $d_n : X_n \to X_{n-1}$ defines a simplicial map

$$u_X : \mathrm{Dec}\,X \to X\ ,$$

natural in X, levelwise surjective, which is the component of the counit of the comonad.

The composition of $d_0 : X_1 \to d(X_0)$ with the retained face maps gives a simplicial map

$$\mathrm{Dec}\, X \to d(X_0)\ ,$$

natural in X, where $d(X_0)$ is the constant simplicial set at X_0. Composition of $s_0 : X_0 \to X_1$ with the degeneracies gives a unique map $s^{(n)} : X_0 \to X_n$ and thus a simplicial map

$$d(X_0) \to \mathrm{Dec}\, X\ ,$$

natural in X, that is a section for $\mathrm{Dec}\, X \to d(X_0)$; it is in fact a contracting homotopy, so $d(X_0)$ is a deformation retract of $\mathrm{Dec}\, X$.

Suppose that X is the nerve of a groupoid. Then $\mathrm{Dec}\, X$ is the equivalence relation corresponding to the surjective map of sets $d_0 : X_1 \to X_0$ (see [33]). If $X_1 \times_{X_0} X_1$ is the pullback of $X_1 \overset{\partial_0}{\to} X_0 \overset{\partial_1}{\leftarrow} X_1$ and $X_1 \times_{d_0} X_1$ is the kernel pair of $d_0 : X_1 \to X_1$, in this case there is an isomorphism

$$X_1 \times_{X_0} X_1 \cong X_1 \times_{d_0} X_1$$

sending $(f, g) \in X_1 \times_{X_0} X_1$ to $(f, g \circ f) \in X_1 \times_{d_0} X_1$; ∂_0, ∂_1 correspond to the two projections $p_1, p_2 : X_1 \times_{d_0} X_1 \to X_1$ while the map $X_1 \to X_1 \times_{X_0} X_1$ sending f to (f, Id) corresponds to the diagonal map $X_1 \to X_1 \times_{d_0} X_1$ sending f to (f, f).

There is also a version of the décalage comonad which omits the first face and degeneracy operators, which we denote by

$$\mathrm{Dec}' : [\Delta^{op}, \mathsf{Set}] \to [\Delta^{op}, \mathsf{Set}]\ .$$

Chapter 3
An Introduction to the Three Segal-Type Models

Abstract This chapter consists of an introduction to the three Segal-type models. We convey the main ideas and intuitions beyond the definitions of our models. We discuss multi-simplicial structures as a natural environment for the development of higher categories, and we introduce one of the central ideas of our models: the notion of weak globularity. We illustrate the main common features of the three Segal-type models and give a summary of the main results. To guide the reader through this book, we discuss its organization and provide diagrammatic summaries of the interconnections between some of the main ideas and results.

In this chapter we give an overview of the approach to weak n-categories developed in this book, which is based on n-fold structures and on the idea of weak globularity. We study three models based on multi-simplicial structures, which we call Segal-type models since the compositions of higher cells is related to the notion of Segal maps. One of these models is due to Tamsamani [126], the other two are new. Our aim in this chapter is to convey the main intuitions and ideas, referring the reader to later chapters for the precise definitions and results.

In Sect. 3.1 we highlight one of the issues that we face when dealing with the general higher categorical case, instead of the higher groupoidal one, when modelling weak higher structures, which is the notion of higher categorical equivalence. In Sect. 3.2 we explain why multi-simplicial structures are a natural environment for the development of models of higher categories, referring to the crucial notion of Segal maps.

In Sect. 3.3 we explain the intuition behind the idea of weak globularity, which is central to this work in developing the new paradigm to weaken higher categorical structures. Finally, in Sect. 3.4 we give a broad overview of the main features of the three Segal-type models of higher categories treated in this book and we give a summary of the main results. We also provide an overview of the organization of this work.

In this chapter we sometimes need to recall some notation and definitions from the background part, in which case we refer the reader to Chap. 2 for further details.

S. Paoli, *Simplicial Methods for Higher Categories*, Algebra and Applications 26,
https://doi.org/10.1007/978-3-030-05674-2_3

3.1 Geometric Versus Higher Categorical Equivalences

An important aspect of the passage from the higher groupoidal case to the general higher categorical case in modelling weak higher structures concerns the notion of equivalence. Let's consider first the case of dimension $n = 1$. The classifying space of a groupoid is the geometric realization of its simplicial nerve, and a weak homotopy equivalence is a map inducing isomorphisms of homotopy groups in all dimensions (see [63]).

In the category of groupoids, categorical and geometric weak equivalences coincide. In fact, let $F : X \to Y$ be a functor between groupoids which is a weak homotopy equivalence. Then $\pi_0(BX) = q(X)$, where q denotes the connected components functor, and similarly for Y. Thus $\pi_0(BF)$ being an isomorphism implies that F is essentially surjective on objects. Given $a, b \in X$, since X and Y are groupoids there are bijections

$$X(a,b) \cong X(a,a), \qquad Y(Fa,Fb) \cong Y(Fa,Fa)\,.$$

On the other hand, $\pi_1(BX, a) = X(a,a)$ and $\pi_1(BY, Fa) = Y(Fa, Fa)$. Thus $\pi_1(BF, a)$ being an isomorphism implies from the above that there is a bijection

$$X(a,b) \cong Y(Fa,Fb)$$

for all $a, b \in X$, that is, F is fully faithful. In conclusion, F is an equivalence of categories.

Conversely, if $F : X \to Y$ is an equivalence of categories, then $q(F)$ is an isomorphism, hence from the above, $\pi_0 B(F)$ is an isomorphism. Also, since F is fully faithful, for each $a \in X$, $\pi_1(BX, a) = X(a,a) \cong Y(Fa, Fa) = \pi_1(BY, Fa)$. That is, $\pi_1(B(F), a)$ is an isomorphism. In conclusion, $B(F)$ is a weak homotopy equivalence, since it induces isomorphisms of all homotopy groups (recall that $\pi_i(BX, a) = 0$ for each $i > 1$ since X is a groupoid).

This relation between categorical and weak homotopy equivalences for groupoidal structures extends to higher dimensions: for instance for weakly globular cat^n-groups in [102] and for weakly globular n-fold groupoids in [29], weak equivalences are defined using classifying spaces, the latter being the geometric realization of a multi-diagonal, and it is shown that these can also be described in terms of higher categorical equivalences.

The general categorical case is more complex: an equivalence of categories is also a weak homotopy equivalence of its simplicial nerves, but not conversely. Thus a notion of higher categorical equivalence needs to be defined alongside notions of weak n-categories. In the case of our Segal-type models, n-equivalences are part of the inductive definition of the structure.

3.2 Multi-Simplicial Structures as an Environment for Higher Categories

Our Segal-type models of weak n-categories are based on multi-simplicial objects, more precisely functors from $\Delta^{n^{op}}$ to Set, where $\Delta^{n^{op}}$ denotes the product of n copies of Δ^{op}. Given $X \in [\Delta^{n^{op}}, \mathsf{Set}]$ and $([k_1], \ldots, [k_n]) \in \Delta^{n^{op}}$, we denote $X([k_1], \ldots, [k_n])$ by $X_{k_1 \ldots k_n}$. The reason why multi-simplicial objects are a good environment for building models of higher categories is the fact that their combinatorics exhibits certain maps, called Segal maps, as natural candidates for the compositions of higher cells.

Let us first illustrate the case $n = 1$. As already recalled in Chap. 2, there is a nerve functor

$$N : \mathsf{Cat} \to [\Delta^{^{op}}, \mathsf{Set}] ,$$

where

$$(NX)_k = \begin{cases} X_0, & k = 0; \\ X_1, & k = 1; \\ X_1 \times_{X_0} \overset{k}{\cdots} \times_{X_0} X_1, & k > 1. \end{cases}$$

It is possible to give a characterization of the simplicial sets that are nerves of small categories by using the notion of Segal maps. As recalled in Definition 2.1.2, for each $k \geq 2$ these are maps

$$\mu_k : X_k \to X_1 \times_{X_0} \overset{k}{\cdots} \times_{X_0} X_1$$

which arise from the commuting diagram

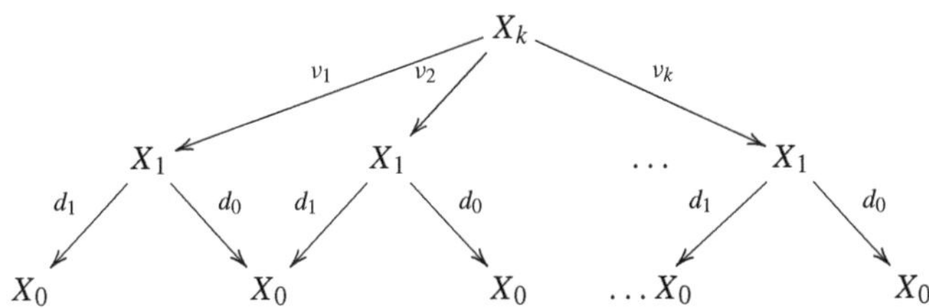

where for each $1 \leq j \leq k$ and $k \geq 2$, the maps $\nu_j : X_k \to X_1$ are induced by the map $[1] \to [k]$ in Δ sending 0 to $j - 1$ and 1 to j.

Then a simplicial set is the nerve of a small category if and only if it satisfies the *Segal condition* that all the Segal maps are isomorphisms. Under these isomorphisms the composition map

$$X_1 \times_{X_0} X_1 \xrightarrow{c} X_1$$

corresponds to the face map $\partial_1 : X_2 \to X_1$ induced by the map $[1] \to [2]$ in Δ sending 0 to 0 and 1 to 2.

Thus given a simplicial set, its simplicial structure together with the Segal maps contain a natural candidate for the composition, and the Segal condition ensures that such composition satisfies the axioms of a categorical composition (that is, it is associative and unital). This paradigm carries on in higher dimensions and it can be used to give notions of strict as well as weak higher categories.

Namely, Segal maps can be defined for any simplicial object $[\Delta^{^{op}}, \mathscr{C}]$ in a category $\mathscr{C}$ with finite limits, and they can be used to characterize nerves of internal categories in $\mathscr{C}$ using the Segal condition that all Segal maps are isomorphisms, in a way formally analogous to the case of Cat (see Chap. 2 for more details). For instance internal categories in Cat (also called double categories) can be described as simplicial objects in Cat such that all the Segal maps are isomorphisms.

Strict 2-categories can be described as simplicial objects X in Cat satisfying the Segal condition as well as the condition that X_0 is a discrete category. Once again, the simplicial structure and the Segal maps give the candidates for the compositions of higher cells while the Segal condition implies the associativity and unitality of the compositions. We refer to Examples 2.5.7 and 2.6.8 in the previous chapter for a pictorial representation of double categories and strict 2-categories.

Another approach to the use of Segal maps is possible: given a simplicial object $X \in [\Delta^{^{op}}, \mathsf{Cat}]$ such that X_0 is a discrete category, we can relax the Segal condition and require that Segal maps for each $k \geq 2$

$$\mu_k : X_k \to X_1 \times_{X_0} \overset{k}{\cdots} \times_{X_0} X_1$$

are not isomorphisms but merely equivalences of categories. We obtain a structure with sets of 0-cells X_{00}, sets of 1-cells X_{10}, and sets of 2-cells X_{11}. We can also define a composition

$$X_1 \times_{X_0} X_1 \xrightarrow{\mu'_2} X_2 \xrightarrow{\partial_1} X_1 \ ,$$

where μ'_2 is a pseudo-inverse to μ_2 and the map $\partial_1 : X_2 \to X_1$ is induced by the map $[1] \to [2]$ in Δ sending 0 to 0 and 1 to 2. Since μ'_2 is no longer an isomorphism, this composition is no longer associative and unital. What we obtain is a Tamsamani 2-category [126]. Lack and the author proved in [82] that Tamsamani 2-categories are suitably equivalent to bicategories.

In dimensions greater than 2, we use the category $[\Delta^{n-1^{op}}, \mathsf{Cat}]$ of multi-simplicial objects in Cat to obtain suitable notions of n-categories. Segal maps can be defined for multi-simplicial objects, and it is easy to see that there are fully faithful multi-nerve functors

$$\mathsf{Cat}^{\mathsf{n}} \to [\Delta^{n-1^{op}}, \mathsf{Cat}] \ ,$$

$$n\text{-}\mathsf{Cat} \to [\Delta^{n-1^{op}}, \mathsf{Cat}]$$

(see Sects. 2.4 and 2.6 for more details). A characterization of the essential image of these functors is also easy to give in terms of Segal maps (see Propositions 2.4.5 and 2.6.4). However, as in the case $n = 2$, we can impose different kinds of Segal conditions to obtain a different behavior of the compositions of higher cells and therefore define models of weak higher categories.

In summary, multi-simplicial objects are a good environment for building models of higher categories since the multi-simplicial maps and the corresponding Segal maps provide suitable candidates for the compositions of higher cells.

To obtain models of weak n-categories using multi-simplicial structures we need to impose extra conditions to encode:

(a) The sets of cells in dimensions $0, \dots, n$.
(b) The behavior of the compositions, giving weak associative and unit laws.
(c) A notion of higher categorical equivalence.

As outlined in Sect. 3.4, we adopt three different approaches that encode this extra structure, giving rise to three different Segal-type models of weak n-categories.

3.3 The Idea of Weak Globularity

Weakly globular n-fold categories form a full subcategory of n-fold categories. They are based on a new paradigm to weaken higher categorical structures: the idea of *weak globularity*.

In a strict n-category, the k-cells, for each $0 \leq k \leq n$, form a set. Equivalently, considering strict n-categories as embedded in $[\Delta^{n-1^{op}}, \mathsf{Cat}]$ via the multinerve functor J_n (see Definition 2.6.3), the k-cells for each $0 \leq k \leq (n-2)$ form a discrete $(n-k-1)$-fold category, that is, one in which all structure maps are identities.

In the weakly globular approach the k cells (for each $0 \leq k \leq (n-2)$) no longer form a set but have a higher categorical structure on their own. More precisely, they form a 'homotopically discrete $(n-k-1)$-fold category': this is an $(n-k-1)$-fold category which is suitably equivalent to a discrete one. We call this property the *weak globularity condition*.

We refer to Chap. 5 for the technical details about homotopically discrete n-fold categories, but a rough intuition about them is as follows. When $n = 1$, a homotopically discrete category is an equivalence relation, that is, a groupoid with no non-trivial loops. Such a groupoid is categorically equivalent to the discrete category on its set of paths components. Given a set B, to build an equivalence relation having B as set of path components we take a surjective map of sets $f : A \to B$ and then build the groupoid

$$(A \times_B A) \times_A (A \times_B A) \xrightarrow{\ m\ } A \times_B A \underset{s}{\overset{p_1, p_2}{\rightrightarrows}} A \tag{3.1}$$

where for each $(a, b) \in A \times_B A$ and $(a, b, b, c) \in (A \times_B A) \times_A (A \times_B A)$, $p_1(a, b) = a$, $p_2(a, b) = b$, $s(a) = (a, a)$, and $m(a, b, b, c) = (a, c)$.

The construction (3.1) can be internalized in any category with finite limits. So if we start with B being itself an equivalence relation, A any category and f a functor then (3.1) affords an internal groupoid in Cat, so in particular a double category. By requiring Nf to be a levelwise surjection (where $N : \mathsf{Cat} \to [\Delta^{op}, \mathsf{Set}]$ is the nerve functor), we obtain precisely a homotopically discrete double category. Notice that the double category (3.1) is groupoidal in only one of the two simplicial directions, and in that direction it is levelwise an equivalence relation. By taking path components in each of these equivalence relations, we recover B. Thus this structure is suitably equivalent to B and so, since the latter is an equivalence relation, it is also equivalent to the discrete category on the set of path components of B.

This process can be iterated: that is, given a homotopically discrete double category, f a map of double categories such that Nf is a levelwise surjection of double nerves, then (3.1) is what we call a homotopically discrete 3-fold category. Continuing in this way, at step n we build the structure from the one at level $n-1$. If an n-fold category X is homotopically discrete, it is suitably equivalent to a discrete n-fold category X^d via a 'discretization map' $\gamma : X \to X^d$.

The weakness in a weakly globular n-fold category is encoded by the weak globularity condition. Further, the weak globularity condition allows us to recover the notion of 'sets of higher cells' in an n-fold category. More precisely, in a weakly globular n-fold category, for each $0 \leq k \leq (n-2)$ there are substructures which are homotopically discrete $(n-k-1)$-fold categories, equivalent to discrete structures: the underlying sets of these discrete structures correspond to the sets of k-cells. The definition of a weakly globular n-fold category also requires several additional conditions to obtain well behaved compositions of higher cells.

The main result of this work is that there is a suitable equivalence after localization between weakly globular n-fold categories and Tamsamani n-categories. To establish this result, we work in the broader context of three Segal-type models: weakly globular n-fold categories, Tamsamani n-categories, and a further new model called weakly globular Tamsamani n-categories, containing the previous two as special cases.

Below we give a summary account of the main features of these three models.

3.4 The Three Segal-Type Models

We identify three multi-simplicial models based on the notion of Segal maps, which we therefore call Segal-type models. The first is the category Ta^n of Tamsamani n-categories introduced by Tamsamani [126] and further studied by Simpson [119]. The second is the category $\mathsf{Cat}^n_{\mathsf{wg}}$ of weakly globular n-fold categories, introduced in [103] when $n = 2$ and in this work for any $n \geq 3$. This is a full subcategory of the category Cat^n of n-fold categories. The third is another new model, the category $\mathsf{Ta}^n_{\mathsf{wg}}$ of weakly globular Tamsamani n-categories.

There is a fourth higher categorical structure which embeds in all three, which is the category n-Cat of strict n-categories. There are full and faithful inclusions

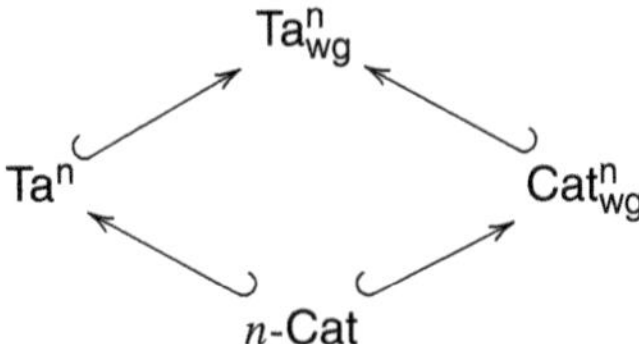

The category n-Cat admits a multi-simplicial description (see Sect. 2.6 for more details) as the full subcategory of $(n-1)$-fold simplicial objects $X \in [\Delta^{n-1^{op}}, \mathsf{Cat}]$ satisfying the following

(i) $X_0 \in [\Delta^{n-2^{op}}, \mathsf{Cat}]$ and $X_{k_1\dots k_r 0} \in [\Delta^{n-r-2^{op}}, \mathsf{Cat}]$ are constant multi-simplicial objects taking values in a discrete category, for all $1 \leq r \leq n-2$ and $(k_1, \dots, k_r) \in \Delta^{r^{op}}$. Here we use Notation 2.1.4.
(ii) The Segal maps (see Definition 2.1.2) in all directions are isomorphisms.

The underlying sets of the discrete structures X_0 (resp. $X_{\underset{r}{1\dots1}0}$) in (i) correspond to the sets of 0-cells (resp. r-cells) for $1 \leq r \leq n-2$; the sets of $(n-1)$ and of n-cells are given by $ob(X_{\underset{n-1}{1\dots1}})$ and $mor(X_{\underset{n-1}{1\dots1}})$ respectively.

The isomorphisms of the Segal maps (condition (ii)) ensures that all compositions of cells are associative and unital.

The discreteness condition (i) is also called the *globularity condition*. The name comes from the fact that it determines the globular shape of the cells in a strict n-category. For instance, when $n = 2$, we can picture 2-cells as globes

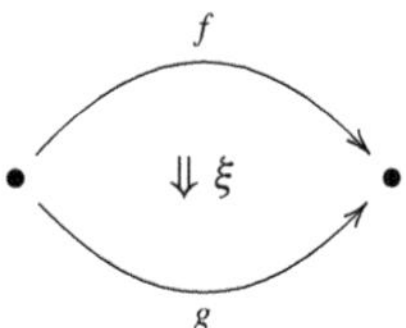

See Examples 2.6.8 and 2.6.9 for more detailed descriptions of the cases $n = 2$ and $n = 3$.

Strict n-categories have found several applications, for example in the groupoidal case where they are equivalent to crossed n-complexes (see [36]). However, they do not satisfy the homotopy hypothesis (see [118] for a counterexample showing that strict 3-groupoids do not model 3-types).

Therefore we must relax the structure to obtain a model of weak n-categories. Using the multi-simplicial framework, we consider three approaches to this weakening:

(a) *Category* Ta^{n} *of Tamsamani n-categories.* In the first approach, we preserve the globularity condition (i) and we relax the Segal map condition (ii) by requiring the Segal maps to be suitably defined higher categorical equivalences. This property means the composition of cells is no longer strictly associative and unital.

(b) *Category* $\mathsf{Cat}^{\mathsf{n}}_{\mathsf{wg}}$ *of weakly globular n-fold categories.* In the second approach condition (ii) is preserved so we obtain a subcategory of n-fold categories. However, the globularity condition (i) is replaced by weak globularity: the substructures $X_0, X_{k_1\dots k_r0}$ $(1 \le r \le n-2)$ are no longer discrete but 'homotopically discrete' in a higher categorical sense that allows iterations: see Chap. 5 for details on the category $\mathsf{Cat}^{\mathsf{n}}_{\mathsf{hd}}$ of homotopically discrete n-fold categories, and Sect. 3.3 for a short intuitive description. The notion of a homotopically discrete n-fold category is a higher order version of equivalence relations.

(c) *Category* $\mathsf{Ta}^{\mathsf{n}}_{\mathsf{wg}}$ *of weakly globular Tamsamani n-categories.* In the third approach, both conditions (i) and (ii) are relaxed.

3.4.1 Notational Conventions

We briefly summarize some notational conventions introduced in the previous chapter which we will need in the next Sect. 3.4.2. A more detailed description of these notational conventions can be found in Sects. 2.2 and 2.5.

(a) We identify $(n-1)$-fold simplicial sets X with those n-fold simplicial sets X for which the simplicial set $X_{k_1\dots k_{n-1}}$ is discrete for all $(k_1 \dots k_{n-1}) \in \Delta^{{n-1}^{op}}$.

(b) Let $q : [\Delta^{^{op}}, \mathsf{Set}] \to \mathsf{Set}$ be the connected component functor and $p : [\Delta^{^{op}}, \mathsf{Set}] \to \mathsf{Set}$ be obtained by applying the homotopy category construction and then taking the set of isomorphism classes of objects.

We inductively define functors

$$p_n, q_n : [\Delta^{n^{op}}, \mathsf{Set}] \to \mathsf{Set}$$

by $p_1 = p$, $q_1 = q$, $p_n = \bar{p} \circ p_{n-1}$, $q_n = \bar{q} \circ q_{n-1}$ for $n > 1$, where $\bar{p}$ and $\bar{q}$ are obtained by applying p and q levelwise (see Definition 2.1.1).

By notational convention (a), p_n, q_n can simply be denoted by

$$p, q : [\Delta^{n^{op}}, \mathsf{Set}] \to \mathsf{Set}\,.$$

In turn, this gives rise to functors

$$p^{(r)}, q^{(r)} : [\Delta^{n^{op}}, \mathsf{Set}] \to [\Delta^{r^{op}}, \mathsf{Set}]$$

for each $0 \leq r \leq n-1$ with $p^{(0)} = p,\ q^{(0)} = q$ and

$$(p^{(r)}X)_{k_1\dots k_r} = pX_{k_1\dots k_r} \qquad (q^{(r)}X)_{k_1\dots k_r} = qX_{k_1\dots k_r}$$

for $1 \leq r \leq n-1,\ (k_1 \dots k_r) \in \Delta^{r^{op}},\ X \in [\Delta^{n^{op}}, \mathsf{Set}]$.

(c) For each $n \geq 1$ there is a fully faithful multinerve functor

$$N_{(n)} : \mathsf{Cat^n} \to [\Delta^{n^{op}}, \mathsf{Set}]\ .$$

We identify $\mathsf{Cat^n}$ with the essential image of the functor $N_{(n)}$. In particular, we identify objects of $[\Delta^{n-1^{op}}, \mathsf{Cat}\,]$ with n-fold simplicial sets X such that for all $(k_1 \dots k_{n-1}) \in \Delta^{n-1^{op}}$, the simplicial set $X_{k_1\dots k_{n-1}}$ is the nerve of a category.

(d) As a consequence of convention (c), if $X \in \mathsf{Cat^n}$ and $N^{(1)}$ is the nerve functor in direction 1, for each $k \geq 0$, we denote $(N^{(1)}X)_k$ by X_k.

3.4.2 Common Features of the Three Segal-Type Models

In this section we describe the main common features of the three models, which we denote collectively by $\mathsf{Seg_n}$. Our purpose is to convey some of the main ideas underpinning the three Segal-type models. We point out, however, that to prove our results, the main structures need to be developed in the order presented in this work: namely we first need to define the category $\mathsf{Cat^n_{hd}}$ of homotopically discrete n-fold categories, and only later the three Segal-type models $\mathsf{Ta^n_{wg}}$, $\mathsf{Cat^n_{wg}}$, $\mathsf{Ta^n}$.

(1) *Inductive definition*

$\mathsf{Seg_n}$ is defined inductively on dimension, starting with $\mathsf{Seg_0} = \mathsf{Set}$ and $\mathsf{Seg_1} = \mathsf{Cat}$. For each $n \geq 1$ we set

$$\mathsf{Seg_n} \subset [\Delta^{^{op}}, \mathsf{Seg_{n-1}}].$$

Unravelling this definition gives embeddings

$$\mathsf{Seg_n} \hookrightarrow [\Delta^{n-1^{op}}, \mathsf{Cat}\,] \hookrightarrow [\Delta^{n^{op}}, \mathsf{Set}]\ .$$

Thus our models have a multi-simplicial structure. The simplicial maps are the candidates for encoding the compositions of higher cells. Under the notational conventions of Sect. 3.4.1, $\mathsf{Seg_{n-1}}$ is a full subcategory of $\mathsf{Seg_n}$, whose objects are discrete in the nth direction.

(2) *Closure properties*

The subcategory $\mathsf{Seg_n} \hookrightarrow [\Delta^{n^{op}}, \mathsf{Set}]$ contains the terminal object and has the following closure properties:

C1 It is replete under isomorphisms; that is, if $A \cong B$ in $[\Delta^{n^{op}}, \mathsf{Set}]$ and $A \in \mathsf{Seg_n}$ then $B \in \mathsf{Seg_n}$.
C2 It is closed under finite products.
C3 It is closed under small coproducts.
C4 If the small coproduct $\coprod_i A_i$ is in $\mathsf{Seg_n}$, then each $A_i \in \mathsf{Seg_n}$.

These closure properties are such that all (co)limits of interest in $\mathsf{Seg_n}$ are computed as they are in $[\Delta^{n^{op}}, \mathsf{Set}]$. An object X of $\mathsf{Seg_n}$ is called *discrete* if X, viewed as an object of $[\Delta^{n-1^{op}}, \mathsf{Cat}]$, is a constant functor taking values in a discrete category.

As a consequence of these closure properties, it is easily shown (see Lemma 2.2.8) that every discrete n-fold simplicial set is an object of $\mathsf{Seg_n}$ and that if $A \xrightarrow{f} X \xleftarrow{g} B$ is a diagram in $\mathsf{Seg_n}$ with X discrete, then $A \times_X B \in \mathsf{Seg_n}$.

(3) *The weak globularity condition*

This condition encodes the sets of higher cells and is based on our new paradigm to weaken higher categorical structures. Namely, if $X \in \mathsf{Seg_n}$, then X_0 is a homotopically discrete $(n-1)$-fold category, and it is discrete if $X \in \mathsf{Ta^n}$.

We refer to Chap. 5 for the details on the notion of a homotopically discrete $(n-1)$-fold category, and to Sect. 3.3 for an intuitive idea about it. If X is homotopically discrete, there is a discretization map

$$\gamma : X \to X^d \tag{3.2}$$

where X^d is discrete (as defined above). Thus, if $X \in \mathsf{Seg_n}$, X_0 and $X_{k_1 \dots k_r 0}$ (for $1 \le r \le n-2$ and $(k_1 \dots k_r) \in \Delta^{r^{op}}$) are homotopically discrete. The sets underlying the discrete structures X_0^d, $X^d_{\underset{r}{1 \dots 1}0}$ correspond to the sets of r-cells for $0 \le r \le n-2$.

The weak globularity condition is used in Definition 6.1.8 (for the category $\mathsf{Ta^n_{wg}}$) and in Definition 6.3.3 (for the category $\mathsf{Cat^n_{wg}}$).

(4) *The truncation functor* $p^{(r)}$

For each $0 \le r < n$ the functor $p^{(r)} : [\Delta^{n^{op}}, \mathsf{Set}] \to [\Delta^{r^{op}}, \mathsf{Set}]$ of Sect. 3.4.1 restricts to functors

$$p^{(r)} : \mathsf{Seg_n} \to \mathsf{Seg_r}\,.$$

See Definition 5.1.2 (for the category $\mathsf{Cat^n_{hd}}$), Definition 6.1.8 (for the category $\mathsf{Ta^n_{wg}}$) and Definition 6.3.3 (for the category $\mathsf{Cat^n_{wg}}$). The functor $p^{(n-1)}$ is a 'truncation functor' that divides out by the highest dimensional invertible cells, and is used to give the notion of n-equivalence in $\mathsf{Seg_n}$.

(5) *Higher categorical equivalences*
Given $X \in \mathsf{Seg_n}$ and $(a, b) \in X_0^d \times X_0^d$, let $X(a, b) \subset X_1$ be the fiber at (a, b) of the map

$$X_1 \xrightarrow{(d_0,d_1)} X_0 \times X_0 \xrightarrow{\gamma\times\gamma} X_0^d \times X_0^d ,$$

where γ is the discretization map as in (3.2). Since $X_1 = \coprod_{a,b\in X_0^d} X(a, b)$, by the closure property **C4** above, $X(a, b) \in \mathsf{Seg_{n-1}}$. Each $X(a, b) \in \mathsf{Seg_{n-1}}$ should be thought of as a hom-$(n-1)$-category.

The 1-equivalences in $\mathsf{Seg_1}$ are equivalences of categories. Inductively, if we have defined $(n-1)$-equivalences in $\mathsf{Seg_{n-1}}$, we define a map $f : X \to Y$ in $\mathsf{Seg_n}$ to be an n-equivalence if the following conditions hold

(i) For all $a, b \in X_0^d$,

$$f(a, b) : X(a, b) \to Y(fa, fb)$$

are $(n-1)$-equivalences.
(ii) $p^{(n-1)} f$ is an $(n-1)$-equivalence.

We denote by $\sim^n$ the class of n-equivalences in $\mathsf{Seg_n}$. This definition is a higher dimensional generalization of a functor which is an equivalence of categories. Indeed the latter can be formulated by saying that a functor $F : X \to Y$ is such that $X(a, b) \cong Y(Fa, Fb)$, for all $a, b \in X$, and $p(F)$ is an isomorphism.
Condition (i) is a higher dimensional generalization of fully faithfulness while condition (ii) generalizes essential surjectivity on objects. As in the case of an equivalence of categories, condition (ii) can be weakened.
The notion of n-equivalence is given in Definition 5.2.5 (for the category $\mathsf{Cat^n_{hd}}$) and it is part of the inductive Definition 6.1.8 of $\mathsf{Ta^n_{wg}}$.

(6) *The induced Segal maps condition*
This condition regulates the behaviour of the compositions. Given $X \in \mathsf{Seg_n}$, since $X \in [\Delta^{op}, \mathsf{Seg_{n-1}}]$ and there is a map $\gamma : X_0 \to X_0^d$, we can consider the induced Segal maps for $k \geq 2$ (see Definition 2.1.3 for more details)

$$\hat{\mu}_k : X_k \to X_1 \times_{X_0^d} \overset{k}{\cdots} \times_{X_0^d} X_1 .$$

In defining $\mathsf{Seg_n}$ we require these maps to be $(n-1)$-equivalences. Note that when $X \in \mathsf{Ta^n}$, $\gamma = \mathrm{Id}$, so $\hat{\mu}_k$ are the same as the Segal maps.
The induced Segal maps condition is shown to hold for the category $\mathsf{Cat^n_{hd}}$ in Proposition 5.2.9, while it is part of the inductive Definition 6.3.3 of $\mathsf{Ta^n_{wg}}$.

(7) *The functor* $q^{(r)}$
For each $0 \le r < n$ the functor $q^{(r)} : [\Delta^{n^{op}}, \mathsf{Set}] \to [\Delta^{r^{op}}, \mathsf{Set}]$ of Sect. 3.4.1 restricts to a functor

$$q^{(r)} : \mathsf{Seg_n} \to \mathsf{Seg_r}\ .$$

There is a map, natural in $X \in \mathsf{Seg_{n-1}}$

$$\gamma^{(r)} : X \to q^{(r)} X.$$

The functor $q^{(n-1)}$ divides out by the (not necessarily invertible) highest dimensional cells and plays an important role in Chap. 9 in the construction of the rigidification functor Q_n.
The existence of the functor $q^{(r)}$ is proved in Proposition 7.1.7 (for the category $\mathsf{Ta^n_{wg}}$) and in Corollary 7.1.8 (for the categories $\mathsf{Cat^n_{hd}}$, $\mathsf{Cat^n_{wg}}$, $\mathsf{Ta^n}$).

(8) *Groupoidal Segal-type models*
The definition of the groupoidal version $\mathsf{GSeg_n}$ of the three Segal-type models is obtained inductively starting from groupoids as follows. We set $\mathsf{GSeg_1} = \mathsf{Gpd}$. Suppose inductively we have defined $\mathsf{GSeg_{n-1}} \subset \mathsf{Seg_{n-1}}$, then $X \in \mathsf{GSeg_n} \subset \mathsf{Seg_n}$ if

(i) $X_k \in \mathsf{GSeg_{n-1}}$ for all $k \ge 0$.
(ii) $p^{(n-1)} X \in \mathsf{GSeg_{n-1}}$.

We show in Corollary 12.3.13 that $\mathsf{GSeg_n}$ is a model of n-types, that is, there is an equivalence of categories

$$\mathsf{GSeg_n}/{\sim^n} \simeq \text{Ho}(n\text{-types})\ ,$$

where $\mathsf{GSeg_n}/\sim^n$ is the localization of $\mathsf{GSeg_n}$ with respect to the n-equivalences. Thus $\mathsf{Seg_n}$ is a model of weak n-categories, satisfying the homotopy hypothesis.

3.4.3 Main Results

The central result of this work is a model comparison between weakly globular n-fold categories and Tamsamani n-categories showing that the two models are suitably equivalent after localization. Our main results are as follows (see Theorems 10.2.1, 12.2.5, 12.2.6 and 12.3.11).

Theorem A *There is a functor* rigidification

$$Q_n : \mathsf{Ta^n_{wg}} \to \mathsf{Cat^n_{wg}}$$

and for each $X \in \mathsf{Ta^n_{wg}}$ *an* n*-equivalence natural in* X

$$s_n(X) : Q_n X \to X.$$

Theorem B *There is a functor* discretization

$$Disc_n : \mathsf{Cat^n_{wg}} \to \mathsf{Ta^n}$$

and, for each $X \in \mathsf{Cat^n_{wg}}$*, a zig-zag of* n*-equivalences in* $\mathsf{Ta^n_{wg}}$ *between* X *and* $Disc_n X$.

Theorem C *The functors*

$$Q_n : \mathsf{Ta^n} \leftrightarrows \mathsf{Cat^n_{wg}} : Disc_n$$

induce an equivalence of categories after localization with respect to the n*-equivalences*

$$\mathsf{Ta^n}/\!\sim^n \;\simeq\; \mathsf{Cat^n_{wg}}/\!\sim^n \ .$$

We also identify a subcategory $\mathsf{GCat^n_{wg}} \subset \mathsf{Cat^n_{wg}}$ of groupoidal weakly globular n-fold categories and we show that it gives an algebraic model of n-types. That is, the category $\mathsf{Cat^n_{wg}}$ satisfies the homotopy hypothesis:

Theorem D *There is an equivalence of categories*

$$\mathsf{GCat^n_{wg}}/\!\sim^n \;\simeq\; \mathrm{Ho}\,(n\text{-}types)\ .$$

In Corollary 12.4.6 we also show that the equivalence of categories of Theorem D can be realized using the fundamental weakly globular n-fold groupoid functor of Blanc and the author [29].

We call the functor Q_n the *rigidification functor* because it replaces a globular and weak structure with an equivalent more rigid structure (an n-fold category), which is no longer globular but it is weakly globular. For $n = 2$ the functor Q_2 was constructed by Pronk and the author in [103]. The construction of Q_n and $Disc_n$ for $n > 2$ is much more complex and it requires several novel ideas and techniques, as presented in this work.

One of the key features in the construction of Q_n is the use of pseudo-functors. Pseudo-functors feature prominently in homotopy theory, for instance in iterated loop space theory [127]. They are also ubiquitous in category theory [32], and can be described with the language of 2-monads and their pseudo-algebras [106].

Weakly globular n-fold categories are a full subcategory of $(n-1)$-fold simplicial objects in Cat, that is, functors $[\Delta^{n-1^{op}}, \mathsf{Cat}\,]$. We consider the pseudo-version of these, that is, pseudo-functors $\mathsf{Ps}[\Delta^{n-1^{op}}, \mathsf{Cat}\,]$.

Crucial to this work is the use of the strictification of pseudo-functors into strict functors

$$St\ :\mathsf{Ps}[\Delta^{n-1^{op}},\mathsf{Cat}\,]\to[\Delta^{n-1^{op}},\mathsf{Cat}\,]$$

left adjoint to the inclusion. There have been many contributions to this topic in category theory, including [122]. We use in this work the elegant formulation of Power [106], further refined by Lack in [81].

We introduce a subcategory

$$\mathsf{SegPs}[\Delta^{n-1^{op}},\mathsf{Cat}\,]\subset\mathsf{Ps}[\Delta^{n-1^{op}},\mathsf{Cat}\,]$$

of *Segalic pseudo-functors*. We show in Theorem 8.2.3 that the strictification functor St restricts to a functor

$$St\ :\mathsf{SegPs}[\Delta^{n-1^{op}},\mathsf{Cat}\,]\to\mathsf{Cat}^{\mathsf{n}}_{\mathsf{wg}}\subset[\Delta^{n-1^{op}},\mathsf{Cat}\,]\,.$$

The rigidification functor factors through the subcategory of Segalic pseudo-functors. That is, Q_n is a composite

$$Q_n:\mathsf{Ta}^{\mathsf{n}}_{\mathsf{wg}}\to\mathsf{SegPs}[\Delta^{n-1^{op}},\mathsf{Cat}\,]\xrightarrow{St}\mathsf{Cat}^{\mathsf{n}}_{\mathsf{wg}}\,.$$

In the case $n=2$, it is easy to build pseudo-functors from $\mathsf{Ta}^2_{\mathsf{wg}}$. More precisely, given $X\in\mathsf{Ta}^2_{\mathsf{wg}}$, define $Tr_2X\in[ob(\Delta^{op}),\mathsf{Cat}\,]$ by

$$(Tr_2X)_k=\begin{cases}X_0^d & k=0\\ X_1 & k=1\\ X_1\times_{X_0^d}\overset{k}{\cdots}\times_{X_0^d}X_1 & k>1\,.\end{cases}\tag{3.3}$$

Since $X\in\mathsf{Ta}^2_{\mathsf{wg}}$, $X_0\in\mathsf{Cat}_{\mathsf{hd}}$, so there are equivalences of categories

$$\begin{aligned}&X_0\simeq X_0^d\\ &X_k\simeq X_1\times_{X_0^d}\overset{k}{\cdots}\times_{X_0^d}X_1\quad\text{for }k>1.\end{aligned}$$

Thus, for all $k\geq 0$ there is an equivalence of categories

$$(Tr_2X)_k\simeq X_k\,.$$

By using transport of structure (more precisely Lemma 4.3.2 with $\mathscr{C}=\Delta^{op}$) we can lift Tr_2X to a pseudo-functor

$$Tr_2X\in\mathsf{Ps}[\Delta^{^{op}},\mathsf{Cat}\,]$$

and by construction $Tr_2 X \in \mathsf{SegPs}[\Delta^{^{op}}, \mathsf{Cat}]$.

Building pseudo-functors from $\mathsf{Ta}^{\mathsf{n}}_{\mathsf{wg}}$ when $n > 2$ is much more complex. The above approach cannot be applied directly because the induced Segal maps, when $n > 2$, are $(n-1)$-equivalences but not in general levelwise equivalence of categories. For this reason we need to introduce an intermediate category $\mathsf{LTa}^{\mathsf{n}}_{\mathsf{wg}} \subset \mathsf{Ta}^{\mathsf{n}}_{\mathsf{wg}}$, from which it is possible to build pseudo-functors using transport of structure. The functor from $\mathsf{Ta}^{\mathsf{n}}_{\mathsf{wg}}$ to Segalic pseudo-functors factorizes as

$$\mathsf{Ta}^{\mathsf{n}}_{\mathsf{wg}} \xrightarrow{P_n} \mathsf{LTa}^{\mathsf{n}}_{\mathsf{wg}} \xrightarrow{Tr_n} \mathsf{SegPs}[\Delta^{n-1^{op}}, \mathsf{Cat}]\,.$$

The functor P_n produces a functorial approximation (up to an n-equivalence) of an object of $\mathsf{Ta}^{\mathsf{n}}_{\mathsf{wg}}$ with an object of $\mathsf{LTa}^{\mathsf{n}}_{\mathsf{wg}}$, while the functor Tr_n is built using transport of structure.

We call the functor $Disc_n$ the *discretization functor* because it replaces a weakly globular structure with a globular one.

The idea of the functor $Disc_n$ is to replace the homotopically discrete substructures in a weakly globular n-fold category by their discretization. This process changes the Segal maps from isomorphisms to higher categorical equivalences.

In the higher groupoidal case, this idea had already appeared in the work by the author in [102] and [29], but further work is needed in the general categorical case to deal with the functoriality of the sections to the discretization maps for the homotopically discrete substructures.

We illustrate this point in the case $n = 2$. Given $X \in \mathsf{Cat}^2_{\mathsf{wg}}$, by definition $X_0 \in \mathsf{Cat}_{\mathsf{hd}}$, so there is a discretization map $\gamma : X_0 \to X^d_0$ which is an equivalence of categories. Given a choice γ' of pseudo-inverse, we have $\gamma\gamma' = \mathrm{Id}$ since X^d_0 is discrete.

We can therefore construct $D_0 X \in [\Delta^{^{op}}, \mathsf{Cat}]$ as follows

$$(D_0 X)_k = \begin{cases} X^d_0, & k = 0 \\ X_k, & k > 0\,. \end{cases}$$

The face maps

$$(D_0 X)_1 \rightrightarrows (D_0 X)_0$$

are given by $\gamma\partial_i \ \ i = 0, 1$ (where $\partial_i : X_1 \rightrightarrows X_0$ are face maps of X) while the degeneracy map

$$(D_0 X)_0 \to (D_0 X)_1$$

is $\sigma_0\gamma'$ (where $\sigma_0 : X_0 \to X_1$ is the degeneracy map of X). All other face and degeneracy maps in $D_0 X$ are as in X. Since $\gamma\gamma' = \mathrm{Id}$, all simplicial identities are satisfied for $D_0 X$. By construction, $(D_0 X)_0$ is discrete while the Segal maps are given, for each $k \geq 2$, by

$$X_1 \times_{X_0} \overset{k}{\cdots} \times_{X_0} X_1 \to X_1 \times_{X_0^d} \overset{k}{\cdots} \times_{X_0^d} X_1$$

and these are equivalences of categories since $X \in \mathsf{Cat}^2_{\mathsf{wg}}$. Thus, by definition, $D_0 X \in \mathsf{Ta}^2$. This construction however does not afford a functor

$$D_0 : \mathsf{Cat}^2_{\mathsf{wg}} \to \mathsf{Ta}^2$$

but only a functor

$$D_0 : \mathsf{Cat}^2_{\mathsf{wg}} \to (\mathsf{Ta}^2)_{\mathsf{ps}} \;,$$

where $(\mathsf{Ta}^2)_{\mathsf{ps}}$ is the full subcategory of $\mathsf{Ps}[\Delta^{^{op}}, \mathsf{Cat}\,]$ whose objects are in Ta^2. In fact, for any morphism $F : X \to Y$ in Ta^2, the diagram in Cat

$$\begin{array}{ccc} X_0^d & \xrightarrow{f^d} & Y_0^d \\ {\scriptstyle \gamma'(X_0)}\downarrow & & \downarrow{\scriptstyle \gamma'(Y_0)} \\ X_0 & \xrightarrow[f]{} & Y_0 \end{array}$$

in general only pseudo-commutes.

To remedy this problem we introduce the category $\mathsf{FCat}^{\mathsf{n}}_{\mathsf{wg}}$ which exhibits functorial sections to the discretization maps of the homotopically discrete substructures in $\mathsf{Cat}^{\mathsf{n}}_{\mathsf{wg}}$. Because of this property of $\mathsf{FCat}^{\mathsf{n}}_{\mathsf{wg}}$, the discretization process can be done functorially, using an iteration of the above idea, via a functor

$$D_n : \mathsf{FCat}^{\mathsf{n}}_{\mathsf{wg}} \to \mathsf{Ta}^{\mathsf{n}}.$$

We show that we can approximate any object of $\mathsf{Cat}^{\mathsf{n}}_{\mathsf{wg}}$ with an n-equivalent object of $\mathsf{FCat}^{\mathsf{n}}_{\mathsf{wg}}$. Namely we prove in Theorem 11.3.6 that there is a functor

$$G_n : \mathsf{Cat}^{\mathsf{n}}_{\mathsf{wg}} \to \mathsf{FCat}^{\mathsf{n}}_{\mathsf{wg}}$$

and an n-equivalence $G_n X \to X$. The discretization functor $Disc_n$ is defined as the composite

$$\mathsf{Cat}^{\mathsf{n}}_{\mathsf{wg}} \xrightarrow{G_n} \mathsf{FCat}^{\mathsf{n}}_{\mathsf{wg}} \xrightarrow{D_n} \mathsf{Ta}^{\mathsf{n}} \,.$$

It is straightforward (Corollary 12.3.9) that Q_n and $Disc_n$ restrict to the groupoidal versions of the models

$$Q_n : \mathsf{GTa}^{\mathsf{n}} \leftrightarrows \mathsf{GCat}^{\mathsf{n}}_{\mathsf{wg}} : Disc_n$$

and that these induce equivalence of categories after localization with respect to the n-equivalences (Proposition 12.3.10). Using the result of Tamsamani [126] it follows (Theorem 12.3.11) that

$$\mathsf{GCat}^{\mathsf{n}}_{\mathsf{wg}}/{\sim^n} \simeq \mathrm{Ho}\,(n\text{-types})\,. \tag{3.4}$$

In Corollary 12.4.6 the equivalence of categories (3.4) is obtained with a different fundamental higher groupoid functor

$$\mathsf{Top} \xrightarrow{\mathscr{H}_n} \mathsf{Gpd}^{\mathsf{n}}_{\mathsf{wg}} \overset{j}{\hookrightarrow} \mathsf{GCat}^{\mathsf{n}}_{\mathsf{wg}}\,,$$

where $\mathscr{H}_n$ is the functor from spaces to weakly globular n-fold groupoids of Blanc and the author in [29] and j is the inclusion. The functor $\mathscr{H}_n$ has the advantage that it is independent of [126] and has a very explicit form, as we illustrate in some low-dimensional examples at the end of Sect. 12.4.

3.4.4 Organization of This Work

We conclude this Part with an account of the overall organization of this book. A more detailed synopsis of the content of each part can be found at the beginning of each, where we also provide some diagrammatic summaries. The core of this work is developed in Parts II, III, IV.

In Part II we introduce some new structures and study their properties: the category $\mathsf{Cat}^{\mathsf{n}}_{\mathsf{hd}}$ of homotopically discrete n-fold categories (Chap. 5), the three Segal-type models $\mathsf{Ta}^{\mathsf{n}}_{\mathsf{wg}}$, Ta^{n}, $\mathsf{Cat}^{\mathsf{n}}_{\mathsf{wg}}$ (Chaps. 6 and 7), and the category of Segalic pseudo-functors $\mathsf{SegPs}[\Delta^{n-1^{op}}, \mathsf{Cat}\,]$ (Chap. 8).

The main result of this Part is Theorem 8.2.3, establishing that the classical strictification of pseudo-functors, when restricted to Segalic pseudo-functors, yields weakly globular n-fold categories; that is, there is a functor

$$St\,:\mathsf{SegPs}[\Delta^{n-1^{op}}, \mathsf{Cat}\,] \to \mathsf{Cat}^{\mathsf{n}}_{\mathsf{wg}}\,.$$

In Part III we prove one of the main results of this work, Theorem 10.2.1, which constructs the rigidification functor

$$Q_n : \mathsf{Ta}^{\mathsf{n}}_{\mathsf{wg}} \to \mathsf{Cat}^{\mathsf{n}}_{\mathsf{wg}}\,.$$

The construction of Q_n makes critical use of the functor St of Theorem 8.2.3 as well as several other intermediate steps, such as the subcategory $\mathsf{LTa}^{\mathsf{n}}_{\mathsf{wg}} \subseteq \mathsf{Ta}^{\mathsf{n}}_{\mathsf{wg}}$ and the functor $Tr_n : \mathsf{LTa}^{\mathsf{n}}_{\mathsf{wg}} \to \mathsf{SegPs}[\Delta^{n-1^{op}}, \mathsf{Cat}\,]$. These are developed in Chap. 9.

In Part IV we establish the main comparison result between Tamsamani n-categories and weakly globular n-fold categories, exhibiting the latter as a model of weak n-categories. One of the main constructions is the discretization functor

$$Disc_n : \mathsf{Cat}^{\mathsf{n}}_{\mathsf{wg}} \to \mathsf{Ta}^{\mathsf{n}}$$

and its properties are established in Theorem 12.2.5. The construction of $Disc_n$ needs several intermediate steps, in particular the new category $\mathsf{FCat}^{\mathsf{n}}_{\mathsf{wg}}$, developed in Chap. 11.

The rigidification and discretization functors lead to the main comparison result, Theorem 12.2.6, on the equivalence of categories

$$\mathsf{Ta}^{\mathsf{n}}/\!\sim^n \;\simeq\; \mathsf{Cat}^{\mathsf{n}}_{\mathsf{wg}}/\!\sim^n \;\; .$$

Chapter 12 concludes with a proof of the homotopy hypothesis, after introducing the higher groupoidal version of the three Segal-type models. Chapter 13 contains a discussion of directions of potential applications and open questions, which will be tackled in future projects. In Figs. 3.1 and 3.2 below we give a schematic account of the main notions and results of this work.

3.4.5 *Informal Discussions*

We have included a number of informal discussions throughout the text to convey the ideas and intuitions behind the main definitions and constructions. A list of the main ideas is as follows:

Section 5.1.1: The category $\mathsf{Cat}^{\mathsf{n}}_{\mathsf{hd}}$ of homotopically discrete n-fold categories.
Section 6.1.1: The category $\mathsf{Ta}^{\mathsf{n}}_{\mathsf{wg}}$ of weakly globular Tamsamani n-categories.
Section 6.3.1: The category $\mathsf{Cat}^{\mathsf{n}}_{\mathsf{wg}}$ of weakly globular n-fold categories.
Section 8.1.2: The category $\mathsf{SegPs}[\Delta^{n^{op}}, \mathsf{Cat}\,]$ of Segalic pseudo-functors.
Section 9.1.1: The category $\mathsf{LTa}^{\mathsf{n}}_{\mathsf{wg}}$.
Section 9.2.1: The main steps in approximating $\mathsf{Ta}^{\mathsf{n}}_{\mathsf{wg}}$ by $\mathsf{LTa}^{\mathsf{n}}_{\mathsf{wg}}$.
Section 10.1.1: The functor $Tr_n : \mathsf{LTa}^{\mathsf{n}}_{\mathsf{wg}} \to \mathsf{SegPs}[\Delta^{n-1^{op}}, \mathsf{Cat}\,]$.
Section 10.2.1: The rigidification functor $Q_n : \mathsf{Ta}^{\mathsf{n}}_{\mathsf{wg}} \to \mathsf{Cat}^{\mathsf{n}}_{\mathsf{wg}}$.
Section 11.1.1: The construction $X(f_0)$.
Section 11.2.1: The functors $V_n : \mathsf{Cat}^{\mathsf{n}}_{\mathsf{hd}} \to \mathsf{Cat}^{\mathsf{n}}_{\mathsf{hd}}$ and $F_n : \mathsf{Cat}^{\mathsf{n}}_{\mathsf{wg}} \to \mathsf{Cat}^{\mathsf{n}}_{\mathsf{wg}}$.
Section 11.3.1: The category $\mathsf{FCat}^{\mathsf{n}}_{\mathsf{wg}}$.
Section 11.3.3: The functor $G_n : \mathsf{Cat}^{\mathsf{n}}_{\mathsf{wg}} \to \mathsf{FCat}^{\mathsf{n}}_{\mathsf{wg}}$.
Section 12.1.1: The functor $D_n : \mathsf{FCat}^{\mathsf{n}}_{\mathsf{wg}} \to \mathsf{Ta}^{\mathsf{n}}$.
Section 12.2.1: The discretization functor $Disc_n : \mathsf{Cat}^{\mathsf{n}}_{\mathsf{wg}} \to \mathsf{Ta}^{\mathsf{n}}$.

A diagrammatic summary of the main connections between these notions is given in Fig. 3.1, and a brief description of these connections follows.

The Comparison Result Our main comparison result Theorem 12.2.6 arises from a pair of functors: the rigidification functor $Q_n : \mathsf{Ta^n} \to \mathsf{Cat^n_{wg}}$ (which is the restriction of $Q_n : \mathsf{Ta^n_{wg}} \to \mathsf{Cat^n_{wg}}$) and the discretization functor $Disc_n : \mathsf{Cat^n_{wg}} \to \mathsf{Ta^n}$. Their construction is based on the following ingredients.

The Categories $\mathsf{Cat^n_{hd}}$, $\mathsf{Ta^n_{wg}}$, $\mathsf{Cat^n_{wg}}$ The category $\mathsf{Cat^n_{hd}}$ of homotopically discrete n-fold categories is used to formulate the weak globularity condition in two of the three Segal-type models: The category $\mathsf{Ta^n_{wg}}$ of weakly globular Tamsamani n-categories and its subcategory $\mathsf{Cat^n_{wg}}$ of weakly globular n-fold categories.

The Functor Tr_n The category $\mathsf{LTa^n_{wg}}$ is intermediate between $\mathsf{Ta^n_{wg}}$ and $\mathsf{Cat^n_{wg}}$: that is, there are embeddings $\mathsf{Cat^n_{wg}} \subset \mathsf{LTa^n_{wg}} \subset \mathsf{Ta^n_{wg}}$, and we give in Theorem 9.2.4 a construction to approximate, up to n-equivalence, objects of $\mathsf{Ta^n_{wg}}$ by objects of $\mathsf{LTa^n_{wg}}$. The category $\mathsf{Cat^n_{wg}}$ also informs the definition of the subcategory

$$\mathsf{SegPs}[\Delta^{n-1^{op}}, \mathsf{Cat}] \subset \mathsf{Ps}[\Delta^{n-1^{op}}, \mathsf{Cat}]$$

of Segalic pseudo-functors. Together with $\mathsf{LTa^n_{wg}}$ this leads (Theorem 10.1.1) to the functor

$$Tr_n : \mathsf{LTa^n_{wg}} \to \mathsf{SegPs}[\Delta^{n-1^{op}}, \mathsf{Cat}].$$

From Tr_n to the Rigidification Functor Q_n Using the above approximation of $\mathsf{Ta^n_{wg}}$ by $\mathsf{LTa^n_{wg}}$ we build (see proof of Theorem 10.2.1) a functor

$$P_n : \mathsf{Ta^n_{wg}} \to \mathsf{LTa^n_{wg}}.$$

On the other hand, we show in Theorem 8.2.3 that the classical strictification of pseudo-functors $St : \mathsf{Ps}[\Delta^{n-1^{op}}, \mathsf{Cat}] \to [\Delta^{n-1^{op}}, \mathsf{Cat}]$ restricts to a functor

$$St : \mathsf{SegPs}[\Delta^{n-1^{op}}, \mathsf{Cat}] \to \mathsf{Cat^n_{wg}}.$$

By pre-composing Tr_n with P_n and post-composing it with St we obtain in Theorem 10.2.1 the rigidification functor

$$Q_n : \mathsf{Ta^n_{wg}} \to \mathsf{Cat^n_{wg}}.$$

The Category $\mathsf{FCat^n_{wg}}$ and the Functor G_n The category $\mathsf{Cat^n_{wg}}$ can be refined to a category $\mathsf{FCat^n_{wg}}$ with better behaved homotopically discrete sub-structures, having functorial sections, and we build in Theorem 11.3.6 a functor

$$G_n : \mathsf{Cat^n_{wg}} \to \mathsf{FCat^n_{wg}}$$

which approximates, up to n-equivalence, any object of $\mathsf{Cat^n_{wg}}$ by one of $\mathsf{FCat^n_{wg}}$.

The definition of G_n uses various ingredients: first a construction on an internal category X and map $f_0 : X'_0 \to X_0$ which produces an internal category $X(f_0)$ and an internal functor $X(f_0) \to X$. We apply this construction to the case where $X \in \mathsf{Cat}^{\mathsf{n}}_{\mathsf{wg}}$ and f_0 is given by a certain map $v_{n-1}(X_0) : V_{n-1}X_0 \to X_0$ arising from a functor

$$V_n : \mathsf{Cat}^{\mathsf{n}}_{\mathsf{hd}} \to \mathsf{Cat}^{\mathsf{n}}_{\mathsf{hd}}.$$

We show that this choice of $f_0 = v_{n-1}(X_0)$ ensures that $X(f_0) \in \mathsf{Cat}^{\mathsf{n}}_{\mathsf{wg}}$ and that the map $X(f_0) \to X$ is an n-equivalence. Thus we obtain the functor

$$F_n : \mathsf{Cat}^{\mathsf{n}}_{\mathsf{wg}} \to \mathsf{Cat}^{\mathsf{n}}_{\mathsf{wg}}$$

given by $F_n X = X(v_{n-1}(X_0))$. We show (Proposition 11.2.5) that for each $X \in \mathsf{Cat}^{\mathsf{n}}_{\mathsf{wg}}$, $(F_n X)_0 \in \mathsf{Cat}^{\mathsf{n-1}}_{\mathsf{hd}}$ admits a functorial section to the discretization map.

The functor G_n is built by iteratively applying F_n (Definition 11.3.4).

The Discretization Functor *Disc*$_n$ The functoriality properties of the homotopically discrete substructures in the category $\mathsf{FCat}^{\mathsf{n}}_{\mathsf{wg}}$ allow us to discretize these substructures and thus produce (Proposition 12.1.4) a functor

$$D_n : \mathsf{FCat}^{\mathsf{n}}_{\mathsf{wg}} \to \mathsf{Ta}^{\mathsf{n}}.$$

By pre-composting the latter with G_n we obtain in Theorem 12.2.5 the discretization functor

$$Disc_n : \mathsf{Cat}^{\mathsf{n}}_{\mathsf{wg}} \to \mathsf{Ta}^{\mathsf{n}}.$$

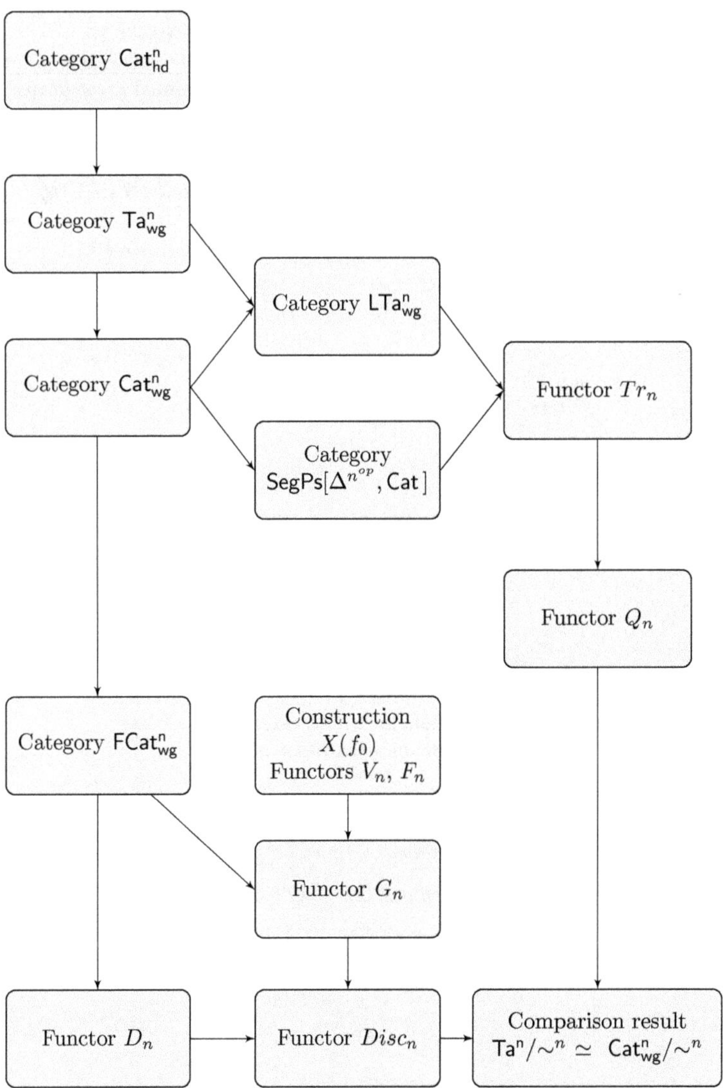

Fig. 3.1 Diagram of connections between the topics in the list of informal discussions

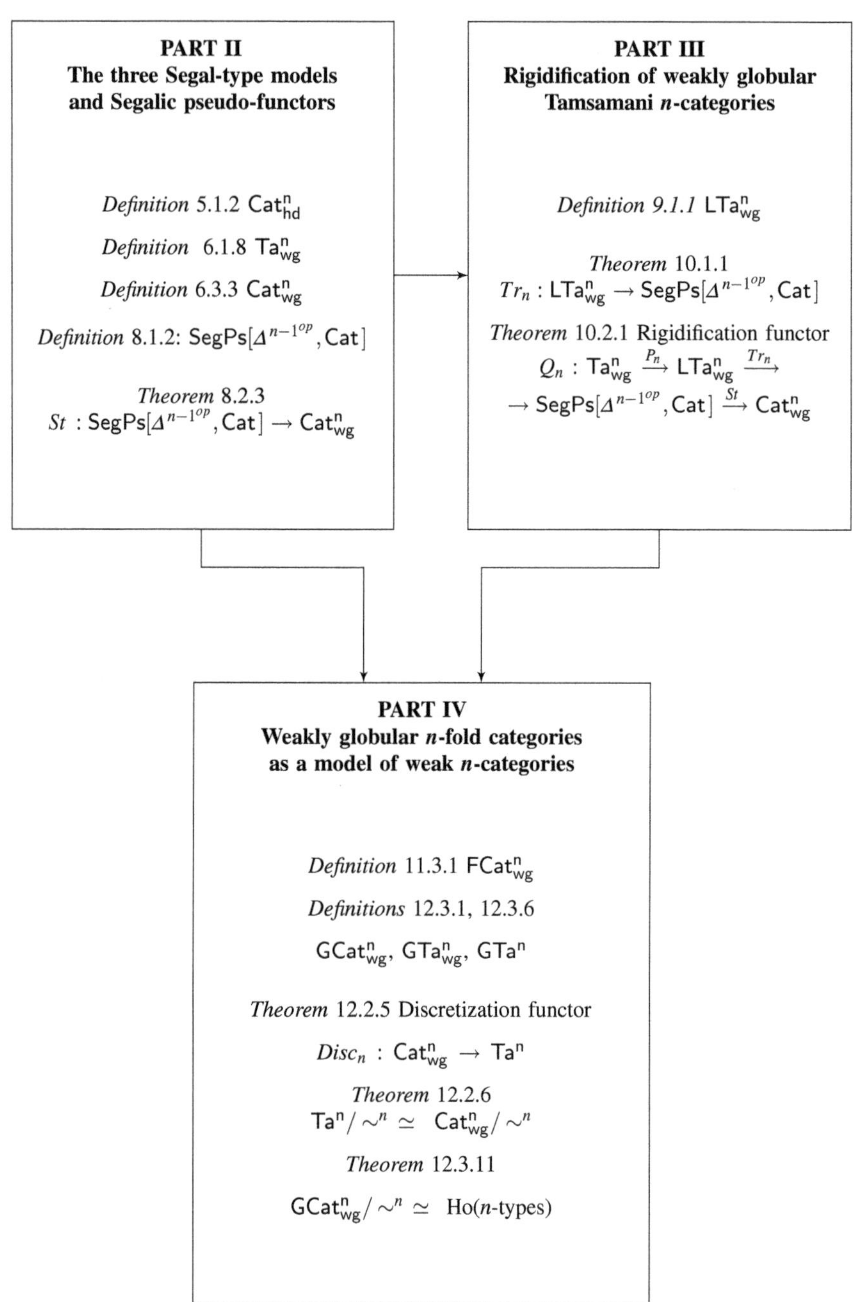

Fig. 3.2 Summary of overall organization and main results

Chapter 4
Techniques from 2-Category Theory

Abstract In this chapter we review another set of techniques crucial to this work, namely techniques from 2-category theory. We recall pseudo-functors and their strictification, as well as a standard technique to produce pseudo-functors. These techniques will be used to build the rigidification functor from weakly globular Tamsamani n-categories to weakly globular n-fold categories. We also review in this chapter the standard categorical notions of pseudo-pullback and isofibration.

In this chapter we cover some categorical background which will be needed in the rest of the book. We first recall two important functors from Cat to Set, associating to a category X the sets of isomorphism classes of objects $p(X)$ and the set of connected components $q(X)$. Their properties are recalled in Sect. 4.1.

The second set of techniques we review in this chapter concern the theory of pseudo-functors. Although pseudo-functors are used in homotopy theory (see for instance [127]), their theory mostly developed within category theory, see for instance [32]. A crucial technique in this work is the use of the strictification functor from pseudo-functors to strict functors. Several versions of strictification exist in the literature (see for instance [122]). Here we review the version due to Power [106] and refined by Lack [81], as these are the forms most suitable for our calculations.

As outlined in Chap. 3 we will show that the rigidification functor from weakly globular Tamsamani n-categories to weakly globular n-fold categories factors through a subcategory of pseudo-functors (called Segalic pseudo-functors); using the form of the strictification functor due to Power [106] and Lack [81] we show that the essential image of the strictification of Segalic pseudo-functors consists of weakly globular n-fold categories.

Another crucial technique we review in this chapter is a way to create pseudo-functors from a small category $\mathscr{C}$ to Cat out of simpler data of a functor from the objects of $\mathscr{C}$ (viewed as a discrete category) to Cat. This is an instance of a more general categorical technique called 'transport of structure along an adjunction' [78] which relies on 2-dimensional monad theory, and it uses the description of pseudo-functors as pseudo-algebras for a 2-monad.

S. Paoli, *Simplicial Methods for Higher Categories*, Algebra and Applications 26,
https://doi.org/10.1007/978-3-030-05674-2_4

4.1 Some Functors on Cat

The connected component functor

$$q : \mathsf{Cat} \to \mathsf{Set}$$

associates to a category its set of path components. This is left adjoint to the discrete category functor

$$d : \mathsf{Set} \to \mathsf{Cat}$$

associating to a set X the discrete category on that set. We denote by

$$\gamma^{(0)} : \mathrm{Id} \Rightarrow dq$$

the unit of the adjunction $q \dashv d$.

Remark 4.1.1 The composite functor

$$\mathsf{Cat} \xrightarrow{N} [\Delta^{op}, \mathsf{Set}] \xrightarrow{q} \mathsf{Set} ,$$

where q is as in Sect. 2.2, coincides with $q : \mathsf{Cat} \to \mathsf{Set}$ as above. Since, by Notational convention 2.5.1, we identify Cat with the essential image of N, it is therefore appropriate to keep the same notation q.

Lemma 4.1.2 *The functor $q : \mathsf{Cat} \to \mathsf{Set}$ preserves fiber products over discrete objects and sends equivalences of categories to isomorphisms.*

Proof We claim that q preserves products; that is, given categories $\mathscr{C}$ and $\mathscr{D}$, there is a bijection

$$q(\mathscr{C} \times \mathscr{D}) = q(\mathscr{C}) \times q(\mathscr{D}) .$$

In fact, given $q(c, d) \in q(\mathscr{C} \times \mathscr{D})$ the map $q(\mathscr{C} \times \mathscr{D}) \to q(\mathscr{C}) \times q(\mathscr{D})$ given by $q(c, d) = (q(c), q(d))$ is well defined and is clearly surjective. On the other hand, this map is also injective: given $q(c, d)$ and $q(c', d')$ with $q(c) = q(c')$ and $q(d) = q(d')$, we have paths in $\mathscr{C}$

$$\begin{array}{ccccc} c & \text{---} & \cdots & \text{---} & c' \\ d & \text{---} & \cdots & \text{---} & d' \end{array}$$

and hence a path in $\mathscr{C} \times \mathscr{D}$

$$(c, d) \;\text{---}\; \cdots \;\text{---}\; (c', d) \;\text{---}\; \cdots \;\text{---}\; (c', d') .$$

Thus $q(c, d) = q(c', d')$ and so the map is also injective, hence it is a bijection, as claimed.

Given a diagram in Cat $\mathscr{C} \xrightarrow[f]{} \mathscr{E} \xleftarrow[g]{} \mathscr{D}$ with $\mathscr{E}$ discrete, we have

$$\mathscr{C} \times_{\mathscr{E}} \mathscr{D} = \coprod_{x \in \mathscr{E}} \mathscr{C}_x \times \, , \mathscr{D}_x \tag{4.1}$$

where $\mathscr{C}_x$, $\mathscr{D}_x$ are the full subcategories of $\mathscr{C}$ and $\mathscr{D}$ with objects c, d such that $f(c) = x = g(d)$. Since q preserves products and (being left adjoint) coproducts, we conclude by (4.1) that

$$q(\mathscr{C} \times_{\mathscr{E}} \mathscr{D}) \cong q(\mathscr{C}) \times_{\mathscr{E}} q(\mathscr{D}) \, .$$

Finally, if $F : \mathscr{C} \simeq \mathscr{D} : G$ is an equivalence of categories, $FG\mathscr{C} \cong \mathscr{C}$ and $FG\mathscr{D} \cong \mathscr{D}$, which implies that $qFqGq\mathscr{C} \cong q\mathscr{C}$ and $qFqGq\mathscr{D} \cong q\mathscr{D}$, so $q\mathscr{C}$ and $q\mathscr{D}$ are isomorphic.

The isomorphism classes of objects functor

$$p : \mathsf{Cat} \to \mathsf{Set}$$

associates to a category the set of isomorphism classes of its objects. Note that if $\mathscr{C}$ is a groupoid, $p\mathscr{C} = q\mathscr{C}$.

Remark 4.1.3 The composite functor

$$\mathsf{Cat} \xrightarrow{N} [\Delta^{op}, \mathsf{Set}] \xrightarrow{p} \mathsf{Set} \, ,$$

where p is as in Sect. 2.2, coincides with $p : \mathsf{Cat} \to \mathsf{Set}$ as above. Since, by Notational convention 2.5.1, we identify Cat with the essential image of N, it is therefore appropriate to keep the same notation p.

Lemma 4.1.4 *The functor p : Cat $\to$ Set preserves pullbacks over discrete objects and sends equivalences of categories to isomorphisms.*

Proof For a category $\mathscr{C}$, let $m\mathscr{C}$ be its maximal sub-groupoid. Then $p\mathscr{C} = qm\mathscr{C}$. Given a diagram in Cat $\mathscr{C} \xrightarrow[f]{} \mathscr{E} \xleftarrow[g]{} \mathscr{D}$ with $\mathscr{E}$ discrete, we have

$$\mathscr{C} \times_{\mathscr{E}} \mathscr{D} = \coprod_{x \in \mathscr{E}} \mathscr{C}_x \times \mathscr{D}_x \, .$$

Since, as easily seen, m commutes with (co)products, and $m\mathscr{E} = \mathscr{E}$, we obtain $m(\mathscr{C} \times_{\mathscr{E}} \mathscr{D}) = m\mathscr{C} \times_{\mathscr{E}} m\mathscr{D}$; so by Lemma 4.1.2,

$$p(\mathscr{C} \times_{\mathscr{E}} \mathscr{D}) = qm(\mathscr{C} \times_{\mathscr{E}} \mathscr{D}) = q(m\mathscr{C} \times_{\mathscr{E}} m\mathscr{D}) = qm\mathscr{C} \times_{q\mathscr{E}} qm\mathscr{D} = p\mathscr{C} \times_{\mathscr{E}} p\mathscr{D} \, .$$

Finally, if $F : \mathscr{C} \simeq \mathscr{D} : G$ is an equivalence of categories, $pF\, pG\, p\mathscr{C} = p\mathscr{C}$ and $pF\, pG\, p\mathscr{D} = p\mathscr{D}$, so $p\mathscr{C}$ and $p\mathscr{D}$ are isomorphic.

Lemma 4.1.5

a) *Let* $X \xrightarrow{f} Z \xleftarrow{g} Y$ *be a diagram in* Cat. *Then*

$$p(X\times_Z Y) \subseteq pX\times_{pZ}\, pY\ .$$

b) *Suppose, further, that* $g_0 = \mathrm{Id}$. *Then*

$$p(X\times_Z Y) \cong pX\times_{pZ}\, pY\ .$$

Proof

a) The map

$$j : p(X\times_Z Y) \to pX\times_{pZ}\, pY$$

is determined by the maps

$$p(X\times_Z Y) \to pX \quad \text{and} \quad p(X\times_Z Y) \to pY$$

induced by the projections

$$X\times_Z Y \to X \quad \text{and} \quad X\times_Z Y \to Y\ .$$

Thus, for each $(a, b) \in X\times_Z Y$,

$$j\, p(a, b) = (p(a), p(b))\ . \tag{4.2}$$

Let $(a, b), (a', b') \in X\times_Z Y$ be such that $jp(a, b) = jp(a', b')$. It follows by (4.2) that $p(a) = p(a')$ and $p(b) = p(b')$. Thus there are isomorphisms $\alpha : a \cong a'$ in X and $\beta : b \cong b'$ in Y and in Z we have

$$f\alpha = g\beta : fa = gb \cong fa' = gb'\ .$$

Thus $(\alpha, \beta) : (a, b) \cong (a', b')$ is an isomorphism in $X\times_Z Y$ and so $p(a, b) = p(a', b')$. This shows that j is injective, proving a).

b) By a) the map j is injective. We now show that, if $g_0 = \mathrm{Id}$, then j is also surjective. Let $(p(x), p(y)) \in pX\times_{pZ}\, pY$ be such that $g_0 = \mathrm{Id}$. Then

$$fp(x) = pf(x) = gp(y) = pg(y) = p(y)\ .$$

Also, $f(x) = g(f(x))$, so that $(x, f(x)) \in X \times_Z Y$. It follows using (4.2) that

$$jp(x, f(x)) = (p(x), pf(x)) = (p(x), p(y))$$

so that j is surjective. Hence j is a bijection.

We recall the notion of a pseudo-pullback [75], which is a special case of a pseudo-limit [32, vol. I, §7.6].

Definition 4.1.6 ([75]) The pseudo-pullback of a pair of functors $g : C \to D$, $f : B \to D$ is the category $C \overset{ps}{\times}_D B$, whose objects are triples (c, ϑ, b) consisting of objects $c \in C$, $b \in B$ and an isomorphism $\vartheta : gc \cong fb$, and whose arrows $(\alpha, \beta) : (c, \vartheta, b) \to (c', \vartheta', b')$ consists of arrows $\alpha : c \to c'$ in C, $\beta : b \to b'$ in D such that the following diagram commutes

$$\begin{array}{ccc} gc & \xrightarrow{\vartheta} & fb \\ {\scriptstyle g\alpha}\downarrow & & \downarrow{\scriptstyle f\beta} \\ gc' & \xrightarrow[\vartheta']{} & fb' \end{array}$$

There is a canonical comparison functor

$$n : C \times_D B \longrightarrow C \overset{ps}{\times}_D B \tag{4.3}$$

$n(c, b) = (c, \mathrm{Id}, b)$, $n(\alpha, \beta) = (\alpha, \beta)$. Clearly n is fully faithful.

Definition 4.1.7 Recall that the functor $F : X \to Y$ is an isofibration if for each $x \in X$ and isomorphism $\alpha : Fx \cong y$ in Y, there is an isomorphism $\beta : x \cong z$ in X with $F\beta = \alpha$.

Theorem 4.1.8 ([75, Theorem 1]) *The canonical comparison functor* (4.3) *is an equivalence of categories for all functors $g : C \to D$ if and only if f is an isofibration.*

Proof Suppose that f is an isofibration. Take $(c, \vartheta, b) \in C \overset{ps}{\times}_D B$ and consider the isomorphism $\vartheta^{-1} : fb \to gc$ to obtain $\phi : b \to b'$ with $fb' = gc$ and $f\phi = \vartheta^{-1}$. This gives an isomorphism $(\mathrm{Id}, \phi) : (c, \vartheta, b) \to n(c, b')$ in $C \overset{ps}{\times}_D B$, showing that n is essentially surjective on objects. Since, as observed above, n is always fully faithful, we conclude that n is an equivalence of categories.

Conversely, suppose that n is an equivalence of categories for all $g : C \to D$. In particular take $g : \mathbb{1} \to D$ $g(1) = d$ and $g(\mathrm{Id}_1) = \mathrm{Id}_d$. Then

$$\mathbb{1} \times_D B \cong \{b \in B \mid fb = d\} ,$$

$$\mathbb{1} \overset{ps}{\times}_D B \cong \{fb \overset{\vartheta}{\cong} d,\ b \in B,\ \vartheta \text{ isomorphism}\} ,$$

while morphisms in $\mathbb{1} \overset{ps}{\times}_D B$ amounts to an isomorphism $\alpha : b \to b'$ making the following diagram commute

$$\begin{array}{ccc} fb & \xrightarrow{\vartheta} & d = g(1) \\ {\scriptstyle f\alpha}\downarrow & & \downarrow{\scriptstyle \mathrm{Id}} \\ fb' & \xrightarrow[\vartheta']{} & d = g(1) \end{array}$$

Thus $n : \mathbb{1} \times_D B \to \mathbb{1} \overset{ps}{\times}_D B$ being essentially surjective on objects means that given $\{fb \overset{\vartheta}{\cong} d\} \in \mathbb{1} \overset{ps}{\times}_D B$, this is isomorphic to $n(b')$; that is, there is an isomorphism $\alpha : b \to b'$ making the following diagram commute

$$\begin{array}{ccc} fb & \xrightarrow{\vartheta} & d \\ {\scriptstyle f\alpha}\downarrow & & \downarrow{\scriptstyle \mathrm{Id}} \\ fb' & \xrightarrow[\mathrm{Id}]{} & d \end{array}$$

Hence $\vartheta = f\alpha$, which by definition means that f is an isofibration. □

Lemma 4.1.9 *Let*

$$\begin{array}{ccc} A & \xrightarrow{g'} & B \\ {\scriptstyle f'}\downarrow & & \downarrow{\scriptstyle f} \\ C & \xrightarrow[g]{} & D \end{array}$$

be a pullback in Cat *with f an isofibration. Then*

a)

$$\begin{array}{ccc} pA & \longrightarrow & pB \\ \downarrow & & \downarrow \\ pC & \longrightarrow & pD \end{array}$$

is a pullback in Set.

b) Suppose that D is a groupoid. Then

$$\begin{array}{ccc} qA & \longrightarrow & qB \\ \downarrow & & \downarrow \\ qC & \longrightarrow & qD \end{array}$$

is a pullback in Set.

Proof

a) Since f is an isofibration, by Theorem 4.1.8, A is equivalent to the pseudo-pullback $A \simeq C \overset{ps}{\times}_D B$. The functor p sends pseudo-pullbacks to pullbacks. In fact, suppose we are given a commuting diagram in Set

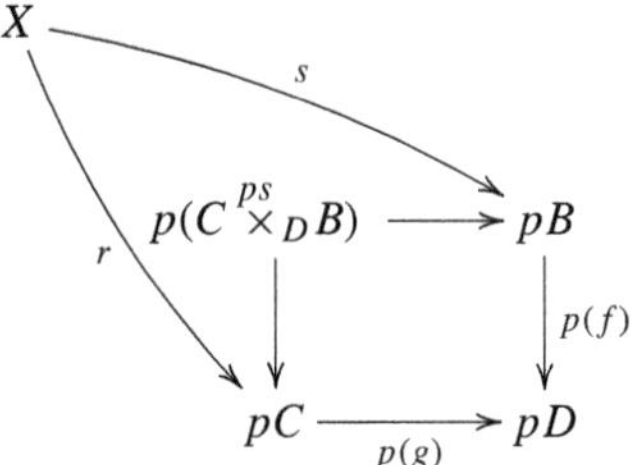

$p(g)r = p(f)s$. If we choose maps $b : dpB \to B$ and $c : dpC \to C$ (so that $p(b) = id$ and $p(c) = id$) we have

$$p(f)p(b)p(d(s)) = p(f)s = p(g)r = p(g)p(c)p(d(r)).$$

It follows that, for each $x \in X$,

$$(fbd(s))(x) \cong (gcd(r))(x) .$$

Therefore, there is a $v : X \to C \overset{ps}{\times}_D B$ such that $g'v = bd(s)$ and $f'v = cd(r)$. Hence

$$p(g')p(v) = p(b)p(d(s)) = s, \qquad p(f')p(v) = p(c)p(d(r)) = r .$$

This shows that

$$p(A) \cong p(C \overset{ps}{\times}_D B) \cong pC \times_{pD} pB .$$

b) Since f is an isofibration, by Joyal and Street [75] A is equivalent to the pseudo-pullback $A \simeq C \overset{ps}{\times}_D B$. The functor q sends pseudo-pullbacks to pullbacks. In fact, suppose we are given a commuting diagram in Set

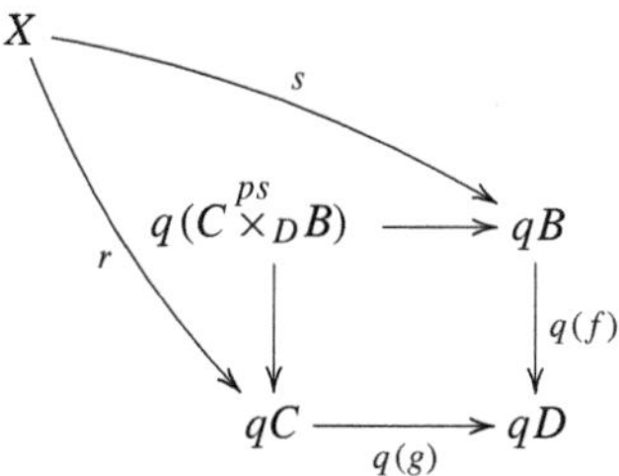

$q(g)r = q(f)s$. If we choose maps $b : dqB \to B$ and $c : dqC \to C$ (so that $q(b) = id$ and $q(c) = id$) we obtain

$$q(f)q(b)q(d(s)) = q(f)s = q(g)r = q(g)q(c)q(d(r)).$$

It follows that, for each $x \in X$, $(fbd(s))(x)$ and $(gcd(r))(x)$ are in the same connected component. Since D is a groupoid, this means that there is an isomorphism

$$(fbd(s))(x) \cong (gcd(r))(x)\ .$$

Therefore, there is a $v : X \to C \overset{ps}{\times}_D B$ such that $g'v = bd(s)$ and $f'v = cd(r)$. Hence

$$q(g')q(v) = q(b)q(d(s)) = s, \qquad q(f')q(v) = q(c)q(d(r)) = r\ .$$

This shows that

$$q(A) \cong q(C \overset{ps}{\times}_D B) \cong qC \times_{qD} qB\ .$$

□

Lemma 4.1.10 *Let*

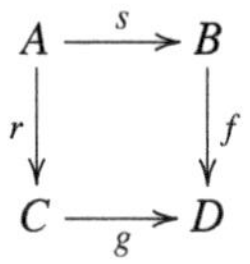

$$\begin{array}{ccc} A & \xrightarrow{s} & B \\ {\scriptstyle r}\downarrow & & \downarrow{\scriptstyle f} \\ C & \xrightarrow[g]{} & D \end{array}$$

be a pullback in Cat, *and suppose that f is fully faithful. Then so is r.*

Proof For all $x, y \in A_0$ there is a pullback in Cat

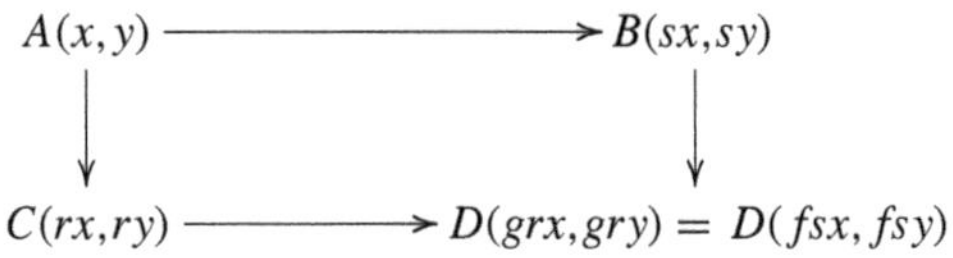

$$\begin{array}{ccc} A(x,y) & \longrightarrow & B(sx,sy) \\ \downarrow & & \downarrow \\ C(rx,ry) & \longrightarrow & D(grx,gry) = D(fsx,fsy) \end{array}$$

Since the right vertical map is a bijection (as f is fully faithful), so is the left vertical map, showing that r is fully faithful.

4.2 Pseudo-Functors and Their Strictification

We recall the notion of pseudo-functor and the classical theory of strictification of pseudo-functors, see [32, 81, 106].

4.2.1 Adjunctions and Equivalences in 2-Categories

We recall some basic 2-categorical notions.

Definition 4.2.1 ([79]) An adjunction in a 2-category $\mathscr{C}$ consists of 1-cells $f : A \to B$ and $u : B \to A$ and 2-cells $\eta : 1_A \to uf$, $\varepsilon : fu \to 1_B$ satisfying the triangle equations

$$B \xrightarrow{u} \underbrace{A \xrightarrow{f} \overbrace{B \xrightarrow{u} A}^{\mathrm{Id}_A,\ \Downarrow\eta}}_{\Downarrow\varepsilon,\ \mathrm{Id}_B} \qquad \equiv \qquad B \xrightarrow{u} A$$

$$A \xrightarrow{f} \overbrace{B \xrightarrow{u} A}^{\mathrm{Id}_A,\ \Downarrow\eta} \xrightarrow{f} B \ \ (\text{with } \Downarrow\varepsilon,\ \mathrm{Id}_B \text{ from } B \text{ to } B) \qquad \equiv \qquad A \xrightarrow{f} B$$

We call f the left adjoint of u and u the right adjoint to f. We call η the unit of the adjunction and ε the counit of the adjunction. If η and ε are invertible 2-cells, the adjunction is called an adjoint equivalence.

For instance, in the 2-category Cat (whose objects are categories, 1-cells are functors and 2-cells are natural transformations), Definition 4.2.1 recovers the usual notion of adjoint functors.

Definition 4.2.2 ([79]) A morphism $f : A \to B$ in a 2-category $\mathscr{C}$ is called an equivalence if there exists a morphism $g : B \to A$ and isomorphisms $gf \cong 1_A$ and $fg \cong 1_B$.

It can be shown [79] that any equivalence in a 2-category can be chosen to be an adjoint equivalence.

4.2.2 The Notion of Pseudo-Functor

Definition 4.2.3 ([32, §7.5]) A pseudo-functor $F : \mathscr{A} \to \mathscr{B}$ between 2-categories $\mathscr{A}$, $\mathscr{B}$ consists of the following data:

(1) For every $A \in \mathscr{A}$, an object $FA \in \mathscr{B}$.
(2) For every pair of objects $A, B \in \mathscr{A}$, a functor

$$F_{AB} : \mathscr{A}(A, B) \to \mathscr{B}(FA, FB) .$$

(3) For every triple of objects $A, B, C \in \mathscr{A}$, a natural isomorphism γ_{ABC}:

$$\begin{array}{ccc} \mathscr{A}(A,B) \times \mathscr{A}(B,C) & \xrightarrow{c_{ABC}} & \mathscr{A}(A,C) \\ {\scriptstyle F_{AB}\times F_{BC}}\downarrow & \overset{\gamma_{ABC}}{\nearrow} & \downarrow{\scriptstyle F_{AC}} \\ \mathscr{B}(FA,FB) \times \mathscr{B}(FB,FC) & \xrightarrow[c_{FA,FB,FC}]{} & \mathscr{B}(FA,FC) \end{array}$$

(4) For every object $A \in \mathscr{A}$, a natural isomorphism δ_A:

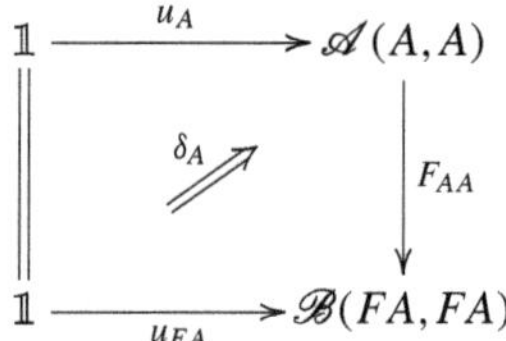

such that the following coherence axioms are satisfied:

(i) Composition axiom: for every triple of arrows

$$A \xrightarrow{f} B \xrightarrow{g} C \xrightarrow{h} D$$

in $\mathscr{A}$, the following equality between 2-cells holds

$$\begin{array}{ccc} Fh \circ Fg \circ Ff & \xrightarrow{i_{Fh} * \gamma_{f,g}} & Fh \circ F(g \circ f) \\ {\scriptstyle \gamma_{g,h} * i_{Ff}}\downarrow & & \downarrow{\scriptstyle \gamma_{g\circ f,h}} \\ F(h \circ g) \circ Ff & \xrightarrow[\gamma_{f,h\circ g}]{} & F(h \circ g \circ f) \end{array}$$

(ii) Unit axiom: for every arrow $f : A \to B$ in $\mathscr{A}$, the following equalities between 2-cells hold

$$\begin{array}{ccc} Ff \circ 1_{FA} & \xrightarrow{i_{Ff} * \delta_A} & Ff \circ F1_A \\ {\scriptstyle i_{Ff}}\downarrow & & \downarrow{\scriptstyle \gamma_{1_A,f}} \\ Ff & \xrightarrow[i_{Ff}]{} & F(f \circ 1_A) \end{array} \qquad \begin{array}{ccc} 1_{FB} \circ Ff & \xrightarrow{\delta_A * i_{Ff}} & F1_B \circ Ff \\ {\scriptstyle i_{Ff}}\downarrow & & \downarrow{\scriptstyle \gamma_{f,1_B}} \\ Ff & \xrightarrow[i_{Ff}]{} & F(1_B \circ f) \end{array}$$

There is also a notion of pseudo-natural transformation between pseudo-functors and modification between pseudo-natural transformations (see [32, §7.5] for more details) and thus a 2-category $Ps[\mathscr{A}, \mathscr{B}]$ of pseudo-functors, pseudo-natural transformations and modifications.

4.2.3 Pseudo T-Algebras

If T is a 2-monad on a 2-category $\mathscr{K}$, a pseudo T-algebra in $\mathscr{K}$ consists of an object A, a morphism $a : TA \to A$ and invertible 2-cells

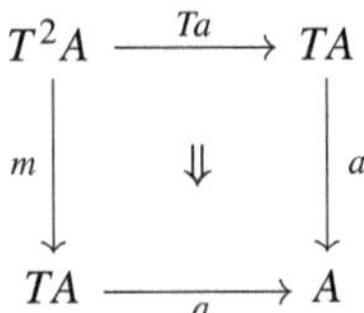

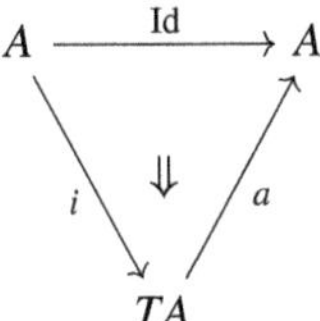

satisfying suitable axioms (see for instance [106] for details), where m is the multiplication of T and i is the unit.

A morphism of pseudo T-algebras is given by the data

$$\begin{array}{ccc} TA & \xrightarrow{Tf} & TB \\ {\scriptstyle a}\downarrow & \Downarrow \tilde{f} & \downarrow{\scriptstyle b} \\ A & \xrightarrow[f]{} & B \end{array}$$

where $\tilde{f}$ is an invertible 2-cell, subject to two coherence conditions (see for instance [106] for details).

An algebra 2-cell $\alpha : (f, \tilde{f}) \Rightarrow (g, \tilde{g})$ in $\mathscr{K}$ is a 2-cell $\alpha : f \Rightarrow g$ in $\mathscr{K}$ such that

$$\begin{array}{ccc} TA & \overset{Tf}{\underset{Tg}{\rightrightarrows}}\ \Downarrow T\alpha & TB \\ {\scriptstyle a}\downarrow & \Downarrow \tilde{g} & \downarrow{\scriptstyle b} \\ A & \xrightarrow[g]{} & B \end{array} \quad = \quad \begin{array}{ccc} TA & \overset{Tf}{\underset{f}{\rightrightarrows}}\ \Downarrow \tilde{f} & TB \\ {\scriptstyle a}\downarrow & \Downarrow \alpha & \downarrow{\scriptstyle b} \\ A & \xrightarrow[g]{} & B \end{array}$$

we denote by Ps-T-alg the 2-category of pseudo T-algebras, morphisms and algebra 2-cells.

4.2.4 Strictification of Pseudo-Functors

The functor 2-category $[\Delta^{n^{op}}, \mathsf{Cat}]$ is 2-monadic over $[ob(\Delta^{n^{op}}), \mathsf{Cat}]$ where $ob(\Delta^{n^{op}})$ is the set of objects of $\Delta^{n^{op}}$. Let

$$U : [\Delta^{n^{op}}, \mathsf{Cat}] \rightarrow [ob(\Delta^{n^{op}}), \mathsf{Cat}]$$

be the forgetful functor given by $(UX)_{\underline{k}} = X_{\underline{k}}$ for each $\underline{k} \in \Delta^{n^{op}}$ and $X \in [\Delta^{n^{op}}, \mathsf{Cat}]$. Its left adjoint F is given on objects by

$$(FY)_{\underline{k}} = \coprod_{\underline{r} \in ob(\Delta^{n^{op}})} \Delta^{n^{op}}(\underline{r}, \underline{k}) \times Y_{\underline{r}}$$

for $Y \in [ob(\Delta^{n^{op}}), \mathsf{Cat}]$, $\underline{k} \in \Delta^{n^{op}}$. If T is the monad corresponding to the adjunction $F \dashv U$, then

$$(TY)_{\underline{k}} = (UFY)_{\underline{k}} = \coprod_{\underline{r} \in ob(\Delta^{n^{op}})} \Delta^{n^{op}}(\underline{r}, \underline{k}) \times Y_{\underline{r}} \, .$$

A pseudo T-algebra is given by $Y \in [ob(\Delta^{n^{op}}), \mathsf{Cat}]$, functors

$$h_{\underline{k}} : \coprod_{\underline{r} \in ob(\Delta^{n^{op}})} \Delta^{n^{op}}(\underline{r}, \underline{k}) \times Y_{\underline{r}} \rightarrow Y_{\underline{k}}$$

and additional data, as described in Sect. 4.2. This amounts precisely to functors from $\Delta^{n^{op}}$ to Cat and the 2-category $\mathsf{Ps\text{-}T\text{-}alg}$ of pseudo T-algebras corresponds to the 2-category $\mathsf{Ps}[\Delta^{n^{op}}, \mathsf{Cat}]$ of pseudo-functors, pseudo-natural transformations and modifications. Note that there is a commuting diagram

$$\begin{array}{ccc} [\Delta^{n^{op}}, \mathsf{Cat}] & \hookrightarrow & \mathsf{Ps}[\Delta^{n^{op}}, \mathsf{Cat}] \\ {\scriptstyle U}\downarrow & \swarrow {\scriptstyle U} & \\ [ob(\Delta^{n^{op}}), \mathsf{Cat}] & & \end{array}$$

Recalling that, if X is a set and $\mathscr{C}$ is a category, $X \times \mathscr{C} \cong \coprod_X \mathscr{C}$, we see that the pseudo T-algebra corresponding to $H \in \mathsf{Ps}[\Delta^{n^{op}}, \mathsf{Cat}]$ has structure map $h : TUH \rightarrow UH$ as follows:

$$(TUH)_{\underline{k}} = \coprod_{\underline{r} \in \Delta^n} \Delta^n(\underline{k}, \underline{r}) \times H_{\underline{r}} = \coprod_{\underline{r} \in \Delta^n} \coprod_{\Delta^n(\underline{k}, \underline{r})} H_{\underline{r}} \, .$$

If $f \in \Delta^n(\underline{k}, \underline{r})$, let

$$i_{\underline{r}} = \coprod_{\Delta^n(\underline{k},\underline{r})} H_{\underline{r}} \to \coprod_{\underline{r}\in\Delta^n} \coprod_{\Delta^n(\underline{k},\underline{r})} H_{\underline{r}} = (TUH)_{\underline{r}} \, ,$$

$$j_f : H_{\underline{r}} \to \coprod_{\Delta^n(\underline{k},\underline{r})} H_{\underline{r}}$$

be the coproduct inclusions, then

$$h_{\underline{k}}\, i_{\underline{r}}\, j_f = H(f) \, . \tag{4.4}$$

The structure map $TUH \to UH$ carries a canonical enrichment to a pseudo-natural transformation $FUH \to H$. The strictification of pseudo-algebras result proved in [106] yields that every pseudo-functor from $\Delta^{n^{op}}$ to Cat is equivalent, in $\mathsf{Ps}[\Delta^{n^{op}}, \mathsf{Cat}\,]$, to a 2-functor, that is, an object of $[\Delta^{n^{op}}, \mathsf{Cat}\,]$.

Given a pseudo T-algebra as above, [106] consider the factorization of $h : TUH \to UH$ as

$$TUH \overset{v}{\to} L \overset{g}{\to} UH$$

with $v_{\underline{k}}$ bijective on objects and $g_{\underline{k}}$ fully faithful, for each $\underline{k} \in \Delta^{n^{op}}$. It is shown in [106] that g is a pseudo-natural transformation and it is possible to give a strict T-algebra structure $TL \to L$ such that (g, Tg) is an equivalence of pseudo T-algebras. It is immediate to see that, for each $\underline{k} \in \Delta^{n^{op}}$, $g_{\underline{k}}$ is an equivalence of categories.

Further, it is shown in [81] that $St : \mathsf{Ps}[\Delta^{n^{op}}, \mathsf{Cat}\,] \to [\Delta^{n^{op}}, \mathsf{Cat}\,]$ as described above is left adjoint to the inclusion

$$J : [\Delta^{n^{op}}, \mathsf{Cat}\,] \to \mathsf{Ps}[\Delta^{n^{op}}, \mathsf{Cat}\,]$$

and that the components of the units are equivalences in $\mathsf{Ps}[\Delta^{n^{op}}, \mathsf{Cat}\,]$.

4.3 Transport of Structure

We now recall a general categorical technique, known as transport of structure along an adjunction, with one of its applications. This technique will be used crucially in the proof of Theorem 10.1.1.

Theorem 4.3.1 ([78, Theorem 6.1]) *Given an equivalence $\eta,\ \varepsilon : f \dashv f^* : A \to B$ in the complete and locally small 2-category $\mathscr{A}$, and an algebra (A, a) for the monad $T = (T, i, m)$ on $\mathscr{A}$, the equivalence enriches to an equivalence*

$$\eta, \varepsilon : (f, \overline{\overline{f}}) \vdash (f^*, \overline{\overline{f^*}}) : (A, a) \to (B, b, \hat{b}, \overline{b})$$

in Ps-T-alg, *where* $\hat{b} = \eta,\ \overline{b} = f^* a{\cdot}T\varepsilon{\cdot}Ta{\cdot}T^2 f,\ \overline{\overline{f}} = \varepsilon^{-1} a{\cdot}Tf,\ \overline{\overline{f^*}} = f^* a{\cdot}T\varepsilon.$

Let $\eta', \varepsilon' : f' \vdash f'^{*} : A' \to B'$ be another equivalence in $\mathscr{A}$ and let $(B', b', \hat{b'}, \overline{b'})$ be the corresponding pseudo-T-algebra as in Theorem 4.3.1. Suppose $g : (A, a) \to (A', a')$ is a morphism in $\mathscr{A}$ and γ is an invertible 2-cell in $\mathscr{A}$

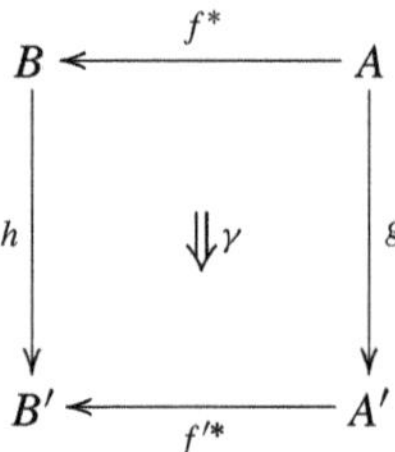

Let $\overline{\gamma}$ be the invertible 2-cell given by the following pasting:

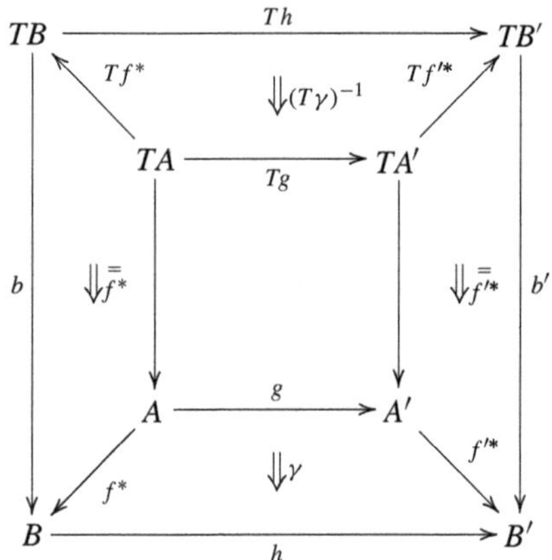

Then it is not difficult to show that $(h, \overline{\gamma}) : (B, b, \hat{b}, \overline{b}) \to (B', b', \hat{b'}, \overline{b'})$ is a pseudo-T-algebra morphism.

The following fact is essentially known and, as sketched in the proof below, it is an instance of Theorem 4.3.1.

Lemma 4.3.2 ([103]) *Let $\mathscr{C}$ be a small 2-category, $F, F' : \mathscr{C} \to$ Cat be 2-functors, and $\alpha : F \to F'$ a 2-natural transformation. Suppose that, for all objects C of $\mathscr{C}$, the following conditions hold:*

i) $G(C)$, $G'(C)$ are objects of Cat *and there are adjoint equivalences of categories $\mu_C \vdash \eta_C$, $\mu'_C \vdash \eta'_C$,*

$$\mu_C : G(C) \rightleftarrows F(C) : \eta_C \qquad \mu'_C : G'(C) \rightleftarrows F'(C) : \eta'_C,$$

ii) there are functors $\beta_C : G(C) \to G'(C)$,

iii) there is an invertible 2-cell

$$\gamma_C : \beta_C\, \eta_C \Rightarrow \eta'_C\, \alpha_C.$$

Then

a) There exists a pseudo-functor $G : \mathscr{C} \to \mathsf{Cat}$ given on objects by $G(C)$, and pseudo-natural transformations $\eta : F \to G$, $\mu : G \to F$ with $\eta(C) = \eta_C$, $\mu(C) = \mu_C$; these are part of an adjoint equivalence $\mu \vdash \eta$ in the 2-category $\mathsf{Ps}[\mathscr{C}, \mathsf{Cat}]$.

b) There is a pseudo-natural transformation $\beta : G \to G'$ with $\beta(C) = \beta_C$ and an invertible 2-cell in $\mathsf{Ps}[\mathscr{C}, \mathsf{Cat}]$, $\gamma : \beta\eta \Rightarrow \eta\alpha$ with $\gamma(C) = \gamma_C$.

Proof Recall [106] that the functor 2-category $[\mathscr{C}, \mathsf{Cat}]$ is 2-monadic over $[ob(\mathscr{C}), \mathsf{Cat}]$, where $ob(\mathscr{C})$ is the set of objects in $\mathscr{C}$. Let

$$\mathscr{U} : [\mathscr{C}, \mathsf{Cat}] \to [ob(\mathscr{C}), \mathsf{Cat}]$$

be the forgetful functor. Let T be the 2-monad; then the pseudo-T-algebras are precisely the pseudo-functors from $\mathscr{C}$ to Cat.

The adjoint equivalences $\mu_C \vdash \eta_C$ amount precisely to an adjoint equivalence in $[ob(\mathscr{C}), \mathsf{Cat}]$, $\mu_0 \vdash \eta_0$, $\mu_0 : G_0 \rightleftarrows \mathscr{U}F : \eta_0$, where $G_0(C) = G(C)$ for all $C \in ob(\mathscr{C})$. By Theorem 4.3.1 this equivalence enriches to an adjoint equivalence $\mu \vdash \eta$ in $\mathsf{Ps}[\mathscr{C}, \mathsf{Cat}]$

$$\mu : G \rightleftarrows F : \eta$$

between F and a pseudo-functor G; we have $\mathscr{U}G = G_0$, $\mathscr{U}\eta = \eta_0$, $\mathscr{U}\mu = \mu_0$; hence on objects G is given by $G(C)$, and $\eta(C) = \mathscr{U}\eta(C) = \eta_C$, $\mu(C) = \mathscr{U}\mu(C) = \mu_C$.

Let $\nu_C : \mathrm{Id}_{G(C)} \Rightarrow \eta_C\mu_C$ and $\varepsilon_C : \mu_C\eta_C \Rightarrow \mathrm{Id}_{F(C)}$ be the unit and counit of the adjunction $\mu_C \vdash \eta_C$. Given a morphism $f : C \to D$ in $\mathscr{C}$, we have

$$G(f) = \eta_D F(f)\mu_C$$

and we have natural isomorphisms:

$$\eta_f : G(f)\eta_C = \eta_D F(f)\mu_C\eta_C \xRightarrow{\eta_D F(f)\varepsilon_C} \eta_D F(f)$$

$$\mu_f : F(f)\mu_C \xRightarrow{\nu_{F(f)}\mu_C} \mu_D\eta_D F(f)\mu_C = \mu_D G(f).$$

Also, the natural isomorphism

$$\beta_f : G'(f)\beta_C \Rightarrow \beta_D G(f)$$

is the result of the following pasting

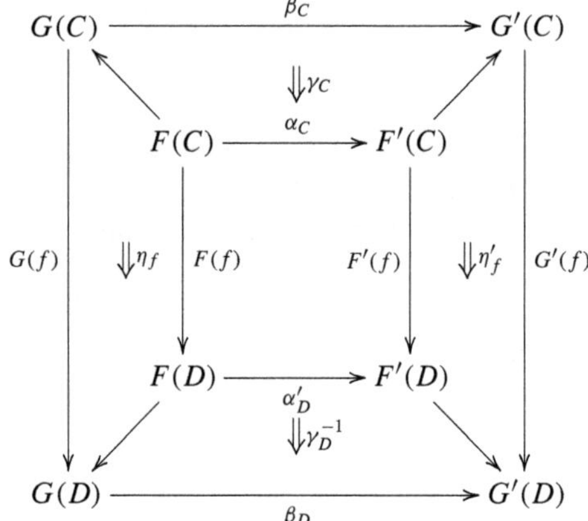

□

Part II
The Three Segal-Type Models and Segalic Pseudo-Functors

In this Part we introduce the three Segal-type models: the category $\mathsf{Ta}^{\mathsf{n}}_{\mathsf{wg}}$ of weakly globular n-fold categories and its subcategories $\mathsf{Cat}^{\mathsf{n}}_{\mathsf{wg}}$ (weakly globular n-fold categories) and Ta^{n} (Tamsamani n-categories). We study the relation between $\mathsf{Cat}^{\mathsf{n}}_{\mathsf{wg}}$ and the special class of pseudo-functors

$$\mathsf{SegPs}[\Delta^{n-1^{op}}, \mathsf{Cat}\,] \subset \mathsf{Ps}[\Delta^{n-1^{op}}, \mathsf{Cat}\,]$$

called Segalic pseudo-functors. The main result of this Part is Theorem 8.2.3, establishing that the classical strictification of pseudo-functors, when restricted to Segalic pseudo-functors, yields weakly globular n-fold categories; that is, there is a functor

$$St\ : \mathsf{SegPs}[\Delta^{n-1^{op}}, \mathsf{Cat}\,] \rightarrow \mathsf{Cat}^{\mathsf{n}}_{\mathsf{wg}}\ .$$

A summary of the main results in this Part is given in Fig. 4.1.

In Chap. 5 we introduce the category $\mathsf{Cat}^{\mathsf{n}}_{\mathsf{hd}}$ of homotopically discrete n-fold categories and we study its properties. The idea of the category $\mathsf{Cat}^{\mathsf{n}}_{\mathsf{hd}}$ is introduced in Sect. 5.1.1 before the formal definition. This category is needed for the precise formulation of the weak globularity condition in the definition of $\mathsf{Ta}^{\mathsf{n}}_{\mathsf{wg}}$. We also introduce n-equivalences of homotopically discrete n-fold categories, and we show in Lemma 5.2.6 that they are detected by isomorphisms of their discretizations.

In Chap. 6 we introduce the three Segal-type models and in Chap. 7 we study their properties. The ideas of the categories $\mathsf{Ta}^{\mathsf{n}}_{\mathsf{wg}}$ and $\mathsf{Cat}^{\mathsf{n}}_{\mathsf{wg}}$ are introduced in Sects. 6.1.1 and 6.3.1 before their formal definition.

In Sect. 7.1 we study the properties of the category $\mathsf{Ta}^{\mathsf{n}}_{\mathsf{wg}}$, which are used throughout the rest of the work: Proposition 7.1.2 establishes important properties of n-equivalences, Proposition 7.1.5 gives a sufficient criterion for an n-equivalence to be a levelwise equivalence of categories. Proposition 7.1.7 establishes the existence of the functor

$$q^{(n-1)} : \mathsf{Ta}^{\mathsf{n}}_{\mathsf{wg}} \rightarrow \mathsf{Ta}^{n-1}_{\mathsf{wg}}\,.$$

The functor $q^{(n-1)}$ divides out by the highest dimensional cells in the structure. The properties of $q^{(n-1)}$ in relation to certain pullbacks are further studied in Sect. 7.1.2, and they play a key role in the proof of Theorem 9.2.4, leading to the rigidification functor Q_n.

In Sect. 7.2 we study the properties of the category $\mathsf{Cat}^{\mathsf{n}}_{\mathsf{wg}}$. The main result of this section is Proposition 7.2.8. Part (a) of this proposition establishes that the nerve functor in direction 2 on the category $\mathsf{Cat}^{\mathsf{n}}_{\mathsf{wg}}$ is levelwise a weakly globular $(n-1)$-fold category: this property will be used on several occasions in the rest of the book, for instance in Corollary 7.1.8 in showing the existence of the functor

$$q^{(n-1)} : \mathsf{Cat}^{\mathsf{n}}_{\mathsf{wg}} \to \mathsf{Cat}^{n-1}_{\mathsf{wg}} ,$$

which plays an important role in the construction of the rigidification functor.

Part (b) of Proposition 7.2.8 gives a sufficient criterion for an n-fold category to be weakly globular. This criterion plays a crucial role in the proof of Proposition 8.2.1, leading to the main result of this Part, Theorem 8.2.3.

In Chap. 8 we introduce the category $\mathsf{SegPs}[\Delta^{n^{op}}, \mathsf{Cat}]$ of Segalic pseudo-functors and study its main properties. The idea of this category is introduced in Sect. 8.1.2 before the formal definition.

In Proposition 8.2.1 we establish that an n-fold category levelwise equivalent via a pseudo-natural transformation to a Segalic pseudo-functor is weakly globular: the proof of this result is based on the criterion of Proposition 7.2.8 (b) together with Corollary 7.2.2, which gives a sufficient condition for a weakly globular n-fold category to be homotopically discrete.

In Lemma 8.2.2 we study the properties of the monad corresponding to Segalic pseudo-functors. Together with Proposition 8.2.1 this leads to the main result, Theorem 8.2.3 on the strictification of Segalic pseudo-functors.

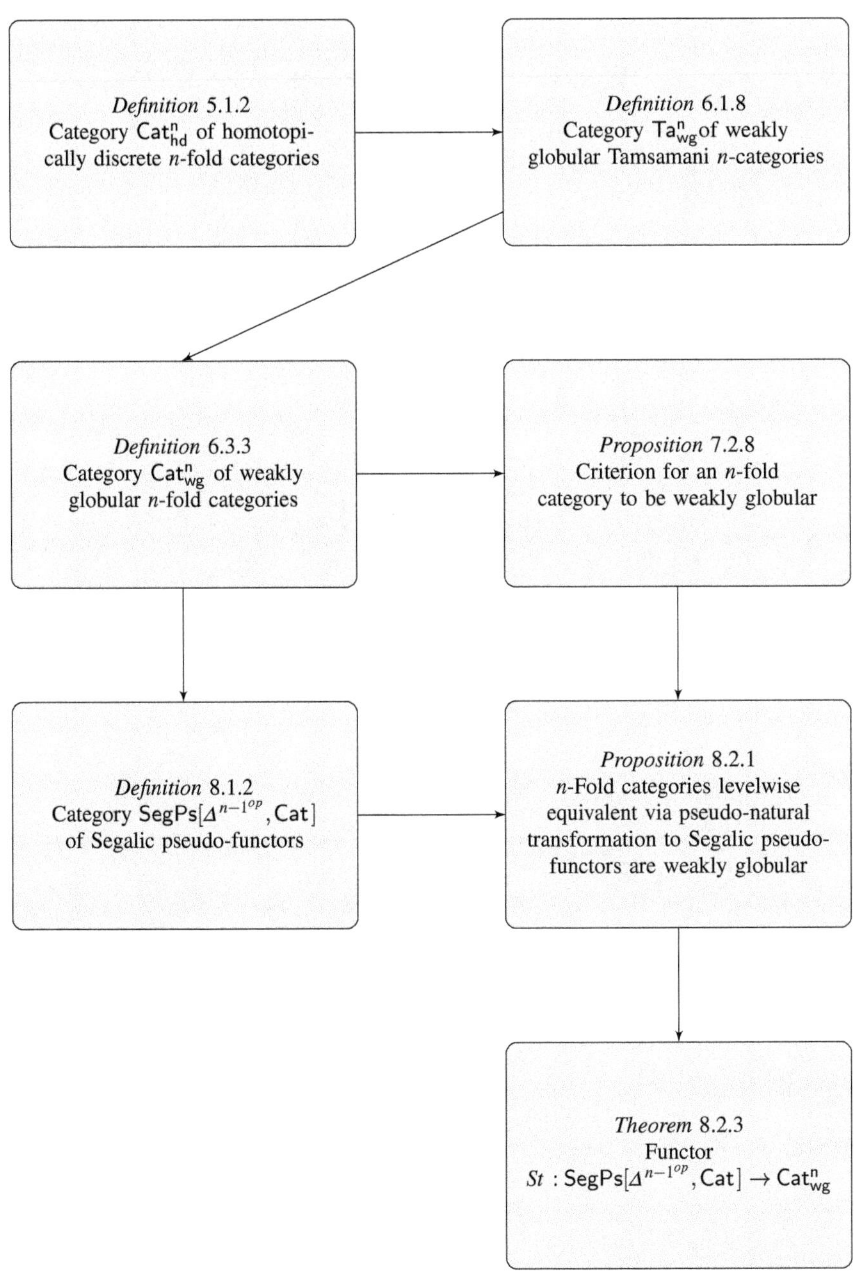

Fig. 4.1 The three Segal-type models and Segalic pseudo-functors

Chapter 5
Homotopically Discrete n-Fold Categories

Abstract In this chapter we introduce in details the first new structure of this work: the category $\mathsf{Cat}^n_{\mathsf{hd}}$ of homotopically discrete n-fold categories. This structure is essential in stating the weak globularity condition in the definition of our Segal-type models. After defining homotopically discrete n-fold categories, we characterize them in terms of internal equivalence relations and we discuss their main properties.

In this chapter we introduce the category $\mathsf{Cat}^n_{\mathsf{hd}}$ of homotopically discrete n-fold categories, which will be needed in Chap. 6 to state the weak globularity condition in the definition of the category of weakly globular Tamsamani n-categories.

Recall (see Definition 2.4.4) that a discrete n-fold category is one in which the multinerve is a constant functor: thus it amounts to a set and identity structure maps in all the simplicial directions. Homotopically discrete n-fold categories are 'discrete up to homotopy' in a specified way. An object $X \in \mathsf{Cat}^n_{\mathsf{hd}}$ comes equipped with a discretization map

$$\gamma_{(n)} : X \to X^d \ ,$$

where X^d is a discrete n-fold category and $\gamma_{(n)}$ is a suitably defined higher categorical equivalence.

In the case $n = 1$, an object of $\mathsf{Cat}_{\mathsf{hd}}$ is a groupoid equivalent to a discrete category, that is, an equivalence relation, which is a groupoid with no non-trivial loops.

As outlined in Sect. 3.3, homotopically discrete k-fold categories (for $1 \leq k \leq n-1$) are sub-structures of our Segal-type models $\mathsf{Cat}^n_{\mathsf{wg}}$ and $\mathsf{Ta}^n_{\mathsf{wg}}$. The sets underlying the discretizations of the homotopically discrete sub-structures in objects of $\mathsf{Cat}^n_{\mathsf{wg}}$ and $\mathsf{Ta}^n_{\mathsf{wg}}$ play the role of sets of higher cells. In the category $\mathsf{Cat}^n_{\mathsf{wg}}$ the weakness of the structure is encoded in these homotopically discrete objects: indeed an object of $\mathsf{Cat}^n_{\mathsf{wg}}$ in which all these homotopically discrete sub-structures are discrete is a strict n-category.

S. Paoli, *Simplicial Methods for Higher Categories*, Algebra and Applications 26,
https://doi.org/10.1007/978-3-030-05674-2_5

Homotopically discrete n-fold categories are higher groupoidal structures to which we can associate a classifying space which is a 0-type (see Proposition 5.3.2), that is, a topological space whose homotopy groups are zero in dimension greater than 0.

After defining homotopically discrete n-fold categories in Definition 5.1.2 we prove in Sect. 5.1.3 that they are iterated internal equivalence relations, offering a more conceptual viewpoint on them. We also show in Proposition 5.2.9 that the induced Segal maps of a homotopically discrete n-fold category are $(n-1)$-equivalences. As we will see in Chap. 6, this makes $\mathsf{Cat}^n_{\mathsf{hd}}$ a full subcategory of the category $\mathsf{Cat}^n_{\mathsf{wg}}$ of weakly globular n-fold categories.

This chapter is organized as follows. In Sect. 5.1 we define the category $\mathsf{Cat}^n_{\mathsf{hd}}$ and the discretization map $\gamma_{(n)} : X \to X^d$ for each $X \in \mathsf{Cat}^n_{\mathsf{hd}}$. In Sect. 5.1.3 we explain how to view homotopically discrete n-fold categories as internal equivalence relations.

In Sect. 5.2 we investigate the properties of the category $\mathsf{Cat}^n_{\mathsf{hd}}$, in particular its closure properties with respect to certain limits and colimits. These properties will be needed to discuss the closure properties of $\mathsf{Ta}^n_{\mathsf{wg}}$ in Chap. 6. In Sect. 5.2.2 we define n-equivalences in $\mathsf{Cat}^n_{\mathsf{hd}}$ and we show in Lemma 5.2.6 that they are detected by isomorphisms of their discretizations. We deduce in Proposition 5.2.9 that the induced Segal maps of objects of $\mathsf{Cat}^n_{\mathsf{hd}}$ are $(n-1)$-equivalences and in Corollary 5.2.7 that the discretization maps are n-equivalences. In Sect. 5.3 we investigate homotopically discrete n-fold categories from a homotopical viewpoint.

5.1 The Definition of Homotopically Discrete n-Fold Categories

In this section we give an inductive definition of the category $\mathsf{Cat}^n_{\mathsf{hd}}$ of homotopically discrete n-fold categories and we discuss in Lemma 5.1.6 a straightforward but useful criterion for an n-fold category to be homotopically discrete. We also define the discretization map for objects of $\mathsf{Cat}^n_{\mathsf{hd}}$.

5.1.1 The Idea of a Homotopically Discrete n-Fold Category

When $n = 1$, a homotopically discrete category is simply a groupoid with no non-trivial loops, in other words an equivalence relation. The idea of a homotopically discrete n-fold category when $n > 1$ is that it is an n-fold category suitably equivalent to a discrete one both 'globally' and in each simplicial dimension. Thus, we require that $X \in \mathsf{Cat}^n_{\mathsf{hd}}$ is a levelwise equivalence relation, that is, for each $(k_1, \ldots, k_{n-1}) \in \Delta^{n-1^{op}}$, $X_{k_1 \ldots k_{n-1}}$ is (the nerve of) an equivalence relation.

We further impose the condition that taking isomorphism classes of objects in each dimension along the last simplicial direction gives a homotopically discrete $(n-1)$-fold category. That is, we have truncation functors

$$\mathsf{Cat}^{\mathsf{n}}_{\mathsf{hd}} \xrightarrow{p^{(n-1)}} \mathsf{Cat}^{\mathsf{n-1}}_{\mathsf{hd}} \xrightarrow{p^{(n-2)}} \cdots \mathsf{Cat}_{\mathsf{hd}} \xrightarrow{p^{(0)}} \mathsf{Set}\,,$$

where the functor $p^{(r)} : [\Delta^{n^{op}}, \mathsf{Set}] \to [\Delta^{r^{op}}, \mathsf{Set}]$ is as in Definition 2.2.5. There are corresponding maps for each $X \in \mathsf{Cat}^{\mathsf{n}}_{\mathsf{hd}}$

$$X \xrightarrow{\gamma^{(n-1)}} p^{(n-1)}X \xrightarrow{\gamma^{(n-2)}} p^{(n-2)}p^{(n-1)}X \to \cdots \xrightarrow{\gamma^{(0)}} X^d\,, \tag{5.1}$$

where

$$X^d = p^{(0)}p^{(2)}\dots p^{(n-1)}X$$

is a discrete n-fold category.

We show in Corollary 5.2.7 that the maps (5.1) and their composite (the discretization map) are n-equivalences, so X is n-equivalent to a discrete category.

In the case $n = 1$, an equivalence relation is a groupoid whose classifying space is a 0-type, since its first homotopy group is zero (as this groupoid has no non-trivial loops). Similarly, for $n > 1$ we build a classifying space functor from $\mathsf{Cat}^{\mathsf{n}}_{\mathsf{hd}}$ to 0-types; so we can think of the category $\mathsf{Cat}^{\mathsf{n}}_{\mathsf{hd}}$ as an appropriate n-fold model of 0-types.

The category $\mathsf{Cat}^{\mathsf{n}}_{\mathsf{hd}}$ is used to formulate the weak globularity condition in two of the three Segal-type models: The category $\mathsf{Ta}^{\mathsf{n}}_{\mathsf{wg}}$ of weakly globular Tamsamani n-categories and its subcategory $\mathsf{Cat}^{\mathsf{n}}_{\mathsf{wg}}$ of weakly globular n-fold categories, which are introduced informally in Sects. 6.1.1 and 6.3.1 respectively.

5.1.2 The Formal Definition of $\mathsf{Cat}^{\mathsf{n}}_{\mathsf{hd}}$

We now formally introduce the category $\mathsf{Cat}^{\mathsf{n}}_{\mathsf{hd}}$ of homotopically discrete n-fold categories.

Definition 5.1.1 An equivalence relation is a groupoid with no non-trivial loops.

In what follows we recall that, according to our notational conventions established in Chap. 2,

(i) $\mathsf{Cat}^{\mathsf{n}}$ is identified with a full subcategory of $[\Delta^{n^{op}}, \mathsf{Set}]$ via the multinerve functor (see Notational Convention 2.5.1). Given $X \in \mathsf{Cat}^{\mathsf{n}}$ and $(k_1, \dots, k_n) \in \Delta^{n^{op}}$, $X_{k_1\dots k_n} \in \mathsf{Set}$ denotes the multinerve of X evaluated at $(k_1, \dots, k_n)$.

(ii) If we denote by $\mathsf{Cat_{hd}}$ the category of equivalence relations, an object $X \in [\Delta^{n^{op}}, \mathsf{Set}]$ is in the subcategory $[\Delta^{n-1^{op}}, \mathsf{Cat_{hd}}]$ if and only if it is a levelwise equivalence relation; that is, for each $(k_1, \ldots, k_{n-1}) \in \Delta^{n-1^{op}}$, $X_{k_1 \ldots k_{n-1}}$ is (the nerve of) an equivalence relation.
(iii) From (i) and (ii) we say that $X \in \mathsf{Cat^n}$ is a levelwise equivalence relation if and only if, for each $(k_1, \ldots, k_{n-1}) \in \Delta^{n-1^{op}}$, $X_{k_1 \ldots k_{n-1}}$ is an equivalence relation.
(iv) For each $0 \leq r < n$ the functors $p^{(r)}, q^{(r)} : [\Delta^{n^{op}}, \mathsf{Set}] \to [\Delta^{r^{op}}, \mathsf{Set}]$ are as in Definition 2.2.5.

We now inductively define the subcategory $\mathsf{Cat^n_{hd}} \subset \mathsf{Cat^n}$ of homotopically discrete n-fold categories.

Definition 5.1.2 Let $\mathsf{Cat^0_{hd}} = \mathsf{Set}$. Suppose, inductively, we have defined the subcategory $\mathsf{Cat^{n-1}_{hd}} \subset \mathsf{Cat^{n-1}}$ of homotopically discrete $(n-1)$-fold categories. We say that the n-fold category $X \in \mathsf{Cat^n}$ is homotopically discrete if:

(a) X is a levelwise equivalence relation.
(b) $p^{(n-1)}X \in \mathsf{Cat^{n-1}_{hd}}$.

When $n = 1$ we denote by $\mathsf{Cat^1_{hd}} = \mathsf{Cat_{hd}}$ the subcategory of Cat consisting of equivalence relations.

Remark 5.1.3 Since, as observed in Lemma 2.2.7, the functor $p^{(r)} : [\Delta^{n^{op}}, \mathsf{Set}] \to [\Delta^{r^{op}}, \mathsf{Set}]$ $(0 \leq r < n)$ of Definition 2.2.5 factors as a composite $p^{(r)} p^{(r+1)} \ldots p^{(n-1)}$ it follows by Definition 5.1.2 that $p^{(r)}$ restricts to a functor

$$p^{(r)} : \mathsf{Cat^n_{hd}} \to \mathsf{Cat^r_{hd}} .$$

Also note that, since p and q agree on equivalence relations (since the latter are in particular groupoids), $p^{(r)} = q^{(r)} : \mathsf{Cat^n_{hd}} \to \mathsf{Cat^r_{hd}}$, where $q^{(r)}$ is as in Definition 2.2.5.

Definition 5.1.4 Let $X \in \mathsf{Cat^n_{hd}}$. Denote by $\gamma_X^{(n-1)} : X \to p^{(n-1)}X$ the morphism given by

$$(\gamma_X^{(n-1)})_{s_1 \ldots s_{n-1}} : X_{s_1 \ldots s_{n-1}} \to qX_{s_1 \ldots s_{n-1}} = pX_{s_1 \ldots s_{n-1}} .$$

Denote by

$$X^d = qX = pX = p^{(0)} p^{(1)} \ldots p^{(n-1)} X$$

and by $\gamma_{(n)}$ the composite

$$X \xrightarrow{\gamma^{(n-1)}} p^{(n-1)}X \xrightarrow{\gamma^{(n-2)}} p^{(n-2)} p^{(n-1)} X \to \cdots \xrightarrow{\gamma^{(0)}} X^d .$$

We call $\gamma_{(n)}$ the discretization map of X.

Remark 5.1.5 Note that, for each $X \in \mathsf{Cat}^n_{\mathsf{hd}}$, the discretization map $\gamma_{(n)} : X \to X^d$ is natural in X.

The following lemma is immediate from the definitions:

Lemma 5.1.6 *Let $X \in \mathsf{Cat}^n$. Then $X \in \mathsf{Cat}^n_{\mathsf{hd}}$ if and only if, for each $k \geq 0$, $X_k \in \mathsf{Cat}^{n-1}_{\mathsf{hd}}$ and $p^{(n-1)}X \in \mathsf{Cat}^{n-1}_{\mathsf{hd}}$.*

Proof By induction on n. The lemma is trivial for $n = 1$. Suppose it holds for $(n-1)$. If $X \in \mathsf{Cat}^n_{\mathsf{hd}}$, by definition $p^{(n-1)}X \in \mathsf{Cat}^{n-1}_{\mathsf{hd}}$, so by the inductive hypothesis $(p^{(n-1)}X)_k = p^{(n-2)}X_k \in \mathsf{Cat}^{n-2}_{\mathsf{hd}}$. Also by definition, X is a levelwise equivalence relation, thus so is X_k. We conclude that $X_k \in \mathsf{Cat}^{n-1}_{\mathsf{hd}}$.

Conversely, let $X \in \mathsf{Cat}^n$ be such that $p^{(n-1)}X \in \mathsf{Cat}^{n-1}_{\mathsf{hd}}$ and $X_k \in \mathsf{Cat}^{n-1}_{\mathsf{hd}}$ for all $k \geq 0$. Then X_k is a levelwise equivalence relation; that is, for each $k_2 \dots k_{n-1}$, $(X_k)_{k_2 \dots k_{n-1}} = X_{kk_2 \dots k_{n-1}}$ is an equivalence relation. This means that X is a levelwise equivalence relation. By definition, we conclude that $X \in \mathsf{Cat}^n_{\mathsf{hd}}$. □

Remark 5.1.7 We observe that $p^{(n-1)}$ commutes with pullbacks over discrete objects. In fact, if $X \to Z \leftarrow Y$ is a diagram in $\mathsf{Cat}^n_{\mathsf{hd}}$ with Z discrete and $X \times_Z Y \in \mathsf{Cat}^n_{\mathsf{hd}}$, by Definition 5.1.2, for all $(s_1 \dots s_{n-1}) \in \Delta^{{n-1}^{op}}$,

$$
\begin{aligned}
&(p^{(n-1)}(X \times_Z Y))_{s_1 \dots s_{n-1}} = p(X_{s_1 \dots s_{n-1}} \times_Z Y_{s_1 \dots s_{n-1}}) \\
&= pX_{s_1 \dots s_{n-1}} \times_{pZ} pY_{s_1 \dots s_{n-1}} = (p^{(n-1)}X \times_{p^{(n-1)}Z} p^{(n-1)}Y)_{s_1 \dots s_{n-1}} ,
\end{aligned}
$$

where we used the fact (Lemma 4.1.4) that p commutes with pullbacks over discrete objects. Since this holds for each $s_1 \dots s_{n-1}$ we conclude that

$$
p^{(n-1)}(X \times_Z Y) \cong p^{(n-1)}X \times_{p^{(n-1)}Z} p^{(n-1)}Y .
$$

5.1.3 Homotopically Discrete n-Fold Categories As Internal Equivalence Relations

In this section we give a different description of homotopically discrete n-fold categories via a notion of internal equivalence relation associated to a morphism $f : A \to B$ in a category $\mathscr{C}$ with finite limits. When $\mathscr{C} = \mathsf{Set}$ and f is surjective, this affords the category $\mathsf{Cat}_{\mathsf{hd}}$ of equivalence relations. In Proposition 5.1.11 we make a choice of $f : A \to B$ in Cat^{n-1} such that the corresponding internal equivalence relation is in $\mathsf{Cat}^n_{\mathsf{hd}}$ and we show that this gives a characterization of homotopically discrete n-fold categories.

Definition 5.1.8 Let $A \to B$ be a morphism in a category $\mathscr{C}$ with finite limits. The diagonal map defines a unique section

$$s : A \to A \times_B A$$

so that $p_1 s = \mathrm{Id}_A = p_2 s$ where $A \times_B A$ is the pullback of $A \xrightarrow{f} B \xleftarrow{f} A$ and $p_1, p_2 : A \times_B A \to A$ are the two projections. The commutative diagram

$$\begin{array}{ccccc} A\times_B A & \xrightarrow{p_1} & A & \xleftarrow{p_2} & A\times_B A \\ {\scriptstyle p_2}\downarrow & & \downarrow{\scriptstyle f} & & \downarrow{\scriptstyle p_1} \\ A & \xrightarrow[f]{} & B & \xleftarrow[f]{} & A \end{array}$$

defines a unique morphism

$$m : (A \times_B A) \times_A (A \times_B A) \to A \times_B A$$

such that $p_2 m = p_2 \pi_2$ and $p_1 m = p_1 \pi_1$, where π_1 and π_2 are the two projections. We denote by $A[f]$ the following object of $\mathsf{Cat}\,(\mathscr{C})$

$$(A\times_B A)\times_A (A\times_B A) \xrightarrow{\quad m \quad} A\times_B A \overset{p_1}{\underset{s}{\overset{\xrightarrow{\quad}}{\underset{\xleftarrow{\quad}}{\xrightarrow{p_2}}}}} A$$

Lemma 5.1.9 *Let $A \to B$ be a morphism in $\mathsf{Cat}\,\mathscr{C}$ with finite limits. Then $A[f]$ (as in Definition 5.1.8) is an internal groupoid in $\mathscr{C}$.*

Proof We need to prove the axioms of an internal groupoid as in Sect. 2.3. Axioms (1) and (2) in Definition 2.3.1 hold by construction. As for (3), denote by

$$\pi_i : (A \times_B A) \times_A (A \times_B A) \to A \times_B A \qquad i = 1, 2$$

the two projections. Then

$$f p_1 m \begin{pmatrix} \mathrm{Id} \\ s p_1 \end{pmatrix} = f p_1 \pi_1 \begin{pmatrix} \mathrm{Id} \\ s p_1 \end{pmatrix} = f p_1 \,\mathrm{Id} = f p_1 \;,$$

$$f p_2 m \begin{pmatrix} \mathrm{Id} \\ s p_1 \end{pmatrix} = f p_2 \pi_2 \begin{pmatrix} \mathrm{Id} \\ s p_2 \end{pmatrix} = f p_2 s p_2 \,\mathrm{Id} = f p_2 \;.$$

Therefore

$$m \circ \begin{pmatrix} \mathrm{Id} \\ s p_1 \end{pmatrix} = \mathrm{Id}_{A \times_B A} \;,$$

which is the first half of axiom (3). The second half is proved similarly.

To show (4) denote by

$$r_i : (A \times_B A) \times_A (A \times_B A) \times_A (A \times_B A) \to (A \times_B A) \qquad i = 1, 2, 3$$

the three projections and by

$$r_{12} : (A \times_B A) \times_A (A \times_B A) \times_A (A \times_B A) \to (A \times_B A) \times_A (A \times_B A)$$
$$r_{23} : (A \times_B A) \times_A (A \times_B A) \times_A (A \times_B A) \to (A \times_B A) \times_A (A \times_B A)$$

the projections

$$r_{12} = r_1 \times r_2 \qquad r_{23} = r_2 \times r_3 \ .$$

To show axiom (3) in Definition 2.3.1, it is enough to prove that

$$f p_2 m(\mathrm{Id} \times m) = f p_2 m(m \times \mathrm{Id}) \ , \tag{5.2}$$

$$f p_1 m(\mathrm{Id} \times m) = f p_1 m(m \times \mathrm{Id}) \ . \tag{5.3}$$

We calculate

$$f p_2 m r_{23} = f p_2 \pi_2 r_{23} = f p_2 r_3 = f p_2 \,\mathrm{Id}\, r_3 \ ,$$

which is (5.2). Also,

$$f p_1 m r_{12} = f p_1 \pi_1 r_{12} = f p_1 r_1 = f p_1 \,\mathrm{Id}\, r_1 \ ,$$

which is (5.3).

Thus $A[f] \in \mathsf{Cat}\,\mathscr{C}$. To show that $A[f]$ is an internal groupoid, consider the inverse map

$$i : A \times_B A \to A \times_B A$$

determined by the diagram

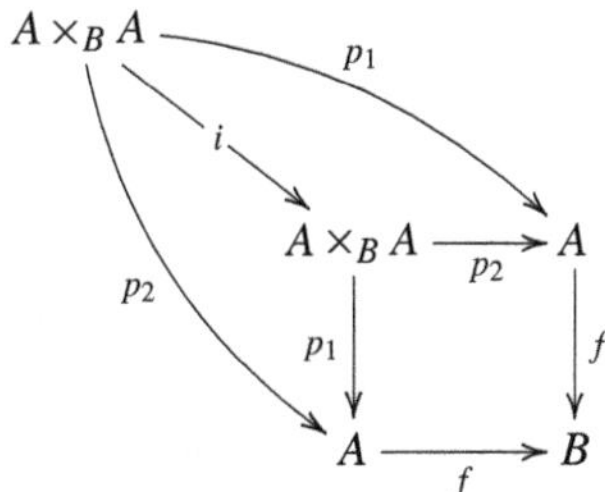

Then $m(\mathrm{Id}, i) = sp_2$ and $m(i, \mathrm{Id}) = sp_1$. □

Remark 5.1.10 If $X \in \mathsf{Cat}_{\mathsf{hd}}$ then $X = X_0[\gamma]$, where $\gamma : X_0 \to pX$ is a surjective map of sets.

Proposition 5.1.11 *Let $X \in \mathsf{Cat^n}$. Then $X \in \mathsf{Cat^n_{hd}}$ if and only if $X = A[f]$ for a morphism $f : A \to B$ in $\mathsf{Cat^{n-1}}$ with $B \in \mathsf{Cat^{n-1}_{hd}}$ and f a levelwise surjection in* Set.

Proof Let $X \in \mathsf{Cat^n_{hd}}$. By definition for each $\underline{k} = (k_1, \dots, k_{n-1}) \in \Delta^{n-1^{op}}$, $X_{\underline{k}}$ is an equivalence relation, thus the map of sets

$$X_{\underline{k}} \to qX_{\underline{k}} = (q^{(n-1)}X)_{\underline{k}} = (p^{(n-1)}X)_{\underline{k}}$$

is surjective, and

$$X_{\underline{k}1} = X_{\underline{k}0} \times_{qX_{\underline{k}}} X_{\underline{k}0} \,.$$

Recall from (2.3) that $(X_0^{\{n\}})_{\underline{k}} = X_{\underline{k}0}$. Thus from the above

$$X = X_0^{\{n\}}[(\gamma^{(n-1)})_0^{\{n\}}] \,,$$

where $(\gamma^{(n-1)})_0^{\{n\}} : X_0^{\{n\}} \to (p^{(n-1)}X)_0^{\{n\}} = p^{(n-1)}X$ is a map in $\mathsf{Cat^{n-1}}$ with $p^{(n-1)}X \in \mathsf{Cat^{n-1}_{hd}}$ and, from the above, is a levelwise surjection.

Conversely, let $f : A \to B$ be a map in $\mathsf{Cat^{n-1}}$ with $B \in \mathsf{Cat^{n-1}_{hd}}$ and such that f is a levelwise surjection. We show that $A[f] \in \mathsf{Cat^n_{hd}}$. Clearly $A[f] \in \mathsf{Cat^n}$ since $A[f]$ is an internal category in $\mathsf{Cat^{n-1}}$. Since pullbacks in $[\Delta^{n^{op}}, \mathsf{Set}]$ are computed pointwise, for each $\underline{k} \in \Delta^{n-1^{op}}$,

$$(A[f])_{\underline{k}} = A_{\underline{k}}[f_{\underline{k}}]$$

is an equivalence relation; since $f_{\underline{k}}$ is surjective

$$q(A[f])_{\underline{k}} = B_{\underline{k}} \,.$$

It follows that

$$q^{(n-1)}(A[f])_{\underline{k}} = q(A[f])_{\underline{k}} = B_{\underline{k}} \,,$$

that is,

$$p^{(n-1)}A[f] = q^{(n-1)}A[f] = B \in \mathsf{Cat^{n-1}_{hd}} \,.$$

By Definition 5.1.2 we conclude that $A[f] \in \mathsf{Cat^n_{hd}}$. □

Example 5.1.12 Let $X \in \mathsf{Cat}^2_{\mathsf{hd}}$; then $p^{(1)}X$ is the equivalence relation associated to the surjective map of sets

$$\gamma : X_0^d = (p^{(1)}X)_0 \to p^{(0)}(p^{(1)}X) = X^d$$

and X has the form

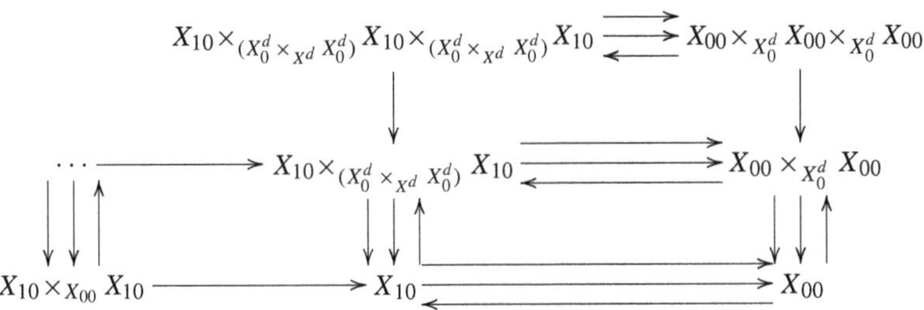

The horizontal nerve $N^{(1)}X \in [\Delta^{op}, \mathsf{Cat}]$ has in each component an equivalence relation so in particular a groupoid. The horizontal structure is not in general groupoidal; however $p^{(1)}X$ is an equivalence relation, so in particular a groupoid. This means that the horizontal arrows in the double category X have inverses after dividing out by the double cells.

Remark 5.1.13 Homotopically discrete n-fold categories are more general than the homotopically discrete n-fold groupoids of [29]. In particular, they are n-fold categories but not in general n-fold groupoids since only some but not all of the n different simplicial directions in the structure are required to be groupoidal. See for instance Example 5.1.12 above.

The way weakly globular n-fold categories arise as strictifications of Segalic pseudo-functors (see Theorem 8.2.3) requires this added generality to the notion of homotopically discrete n-fold categories compared to the homotopically discrete n-fold groupoids of [29].

5.2 Properties of Homotopically Discrete n-Fold Categories

In this section we establish the main properties of homotopically discrete n-fold categories. We investigate in Sect. 5.2.1 some closure properties of the category $\mathsf{Cat}^n_{\mathsf{hd}}$ with respect to some limits and colimits. These closure properties will be used in Chap. 6 to discuss corresponding closure properties of the three Segal-type models, which are an important part of their definition.

We also explore some consequences of the closure properties of $\mathsf{Cat}^n_{\mathsf{hd}}$ in Corollary 5.2.2. This allows us to define the hom-$(n-1)$-category $X(a,b) \in \mathsf{Cat}^{n-1}_{\mathsf{hd}}$ of any $X \in \mathsf{Cat}^n_{\mathsf{hd}}$. This is used crucially to give a notion of n-equivalence in $\mathsf{Cat}^n_{\mathsf{hd}}$ (Definition 5.2.5); the latter is a higher dimensional generalization of a functor which is fully faithful and essentially surjective on objects. In Sect. 5.2.2 we show that every homotopically discrete n-fold category X is n-equivalent to the discrete n-fold category X^d via the discretization map $\gamma_{(n)}$ of Definition 5.1.4. In addition, homotopically discrete n-fold categories commute with pullbacks over discrete objects (Lemma 5.2.2). These properties imply that the induced Segal maps in a homotopically discrete n-fold category are $(n-1)$-equivalences (Proposition 5.2.9). As a consequence we will see in Chap. 6 that homotopically discrete n-fold categories are a subcategory of weakly globular n-fold categories.

5.2.1 Closure Properties of $\mathsf{Cat}^{\mathsf{n}}_{\mathsf{hd}}$

We illustrate some closure properties of $\mathsf{Cat}^{\mathsf{n}}_{\mathsf{hd}}$ and their consequences. These follow the general pattern for the Segal-type models summarized in Sect. 3.4.2.

Note that $\mathsf{Cat}^{\mathsf{n}}_{\mathsf{hd}}$ contains the terminal object since the latter is in particular a discrete n-fold category and discrete n-fold categories are homotopically discrete.

Lemma 5.2.1 *The subcategory* $\mathsf{Cat}^{\mathsf{n}}_{\mathsf{hd}}$ *of* $[\Delta^{n^{op}}, \mathsf{Set}]$ *has the following closure properties:*

a) It is replete under isomorphisms.
b) It is closed under finite products.
c) It is closed under small coproducts.
d) If the small coproduct $\coprod_i A_i$ *is in* $\mathsf{Cat}^{\mathsf{n}}_{\mathsf{hd}}$*, then each* $A_i \in \mathsf{Cat}^{\mathsf{n}}_{\mathsf{hd}}$*.*

Proof

a) By induction on n. It is clear for $n = 1$. Suppose it holds for $(n - 1)$ and let $f : A \to B$ be an isomorphism in $[\Delta^{n^{op}}, \mathsf{Set}]$ with $A \in \mathsf{Cat}^{\mathsf{n}}_{\mathsf{hd}}$. Then all Segal maps of B are isomorphisms (since the same is true for all Segal maps of A), thus $B \in \mathsf{Cat}^{\mathsf{n}}$. For each $\underline{k} \in \Delta^{n^{op}}$, $A_{\underline{k}} \cong B_{\underline{k}}$, and since $A_{\underline{k}}$ is an equivalence relation, so is $B_{\underline{k}}$. Also, $p^{(n-1)}A \cong p^{(n-1)}B$, so by the induction hypothesis $p^{(n-1)}B \in \mathsf{Cat}^{\mathsf{n-1}}_{\mathsf{hd}}$. By definition we conclude that $B \in \mathsf{Cat}^{\mathsf{n}}_{\mathsf{hd}}$.

b) By induction on n. It is clear for $n = 1$; suppose it holds for $n - 1$. Then for each $s \geq 0$, by the induction hypothesis

$$(X \times Y)_s = X_s \times Y_s \in \mathsf{Cat}^{\mathsf{n-1}}_{\mathsf{hd}}\ .$$

Since $p^{(n-1)}$ commutes with pullbacks over discrete objects (see Remark 5.1.7) and therefore with products, by the induction hypothesis

$$p^{(n-1)}(X \times Y) = p^{(n-1)}X \times p^{(n-1)}Y \in \mathsf{Cat}^{\mathsf{n-1}}_{\mathsf{hd}}\ .$$

By Lemma 5.1.6 this proves that $X \times Y \in \mathsf{Cat}^{\mathsf{n}}_{\mathsf{hd}}$.

c) By induction on n. It is clear for $n = 1$. Suppose it holds for $n-1$ and let $X_i \in \mathsf{Cat}^n_{\mathsf{hd}}$. Since $\mathsf{Cat}^n_{\mathsf{hd}} \subset [\Delta^{op}, \mathsf{Cat}^{n-1}_{\mathsf{hd}}]$ and coproducts in functor categories are computed pointwise, for each $s \geq 0$ we have, by the induction hypothesis

$$(\coprod_i X_i)_s = \coprod_i (X_i)_s \in \mathsf{Cat}^{n-1}_{\mathsf{hd}} .$$

Since p commutes with coproducts, the same holds for $p^{(n-1)}$, thus by the induction hypothesis

$$p^{(n-1)}(\coprod_i X_i) = \coprod_i p^{(n-1)} X_i \in \mathsf{Cat}^{n-1}_{\mathsf{hd}} .$$

By Lemma 5.1.6 this proves that $\coprod_i X_i \in \mathsf{Cat}^n_{\mathsf{hd}}$.

d) By induction on n. When $n = 1$, let $\coprod_i A_i$ be an equivalence relation. Thus $\coprod_i A_i$ is a groupoid, hence so is A_i for each i. For each $x \in \coprod_i A_{i0}$, $(\coprod_i A_i)(x,x) = \coprod_i A_i(x,x)$ is trivial, thus each $A_i(x,x)$ is trivial. In conclusion, each A_i is a groupoid with no non-trivial loops, hence an equivalence relation.

Suppose the lemma holds for $(n-1)$ and let $\coprod_i A_i \in \mathsf{Cat}^n_{\mathsf{hd}}$, so for each $k \geq 0$, $\coprod_i A_{ik} \in \mathsf{Cat}^{n-1}_{\mathsf{hd}}$, thus by the inductive hypothesis $(A_i)_k \in \mathsf{Cat}^{n-1}_{\mathsf{hd}}$. Also,

$$p^{(n-1)}(\coprod_i A_i) = \coprod_i p^{(n-1)} A_i \in \mathsf{Cat}^{n-1}_{\mathsf{hd}} ,$$

thus by the inductive hypothesis $p^{(n-1)} A_i \in \mathsf{Cat}^{n-1}_{\mathsf{hd}}$. By Lemma 5.1.6 we conclude that each $A_i \in \mathsf{Cat}^{n-1}_{\mathsf{hd}}$.

□

Corollary 5.2.2

a) If $f : E \to X$ is a map in $\mathsf{Cat}^{n-1}_{\mathsf{hd}}$ with X discrete, the fiber E_x of f at $x \in X$ is in $\mathsf{Cat}^{n-1}_{\mathsf{hd}}$.

b) Let $X \xrightarrow{f} Z \xleftarrow{g} Y$ be a diagram in $\mathsf{Cat}^n_{\mathsf{hd}}$ with Z discrete. Then $X \times_Z Y \in \mathsf{Cat}^n_{\mathsf{hd}}$ and $(X \times_Z Y)^d = X^d \times_Z Y^d$.

Proof

a) Since $E = \coprod_{x \in X} E_x$, this follows from Lemma 5.2.1 part d).

b) Since Z is discrete,

$$X \times_Z Y = \coprod_{c \in Z} X_c \times Y_c ,$$

where X_c (resp. Y_c) is the fiber of f (resp. g) at c. Since by a) $X_c,\ Y_c \in \mathsf{Cat}^n_{\mathsf{hd}}$, from Lemma 5.2.1 it follows that $X\times_Z Y \in \mathsf{Cat}^n_{\mathsf{hd}}$. Since by Remark 5.1.7 $p^{(n-1)}$ commutes with pullbacks over discrete objects for all n, we have

$$(X\times_Z Y)^d = p^{(0)}\cdots p^{(n-1)}(X\times_Z Y)$$
$$= p^{(0)}\cdots p^{(n-1)}X\times_{p^{(0)}\dots p^{(n-1)}Z}\, p^{(0)}\cdots p^{(n-1)}Y = X^d\times_Z Y^d\ .$$

□

Lemma 5.2.3 *Let*

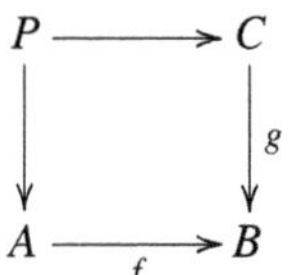

be a pullback in Cat *with* $A, B, C \in \mathsf{Cat}_{\mathsf{hd}}$ *and* g *an isofibration. Then* $P \in \mathsf{Cat}_{\mathsf{hd}}$.

Proof Consider the diagram in Cat in which the vertical maps are discretization maps (which are natural by Remark 5.1.5)

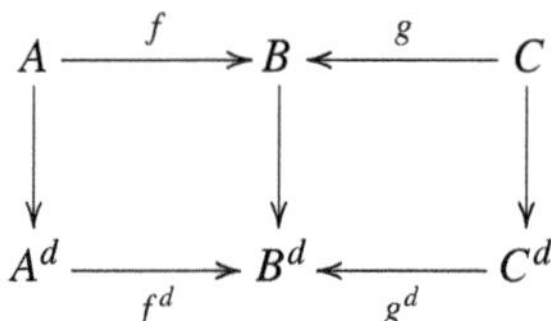

In this diagram the vertical maps are equivalences of categories since $A, B, C \in \mathsf{Cat}_{\mathsf{hd}}$; by hypothesis g is an isofibration, and g^d is also an isofibration (it is a map in Set). Therefore the induced map of pullbacks

$$P \to A^d\times_{B^d} C^d$$

is an equivalence of categories. Since $A^d\times_{B^d} C^d$ is discrete, this means that $P \in \mathsf{Cat}_{\mathsf{hd}}$. □

5.2.2 *n-Equivalences in* $\mathsf{Cat}^n_{\mathsf{hd}}$

Definition 5.2.4 Given $X \in \mathsf{Cat}^n_{\mathsf{hd}}$, for each $a, b \in X_0^d$ denote by $X(a, b)$ the fiber at (a, b) of the map

$$X_1 \xrightarrow{(d_0,d_1)} X_0 \times X_0 \xrightarrow{\gamma_{(n)}\times\gamma_{(n)}} X_0^d \times X_0^d\ .$$

By Corollary 5.2.2, $X(a,b) \in \mathsf{Cat}^{n-1}_{\mathsf{hd}}$. This should be thought of as a hom-$(n-1)$-category.

Definition 5.2.5 Define inductively n-equivalences in $\mathsf{Cat}^{n}_{\mathsf{hd}}$. For $n = 1$, a 1-equivalence is an equivalence of categories. Suppose we have defined $(n-1)$-equivalences in $\mathsf{Cat}^{n-1}_{\mathsf{hd}}$. Then a map $f : X \to Y$ in $\mathsf{Cat}^{n}_{\mathsf{hd}}$ is an n-equivalence if

a) For all $a, b \in X^d_0$,

$$f(a,b) : X(a,b) \to Y(fa, fb)$$

is an $(n-1)$-equivalence.

b) $p^{(n-1)} f$ is an $(n-1)$-equivalence.

Lemma 5.2.6 *A map $f : X \to Y$ in $\mathsf{Cat}^{n}_{\mathsf{hd}}$ is an n-equivalence if and only if it induces an isomorphism $X^d \cong Y^d$.*

Proof By induction on n. For $n = 1$, if f is an equivalence of categories, then $X^d \cong Y^d$. Conversely, if $X^d \cong Y^d$, then f is essentially surjective on objects. It is also fully faithful, since it is a functor between equivalence relations, hence f is an equivalence of categories.

Suppose the lemma holds for $(n-1)$ and let $f : X \to Y$ be an n-equivalence in $\mathsf{Cat}^{n}_{\mathsf{hd}}$. Then by definition $p^{(n-1)} f$ is an $(n-1)$-equivalence; therefore by the induction hypothesis

$$X^d = (p^{(n-1)} X)^d \cong (p^{(n-1)} Y)^d = Y^d \ .$$

Conversely, suppose that $f : X \to Y$ is such that $X^d \cong Y^d$. This is the same as $(p^{(n-1)} X)^d = (p^{(n-1)} Y)^d$, so, by induction, $p^{(n-1)} f$ is an $(n-1)$-equivalence. This implies that, for each $a, b \in X^d$, $(p^{(n-1)} f)(a,b)$ is an $(n-2)$-equivalence. But

$$(p^{(n-1)} f)(a,b) = (p^{(n-2)} f)(a,b)$$

so $(p^{(n-2)} f)(a,b)$ is an $(n-2)$-equivalence. This implies that

$$X(a,b)^d = (p^{(n-2)} X(a,b))^d \cong (p^{(n-2)} Y(fa, fb))^d = Y(fa, fb)^d \ .$$

By the induction hypothesis, we deduce that

$$f_{(a,b)} : X(a,b) \to Y(fa, fb)$$

is an $(n-1)$-equivalence. We conclude that f is an n-equivalence. □

Corollary 5.2.7 *Let $X \in \mathsf{Cat}^{n}_{\mathsf{hd}}$. Then the maps $\gamma^{(n-1)} : X \to p^{(n-1)} X$ and $\gamma_{(n)} : X \to X^d$ are n-equivalences.*

Proof This follows from Lemma 5.2.6 since

$$X^d \cong (p^{(n-1)}X)^d \cong (X^d)^d.$$

□

Remark 5.2.8 It follows from Lemma 5.2.6 that n-equivalences in $\mathsf{Cat}^{\mathsf{n}}_{\mathsf{hd}}$ have the 2-out-of-3 property.

Given $X \in \mathsf{Cat}^{\mathsf{n}}_{\mathsf{hd}}$, since X_0^d is discrete and $X_1 \in \mathsf{Cat}^{\mathsf{n-1}}_{\mathsf{hd}}$, by Corollary 5.2.2, for all $s \geq 2$,

$$X_1 \times_{X_0^d} \overset{s}{\cdots} \times_{X_0^d} X_1 \in \mathsf{Cat}^{\mathsf{n-1}}_{\mathsf{hd}}\ .$$

We can therefore consider the induced Segal maps in $\mathsf{Cat}^{\mathsf{n-1}}_{\mathsf{hd}}$

$$\hat{\mu}_s : X_s = X_1 \times_{X_0} \overset{s}{\cdots} \times_{X_0} X_1 \to X_1 \times_{X_0^d} \overset{s}{\cdots} \times_{X_0^d} X_1$$

(see Definition 2.1.3). Using Lemma 5.2.6 we show next that these maps are $(n-1)$-equivalences.

Proposition 5.2.9 *Let $X \in \mathsf{Cat}^{\mathsf{n}}_{\mathsf{hd}}$. For each $s \geq 2$ the induced Segal maps*

$$\hat{\mu}_s : X_1 \times_{X_0} \overset{s}{\cdots} \times_{X_0} X_1 \to X_1 \times_{X_0^d} \overset{s}{\cdots} \times_{X_0^d} X_1$$

are $(n-1)$-equivalences.

Proof We show this for $s = 2$, the case $s > 2$ being similar. By Lemma 5.2.6 it is enough to show that

$$(X_1 \times_{X_0} X_1)^d \cong X_1^d \times_{X_0^d} X_1^d\ . \tag{5.4}$$

We claim that for each $1 \leq j \leq n-1$ and $X \in \mathsf{Cat}^{\mathsf{n-1}}_{\mathsf{hd}}$

$$p^{(j-1)}(X_1 \times_{X_0} X_1) \cong p^{(j-1)}X_1 \times_{p^{(j-1)}X_0} p^{(j-1)}X_1\ . \tag{5.5}$$

In fact we have

$$(p^{(j)}X)_2 = p^{(j-1)}X_2 = p^{(j-1)}(X_1 \times_{X_0} X_1)\ . \tag{5.6}$$

Since by Remark 5.1.3 $p^{(j)} : \mathsf{Cat}^{\mathsf{n}}_{\mathsf{hd}} \to \mathsf{Cat}^{\mathsf{j}}_{\mathsf{hd}}$, the Segal maps of $p^{(j)}X$ are isomorphisms, so

$$(p^{(j)}X)_2 \cong (p^{(j)}X)_1 \times_{(p^{(j)}X)_0} (p^{(j)}X)_1 = p^{(j-1)}X_1 \times_{p^{(j-1)}X_0} p^{(j-1)}X_1 \,. \tag{5.7}$$

By (5.6) and (5.7), (5.5) follows. In the case $j = 1$ this gives (5.4). □

5.3 Homotopically Discrete n-Fold Categories and 0-Types

In this section we discuss the homotopical significance of the category $\mathsf{Cat}^{\mathsf{n}}_{\mathsf{hd}}$. We introduce a classifying space functor from $\mathsf{Cat}^{\mathsf{n}}_{\mathsf{hd}}$ to spaces and we show that the classifying space of a homotopically discrete n-fold category is a 0-type.

Definition 5.3.1 The classifying space functor is the composite

$$B : \mathsf{Cat}^{\mathsf{n}}_{\mathsf{hd}} \hookrightarrow [\Delta^{n^{op}}, \mathsf{Set}] \xrightarrow{Diag_n} [\Delta^{op}, \mathsf{Set}] \,,$$

where $Diag_n$ denotes the multi-diagonal defined by

$$(Diag_n Y)_k = Y_{k\overset{n}{\dots}k}$$

for $Y \in [\Delta^{n^{op}}, \mathsf{Set}]$ and $k \geq 0$.

Proposition 5.3.2 *If $X \in \mathsf{Cat}^{\mathsf{n}}_{\mathsf{hd}}$, $B\gamma_{(n)} : BX \to BX^d$ is a weak homotopy equivalence. In particular, BX is a 0-type with $\pi_i(BX, x) = 0$ for $i > 0$ and $\pi_0 BX = UX^d$, where UX^d is the set underlying the discrete n-fold category X^d.*

Proof By induction on n. For $n = 1$, X is a groupoid with no non-trivial loops, hence, since groupoids are models of 1-types, $\pi_i(BX, x) = 0$ for $i > 0$ while $\pi_0 BX = UX^d$; suppose the statement holds for $(n-1)$.

The functor B is also the composite

$$B : \mathsf{Cat}^{\mathsf{n}}_{\mathsf{hd}} \xrightarrow{N_1} [\Delta^{op}, \mathsf{Cat}^{\mathsf{n-1}}_{\mathsf{hd}}] \xrightarrow{\overline{B}} [\Delta^{op}, [\Delta^{op}, \mathsf{Set}]] \xrightarrow{Diag_2} [\Delta^{op}, \mathsf{Set}] \,,$$

where the notation $\overline{B}$ is as in Definition 2.1.1. Thus $B\gamma_{(n)}$ is obtained by applying $Diag_2$ to the map of bisimplicial sets $N_1\overline{B}\gamma_{(n)}$. For each $s \geq 0$ the latter is given by

$$(N_1\overline{B}\gamma_{(n)})_s = B(\gamma_{(n)})_s : BX_s \to BX^d_s$$

and this is a weak homotopy equivalence by the induction hypothesis.

A map of bisimplicial sets which is a levelwise weak homotopy equivalence induces a weak homotopy equivalence of diagonals (see [63, Proposition 1.7]).

Hence

$$Diag_2 N_1 \overline{B}\gamma_{(n)} = B\gamma_{(n)}$$

is a weak homotopy equivalence, as required. Thus BX is weakly homotopy equivalent to $B(p^{(1)} \dots p^{(n-1)}X)$, which is a 0-type since $p^{(1)} \dots p^{(n-1)}X \in \mathsf{Cat}_{\mathsf{hd}}$. Further,

$$\pi_0 BX \cong \pi_0 B(p^{(1)} \dots p^{(n-1)}X) \cong Up^{(0)} \dots p^{(n-1)}X \cong UX^d.$$

□

Chapter 6
The Definition of the Three Segal-Type Models

Abstract In this chapter we introduce the three Segal-type models: the category $\mathsf{Ta}^{\mathsf{n}}_{\mathsf{wg}}$ of weakly globular Tamsamani n-categories and its subcategories Ta^{n} (Tamsamani n-categories) and $\mathsf{Cat}^{\mathsf{n}}_{\mathsf{wg}}$ (weakly globular n-fold categories). We also define the notion of n-equivalence in these models, as part of their inductive definition. We discuss some elementary properties of the three models, and we illustrate some examples in low dimensions.

In this chapter we introduce the three Segal-type models of this work: the category $\mathsf{Ta}^{\mathsf{n}}_{\mathsf{wg}}$ of weakly globular Tamsamani n-categories, and its subcategories Ta^{n} of Tamsamani n-categories and $\mathsf{Cat}^{\mathsf{n}}_{\mathsf{wg}}$ of weakly globular n-fold categories. We refer the reader to Sect. 3.4.2 for an overview of the common features of the three Segal type models.

In Sect. 6.1 we define the most general structure, the category $\mathsf{Ta}^{\mathsf{n}}_{\mathsf{wg}}$: in this category the weakening occurs both via the weak globularity condition and via the induced Segal map condition. The idea of the category $\mathsf{Ta}^{\mathsf{n}}_{\mathsf{wg}}$ is discussed in Sect. 6.1.1, before the formal Definition 6.1.8.

The category $\mathsf{Ta}^{\mathsf{n}}_{\mathsf{wg}}$ is defined by induction on dimension, and this necessitates a careful treatment of some closure properties of certain subcategories of $[\Delta^{n^{op}}, \mathsf{Set}]$, see Sect. 6.1.2.

In Sect. 6.2 we specialize to the subcategory Ta^{n} of Tamsamani n-categories (see Definition 6.2.1). In this subcategory we impose the strict globularity condition, so the weakness of the structure is encoded by the Segal maps (which coincide with the induced Segal maps).

Finally, in Sect. 6.3 we introduce the subcategory $\mathsf{Cat}^{\mathsf{n}}_{\mathsf{wg}}$ of weakly globular n-fold categories. In this subcategory the weakness of the structure is encoded via the weak globularity condition while the induced Segal map condition regulates the behaviour of the compositions; weakly globular n-fold categories are more rigid structures than weakly globular Tamsamani n-categories, because they are n-fold categories and, for all $0 \leq r < n$, they give rise to r-fold categories after applying the functor truncation $p^{(r)}$. That is, the functor $p^{(r)} : \mathsf{Ta}^{\mathsf{n}}_{\mathsf{wg}} \to \mathsf{Ta}^{\mathsf{r}}_{\mathsf{wg}}$ (which is part of the definition of $\mathsf{Ta}^{\mathsf{n}}_{\mathsf{wg}}$) restricts to the functor $p^{(r)} : \mathsf{Cat}^{\mathsf{n}}_{\mathsf{wg}} \to \mathsf{Cat}^{\mathsf{r}}_{\mathsf{wg}}$.

S. Paoli, *Simplicial Methods for Higher Categories*, Algebra and Applications 26,
https://doi.org/10.1007/978-3-030-05674-2_6

We also discuss in Sect. 6.3 some elementary properties of weakly globular n-fold categories, while the more technical properties are treated in Chap. 7.

6.1 Weakly Globular Tamsamani n-Categories

6.1.1 The Idea of Weakly Globular Tamsamani n-Categories

The definition of the category $\mathsf{Ta}^{\mathsf{n}}_{\mathsf{wg}}$ is inductive on dimension, starting with Set when $n = 0$ (with 0-equivalences being isomorphisms) and Cat when $n = 1$ (with 1-equivalences being equivalences of categories). In dimension $n > 1$, a weakly globular Tamsamani n-category X is a simplicial object in the category $\mathsf{Ta}^{\mathsf{n-1}}_{\mathsf{wg}}$ satisfying additional conditions. By unravelling the inductive definition, one obtains an embedding

$$\mathsf{Ta}^{\mathsf{n}}_{\mathsf{wg}} \hookrightarrow [\Delta^{n^{op}}, \mathsf{Set}] \; .$$

The additional conditions in the definition of $\mathsf{Ta}^{\mathsf{n}}_{\mathsf{wg}}$ encode the weakness of the structure in two ways: the first is the weak globularity condition, requiring X_0 to be a homotopically discrete $(n-1)$-fold category, in the sense of Definition 5.1.2. The second is the induced Segal map condition, requiring the maps

$$\hat{\mu}_k : X_k \to X_1 \times_{X_0^d} \overset{k}{\cdots} \times_{X_0^d} X_1 \tag{6.1}$$

to be $(n-1)$-equivalences for each $k \geq 2$. For this to make sense, we need to impose conditions to ensure that

$$X_1 \times_{X_0^d} \overset{k}{\cdots} \times_{X_0^d} X_1 \in \mathsf{Ta}^{\mathsf{n-1}}_{\mathsf{wg}} \; . \tag{6.2}$$

This is done inductively by requiring the subcategory $\mathsf{Ta}^{\mathsf{n-1}}_{\mathsf{wg}}$ of $[\Delta^{n-1^{op}}, \mathsf{Set}]$ to satisfy the closure properties **C1–C4** of page 27 from which (6.2) follows by Lemma 2.2.8.

The inductive Definition 6.1.8 of the category $\mathsf{Ta}^{\mathsf{n}}_{\mathsf{wg}}$ also requires the existence of a truncation functor

$$p^{(n-1)} : \mathsf{Ta}^{\mathsf{n}}_{\mathsf{wg}} \to \mathsf{Ta}^{\mathsf{n-1}}_{\mathsf{wg}}$$

obtained by restriction of $p^{(n-1)} : [\Delta^{n^{op}}, \mathsf{Set}] \to [\Delta^{n-1^{op}}, \mathsf{Set}]$ as in Definition 2.2.5. This truncation functor is used to define n-equivalences in $\mathsf{Ta}^{\mathsf{n}}_{\mathsf{wg}}$. This notion is a higher dimensional generalization of a functor being fully faithful and essentially surjective on objects.

In order for the closure properties **C1–C4** to hold at step n (thus completing the inductive step of the definition of $\mathsf{Ta}^{\mathsf{n}}_{\mathsf{wg}}$) we need to require analogous closure properties to hold, inductively, for $(n-1)$-equivalences.

We treat the closure properties of n-equivalences in Sect. 6.1.2, where we establish them separately in a general setting (see Lemma 6.1.3 and Proposition 6.1.6). These results are then applied in Definition 6.1.8 in stating the inductive hypothesis and in completing the inductive step. The reader may want at first reading to jump directly to Definition 6.1.8, and only later come back to Sect. 6.1.2 for the details of the proof of Proposition 6.1.6.

6.1.2 Closure Properties

We discuss some closure properties of certain subcategories of $[\Delta^{n^{op}}, \mathsf{Set}]$. These will be used crucially in the next section in giving the definition of the category $\mathsf{Ta}^{\mathsf{n}}_{\mathsf{wg}}$. More precisely, in the definition of $\mathsf{Ta}^{\mathsf{n}}_{\mathsf{wg}}$ the closure properties **E1–E4** as in Lemma 6.1.3 are used in the inductive hypothesis, while Proposition 6.1.6 will be applied to the case where $\mathscr{C}_i = \mathsf{Ta}^{\mathsf{i}}_{\mathsf{wg}}$, $\mathscr{W}_i$ are i-equivalences ($i = n-1, n$) in order to complete the inductive step.

Given $Y \in [\Delta^{n^{op}}, \mathsf{Set}]$ with $Y_0 \in \mathsf{Cat}^{\mathsf{n-1}}_{\mathsf{hd}}$, recall (see Definition 2.1.3) that the induced Segal maps of Y (induced by the map $\gamma : Y_0 \to Y_0^d$) are given, for all $s \geq 2$, by

$$Y_s \to Y_1 \times_{Y_0^d} \overset{s}{\cdots} \times_{Y_0^d} Y_1 \ .$$

Also, for each $a, b \in Y_0^d$ we denote by $Y(a, b)$ the fiber at (a, b) of the map

$$Y_1 \xrightarrow{(\partial_0, \partial_1)} Y_0 \times Y_0 \xrightarrow{\gamma \times \gamma} Y_0^d \times Y_0^d \ .$$

Definition 6.1.1 Let $(\mathscr{C}_{n-1}, \mathscr{W}_{n-1})$ consist of a full subcategory $\mathscr{C}_{n-1}$ of $[\Delta^{{n-1}^{op}}, \mathsf{Set}]$ containing the terminal object together with a class of maps $\mathscr{W}_{n-1}$ called $(n-1)$-equivalences. Let $\mathscr{C}_n$ be the full subcategory of $[\Delta^{n^{op}}, \mathsf{Set}]$ whose objects X are such that

i) $X_0 \in \mathsf{Cat}^{\mathsf{n-1}}_{\mathsf{hd}}$.
ii) $X_k \in \mathscr{C}_{n-1}$ for all $k > 0$.
iii) For all $s \geq 2$ the induced Segal maps $X_s \to X_1 \times_{X_0^d} \overset{s}{\cdots} \times_{X_0^d} X_1$ are in $\mathscr{W}_{n-1}$.

Remark 6.1.2 In Definition 6.1.1 note that, since $\mathsf{Cat}^{\mathsf{n-1}}_{\mathsf{hd}}$ and $\mathscr{C}_{n-1}$ contain the terminal object, so does $\mathscr{C}_n$.

Lemma 6.1.3 *Let* $(\mathcal{C}_{n-1}, \mathcal{W}_{n-1})$ *and* $\mathcal{C}_n$ *be as in Definition 6.1.1. Suppose that* $\mathcal{C}_{n-1}$ *satisfies the closure properties* **C1–C4** *as on page 27 and suppose that* $\mathcal{W}_{n-1}$ *satisfies the following closure properties:*

E1 $\mathcal{W}_{n-1}$ *is closed under composition with isomorphisms.*
E2 $\mathcal{W}_{n-1}$ *is closed under finite products.*
E3 $\mathcal{W}_{n-1}$ *is closed under small colimits.*
E4 *If the small colimit* $\coprod_i f_i$ *of maps in* $\mathcal{C}_{n-1}$ *is in* $\mathcal{W}_{n-1}$*, then each* $f_i \in \mathcal{W}_{n-1}$.

Then $\mathcal{C}_n$ *has the closure properties* **C1–C4***.*

Proof

C1) Let $A \cong B$ be an isomorphism in $[\Delta^{n^{op}}, \mathsf{Set}]$ with $A \in \mathcal{C}_n$. Then $A_0 \cong B_0$ with $A_0 \in \mathsf{Cat}^{\mathsf{n-1}}_{\mathsf{hd}}$, so by Lemma 5.2.1 $B_0 \in \mathsf{Cat}^{\mathsf{n-1}}_{\mathsf{hd}}$; also $A_k \cong B_k$ for $k > 0$ hence, since by hypothesis $\mathcal{C}_{n-1}$ satisfies **C1** and $A_k \in \mathcal{C}_{n-1}$, we have $B_k \in \mathcal{C}_{n-1}$. From the commuting diagram for $s \geq 2$

$$\begin{array}{ccc} A_s & \longrightarrow & A_1 \times_{A_0^d} \overset{s}{\cdots} \times_{A_0^d} A_1 \\ \downarrow \wr\Vert & & \downarrow \wr\Vert \\ B_s & \longrightarrow & B_1 \times_{B_0^d} \overset{s}{\cdots} \times_{B_0^d} B_1 \end{array}$$

we see that the induced Segal maps of B are obtained by composing the induced Segal maps of A with isomorphisms. Using hypothesis **E1** we conclude that the induced Segal maps of B are in $\mathcal{W}_{n-1}$. Therefore $B \in \mathcal{C}_n$.

C2) Let $X, Y \in \mathcal{C}_n$. By Lemma 5.2.1 $(X \times Y)_0 = X_0 \times Y_0 \in \mathsf{Cat}^{\mathsf{n-1}}_{\mathsf{hd}}$. Also, for each $k > 0$, by hypothesis **C2** for $\mathcal{C}_{n-1}$, $(X \times Y)_k = X_k \times Y_k \in \mathcal{C}_{n-1}$. Consider the induced Segal map $\hat{\mu}_2(X \times Y)$. Since (see Corollary 5.2.2) $(X_0 \times Y_0)^d = X_0^d \times Y_0^d$, this is given by

$$X_2 \times Y_2 \to (X_1 \times Y_1) \times_{(X_0^d \times Y_0^d)} (X_1 \times Y_1) \cong (X_1 \times_{X_0^d} X_1) \times (Y_1 \times_{Y_0^d} Y_1)\,.$$

Hence $\hat{\mu}_2(X \times Y) = \hat{\mu}_2(X) \times \hat{\mu}_2(Y)$ and by hypothesis **E2**, $\hat{\mu}_2(X \times Y) \in \mathcal{W}_{n-1}$. The proof for $\hat{\mu}_s(X \times Y)$ when $s > 2$ is similar. In conclusion, $X \times Y \in \mathcal{C}_n$.

C3) Consider the small coproduct $\coprod_i X_i$ with $X_i \in \mathcal{C}_n$. By Lemma 5.2.1

$$(\coprod_i X_i)_0 = \coprod_i X_{i0} \in \mathsf{Cat}^{\mathsf{n-1}}_{\mathsf{hd}}$$

and for each $k > 0$, using hypothesis **C3** for $\mathcal{C}_{n-1}$

$$(\coprod_i X_i)_k = \coprod_i X_{ik} \in \mathcal{C}_{n-1}\,.$$

Since $(\coprod_i X_{i0})^d = \coprod_i X^d_{i0}$, the induced Segal map $\hat{\mu}_2(\coprod_i X_i)$ of $\coprod_i X_i$ is given by

$$\coprod_i X_{i2} \to (\coprod_i X_{i1}) \times_{\coprod_i X^d_{i0}} (\coprod_i X_{i1}) \cong \coprod_i (X_{i1} \times_{X^d_{i0}} X_{i1}) \,.$$

That is, $\hat{\mu}_2(\coprod_i X_i) = \coprod_i \hat{\mu}_2(X_i)$, so by hypothesis **E3**, $\hat{\mu}_2 \in \mathcal{W}_{n-1}$. Similarly, $\hat{\mu}_s \in \mathcal{W}_{n-1}$ for all $s \geq 2$. In conclusion, $\coprod_i X_i \in \mathcal{C}_n$.

C4) Suppose that $\coprod_i X_i \in \mathcal{C}_n$. Then $(\coprod_i X_i)_0 \cong \coprod_i X_{i0} \in \mathsf{Cat}^{n-1}_{\mathsf{hd}}$ so by Lemma 5.2.1 $X_{i0} \in \mathsf{Cat}^{n-1}_{\mathsf{hd}}$ for all i. Similarly when $k > 0$, using hypothesis **C4** for $\mathcal{C}_{n-1}$ we see that, since $(\coprod_i X_i)_k = \coprod_i X_{ik} \in \mathcal{C}_{n-1}$, we have $X_{ik} \in \mathcal{C}_{n-1}$ for all k. Reasoning as in the proof of **C3** above, the induced Segal maps of $\coprod_i X_i$ are the small coproduct of the induced Segal maps of X_i. Hence, by hypothesis **E4**, the induced Segal maps of X_i are in $\mathcal{W}_{n-1}$ for all i. In conclusion, $X_i \in \mathcal{C}_n$ for all i. □

Lemma 6.1.4 *Let $(\mathcal{C}_{n-1}, \mathcal{W}_{n-1})$ be as in Lemma 6.1.3, that is, $\mathcal{C}_{n-1}$ satisfies **C1–C4** and $\mathcal{W}_{n-1}$ satisfies **E1–E4**. Consider the diagram in $\mathcal{C}_{n-1}$*

$$\begin{array}{ccccc} X & \xrightarrow{f} & Z & \xleftarrow{g} & Y \\ {\scriptstyle\alpha}\downarrow & & \| & & \downarrow{\scriptstyle\beta} \\ X' & \xrightarrow[f']{} & Z & \xleftarrow[g']{} & Y' \end{array}$$

with Z discrete. Then, if $\alpha, \beta \in \mathcal{W}_{n-1}$, the morphism

$$(\alpha, \beta) : X \times_Z Y \to X' \times_Z Y'$$

is also in $\mathcal{W}_{n-1}$.

Proof Since Z is discrete

$$(\alpha, \beta) : X \times_Z Y = \coprod_{a \in Z} X_a \times Y_a \to \coprod_{a \in Z} X'_a \times Y'_a \,,$$

therefore

$$(\alpha, \beta) = (\coprod_{a \in Z} \alpha_a, \coprod_{a \in Z} \beta_a) = \coprod_{a \in Z} (\alpha_a, \beta_a) \,, \tag{6.3}$$

where $\alpha_a : X_a \to X'_a$ and $\beta_a : Y_a \to Y'_a$. By property **E4** for $\mathcal{W}_{n-1}$, $\alpha_a, \beta_a \in \mathcal{W}_{n-1}$ for all $a \in Z$. Hence, by property **E2**, $(\alpha_a, \beta_a) \in \mathcal{W}_{n-1}$ and by property **E3**, $\coprod_{a \in Z} (\alpha_a, \beta_a) \in \mathcal{W}_{n-1}$. Thus, by (6.3), $(\alpha, \beta) \in \mathcal{W}_{n-1}$. □

Definition 6.1.5 Let $(\mathscr{C}_{n-1}, \mathscr{W}_{n-1})$ and $\mathscr{C}_n$ be as in Definition 6.1.1. Suppose, further, that for each $X \in \mathscr{C}_n$ and $\underline{k} \in \Delta^{n-1^{op}}$, $X_{\underline{k}} \in \mathsf{Cat} \hookrightarrow [\Delta^{^{op}}, \mathsf{Set}]$ and suppose that the functor $p^{(n-1)} : [\Delta^{n^{op}}, \mathsf{Set}] \to [\Delta^{n-1^{op}}, \mathsf{Set}]$ as in Definition 2.2.5 restricts to a functor $p^{(n-1)} : \mathscr{C}_n \to \mathscr{C}_{n-1}$. We say that a map $f : X \to Y$ in $\mathscr{C}_n$ is an n-equivalence if:

a) For all $a, b \in X_0^d$, $f(a,b) : X(a,b) \to Y(fa, fb)$ is in $\mathscr{W}_{n-1}$.
b) $p^{(n-1)} f \in \mathscr{W}_{n-1}$.

We denote by $\mathscr{W}_n$ the class of n-equivalences in $\mathscr{C}_n$.

Proposition 6.1.6 *Let $(\mathscr{C}_{n-1}, \mathscr{W}_{n-1})$ and $(\mathscr{C}_n, \mathscr{W}_n)$ be as in Definition 6.1.5. Suppose that $\mathscr{C}_{n-1}$ has the closure properties **C1–C4** and $\mathscr{W}_{n-1}$ has the closure properties **E1–E4**. Then $\mathscr{C}_n$ has the closure properties **C1–C4** and $\mathscr{W}_n$ has the closure properties **E1–E4**.*

Proof The fact that $\mathscr{C}_n$ satisfies **C1–C4** is proved in Lemma 6.1.3. We now show that $\mathscr{W}_n$ satisfies **E1–E4**.

E1) Let $X \xrightarrow{f} Y \xrightarrow{g} Z$ be maps in $\mathscr{C}_{n-1}$ with $f \in \mathscr{W}_{n-1}$ and g an isomorphism. Then, for all $a, b \in X_0^d$, the composite

$$X(a,b) \xrightarrow{f(a,b)} Y(fa, fb) \xrightarrow{g(fa,fb)} Z(gf(a), gf(b))$$

has $f(a,b) \in \mathscr{W}_{n-1}$ and $g(fa, fb)$ an isomorphism. By hypothesis **E1** for $\mathscr{W}_{n-1}$, this composite is in $\mathscr{W}_{n-1}$. Also, $p^{(n-1)}(gf) = (p^{(n-1)}g)(p^{(n-1)}f)$ has $p^{(n-1)}f \in \mathscr{W}_{n-1}$ and $p^{(n-1)}g$ an isomorphism. Hence by hypothesis **E1** for $\mathscr{W}_{n-1}$, $p^{(n-1)}(gf) \in \mathscr{W}_{n-1}$. In conclusion, $gf \in \mathscr{W}_n$.

E2) Let $f_1 : X_1 \to Y_1$, $f_2 : X_2 \to Y_2$ be in $\mathscr{W}_n$. Then for each $(a_1, a_2), (b_1, b_2) \in (X_1 \times X_2)_0^d \cong X_{10}^d \times X_{20}^d$ and by hypothesis **E2** for $\mathscr{W}_{n-1}$

$$(f_1 \times f_2)((a_1, a_2), (b_1, b_2)) = f_1(a_1, b_1) \times f_2(a_2, b_2) \in \mathscr{W}_{n-1} .$$

We note that $p^{(n-1)}$ preserves finite products. In fact, since (Lemma 4.1.2) p preserves finite products, for each $\underline{k} \in \Delta^{n-1^{op}}$,

$$\begin{aligned}&(p^{(n-1)}(X \times Y))_{\underline{k}} = p(X \times Y)_{\underline{k}} = p(X_{\underline{k}} \times Y_{\underline{k}}) = pX_{\underline{k}} \times pY_{\underline{k}} \\ &= (p^{(n-1)}X)_{\underline{k}} \times (p^{(n-1)}Y)_{\underline{k}}\end{aligned}$$

so that $p^{(n-1)}(X \times Y) = p^{(n-1)}X \times p^{(n-1)}Y$. It follows that $p^{(n-1)}(f_1 \times f_2) = p^{(n-1)} f_1 \times p^{(n-1)} f_2$. Thus by hypothesis **E2** for $\mathscr{W}_{n-1}$ we deduce that $p^{(n-1)} (f_1 \times f_2) \in \mathscr{W}_{n-1}$. In conclusion, $f_1 \times f_2 \in \mathscr{W}_n$.

E3) Let $f_i : X_i \to Y_i$ be in $\mathscr{W}_{n-1}$ and consider the small coproduct $\coprod\limits_i f_i$. For each $a_i, b_i \in X_{i0}^d$, $f_i(a_i, b_i) \in \mathscr{W}_{n-1}$ so by hypothesis **E3** for $\mathscr{W}_{n-1}$,

$$(\coprod_i f_i)(\coprod_i (a_i, b_i) = \coprod_i f_i(a_i, b_i) \in \mathscr{W}_{n-1} .$$

Note that, since p preserves coproducts, so does $p^{(n-1)}$. Thus by hypothesis **E3** for $\mathcal{W}_{n-1}$

$$p^{(n-1)}(\coprod_i f_i) = \coprod_i (p^{(n-1)} f_i) \in \mathcal{W}_{n-1} \,.$$

In conclusion, $\coprod_i f_i \in \mathcal{W}_n$.

E4) Suppose that $\coprod_i f_i \in \mathcal{W}_n$. Then for each $a_i, b_i \in X^d_{i0}$, $\coprod_i f_i(a_i, b_i) \in \mathcal{W}_{n-1}$ so by hypothesis **E4** for $\mathcal{W}_{n-1}$, $f_i(a_i, b_i) \in \mathcal{W}_{n-1}$ for all i. Also, $p^{(n-1)}(\coprod_i f_i) = \coprod_i (p^{(n-1)} f_i) \in \mathcal{W}_{n-1}$ so $p^{(n-1)} f_i \in \mathcal{W}_{n-1}$ for all i. In conclusion, $f_i \in \mathcal{W}_n$ for all i. □

Remark 6.1.7 Under the hypotheses of Proposition 6.1.6 since, as shown there, $\mathcal{W}_n$ has properties **E1–E4**, reasoning as in Lemma 6.1.4 we see that the following holds: Given a diagram in $\mathcal{C}_n$

$$\begin{array}{ccccc} X & \xrightarrow{f} & Z & \xleftarrow{g} & Y \\ {\scriptstyle\alpha}\downarrow & & \| & & \downarrow{\scriptstyle\beta} \\ X' & \xrightarrow[f']{} & Z & \xleftarrow[g']{} & Y' \end{array}$$

with Z discrete, if $\alpha, \beta \in \mathcal{W}_n$ then $(\alpha, \beta) : X \times_Z Y \to X' \times_Z Y'$ is also in $\mathcal{W}_n$.

6.1.3 *The Formal Definition of the Category* $\mathsf{Ta}^{\mathsf{n}}_{\mathsf{wg}}$

Definition 6.1.8 We define the category $\mathsf{Ta}^{\mathsf{n}}_{\mathsf{wg}} \subset [\Delta^{n^{op}}, \mathsf{Set}]$ and n-equivalences by induction on n. When $n = 0$, $\mathsf{Ta}^{\mathsf{0}}_{\mathsf{wg}} = \mathsf{Set}$ and 0-equivalences are isomorphisms. When $n = 1$, $\mathsf{Ta}^{\mathsf{1}}_{\mathsf{wg}} = \mathsf{Cat} \hookrightarrow [\Delta^{op}, \mathsf{Set}]$ and 1-equivalences are equivalences of categories.

Inductive Hypothesis

Suppose, inductively, that we defined for each $1 \le k \le n-1$ a subcategory

$$\mathsf{Ta}^{\mathsf{k}}_{\mathsf{wg}} \subset [\Delta^{k-1^{op}}, \mathsf{Cat}\,] \subset [\Delta^{k^{op}}, \mathsf{Set}]$$

containing the terminal object and a class $\mathcal{W}_k$ of maps in $\mathsf{Ta}^{\mathsf{k}}_{\mathsf{wg}}$ (called k-equivalences) such that

I1) $\mathsf{Ta}^{\mathsf{k}}_{\mathsf{wg}}$ satisfies the closure properties **C1–C4** (see page 27).

I2) The functor $p^{(k-1)} : [\Delta^{k^{op}}, \mathsf{Set}] \to [\Delta^{k-1^{op}}, \mathsf{Set}]$ of Definition 2.2.5 restricts to a functor $p^{(k-1)} : \mathsf{Ta}^{\mathsf{k}}_{\mathsf{wg}} \to \mathsf{Ta}^{\mathsf{k-1}}_{\mathsf{wg}}$ which sends k-equivalences to $(k-1)$-equivalences.

I3) $\mathscr{W}_k$ satisfies the closure properties **E1–E4** (as in Lemma 6.1.3).

Defining $\mathsf{Ta}^{\mathsf{n}}_{\mathsf{wg}}$

An object X of $[\Delta^{n-1^{op}}, \mathsf{Cat}] \subset [\Delta^{n^{op}}, \mathsf{Set}]$ is a weakly globular Tamsamani n-category if:

a) Weak globularity condition: $X_0 \in \mathsf{Cat}^{\mathsf{n-1}}_{\mathsf{hd}}$.
b) $X_k \in \mathsf{Ta}^{\mathsf{n-1}}_{\mathsf{wg}}$ for all $k > 0$.
c) Induced Segal maps condition. For all $s \geq 2$ the induced Segal maps

$$X_s \to X_1 \times_{X_0^d} \overset{s}{\cdots} \times_{X_0^d} X_1$$

(induced by the map $\gamma : X_0 \to X_0^d$) are $(n-1)$-equivalences.

Note that, since $\mathsf{Cat}^{\mathsf{n-1}}_{\mathsf{hd}}$ and $\mathsf{Ta}^{\mathsf{n-1}}_{\mathsf{wg}}$ contain the terminal object, so does $\mathsf{Ta}^{\mathsf{n}}_{\mathsf{wg}}$. Also note that since by induction $\mathsf{Ta}^{\mathsf{n-1}}_{\mathsf{wg}}$ satisfies the closure properties **C1–C4**, by Lemma 2.2.8,

$$X_1 \times_{X_0^d} \overset{s}{\cdots} \times_{X_0^d} X_1 \in \mathsf{Ta}^{\mathsf{n-1}}_{\mathsf{wg}} .$$

To satisfy the inductive hypothesis, we check that the functor $p^{(n-1)}$: $[\Delta^{n^{op}}, \mathsf{Set}] \to [\Delta^{n-1^{op}}, \mathsf{Set}]$ of Definition 2.2.5 restricts to a functor $p^{(n-1)}$: $\mathsf{Ta}^{\mathsf{n}}_{\mathsf{wg}} \to \mathsf{Ta}^{\mathsf{n-1}}_{\mathsf{wg}}$.

Since $X_0 \in \mathsf{Cat}^{\mathsf{n-1}}_{\mathsf{hd}}$, $(p^{(n-1)}X)_0 = p^{(n-2)}X_0 \in \mathsf{Cat}^{\mathsf{n-2}}_{\mathsf{hd}}$. Further, $p^{(n-2)}$ preserves pullbacks over discrete objects (as the same is true for p, see Lemma 4.1.4) so that

$$p^{(n-2)}(X_1 \times_{X_0^d} \overset{s}{\cdots} \times_{X_0^d} X_1) \cong p^{(n-2)}X_1 \times_{(p^{(n-2)}X_0^d)} \overset{s}{\cdots} \times_{(p^{(n-2)}X_0^d)} p^{(n-2)}X_1 .$$

Further,

$$p^{(n-2)}X_0^d = (p^{(n-2)}X_0)^d$$

and, by the inductive hypothesis, $p^{(n-2)}$ sends $(n-1)$-equivalences to $(n-2)$-equivalences.

Therefore, the induced Segal maps for $s \geq 2$

$$X_s \to X_1 \times_{X_0^d} \overset{s}{\cdots} \times_{X_0^d} X_1 ,$$

being $(n-1)$-equivalences, give rise to $(n-2)$-equivalences

$$p^{(n-2)}X_s \to p^{(n-2)}X_1 \times_{(p^{(n-2)}X_0)^d} \overset{s}{\cdots} \times_{(p^{(n-2)}X_0)^d} p^{(n-2)}X_1 .$$

This shows that $p^{(n-1)}X \in \mathsf{Ta}^{\mathsf{n-1}}_{\mathsf{wg}}$.

Defining n-Equivalences

Given $a, b \in X_0^d$, denote by $X(a, b)$ the fiber at (a, b) of the map

$$X_1 \xrightarrow{(\partial_0, \partial_1)} X_0 \times X_0 \xrightarrow{\gamma \times \gamma} X_0^d \times X_0^d .$$

Since, by the inductive hypothesis, $\mathsf{Ta}_{\mathsf{wg}}^{\mathsf{n-1}}$ satisfies **C1–C4**, by Lemma 2.2.8 $X(a, b) \in \mathsf{Ta}_{\mathsf{wg}}^{\mathsf{n-1}}$. One should think of $X(a, b) \in \mathsf{Ta}_{\mathsf{wg}}^{\mathsf{n-1}}$ as a hom-$(n-1)$-category. We define a map $f : X \to Y$ in $\mathsf{Ta}_{\mathsf{wg}}^{\mathsf{n}}$ to be an n-equivalence if

i) For all $a, b \in X_0^d$

$$f(a, b) : X(a, b) \to Y(fa, fb)$$

is an $(n-1)$-equivalence.

ii) $p^{(n-1)} f$ is an $(n-1)$-equivalence.

Completing the Inductive Step

To complete the inductive step in the definition of $\mathsf{Ta}_{\mathsf{wg}}^{\mathsf{n}}$ we need to check that $\mathsf{Ta}_{\mathsf{wg}}^{\mathsf{n}}$ satisfies the inductive hypotheses I1)–I3) at step n.

The property I2) has already been checked above, and the fact that $p^{(n-1)}$ sends n-equivalences to $(n-1)$-equivalences is part of the definition of n-equivalences in $\mathsf{Ta}_{\mathsf{wg}}^{\mathsf{n}}$.

Finally, I1) and I3) follow by Proposition 6.1.6, taking $\mathcal{C}_{n-1} = \mathsf{Ta}_{\mathsf{wg}}^{\mathsf{n-1}}$ with $\mathcal{W}_{n-1}$ the $(n-1)$-equivalences and $\mathcal{C}_n = \mathsf{Ta}_{\mathsf{wg}}^{\mathsf{n}}$ with $\mathcal{W}_n$ the n-equivalences.

Remark 6.1.9 We note that, for each $X \in \mathsf{Ta}_{\mathsf{wg}}^{\mathsf{n}}$, $1 \le r \le n-1$ and $a, b \in X_0^d$,

$$(p^{(r)} X)(a, b) = p^{(r-1)}(X(a, b)) . \tag{6.4}$$

In fact, since $X_1 = \coprod_{a,b \in X_0^d} X(a, b)$ and $p^{(r-1)}$ preserves coproducts,

$$(p^{(r)} X)_1 = p^{(r-1)} X_1 = \coprod_{a,b \in X_0^d} p^{(r-1)}(X(a, b)) . \tag{6.5}$$

On the other hand,

$$(p^{(r)} X)_1 = \coprod_{a,b \in X_0^d} (p^{(r)} X)(a, b) . \tag{6.6}$$

From (6.5) and (6.6), (6.4) follows.

Example 6.1.10 Weakly globular Tamsamani 2-categories.
From the definition, $X \in \mathsf{Ta}^2_{\mathsf{wg}}$ consists of a simplicial object $X \in [\Delta^{^{op}}, \mathsf{Cat}]$ such that $X_0 \in \mathsf{Cat}_{\mathsf{hd}}$ and the induced Segal maps

$$\hat{\mu}_k : X_k \to X_1 \times_{X_0^d} \overset{k}{\cdots} \times_{X_0^d} X_1$$

are equivalences of categories. The functor $p^{(1)} : \mathsf{Ta}^2_{\mathsf{wg}} \to \mathsf{Cat} \subset [\Delta^{^{op}}, \mathsf{Set}]$ associates to $X \in \mathsf{Ta}^2_{\mathsf{wg}}$ the simplicial set taking $k \in \Delta^{op}$ to $p(X_k)$; this simplicial set is the nerve of a category since, for each $k \geq 2$,

$$\begin{aligned} pX_k &\cong p(X_1 \times_{X_0^d} \overset{k}{\cdots} \times_{X_0^d} X_1) \cong pX_1 \times_{pX_0^d} \overset{k}{\cdots} \times_{pX_0^d} pX_1 \\ &\cong pX_1 \times_{pX_0} \overset{k}{\cdots} \times_{pX_0} pX_1 \; . \end{aligned}$$

The following are elementary properties of $\mathsf{Ta}^n_{\mathsf{wg}}$.

Lemma 6.1.11 *For each $n \geq 1$*

a) *There is an inclusion $\mathsf{Ta}^{n-1}_{\mathsf{wg}} \hookrightarrow \mathsf{Ta}^n_{\mathsf{wg}}$ making the following diagram commute*

$$\begin{array}{ccc} \mathsf{Ta}^{n-1}_{\mathsf{wg}} & \hookrightarrow & [\Delta^{n-1^{op}}, \mathsf{Set}] \\ \downarrow & & \downarrow \\ \mathsf{Ta}^n_{\mathsf{wg}} & \hookrightarrow & [\Delta^{n^{op}}, \mathsf{Set}] \end{array} \tag{6.7}$$

where the inclusion $[\Delta^{n-1^{op}}, \mathsf{Set}] \hookrightarrow [\Delta^{n^{op}}, \mathsf{Set}]$ on the right side of (6.7) is as in Notational Convention 2.2.1.

b) *A morphism f in $\mathsf{Ta}^{n-1}_{\mathsf{wg}}$ is an $(n-1)$-equivalence if and only if it is an n-equivalence in $\mathsf{Ta}^n_{\mathsf{wg}}$.*

Proof By induction on n. When $n = 1$, (6.7) becomes

$$\begin{array}{ccc} \mathsf{Set} & \xrightarrow{\mathrm{Id}} & \mathsf{Set} \\ \downarrow & & \downarrow \\ \mathsf{Cat} & \longrightarrow & [\Delta^{^{op}}, \mathsf{Set}] \end{array}$$

and this commutes by Notational Convention 2.5.1. Also, a morphism f in Set is a 0-equivalence (that is, an isomorphism) if and only if it is an equivalence of the corresponding discrete categories. Suppose, inductively, that the lemma holds for $(n-1)$.

a) Let $X \in \mathsf{Ta}^{\mathsf{n-1}}_{\mathsf{wg}}$. Under Notational Convention 2.2.1 $X \in [\Delta^{n^{op}}, \mathsf{Set}]$ with $X_{\underline{k}}$ a discrete simplicial set for all $\underline{k} \in \Delta^{n-1^{op}}$. Then $X \in \mathsf{Ta}^{\mathsf{n}}_{\mathsf{wg}}$ since $X_k \in \mathsf{Ta}^{\mathsf{n-1}}_{\mathsf{wg}}$ for all $k > 0$ (by the inductive hypothesis applied to X_k), $X_0 \in \mathsf{Cat}^{\mathsf{n-1}}_{\mathsf{hd}}$ and the induced Segal maps are $(n-1)$-equivalences (by inductive hypothesis b) applied to the induced Segal maps of $X \in \mathsf{Ta}^{\mathsf{n-1}}_{\mathsf{wg}}$). Thus (6.7) commutes.
b) Let $f \in \mathsf{Ta}^{\mathsf{n-1}}_{\mathsf{wg}}$ be an $(n-1)$-equivalence. Then for all $a, b \in X_0^d$, $f(a,b)$ and $p^{(n-2)}f$ are $(n-2)$-equivalences. Thus, by inductive hypothesis b), they are $(n-1)$-equivalences when viewing $f \in \mathsf{Ta}^{\mathsf{n}}_{\mathsf{wg}}$. So f is an n-equivalence. Similarly for the converse. □

Corollary 6.1.12 *Consider the diagram in* $\mathsf{Ta}^{\mathsf{n}}_{\mathsf{wg}}$

$$\begin{array}{ccccc} X & \xrightarrow{f} & Z & \xleftarrow{g} & Y \\ {\scriptstyle\alpha}\downarrow & & \| & & \downarrow{\scriptstyle\beta} \\ X' & \xrightarrow[f']{} & Z & \xleftarrow[g']{} & Y' \end{array}$$

with Z discrete. If α and β are n-equivalences, so is $(\alpha, \beta) : X\times_Z Y \to X'\times_Z Y'$.

Proof This follows from Remark 6.1.7, taking $\mathscr{C}_n = \mathsf{Ta}^{\mathsf{n}}_{\mathsf{wg}}$ and $\mathscr{W}_n$ the class of n-equivalences. □

6.2 Tamsamani n-Categories

Definition 6.2.1 For each $n \geq 0$ the category Ta^{n} of Tamsamani n-categories is the full subcategory of $\mathsf{Ta}^{\mathsf{n}}_{\mathsf{wg}}$ whose objects X are such that X_0 and $X_{k_1 \dots k_r\, 0}$ are discrete for all $(k_1 \dots k_r) \in \Delta^{r^{op}}$, $1 \leq r \leq n-2$. A morphism in Ta^{n} is an n-equivalence if it is an n-equivalence as a morphism in $\mathsf{Ta}^{\mathsf{n}}_{\mathsf{wg}}$.

This recovers the original definition of Tamsamani's weak n-category [126].

Remark 6.2.2 If $X \in \mathsf{Ta}^{\mathsf{n}}$, X_0 is discrete so for each $s \geq 2$ the induced Segal maps of X coincide with the Segal maps

$$\nu_s : X_s \to X_1 \times_{X_0} \overset{s}{\cdots} \times_{X_0} X_1 \, .$$

The following property is immediate from the definition.

Lemma 6.2.3

a) *Let $X \in \mathsf{Ta}^{\mathsf{n}}_{\mathsf{wg}}$. Then $X \in \mathsf{Ta}^{\mathsf{n}}$ if and only if $X_s \in \mathsf{Ta}^{\mathsf{n-1}}$ for $s \geq 0$ and X_0 is discrete.*

b) The functor $p^{(n-1)} : \mathsf{Ta}^{\mathsf{n}}_{\mathsf{wg}} \to \mathsf{Ta}^{\mathsf{n-1}}_{\mathsf{wg}}$ *restricts to a functor* $p^{(n-1)} : \mathsf{Ta}^{\mathsf{n}} \to \mathsf{Ta}^{\mathsf{n-1}}$.

c) Ta^{n} *contains the terminal object and has the closure properties* ***C1–C4***.

Proof

a) Let $X \in \mathsf{Ta}^{\mathsf{n}}$. Then X_0 is discrete and, for each $s > 0$, $(k_2 \dots k_r) \in \Delta^{r-2^{op}}$, $(X_s)_{k_2 \dots k_r 0} = X_{s\, k_2 \dots k_r 0}$ is discrete. Thus, by definition, $X_s \in \mathsf{Ta}^{\mathsf{n-1}}$. Conversely, if $X \in \mathsf{Ta}^{\mathsf{n}}_{\mathsf{wg}}$ satisfies the hypotheses, X_0 and $X_{k_1 \dots k_r 0} = (X_{k_1})_{k_2 \dots k_r 0}$ are discrete, so $X \in \mathsf{Ta}^{\mathsf{n}}$.

b) Let $X \in \mathsf{Ta}^{\mathsf{n}}$. Then $(p^{(n-1)}X)_0 = p^{(n-2)}X_0$ is discrete (since X_0 is discrete) and $(p^{(n-1)}X)_{k_1 \dots k_r 0} = p^{(n-r-2)}X_{k_1 \dots k_r 0}$ is discrete, since $X_{k_1 \dots k_r 0}$ is. Thus $p^{(n-1)}X \in \mathsf{Ta}^{\mathsf{n-1}}$.

c) Since the terminal object is in particular discrete, by construction it is in Ta^{n}. We prove the rest by induction on n. It is clear for $n = 0$. Suppose it holds for $(n-1)$. In Lemma 6.1.3 let $\mathscr{C}_{n-1} = \mathsf{Ta}^{\mathsf{n-1}}$ and $\mathscr{W}_{n-1}$ consists of $(n-1)$-equivalences. By the inductive hypothesis the properties **C1–C4** hold for $\mathscr{C}_{n-1}$ and **E1–E4** hold for $\mathscr{W}_{n-1}$ as they hold for $(n-1)$-equivalences in $\mathsf{Ta}^{\mathsf{n-1}}_{\mathsf{wg}}$. So the hypotheses of Lemma 6.1.3 are satisfied and we conclude that Ta^{n} has the properties **C1–C4**. □

Remark 6.2.4 It is elementary to see that the inclusion $\mathsf{Ta}^{\mathsf{n-1}}_{\mathsf{wg}} \hookrightarrow \mathsf{Ta}^{\mathsf{n}}_{\mathsf{wg}}$ also restricts to an inclusion $\mathsf{Ta}^{\mathsf{n-1}} \hookrightarrow \mathsf{Ta}^{\mathsf{n}}$.

Example 6.2.5 (Tamsamani 2-Categories) From the definition, $X \in \mathsf{Ta}^2$ consists of $X \in [\Delta^{op}, \mathsf{Cat}]$ such that X_0 is discrete and the Segal maps $X_k \to X_1 \times_{X_0} \overset{k}{\cdots} \times_{X_0} X_1$ are equivalences of categories for all $k \geq 2$. As in the case of $\mathsf{Ta}^2_{\mathsf{wg}}$ (see Example 6.1.10) we automatically have the functor $p^{(1)} : \mathsf{Ta}^2 \to \mathsf{Cat} \hookrightarrow [\Delta^{op}, \mathsf{Set}]$.

In the case $n = 2$, the relation between Tamsamani 2-categories and bicategories was shown by Lack and Paoli [82], who introduced a 2-nerve functor from the 2-category of bicategories, normal homomorphism and icons to the category Ta^2.

6.3 Weakly Globular *n*-Fold Categories

6.3.1 *The Idea of Weakly Globular n-Fold Categories*

Weakly globular n-fold categories form a full subcategory of the category $\mathsf{Cat}^{\mathsf{n}}$ of n-fold categories; they are therefore rigid structures in which there are compositions in n different directions and all these compositions are associative and unital. Weakly globular n-fold categories also form a full subcategory of the category of weakly globular Tamsamani n-categories, which was informally introduced in Sect. 6.1.1;

thus as in $\mathsf{Ta}^{\mathsf{n}}_{\mathsf{wg}}$, the weakness in a weakly globular n-fold category is encoded by the weak globularity condition.

The underlying sets of the discretizations of the homotopically discrete substructures in a weakly globular n-fold category play the role of sets of cells in the respective dimensions. As for $\mathsf{Ta}^{\mathsf{n}}_{\mathsf{wg}}$, the induced Segal maps condition regulates the behaviour of the compositions of higher cells. Finally, we require the functor

$$p^{(n-1)} : \mathsf{Ta}^{\mathsf{n}}_{\mathsf{wg}} \to \mathsf{Ta}^{\mathsf{n-1}}_{\mathsf{wg}}$$

to restrict to a functor

$$p^{(n-1)} : \mathsf{Cat}^{\mathsf{n}}_{\mathsf{wg}} \to \mathsf{Cat}^{\mathsf{n-1}}_{\mathsf{wg}} .$$

To handle most efficiently the definition of weakly globular n-fold categories, we first define the subcategory $\mathsf{Cat}^{\mathsf{n}}_{\mathsf{t}} \subset \mathsf{Cat}^{\mathsf{n}}$ of truncatable n-fold categories such that $\mathsf{Cat}^{\mathsf{0}}_{\mathsf{t}} = \mathsf{Set}$ and

$$p^{(r)} : \mathsf{Cat}^{\mathsf{n}}_{\mathsf{t}} \to \mathsf{Cat}^{\mathsf{r}}_{\mathsf{t}}$$

for all $0 \le r < n$. Then $X \in \mathsf{Cat}^{\mathsf{n}}_{\mathsf{wg}}$ if $X \in \mathsf{Cat}^{\mathsf{n}}_{\mathsf{t}}$ and $X \in \mathsf{Ta}^{\mathsf{n}}_{\mathsf{wg}}$. In the case $n = 2$, weakly globular double categories were introduced in joint work by Pronk and the author in [103] and shown to be biequivalent to bicategories. The generalization to the case $n > 2$ is much more complex and new to this work. The category $\mathsf{Cat}^{\mathsf{n}}_{\mathsf{wg}}$ also informs the definitions of the category $\mathsf{SegPs}[\Delta^{n-1^{op}}, \mathsf{Cat}\,]$ of Segalic pseudo-functors, discussed informally in Sect. 8.1.2, and of the category $\mathsf{LTa}^{\mathsf{n}}_{\mathsf{wg}}$, discussed informally in Sect. 9.1.1.

6.3.2 *The Formal Definition of the Category* $\mathsf{Cat}^{\mathsf{n}}_{\mathsf{wg}}$

Definition 6.3.1 We say that an n-fold category $X \in \mathsf{Cat}^{\mathsf{n}}$ is truncatable if for all $0 \le r < n$, $p^{(r)}X \in [\Delta^{r^{op}}, \mathsf{Set}]$ is an r-fold category. We denote by $\mathsf{Cat}^{\mathsf{n}}_{\mathsf{t}}$ the full subcategory of $\mathsf{Cat}^{\mathsf{n}}$ consisting of truncatable n-fold categories, with $\mathsf{Cat}^{\mathsf{0}}_{\mathsf{t}} = \mathsf{Set}$.

Remark 6.3.2 From the definition we see that for all $0 \le r < n$, the truncation functor $p^{(r)} : [\Delta^{n^{op}}, \mathsf{Set}] \to [\Delta^{r^{op}}, \mathsf{Set}]$ restricts to give a functor $p^{(r)} : \mathsf{Cat}^{\mathsf{n}}_{\mathsf{t}} \to \mathsf{Cat}^{\mathsf{r}}_{\mathsf{t}}$. Also, this sequence of subcategories is maximal among sequences of full subcategories $\mathscr{D}^n \subset \mathsf{Cat}^{\mathsf{n}}$ for which $p^{(r)}$ restricts to $p^{(r)} : \mathscr{D}^n \to \mathscr{D}^r$.

Definition 6.3.3 We say that $X \in [\Delta^{n^{op}}, \mathsf{Set}]$ is a weakly globular n-fold category if $X \in \mathsf{Cat}^{\mathsf{n}}_{\mathsf{t}}$ and $X \in \mathsf{Ta}^{\mathsf{n}}_{\mathsf{wg}}$. We denote by $\mathsf{Cat}^{\mathsf{n}}_{\mathsf{wg}}$ the category of weakly globular n-fold categories.

Remark 6.3.4 Note that Cat^n_{wg} contains the terminal object, since the latter is a discrete n-fold category, and thus both in Cat^n_t and Ta^n_{wg}.

The following lemma is a straightforward consequence of the definitions.

Lemma 6.3.5

a) *For all $0 \le r < n$ the functor $p^{(r)} : [\Delta^{n^{op}}, \mathsf{Set}] \to [\Delta^{r^{op}}, \mathsf{Set}]$ restricts to a functor $p^{(r)} : \mathsf{Cat}^n_{wg} \to \mathsf{Cat}^r_{wg}$.*
b) *Let $X \in \mathsf{Cat}^n$. Then $X \in \mathsf{Cat}^n_{wg}$ if and only if*

   i) *$X_0 \in \mathsf{Cat}^{n-1}_{hd}$,*
   ii) *$X_k \in \mathsf{Cat}^{n-1}_{wg}$ for all $k \ge 0$,*
   iii) *For each $s \ge 2$ the induced Segal maps*

$$X_1 \times_{X_0} \overset{s}{\cdots} \times_{X_0} X_1 \to X_1 \times_{X_0^d} \overset{s}{\cdots} \times_{X_0^d} X_1$$

   are $(n-1)$-equivalences,
   iv) *$p^{(n-1)} X \in \mathsf{Cat}^{n-1}_{wg}$.*

Proof a) is straightforward from the definitions. As for b), let $X \in \mathsf{Cat}^n$ satisfy i)–iv). Then, since $\mathsf{Cat}^{n-1}_{wg} \subset \mathsf{Ta}^{n-1}_{wg}$, $X \in \mathsf{Ta}^n_{wg}$. Together with iv), this implies that $X \in \mathsf{Cat}^n_{wg}$. Conversely, if $X \in \mathsf{Cat}^n_{wg}$, iv) holds by a); since $X \in \mathsf{Ta}^n_{wg}$, iii) holds. Finally i) and ii) hold since $X_0 \in \mathsf{Cat}^{n-1}_{hd}$, $X_k \in \mathsf{Cat}^{n-1}_t$ and $X_k \in \mathsf{Ta}^{n-1}_{wg}$ for all $k \ge 0$. □

Remark 6.3.6 It follows from Lemma 5.1.6, Proposition 5.2.9 and Lemma 6.3.5 b) that $\mathsf{Cat}^n_{hd} \subset \mathsf{Cat}^n_{wg}$.

Lemma 6.3.7 *The category Cat^n_{wg} has the closure properties **C1–C4**.*

Proof By induction on n. It is clear for $n = 0$. Suppose it holds for $(n-1)$. In Lemma 6.1.3 let $\mathscr{C}_{n-1} = \mathsf{Cat}^{n-1}_{wg}$ and $\mathscr{W}_{n-1}$ consist of $(n-1)$-equivalences. Then by the inductive hypothesis, the properties **C1–C4** hold for $\mathscr{C}_{n-1}$ and **E1–E4** hold for $\mathscr{W}_{n-1}$ as they hold for $(n-1)$-equivalences in Ta^{n-1}_{wg}. So the hypotheses of Lemma 6.1.3 are satisfied and we conclude that Cat^n_{wg} has the properties **C1–C4**. □

Lemma 6.3.8 *For each $X \in \mathsf{Cat}^n_{wg}$, $0 \le j < n-1$ and $s \ge 2$ we have*

$$p^{(j)} X_s \cong p^{(j)}(X_1 \times_{X_0} \overset{s}{\cdots} \times_{X_0} X_1) = p^{(j)} X_1 \times_{p^{(j)} X_0} \overset{s}{\cdots} \times_{p^{(j)} X_0} p^{(j)} X_1 \,. \tag{6.8}$$

Proof Since $X \in \mathsf{Cat}^n_{wg}$, by Lemma 6.3.5 $p^{(j+1)} X \in \mathsf{Cat}^{j+1}_{wg}$ for each $1 \le j < n-2$. Since $(p^{(j+1)} X)_s = p^{(j)} X_s$ for each $s \ge 0$, (6.8) follows. □

Remark 6.3.9 It follows immediately from Lemma 6.3.8 that if $X \in \mathsf{Cat}^n_{wg}$, for all $s \ge 2$

$$X^d_{s0} = (X_{10} \times_{X_{00}} \overset{s}{\cdots} \times_{X_{00}} X_{10})^d = X^d_{10} \times_{X^d_{00}} \overset{s}{\cdots} \times_{X^d_{00}} X^d_{10} \,. \tag{6.9}$$

In fact, by (6.8) in the case $j = 1$, taking the 0-component, we obtain

$$p^{(0)}(X_{10}\times_{X_{00}}\overset{s}{\cdots}\times_{X_{00}} X_{10}) = p^{(0)}X_{s0} = (p^{(1)}X_s)_0$$
$$= (p^{(1)}X_1\times_{p^{(1)}X_0}\overset{s}{\cdots}\times_{p^{(1)}X_0} p^{(1)}X_1)_0$$
$$= p^{(0)}X_{10}\times_{p^{(0)}X_{00}}\overset{s}{\cdots}\times_{p^{(0)}X_{00}} p^{(0)}X_{10}\,,$$

which is the same as (6.9).

Example 6.3.10 (Weakly Globular Double Categories) Let $X \in \mathsf{Cat}^2_{\mathsf{wg}}$. Then, by definition, $X \in [\Delta^{op}, \mathsf{Cat}]$ is such that

a) $X_0 \in \mathsf{Cat}_{\mathsf{hd}}$.
b) For all $k \geq 2$ $X_k \cong X_1\times_{X_0}\overset{k}{\cdots}\times_{X_0} X_1$.
c) For all $k \geq 2$ the induced Segal maps

$$\hat{\mu}_k : X_k \to X_1\times_{X_0^d}\overset{k}{\cdots}\times_{X_0^d} X_1$$

are equivalences of categories.
d) For each $X \in \mathsf{Cat}^2_{\mathsf{wg}}$, $p^{(1)}X \in \mathsf{Cat}$.

Note that in the case $n = 2$ condition d) is redundant. In fact, by b) and c) and the fact (Lemma 4.1.4) that p sends equivalences of categories to isomorphisms and commutes with pullbacks over discrete objects, given $X \in [\Delta^{op}, \mathsf{Cat}]$ satisfying conditions a), b), and c), for each $k \geq 2$

$$pX_k \cong p(X_1\times_{X_0^d}\overset{k}{\cdots}\times_{X_0^d} X_1) = pX_1\times_{X_0^d}\overset{k}{\cdots}\times_{X_0^d} pX_1$$
$$\cong pX_1\times_{pX_0^d}\overset{k}{\cdots}\times_{pX_0^d} pX_1\,.$$

Hence d) holds.

On page 122, Fig. 6.1 is a picture of the corner of $X \in \mathsf{Cat}^2_{\mathsf{wg}}$, where the red structure is homotopically discrete. The corresponding geometric picture is on page 122, Fig. 6.2.

Condition c) also has the following geometric interpretation. Given $X \in [\Delta^{op}, \mathsf{Cat}]$ satisfying a) and b), the induced Segal maps

$$\hat{\mu}_k : X_1\times_{X_0}\overset{k}{\cdots}\times_{X_0} X_1 \to X_1\times_{X_0^d}\overset{k}{\cdots}\times_{X_0^d} X_1$$

are fully faithful. We show this for $k = 2$, the case $k > 2$ being similar. Given $(a, b), (c, d) \in X_{10} \times_{X_{00}} X_{10}$ we have

$$(X_1 \times_{X_0} X_1)\{(a, b), (c, d)\} \cong X_{11}(a, c)\times_{X_{01}(\partial_0 a,\partial_0 c)} X_{11}(b, d)$$
$$\cong X_{11}(a, c) \times X_{11}(b, d)$$

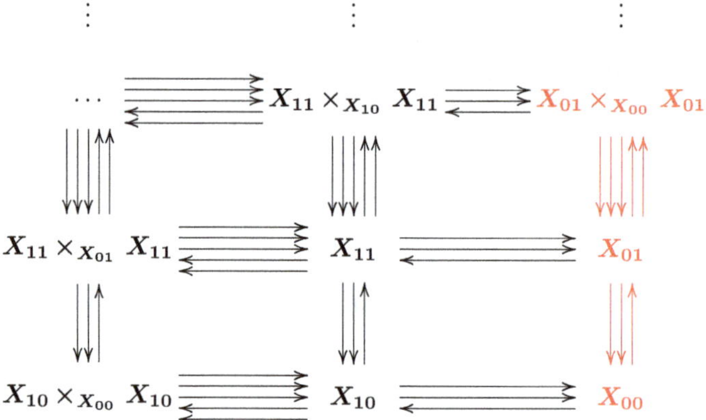

Fig. 6.1 Corner of the double nerve of a weakly globular double category X

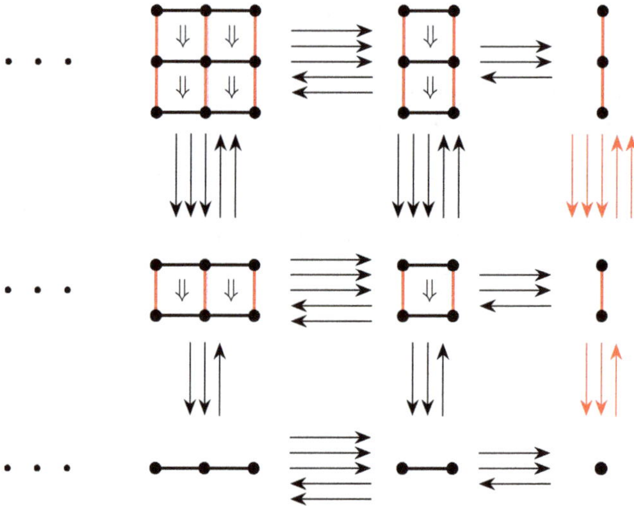

Fig. 6.2 Geometric picture of the corner of the double nerve of a weakly globular double category

since $X_{01}(\partial_0 a, \partial_0 c) = \{\cdot\}$ as $X_0 \in \mathsf{Cat}_{\mathsf{hd}}$. Hence

$$\begin{aligned}(X_1 \times_{X_0} X_1)\{(a, b), (c, d)\} &\cong X_{11}(a, c) \times X_{11}(b, d) \\ &\cong (X_1 \times_{X_0^d} X_1)\{\hat{\mu}_k(a, b), \hat{\mu}_k(c, d)\}\,.\end{aligned}$$

Since, when conditions a) and b) hold, $\hat{\mu}_k$ is always fully and faithful, condition c) that $\hat{\mu}_k$ is an equivalence of categories is equivalent to the requirement that it is essentially surjective on objects.

An object of $X_1 \times_{X_0^d} \overset{k}{\cdots} \times_{X_0^d} X_1$ is a staircase of length k of horizontal arrows whose source and targets match up in the vertical connected component, as in the following picture:

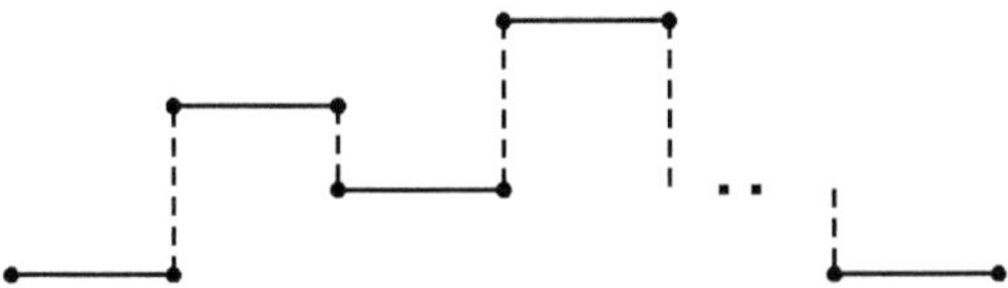

Essential surjectivity of $\hat{\mu}_k$ means that this staircase can be lifted to a sequence of horizontally composable arrows through vertically invertible squares:

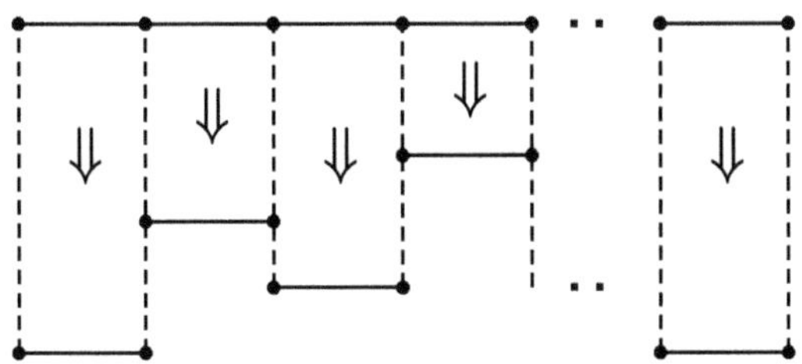

Example 6.3.11 (Weakly Globular 3-Fold Categories) A weakly globular 3-fold category $X \in \mathsf{Cat}^3_{\mathsf{wg}}$ is given by $X \in [\Delta^{op}, \mathsf{Cat}^2_{\mathsf{wg}}]$ such that

a) $X_0 \in \mathsf{Cat}^2_{\mathsf{hd}}$.
b) For each $k \geq 2$, $X_k \cong X_1 \times_{X_0} \overset{k}{\cdots} \times_{X_0} X_1$.
c) For each $k \geq 2$ the induced Segal maps

$$\hat{\mu}_k : X_k \to X_1 \times_{X_0^d} \overset{k}{\cdots} \times_{X_0^d} X_1$$

are 2-equivalences in $\mathsf{Cat}^2_{\mathsf{wg}}$.

d) For each $X \in \mathsf{Cat}^3_{\mathsf{wg}}$, $p^{(2)} X \in \mathsf{Cat}^2_{\mathsf{wg}}$.

It follows from the definition that $X_{k0} \in \mathsf{Cat}_{\mathsf{hd}}$ for all $k \geq 0$. On page 124, Fig. 6.3 is a picture of the corner of $X \in [\Delta^{3^{op}}, \mathsf{Set}]$, where we omitted drawing the degeneracy operators for simplicity. The structures in red are homotopically discrete. A corresponding geometric picture (again with omitted degeneracy operators) is in Fig. 6.4 on page 124.

In the following picture, for all $i, j, k \in \Delta^{op}$
$X_{2jk} \cong X_{1jk} \times_{X_{0jk}} X_{1jk}, \; X_{i2k} \cong X_{i1k} \times_{X_{i0k}} X_{i1k}, \; X_{ij2} \cong X_{ij1} \times_{X_{ij0}} X_{ij1}$.

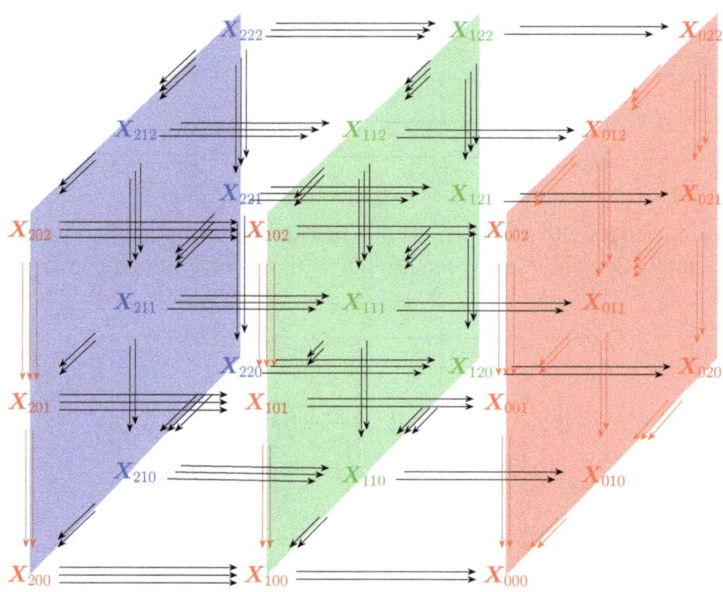

Fig. 6.3 Corner of the multinerve of a weakly globular 3-fold category X

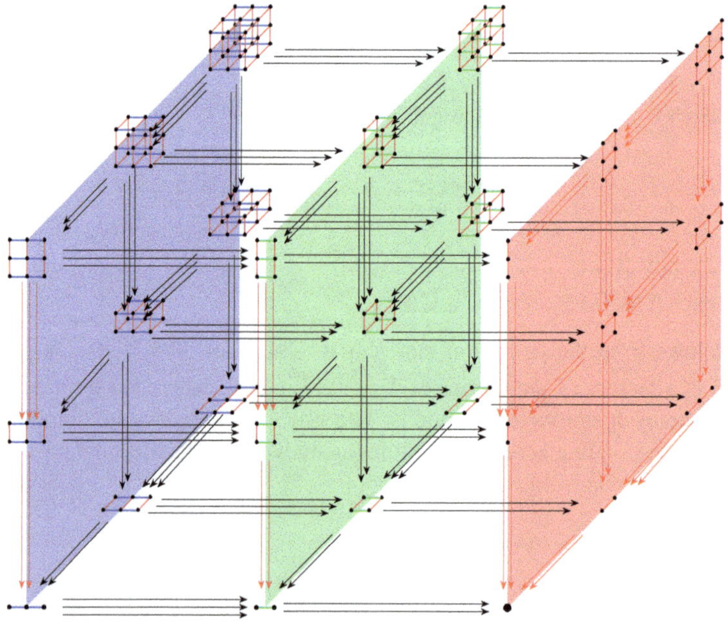

Fig. 6.4 Geometric picture of the corner of the multinerve of a weakly globular 3-fold category

Remark 6.3.12 The previous Examples 6.3.10 and 6.3.11 illustrate an important difference between the cases $n = 2$ and $n > 2$ in the definition of $\mathsf{Cat}^{\mathsf{n}}_{\mathsf{wg}}$. Namely, while in the case $n = 2$, the fact that the functor $p^{(n-1)} : \mathsf{Ta}^{\mathsf{n}}_{\mathsf{wg}} \to \mathsf{Ta}^{\mathsf{n-1}}_{\mathsf{wg}}$ restricts to $p^{(n-1)} : \mathsf{Cat}^{\mathsf{n}}_{\mathsf{wg}} \to \mathsf{Cat}^{\mathsf{n-1}}_{\mathsf{wg}}$ is implied automatically by the condition that $X_0 \in \mathsf{Cat}^{\mathsf{n-1}}_{\mathsf{hd}}$ and by the induced Segal map condition, this is no longer the case when $n > 2$, so this property needs to be part of the definition of $\mathsf{Cat}^{\mathsf{n}}_{\mathsf{wg}}$.

The difference between the case $n = 2$ and $n > 2$ is also reflected in the fact that the rigidification functor Q_n is substantially more complex to define when $n > 2$ than when $n = 2$. We refer the reader to the introduction to Chap. 10 for a detailed discussion about this point.

Chapter 7
Properties of the Segal-Type Models

Abstract In this chapter we discuss the more advanced properties of the three Segal-type models introduced in Chap. 6. We discuss the behaviour of the n-equivalences and certain sufficient criteria for the latter to be levelwise equivalences of categories. We introduce the functor $q^{(n-1)} : \mathsf{Ta}^{\mathsf{n}}_{\mathsf{wg}} \to \mathsf{Ta}^{\mathsf{n-1}}_{\mathsf{wg}}$ and we discuss its properties. We prove a useful criterion for an n-fold category to be weakly globular. This criterion plays a crucial role in Chap. 8 in showing how n-fold categories arise as strictification of a certain class of pseudo-functors.

In this chapter we discuss some important properties of the three Segal-type models. These will be used throughout the rest of this work.

We start in Sect. 7.1 with the properties of $\mathsf{Ta}^{\mathsf{n}}_{\mathsf{wg}}$. In Proposition 7.1.2 we give a useful sufficient criterion for a morphism in $\mathsf{Ta}^{\mathsf{n}}_{\mathsf{wg}}$ to be an n-equivalence and we prove several cases of the 2-out-of-3 property of n-equivalences; a complete proof of the latter will be given in Corollary 12.1.6. We also give in Proposition 7.1.5 a criterion for an n-equivalence to be a levelwise equivalence of categories, which will be used crucially in Chap. 9.

In Sect. 7.1.2 we introduce the functor

$$q^{(n-1)} : \mathsf{Ta}^{\mathsf{n}}_{\mathsf{wg}} \to \mathsf{Ta}^{\mathsf{n-1}}_{\mathsf{wg}} \tag{7.1}$$

and we discuss in Sect. 7.1.3 the properties of pullbacks along the map $\gamma^{(n-1)} : X \to q^{(n-1)}X$ for each $X \in \mathsf{Ta}^{\mathsf{n}}_{\mathsf{wg}}$. These properties will be used in Chap. 9 in the proof of the main Theorem 9.2.4, which gives a method for approximating objects of $\mathsf{Ta}^{\mathsf{n}}_{\mathsf{wg}}$ by objects from the simpler category $\mathsf{LTa}^{\mathsf{n}}_{\mathsf{wg}}$.

Section 7.2 discusses some properties of the category $\mathsf{Cat}^{\mathsf{n}}_{\mathsf{wg}}$. The main result of this section, Proposition 7.2.8, states that if $X \in \mathsf{Cat}^{\mathsf{n}}_{\mathsf{wg}}$, then $X^{\{2\}}_k \in \mathsf{Cat}^{\mathsf{n-1}}_{\mathsf{wg}}$ for all $k \geq 0$, where $X^{\{2\}}_k$ is the n-fold category X viewed as an internal category in $\mathsf{Cat}^{\mathsf{n-1}}$ along direction 2 (see Sect. 2.5). Further, it gives a criterion for an n-fold category to be weakly globular. This criterion will be used crucially in the proof of

S. Paoli, *Simplicial Methods for Higher Categories*, Algebra and Applications 26,
https://doi.org/10.1007/978-3-030-05674-2_7

Proposition 8.2.1 to show that n-fold categories levelwise equivalent via a pseudo-natural transformation to Segalic pseudo-functors are weakly globular. This result leads to the main Theorem 8.2.3 on the strictification of Segalic pseudo-functors.

Using Proposition 7.2.8 we also deduce, in Proposition 7.2.12, that the functor $q^{(n-1)} : \mathsf{Ta}^{\mathsf{n}}_{\mathsf{wg}} \to \mathsf{Ta}^{\mathsf{n-1}}_{\mathsf{wg}}$ of Sect. 7.1 restricts to a functor

$$q^{(n-1)} : \mathsf{Cat}^{\mathsf{n}}_{\mathsf{wg}} \to \mathsf{Cat}^{\mathsf{n-1}}_{\mathsf{wg}}.$$

This functor will be used throughout this work. We also deduce in Corollary 7.2.14 and Remark 7.2.15 a property of $\mathsf{Cat}^{\mathsf{n}}_{\mathsf{wg}}$ that will motivate, in Sect. 9.1, the definition of the category $\mathsf{LTa}^{\mathsf{n}}_{\mathsf{wg}}$, and we discuss a corresponding geometric interpretation.

7.1 Properties of Weakly Globular Tamsamani n-Categories

In this section we establish some important properties of the category $\mathsf{Ta}^{\mathsf{n}}_{\mathsf{wg}}$ of weakly globular Tamsamani n-categories. In Proposition 7.1.2 we prove useful properties of n-equivalences in $\mathsf{Ta}^{\mathsf{n}}_{\mathsf{wg}}$, which will be used throughout the rest of this work. Proposition 7.1.5 is a criterion for an n-equivalence in $\mathsf{Ta}^{\mathsf{n}}_{\mathsf{wg}}$ to be a levelwise equivalence of categories, and it will play an important role in the proofs of the main results of Chap. 9.

7.1.1 Properties of n-Equivalences

In the following proposition, we describe some useful properties of n-equivalences in $\mathsf{Ta}^{\mathsf{n}}_{\mathsf{wg}}$ which include several cases of the 2-out-of-3 property. The complete proof of the latter will be given in Corollary 12.1.6.

Definition 7.1.1 A morphism $f : X \to Y$ in $\mathsf{Ta}^{\mathsf{n}}_{\mathsf{wg}}$ is said to be a local $(n-1)$-equivalence if for all $a, b \in X^d_0$,

$$f(a,b) : X(a,b) \to Y(fa, fb)$$

is an $(n-1)$-equivalence in $\mathsf{Ta}^{\mathsf{n-1}}_{\mathsf{wg}}$.

In what follows $p = p^{(0)} : \mathsf{Ta}^{\mathsf{n}}_{\mathsf{wg}} \to \mathsf{Set}$ is as in Definition 6.1.8.

Proposition 7.1.2

a) Let f be an n-equivalence in $\mathsf{Ta}^{\mathsf{n}}_{\mathsf{wg}}$. Then f is a local $(n-1)$-equivalence and pf is an isomorphism.

b) Let f be a local $(n-1)$-equivalence in $\mathsf{Ta}^{\mathsf{n}}_{\mathsf{wg}}$ with pf surjective. Then f is an n-equivalence.

c) *Let* $X \xrightarrow{g} Z \xrightarrow{h} Y$ *be morphisms in* $\mathsf{Ta}^{\mathsf{n}}_{\mathsf{wg}}$, $f = hg$ *and suppose that* f *and* h *are* n*-equivalences. Then* g *is an* n*-equivalence.*
d) *Let* $X \xrightarrow{g} Z \xrightarrow{h} Y$ *be morphisms in* $\mathsf{Ta}^{\mathsf{n}}_{\mathsf{wg}}$, $f = hg$ *and suppose that* g *and* h *are* n*-equivalences. Then* f *is an* n*-equivalence.*
e) *Let* $X \xrightarrow{g} Z \xrightarrow{h} Y$ *be morphisms in* $\mathsf{Ta}^{\mathsf{n}}_{\mathsf{wg}}$, $f = hg$ *and let* $g_0^d : X_0^d \to Z_0^d$ *be surjective; suppose that* f *and* g *are* n*-equivalences. Then* h *is an* n*-equivalence.*

Proof By induction on n. Consider the case $n = 1$. If f is an equivalence of categories, it is fully faithful (hence a local 0-equivalence) and pf is an isomorphism, showing a). If f is a local 0-equivalence, it is fully faithful, while pf being surjective means that f is essentially surjective on objects. Thus f is an equivalence of categories, proving b). Let f, g, h be as in c). Then for all $a, b \in X_0$

$$f(a,b) = h(ga, gb)g(a,b) \tag{7.2}$$

with $f(a,b)$ and $h(ga, gb)$ isomorphisms. Hence $g(a,b)$ is also an isomorphism, that is, g is fully faithful. Since

$$pf = (ph)(pg) \tag{7.3}$$

and pf, ph are isomorphisms, so is pg. Hence g is essentially surjective on objects. In conclusion, g is an equivalence of categories, proving c). Let f, g, h be as in d). By (7.2) f is fully faithful and by (7.3) pf is an isomorphism, so f is essentially surjective on objects; in conclusion, f is an equivalence of categories, proving d).

Let f, g, h be as in e). By hypothesis, for each $a', b' \in Z_0$, $a' = ga$, $b' = gb$ for $a, b \in X_0$. Therefore $h(a', b') = h(ga, gb)$. Since by (7.2) $h(ga, gb)$ is an isomorphism, so is $h(a', b')$, and so h is fully faithful. Also, by (7.3), ph is an isomorphism, so h is essentially surjective on objects. In conclusion, h is an equivalence of categories.

Suppose the proposition is true for $n - 1$.

a) Let $f : X \to Y$ be an n-equivalence in $\mathsf{Ta}^{\mathsf{n}}_{\mathsf{wg}}$. Then, by definition, f is a local $(n-1)$-equivalence and $p^{(n-1)} f$ is an $(n-1)$-equivalence. Therefore, by the induction hypothesis applied to $p^{(n-1)} f$, pf is an isomorphism.
b) Suppose that $f : X \to Y$ is a local $(n-1)$-equivalence in $\mathsf{Ta}^{\mathsf{n}}_{\mathsf{wg}}$ and pf is surjective. To show that f is an n-equivalence we need to show that $p^{(n-1)} f$ is an $(n-1)$-equivalence. For each $a, b \in X_0^d$

$$(p^{(n-1)} f)(a,b) = p^{(n-2)} f(a,b) \ .$$

Since $f(a,b)$ is an $(n-1)$-equivalence, $p^{(n-2)} f(a,b)$ is an $(n-2)$-equivalence; that is, $p^{(n-1)} f$ is a local $(n-2)$-equivalence.

Since $pf = pp^{(n-1)} f$ is surjective, by the inductive hypothesis applied to $p^{(n-1)} f$ we conclude that $p^{(n-1)} f$ is a $(n-1)$-equivalence as required.

c) For all $a, b \in X_0^d$,

$$f(a,b) = h(ga, gb)g(a,b) \tag{7.4}$$

with $f(a,b)$ and $h(ga, gb)$ $(n-1)$-equivalences. By the inductive hypothesis, $g(a,b)$ is therefore an $(n-1)$-equivalence.

By hypothesis and by part a), pf and ph are isomorphisms. Since

$$pf = (ph)(pg) \tag{7.5}$$

it follows that pg is an isomorphism, hence in particular it is surjective. By part b), this implies that g is an $(n-1)$-equivalence.

d) Suppose that h and g are n-equivalences, then $h(ga, gb)$ and $g(a,b)$ are $(n-1)$-equivalences. By (7.4) and the inductive hypothesis, f is therefore a local $(n-1)$-equivalence and by (7.5) pf is an isomorphism. By b), f is thus an n-equivalence.

e) By hypothesis, for each $a', b' \in Z_0^d$, $a' = ga$, $b' = gb$ for $a, b \in X_0^d$. It follows that $h(a', b') = h(ga, gb)$. Since, by the induction hypothesis and by (7.4), $h(ga, gb)$ is an $(n-1)$-equivalence, it follows that $h(a', b')$ is also an $(n-1)$-equivalence. That is, h is a local equivalence.

By hypothesis and by part a), pf and pg are isomorphisms, so by (7.5), so is ph. We conclude by part b) that h is an n-equivalence.

□

Lemma 7.1.3 *$f : X \to Y$ in $\mathsf{Ta}^n_{\mathsf{wg}}$ is a levelwise $(n-1)$-equivalence in $\mathsf{Ta}^{n-1}_{\mathsf{wg}}$ if and only if it is an n-equivalence and $pf_0 : pX_0 \to pY_0$ is a bijection.*

Proof Let f be a levelwise $(n-1)$-equivalence. We show that f is an n-equivalence by induction on n. Let $n = 2$. If f_0 is an equivalence of categories, $X_0^d \cong Y_0^d$. Hence

$$Y_1 = \coprod_{a',b' \in Y_0^d} Y(a',b') \cong \coprod_{\substack{a,b \in X_0^d \\ fa=a' \\ fb=b'}} Y(fa, fb)\,, \tag{7.6}$$

$$X_1 = \coprod_{a,b \in X_0^d} X(a,b)\,. \tag{7.7}$$

Since f_1 is an equivalence of categories it follows from (7.6) and (7.7) that $f(a,b)$ is an equivalence of categories. Further, f_k is an equivalence of categories for all $k \geq 0$ so that $pf_k = (p^{(1)}f)_k$ is an isomorphism, hence $p^{(1)}f$ is an isomorphism, so in particular $p(p^{(1)}f)$ is surjective; by Proposition 7.1.2 b) we conclude that f is a 2-equivalence.

Suppose the lemma holds for $(n-1)$ and let f be as in the hypothesis. Since f_0 is an $(n-1)$-equivalence in $\mathsf{Cat}^{n-1}_{\mathsf{hd}}$, $X_0^d \cong Y_0^d$, so that (7.6) holds. Since f_1 is an $(n-1)$-equivalence, it follows from (7.6) and (7.7) that $f(a,b)$ is an $(n-1)$-equivalence for all $a,b \in X_0^d$.

Since f_k is an $(n-1)$-equivalence for all $k \geq 0$, $p^{(n-2)} f_k = (p^{(n-1)} f)_k$ is an $(n-2)$-equivalence. So $p^{(n-1)} f$ satisfies the induction hypothesis and is therefore an $(n-1)$-equivalence. In conclusion, f is an n-equivalence.

Conversely, suppose that f is an n-equivalence and pf_0 is an isomorphism, that is, $X_0^d \cong Y_0^d$. Then

$$
\begin{aligned}
Y_1 &= \coprod_{a',b' \in Y_0^d} Y(a',b') \cong \coprod_{\substack{a,b \in X_0^d \\ fa=a' \\ fb=b'}} Y(fa,fb)\,, \\
X_1 &= \coprod_{a,b \in X_0^d} X(a,b)\,.
\end{aligned}
\tag{7.8}
$$

Since f is an n-equivalence, $X(a,b) \to Y(fa,fb)$ is an $(n-1)$-equivalence, so by (7.8), $f_1 : X_1 \to Y_1$ is an $(n-1)$-equivalence. For each $k \geq 2$ there is a commuting diagram

$$
\begin{array}{ccc}
X_k & \xrightarrow{\hat{\mu}_k} & X_1 \times_{X_0^d} \overset{k}{\cdots} \times_{X_0^d} X_1 \\
{\scriptstyle f_k}\big\downarrow & & \big\downarrow{\scriptstyle (f_1,\ldots,f_1)} \\
Y_k & \xrightarrow[\hat{\mu}_k]{} & Y_1 \times_{Y_0^d} \overset{k}{\cdots} \times_{Y_0^d} Y_1
\end{array}
$$

where the horizontal induced Segal maps are $(n-1)$-equivalences since $X, Y \in \mathsf{Ta}^n_{\mathsf{wg}}$ and the right vertical map is an $(n-1)$-equivalence since f_1 is also an $(n-1)$-equivalence. By Proposition 7.1.2 c) it follows that f_k is also an $(n-1)$-equivalence. □

Remark 7.1.4 Applying Lemma 7.1.3 repeatedly it follows immediately that if a morphism f in $\mathsf{Ta}^n_{\mathsf{wg}}$ is a levelwise equivalence of categories, then f is an n-equivalence.

The following is a useful criterion for an n-equivalence in $\mathsf{Ta}^n_{\mathsf{wg}}$ to be a levelwise equivalence of categories. It will be used in the rest of this chapter and in Chap. 9.

Proposition 7.1.5 *Let $f : X \to Y$ be a morphism in $\mathsf{Ta}^n_{\mathsf{wg}}$ with $n \geq 2$, such that*

a) f is an n-equivalence,
b) $p^{(n-2)} X_0 \cong p^{(n-2)} Y_0$,
c) For each $1 \leq r < n-1$ and all $k_1, \ldots, k_r \geq 0$,

$$p^{(n-r-2)} X_{k_1,\ldots,k_r,0} \cong p^{(n-r-2)} Y_{k_1,\ldots,k_r,0}\,.$$

Then f is a levelwise equivalence of categories.

Proof By induction on n. Let $f : X \to Y$ be a 2-equivalence in $\mathsf{Ta}^2_{\mathsf{wg}}$ such that

$$X_0^d = pX_0 \cong pY_0 = Y_0^d.$$

Since $f(a,b)$ is an equivalence of categories for all $a, b \in X_0^d$, we deduce that there is an equivalence of categories

$$f_1 : X_1 = \coprod_{a,b\in X_0^d} X(a,b) \to Y_1 = \coprod_{a',b'\in Y_0^d} Y(a',b') = \coprod_{\substack{fa,fb\in Y_0^d\\ fa=a',\\ fb=b'}} Y(fa,fb).$$

Hence there are equivalences of categories for $k \geq 2$

$$X_k \sim X_1 \times_{X_0^d} \overset{k}{\cdots} \times_{X_0^d} X_1 \sim Y_1 \times_{Y_0^d} \overset{k}{\cdots} \times_{Y_0^d} Y_1 \sim Y_k\ .$$

In conclusion, $X_k \sim Y_k$ for all $k \geq 0$.

Suppose, inductively, that the statement holds for $(n-1)$ and let $f : X \to Y$ be as in the hypothesis. We show that f_k is a levelwise equivalence of categories for each $k \geq 0$ by showing that f_k satisfies the inductive hypothesis. It then follows that f is a levelwise equivalence of categories since

$$(f)_{k_1\ldots k_{n-1}} = (f_{k_1})_{k_2\ldots k_{n-1}}\ .$$

Since $X_0 \in \mathsf{Cat}^{n-1}_{\mathsf{hd}}$, from hypothesis b) and Lemma 5.2.6 we obtain

$$X_0^d = pX_0 \cong pY_0 = Y_0^d\ . \tag{7.9}$$

Thus, by Lemma 5.2.6 again, $f_0 : X_0 \to Y_0$ is an $(n-1)$-equivalence. Further, by hypothesis c),

$$\begin{aligned} p^{(n-3)}X_{00} &\cong p^{(n-3)}Y_{00}\ , \\ p^{(n-r-3)}X_{0\,k_1\ldots k_r\,0} &\cong p^{(n-r-3)}Y_{0\,k_1\ldots k_r\,0} \end{aligned}$$

for each $1 \leq r < n-2$ and all $k_1 \ldots k_r$. Thus $f_0 : X_0 \to Y_0$ satisfies the inductive hypothesis and we conclude that f_0 is a levelwise equivalence of categories. By (7.9) we also have

$$f_1 : \coprod_{a,b\in X_0^d} X(a,b) \to Y_1 = \coprod_{a',b'\in Y_0^d} Y(a',b') = \coprod_{\substack{fa,fb\in Y_0^d\\ fa=a',\\ fb=b'}} Y(fa,fb)\ .$$

Since f is a local $(n-1)$-equivalence, it follows that $f_1 : X_1 \to Y_1$ is an $(n-1)$-equivalence. Further, by hypothesis c),

$$\begin{aligned} p^{(n-3)} X_{10} &\cong p^{(n-3)} Y_{10} \,, \\ p^{(n-r-3)} X_{1\, k_1 \dots k_r\, 0} &\cong p^{(n-r-3)} Y_{1\, k_1 \dots k_r\, 0} \end{aligned}$$

for all $1 \leq r < n-2$. Thus f_1 satisfies the inductive hypothesis, and is therefore a levelwise equivalence of categories.

For each $k \geq 2$ consider the map

$$(f_1, \overset{k}{\dots}, f_1) : X_1 \times_{X_0^d} \overset{k}{\cdots} \times_{X_0^d} X_1 \to Y_1 \times_{Y_0^d} \overset{k}{\cdots} \times_{Y_0^d} Y_1 \,.$$

Since $X_0^d \cong Y_0^d$ and, from above, f_1 is an $(n-1)$-equivalence, then $(f_1, \overset{k}{\dots}, f_1)$ is also an $(n-1)$-equivalence (using the closure properties of $(n-1)$-equivalences, see Definition 6.1.8 and Remark 6.1.7).

There is a commutative diagram in $\mathsf{Ta}_{\mathsf{wg}}^{\mathsf{n-1}}$

$$\begin{array}{ccc} X_k & \xrightarrow{\hat{\mu}_k} & X_1 \times_{X_0^d} \overset{k}{\cdots} \times_{X_0^d} X_1 \\ {\scriptstyle f_k} \downarrow & & \downarrow {\scriptstyle (f_1, \overset{k}{\dots}, f_1)} \\ Y_k & \xrightarrow[\hat{\mu}_k]{} & Y_1 \times_{Y_0^d} \overset{k}{\cdots} \times_{Y_0^d} Y_1 \end{array}$$

where the horizontal induced Segal maps are $(n-1)$-equivalences since $X, Y \in \mathsf{Ta}_{\mathsf{wg}}^{\mathsf{n}}$ and the right vertical map is an $(n-1)$-equivalence from above. It follows from Proposition 7.1.2 c) and d) that f_k is an $(n-1)$-equivalence. Further, from hypothesis c),

$$\begin{aligned} p^{(n-3)} X_{k0} &\cong p^{(n-3)} Y_{k0} \,, \\ p^{(n-r-3)} X_{k\, k_1 \dots k_r\, 0} &\cong p^{(n-r-3)} Y_{k\, k_1 \dots k_r\, 0} \,. \end{aligned}$$

Thus f_k satisfies the induction hypothesis and we conclude that f_k is a levelwise equivalence of categories.

In conclusion, f_k is a levelwise equivalence of categories for all $k \geq 0$, which implies that f is a levelwise equivalence of categories. □

For completeness, we note that Proposition 7.1.5 can be generalized as follows. Let $1 \leq m < n$. Recall that $f : X \to Y$ in $\mathsf{Ta}_{\mathsf{wg}}^{\mathsf{n}}$ is a levelwise $(n-m)$-equivalence if for each $(k_1, \dots, k_m) \in \Delta^{m^{op}}$, $f_{k_1 \dots k_m} \in \mathsf{Ta}_{\mathsf{wg}}^{\mathsf{n-m}}$ is an $(n-m)$-equivalence.

Proposition 7.1.6 *Let $1 \leq m < n$. Then $f : X \to Y$ in $\mathsf{Ta}^{\mathsf{n}}_{\mathsf{wg}}$ is a levelwise $(n-m)$-equivalence if and only if it is an n-equivalence and*

a) The map $pf_0 : pX_0 = X_0^d \to pY_0 = Y_0^d$ is an isomorphism.
b) For each $1 \leq r < m$ and $k_1, \ldots, k_r \in \Delta$, the maps

$$pf_{k_1\ldots k_r 0} : pX_{k_1\ldots k_r 0} \to pY_{k_1\ldots k_r 0}$$

are isomorphisms.

Proof Suppose that f is a levelwise $(n-m)$-equivalence. We proceed by induction on m. When $m = 1$, this is Lemma 7.1.3. Suppose, inductively, that the lemma holds for $(m-1)$ and let $f : X \to Y$ be a levelwise $(n-m)$-equivalence. Then for each $k_1 \geq 0$, $f_{k_1} : X_{k_1} \to Y_{k_1}$ is a levelwise $(n-m+1)$-equivalence. So by the inductive hypothesis applied to f_{k_1}, $pf_{k_1\ldots k_r 0}$ is an isomorphism, that is, b) holds.

When $k_1 = 0$, f_0 being a levelwise $(n-m+1)$-equivalence implies by the inductive hypothesis that f_0 is an $(n-1)$-equivalence in $\mathsf{Cat}^{\mathsf{n-1}}_{\mathsf{hd}}$ and thus by Lemma 5.2.6 pf_0 is an isomorphism, that is, a) holds.

It remains to prove that f is an n-equivalence. By the inductive hypothesis applied to f_1, f_1 is an $(n-1)$-equivalence. Since

$$\begin{aligned} Y_1 &= \coprod_{a',b'\in Y_0^d} Y(a',b') = \coprod_{\substack{a,b\in X_0^d \\ fa=a' \\ fb=b'}} Y(fa,fb)\,, \\ X_1 &= \coprod_{a,b\in X_0^d} X(a,b) \end{aligned} \tag{7.10}$$

it follows that $X(a,b) \to Y(fa,fb)$ is an $(n-1)$-equivalence, that is, f is a local equivalence.

To show that $p^{(n-1)}f$ is an $(n-1)$-equivalence observe that the hypothesis that f is a levelwise $(n-m)$-equivalence implies in particular that for each $(k_1,\ldots,k_{m-1}) \in \Delta^{m-1^{op}}$, $f_{k_1\ldots k_{m-1}0}$ is an $(n-m)$-equivalence in $\mathsf{Cat}^{\mathsf{n-m}}_{\mathsf{hd}}$, which implies by Lemma 5.2.6 that $pf_{k_1\ldots k_m 0}$ is a bijection. By Lemma 7.1.3 it follows that $f_{k_1\ldots k_{m-1}}$ is an $(n-m+1)$-equivalence. Thus

$$p^{(n-m)}(f_{k_1\ldots k_{m-1}}) = (p^{(n-1)}f)_{k_1\ldots k_{m-1}}$$

is an $(n-m)$-equivalence. That is, $p^{(n-1)}f$ is a levelwise $(n-m+1)$-equivalence. By the induction hypothesis applied to $p^{(n-1)}f$ this implies it is an $(n-1)$-equivalence, as required. In conclusion, f is an n-equivalence.

Conversely, let f be an n-equivalence in $\mathsf{Ta}^{\mathsf{n}}_{\mathsf{wg}}$ satisfying a) and b). In particular, f is a local $(n-1)$-equivalence and (7.10) holds since $X_0^d \cong Y_0^d$. Thus f_1 is an

$(n-1)$-equivalence. For each $k \geq 2$ we have the commuting diagram

$$\begin{array}{ccc} X_k & \xrightarrow{\hat{\mu}_k} & X_1 \times_{X_0^d} \overset{k}{\cdots} \times_{X_0^d} X_1 \\ {\scriptstyle f_k}\downarrow & & \downarrow{\scriptstyle (f_1,\dots,f_1)} \\ Y_k & \xrightarrow[\hat{\mu}_k]{} & Y_1 \times_{Y_0^d} \overset{k}{\cdots} \times_{Y_0^d} Y_1 \end{array}$$

By Proposition 7.1.2 c) it follows that f_k is an $(n-1)$-equivalence. It is immediate that f_k satisfies hypotheses a) and b), thus by the inductive hypothesis applied to f_k we conclude that f_k is a levelwise $(n-m+1)$-equivalence for all k, and thus f is a levelwise $(n-m)$-equivalence. □

7.1.2 The Functor $q^{(n-1)}$

This section introduces the functor

$$q^{(n-1)} : \mathsf{Ta}^{\mathsf{n}}_{\mathsf{wg}} \to \mathsf{Ta}^{\mathsf{n-1}}_{\mathsf{wg}} .$$

This functor is a higher dimensional generalization of the connected component functor $q : \mathsf{Cat} \to \mathsf{Set}$ and comes equipped with a morphism

$$\gamma^{(n-1)} : X \to q^{(n-1)} X ,$$

natural in $X \in \mathsf{Ta}^{\mathsf{n}}_{\mathsf{wg}}$. It will be used crucially in Sect. 9.1 to replace a weakly globular n-fold category X with a simpler one (Theorem 9.2.4).

Proposition 7.1.7 *The functor $q^{(n-1)} : [\Delta^{n^{op}}, \mathsf{Set}] \to [\Delta^{n-1^{op}}, \mathsf{Set}]$ as in Definition 2.2.5 restricts to a functor $q^{(n-1)} : \mathsf{Ta}^{\mathsf{n}}_{\mathsf{wg}} \to \mathsf{Ta}^{\mathsf{n-1}}_{\mathsf{wg}}$. The functor $q^{(n-1)}$ sends n-equivalences to $(n-1)$-equivalences and preserves pullbacks over discrete objects. If $X \in \mathsf{Cat}^{\mathsf{n}}_{\mathsf{hd}}$, then $q^{(n-1)}X = p^{(n-1)}X$; further, for each $X \in \mathsf{Ta}^{\mathsf{n}}_{\mathsf{wg}}$, there is a map, natural in X,*

$$\gamma^{(n-1)} : X \to q^{(n-1)} X .$$

Proof By induction on n; for $n = 1$, $q^{(0)} = q : \mathsf{Cat} \to \mathsf{Set}$ is the connected components functor which, by Lemma 4.1.2, has the desired properties. The map $\gamma^{(1)}$ is the unit of the adjunction $q \dashv d$. If $X \in \mathsf{Cat}_{\mathsf{hd}}$, in particular X is a groupoid, so $pX = qX$.

Suppose we have defined $q^{(n-2)}$ with the desired properties and let $X \in \mathsf{Ta}^{\mathsf{n}}_{\mathsf{wg}}$. For each $k \geq 0$, by the inductive hypothesis applied to $X_k \in \mathsf{Ta}^{\mathsf{n-1}}_{\mathsf{wg}}$, $(q^{(n-1)}X)_k = q^{(n-2)}X_k \in \mathsf{Ta}^{\mathsf{n-2}}_{\mathsf{wg}}$.

Since $X_0 \in \mathsf{Cat}^{\mathsf{n-1}}_{\mathsf{hd}}$, $(q^{(n-1)}X)_0 = q^{(n-2)}X_0 \in \mathsf{Cat}^{\mathsf{n-2}}_{\mathsf{hd}}$ and $q^{(n-2)}X_0^d = (q^{(n-2)}X_0)^d$. Further, by the inductive hypothesis $q^{(n-2)}$ preserves pullbacks over discrete objects so that

$$q^{(n-2)}(X_1 \times_{X_0^d} \overset{s}{\cdots} \times_{X_0^d} X_1) \cong q^{(n-2)}X_1 \times_{q^{(n-2)}X_0^d} \overset{s}{\cdots} \times_{q^{(n-2)}X_0^d} q^{(n-2)}X_1 \; .$$

By the induction hypothesis, $q^{(n-2)}$ sends $(n-1)$-equivalences to $(n-2)$-equivalences. Therefore the induced Segal maps for $s \geq 2$

$$X_s \to X_1 \times_{X_0^d} \overset{s}{\cdots} \times_{X_0^d} X_1 \; ,$$

being $(n-1)$-equivalences, give rise to $(n-2)$-equivalences

$$q^{(n-2)}X_s \to q^{(n-2)}X_1 \times_{(q^{(n-2)}X_0)^d} \overset{s}{\cdots} \times_{(q^{(n-2)}X_0)^d} q^{(n-2)}X_1 \; .$$

This shows that $q^{(n-1)}X \in \mathsf{Ta}^{\mathsf{n-1}}_{\mathsf{wg}}$.

If $X \in \mathsf{Cat}^{\mathsf{n}}_{\mathsf{hd}}$, by definition $X_k \in \mathsf{Cat}^{\mathsf{n-1}}_{\mathsf{hd}}$ for each k, so by the induction hypothesis $p^{(n-2)}X_k = q^{(n-2)}X_k$. It follows that $(p^{(n-1)}X)_k = (q^{(n-1)}X)_k$ for all k. That is $p^{(n-1)}X = q^{(n-1)}X$.

Let $f : X \to Y$ be an n-equivalence in $\mathsf{Ta}^{\mathsf{n}}_{\mathsf{wg}}$ and let $a, b \in X_0^d$. Then from the definitions

$$(q^{(n-1)}f)(a,b) = q^{(n-2)}f(a,b) \; .$$

Since $f(a,b)$ is an $(n-1)$-equivalence, by the induction hypothesis $q^{(n-2)}f(a,b)$ is an $(n-2)$-equivalence. By Proposition 7.1.2, to prove that $q^{(n-1)}f$ is an $(n-1)$-equivalence, it is enough to show that $pq^{(n-1)}f$ is surjective.

Recall that for any category $\mathscr{C}$ there is a surjective map $p\mathscr{C} \to q\mathscr{C}$, natural in $\mathscr{C}$. Applying this map levelwise to $X \in \mathsf{Ta}^{\mathsf{n}}_{\mathsf{wg}} \hookrightarrow [\Delta^{n-1^{op}}, \mathsf{Cat}]$ we obtain a map

$$\alpha_X^{(n-1)} : p^{(n-1)}X \to q^{(n-1)}X \; ,$$

natural in X. The map $\alpha^{(n-1)}$ induces a functor

$$p^{(1)}\alpha_X^{(n-1)} : p^{(1)}p^{(n-1)}X = p^{(1)}X \to p^{(1)}q^{(n-1)}X \; , \tag{7.11}$$

which is the identity on objects. In fact, on objects this map is given by

$$p^{(0)}p^{(n-2)}X_0 \to p^{(0)}q^{(n-2)}X_0$$

and since $X_0 \in \mathsf{Cat}^{n-1}_{\mathsf{hd}}$, $p^{(n-2)} X_0 = q^{(n-2)} X_0$, so this map is the identity. It follows that the map in Set

$$p^{(0)} \alpha_X^{(n-1)} : p^{(0)} p^{(n-1)} X \to p^{(0)} q^{(n-1)} X$$

is surjective. We thus have a commuting diagram

$$\begin{array}{ccc} p^{(0)} p^{(n-1)} X & \xrightarrow{p^{(0)} p^{(n-1)} f} & p^{(0)} p^{(n-1)} Y \\ {\scriptstyle p^{(0)} \alpha_X^{(n-1)}} \downarrow & & \downarrow {\scriptstyle p^{(0)} \alpha_Y^{(n-1)}} \\ p^{(0)} q^{(n-1)} X & \xrightarrow{p^{(0)} q^{(n-1)} f} & p^{(0)} q^{(n-1)} Y \end{array}$$

in which the top arrow is an isomorphism (by Proposition 7.1.2) and from the above the vertical arrows are surjective. It follows that the bottom map is also surjective. By Proposition 7.1.2 b) we conclude that $q^{(n-1)} f$ is an $(n-1)$-equivalence.

Finally, the map $\gamma^{(n-1)} : X \to q^{(n-1)} X$ is given levelwise by the maps $\gamma^{(n-2)} : X_s \to q^{(n-2)} X_s$, which exist by the induction hypothesis. Since, by induction, each $X_s \to q^{(n-1)} X_s$ is natural in X, so is $\gamma^{(n-1)}$. □

Corollary 7.1.8 *The functor*

$$q^{(n-1)} : \mathsf{Ta}^{n}_{\mathsf{wg}} \to \mathsf{Ta}^{n-1}_{\mathsf{wg}}$$

restricts to functors

$$q^{(n-1)} : \mathsf{Cat}^{n}_{\mathsf{hd}} \to \mathsf{Cat}^{n-1}_{\mathsf{hd}} \qquad q^{(n-1)} : \mathsf{Ta}^{n} \to \mathsf{Ta}^{n-1} .$$

Proof If $X \in \mathsf{Cat}^{n}_{\mathsf{hd}}$ by Proposition 7.1.7, $q^{(n-1)} X = p^{(n-1)} X \in \mathsf{Cat}^{n-1}_{\mathsf{hd}}$.

Let $X \in \mathsf{Ta}^{n}$. We show by induction that $q^{(n-1)} X \in \mathsf{Ta}^{n-1}$ and that $q^{(n-1)} X$ is discrete if X is discrete. This is clear for $n = 2$, since $q^{(1)} X \in \mathsf{Cat}$.

Inductively, if $X \in \mathsf{Ta}^{n}$, $X_{k-1} \in \mathsf{Ta}^{n-1}$ so by the inductive hypothesis

$$(q^{(n-1)} X)_k = q^{(n-2)} X_k \in \mathsf{Ta}^{n-2}$$

with $(q^{(n-1)} X)_0 = q^{(n-2)} X_0$ discrete since X_0 is discrete (using the inductive hypothesis). It follows that $(q^{(n-1)} X)_0$ and $(q^{(n-1)} X)_{k_1 \dots k_r 0}$ are discrete for all $1 \leq r \leq n-2$ and $(k_1, \dots, k_r) \in \Delta^{op}$, and thus by Definition 6.2.1 $q^{(n-1)} X \in \mathsf{Ta}^{n-1}$. It is straightforward that if X is discrete then so is $q^{(n-1)} X$. □

Remark 7.1.9 From Proposition 7.1.7 and Corollary 7.1.8 it is immediate that for each $1 \leq r \leq n-1$ the functor $q^{(r)} : [\Delta^{n^{op}}, \mathsf{Set}] \to [\Delta^{r^{op}}, \mathsf{Set}]$ restricts to functors

$$q^{(r)} : \mathsf{Ta}^{n}_{\mathsf{wg}} \to \mathsf{Ta}^{r}_{\mathsf{wg}} \qquad q^{(r)} : \mathsf{Ta}^{n} \to \mathsf{Ta}^{r} \qquad q^{(r)} : \mathsf{Cat}^{n}_{\mathsf{hd}} \to \mathsf{Cat}^{r}_{\mathsf{hd}}$$

and for each $X \in \mathsf{Ta}^{n}_{\mathsf{wg}}$ there are maps $\gamma^{(r)} : X \to q^{(r)} X$, natural in X.

7.1.3 Pullback Constructions Using $q^{(n-1)}$

In Sect. 9.1, Theorem 9.2.4, we will replace a weakly globular Tamsamani n-category X with a simpler one, and this result will play a crucial role in the construction of the rigidification functor Q_n. The proof of Theorem 9.2.4 will use pullbacks along the map $\gamma^{(n)}$ of Proposition 7.1.7. In this section we establish several properties of these pullbacks.

Lemma 7.1.10 *Let $X \in \mathsf{Cat}^{\mathsf{n}}_{\mathsf{hd}}$, $Z \in \mathsf{Cat}^{\mathsf{n-1}}_{\mathsf{hd}}$, $r : Z \to q^{(n-1)}X$. Consider the pullback in* $[\Delta^{n-1^{op}}, \mathsf{Cat}]$

$$\begin{array}{ccc} P & \longrightarrow & X \\ \downarrow & & \downarrow \gamma_X^{(n-1)} \\ Z & \longrightarrow & q^{(n-1)}X \end{array}$$

then $P \in \mathsf{Cat}^{\mathsf{n}}_{\mathsf{hd}}$ and $p^{(n-1)}P = Z$.

Proof By induction on n. For $n = 1$, since $q^{(0)}X = qX$ is discrete, the map

$$\gamma^{(0)} : X \to q^{(0)}X = X^d$$

is an isofibration. Therefore, since $\gamma^{(0)}$ is an equivalence of categories (as $X \in \mathsf{Cat}_{\mathsf{hd}}$) we have an equivalence of categories

$$P = Z \times_{q^{(0)}X} X \simeq Z \times_{q^{(0)}X} q^{(0)}X = Z\ .$$

Thus $P \in \mathsf{Cat}_{\mathsf{hd}}$ and $pP = Z$.

Suppose, inductively, that the lemma holds for $n - 1$ and let P be as in the hypothesis. Since pullbacks in $[\Delta^{n-1^{op}}, \mathsf{Cat}]$ are computed pointwise, for each $k \geq 0$ we have a pullback in $[\Delta^{n-2^{op}}, \mathsf{Cat}]$

$$\begin{array}{ccc} P_k & \longrightarrow & X_k \\ \downarrow & & \downarrow \gamma_{X_k}^{(n-2)} \\ Z_k & \longrightarrow & q^{(n-2)}X_k \end{array}$$

where $X_k \in \mathsf{Cat}^{\mathsf{n-1}}_{\mathsf{hd}}$ (since $X \in \mathsf{Cat}^{\mathsf{n}}_{\mathsf{hd}}$) and $Z_k \in \mathsf{Cat}^{\mathsf{n-2}}_{\mathsf{hd}}$ (since $Z \in \mathsf{Cat}^{\mathsf{n-1}}_{\mathsf{hd}}$). By the induction hypothesis, we conclude that $P_k \in \mathsf{Cat}^{\mathsf{n-1}}_{\mathsf{hd}}$.

We now show that, for each $k \geq 2$,

$$P_k \cong P_1 \times_{P_0} \overset{k}{\cdots} \times_{P_0} P_1\ . \tag{7.12}$$

We illustrate this for $k = 2$, the case $k > 2$ being similar. Since $X \in \mathsf{Cat}^{\mathsf{n}}_{\mathsf{hd}}$, $q^{(n-1)}X = p^{(n-1)}X \in \mathsf{Cat}^{\mathsf{n-1}}_{\mathsf{hd}}$, so

$$\begin{aligned} q^{(n-2)}X_2 &= p^{(n-2)}X_2 = p^{(n-2)}(X_1 \times_{X_0} X_1) \\ &= p^{(n-2)}X_1 \times_{p^{(n-2)}X_0} p^{(n-2)}X_1 = q^{(n-2)}X_1 \times_{q^{(n-2)}X_0} q^{(n-2)}X_1 \,. \end{aligned}$$

Since $X_2 \cong X_1 \times_{X_0} X_1$ and $Z_2 \cong Z_1 \times_{Z_0} Z_1$, it follows from Remark 2.4.2 that

$$P_2 \cong P_1 \times_{P_0} P_1 .$$

To prove that $P \in \mathsf{Cat}^{\mathsf{n}}_{\mathsf{hd}}$ it remains to show that $p^{(n-1)}P \in \mathsf{Cat}^{\mathsf{n-1}}_{\mathsf{hd}}$. Since p commutes with fiber products over discrete objects, for each $\underline{s} \in \Delta^{{n-1}^{op}}$ we have

$$(p^{(n-1)}P)_{\underline{s}} = p\, P_{\underline{s}} = p(Z_{\underline{s}} \times_{qX_{\underline{s}}} X_{\underline{s}}) = Z_{\underline{s}} \times_{qX_{\underline{s}}} pX_{\underline{s}} = Z_{\underline{s}} \;,$$

where we used the fact that, since $X_{\underline{s}}$ is a groupoid, $pX_{\underline{s}} = qX_{\underline{s}}$. Since this holds for each $\underline{s}$ we conclude that $p^{(n-1)}P = Z \in \mathsf{Cat}^{\mathsf{n-1}}_{\mathsf{hd}}$ as required. □

Lemma 7.1.11 *Let $Y \in \mathsf{Ta}^{\mathsf{n}}_{\mathsf{wg}}$ and let*

$$X \to q^{(n-1)}Y \leftarrow p^{(n-1)}Y \tag{7.13}$$

be a diagram in $\mathsf{Ta}^{\mathsf{n}}_{\mathsf{wg}}$ such that $X \times_{q^{(n-1)}Y} p^{(n-1)}Y \in \mathsf{Ta}^{\mathsf{n}}_{\mathsf{wg}}$. Then for all $0 \leq j \leq n-1$

$$p^{(j)}(X \times_{q^{(n-1)}Y} p^{(n-1)}Y) = p^{(j)}X \times_{p^{(j)}q^{(n-1)}Y} p^{(j)}p^{(n-1)}Y \,.$$

Proof By induction on n. For $n = 2$ and $j = 0$, the functor $p^{(1)}Y \to q^{(1)}Y$ is the identity on objects, therefore by Lemma 4.1.5

$$p(X \times_{q^{(1)}Y} p^{(1)}Y) = pX \times_{pq^{(1)}Y} pp^{(1)}Y \,.$$

When $n = 2$ and $j = 1$, since p commutes with pullbacks over discrete objects, for each $k \geq 0$,

$$\{p^{(1)}(X \times_{q^{(1)}Y} p^{(1)}Y\}_k = p(X_k \times_{qY_k} pY_k) = pX_k \times_{qY_k} pY_k = \{p^{(1)}X \times_{p^{(1)}q^{(1)}Y} p^{(1)}Y\}_k,$$

so that

$$p^{(1)}(X \times_{q^{(n-1)}Y} p^{(n-1)}Y) = p^{(1)}X \times_{p^{(1)}q^{(n-1)}Y} p^{(1)}p^{(n-1)}Y \,.$$

Suppose, inductively, that the lemma holds for $n-1$. Then for each $k \geq 0$, using the inductive hypothesis,

$$\begin{aligned}&(p^{(j)}(X\times_{q^{(n-1)}Y} p^{(n-1)}Y))_k = p^{(j-1)}(X_k\times_{q^{(n-2)}Y_k} p^{(n-2)}Y_k)\\ &= p^{(j-1)}X_k\times_{p^{(j-1)}q^{(n-2)}Y_k} p^{(j-1)}p^{(n-2)}Y_k = (p^{(j)}X)_k\times_{(p^{(j)}q^{(n-2)}Y)_k} (p^{(j)}p^{(n-2)}Y)_k\,.\end{aligned} \tag{7.14}$$

Since this holds for each $k \geq 0$, the lemma follows. □

Proposition 7.1.12 *Let*

$$A \xrightarrow{f} q^{(n-1)}C \xleftarrow{\gamma^{(n-1)}} C$$

be a diagram in $\mathsf{Ta}^{\mathsf{n}}_{\mathsf{wg}}$ *where* $f : A \to q^{(n-1)}C$ *is a morphism in* $\mathsf{Ta}^{\mathsf{n-1}}_{\mathsf{wg}}$ *and consider the pullback in* $[\Delta^{n-1^{op}}, \mathsf{Cat}]$

$$\begin{array}{ccc} P & \xrightarrow{w} & C \\ \downarrow & & \downarrow{\scriptstyle \gamma_C^{(n-1)}} \\ A & \xrightarrow[f]{} & q^{(n-1)}C \end{array}$$

a) Then $P \in \mathsf{Ta}^{\mathsf{n}}_{\mathsf{wg}}$ *and*

$$p^{(0)}P = p^{(0)}A\times_{p^{(0)}q^{(n-1)}C} p^{(0)}C\,. \tag{7.15}$$

b) Consider the commutative diagram in $\mathsf{Ta}^{\mathsf{n}}_{\mathsf{wg}}$

$$\begin{array}{ccccc} A & \xrightarrow{f} & q^{(n-1)}C & \xleftarrow{\gamma_C^{(n-1)}} & C \\ {\scriptstyle a}\downarrow & & \downarrow{\scriptstyle b} & & \downarrow{\scriptstyle c} \\ D & \xrightarrow[h]{} & q^{(n-1)}F & \xleftarrow[\gamma_F^{(n-1)}]{} & F \end{array} \tag{7.16}$$

where a, b, c *are* n*-equivalences and* f, h *are morphisms in* $\mathsf{Ta}^{\mathsf{n-1}}_{\mathsf{wg}}$*. Then the induced map of pullbacks*

$$(a, c) : A\times_{q^{(n-1)}C} C \to D\times_{q^{(n-1)}F} F$$

is an n*-equivalence in* $\mathsf{Ta}^{\mathsf{n}}_{\mathsf{wg}}$.

c) If f *is an* $(n-1)$*-equivalence,* $P \xrightarrow{w} C$ *is an* n*-equivalence.*

Proof By induction on n. When $n = 1$ a) and c) are trivial. For b), the maps $f, \gamma_C^{(n-1)}, h, \gamma_F^{(n-1)}$ are isofibrations since their targets are discrete categories. Therefore, since a, b, c are equivalences of categories, the induced map of pullbacks

$$(a, c) : A\times_{qC} C \to D\times_{qF} F$$

is an equivalence of categories and

$$p(A\times_{qC} C) = A\times_{qC} pC.$$

Suppose, inductively, that the proposition holds for $n-1$.

a) We have

$$P_0 = (A\times_{q^{(n-1)}C} C)_0 = A_0\times_{q^{(n-2)}C_0} C_0 \ .$$

Since, by hypothesis, $A \in \mathsf{Ta}^{\mathsf{n-1}}_{\mathsf{wg}}$ and $C \in \mathsf{Ta}^{\mathsf{n}}_{\mathsf{wg}}$, by definition $A_0 \in \mathsf{Cat}^{\mathsf{n-2}}_{\mathsf{hd}}$ and $C_0 \in \mathsf{Cat}^{\mathsf{n-1}}_{\mathsf{hd}}$. Therefore by Lemma 7.1.10

$$A_0\times_{q^{(n-2)}C_0} C_0 \in \mathsf{Cat}^{\mathsf{n-1}}_{\mathsf{hd}} \ .$$

Further, for each $k \geq 1$,

$$P_k = (A\times_{q^{(n-1)}C} C)_k = A_k\times_{q^{(n-2)}C_k} C_k \ ,$$

where, by hypothesis, $A_k \in \mathsf{Ta}^{\mathsf{n-2}}_{\mathsf{wg}}$ and $C_k \in \mathsf{Ta}^{\mathsf{n-1}}_{\mathsf{wg}}$. It follows by the inductive hypothesis that

$$A_k\times_{q^{(n-2)}C_k} C_k \in \mathsf{Ta}^{\mathsf{n-1}}_{\mathsf{wg}} \ .$$

To prove that $A\times_{q^{(n-1)}C} C \in \mathsf{Ta}^{\mathsf{n}}_{\mathsf{wg}}$, it remains to show that its induced Segal maps $\hat{\mu}_s$ are $(n-1)$-equivalences for all $s \geq 2$. We show this for $s = 2$, the case $s > 2$ being similar. We have

$$\hat{\mu}_2 : (A\times_{q^{(n-1)}C} C)_2 \to (A\times_{q^{(n-1)}C} C)_1 \times_{(A\times_{q^{(n-1)}C} C)_0^d} (A\times_{q^{(n-1)}C} C)_1 \ .$$

By Lemma 7.1.10,

$$(A\times_{q^{(n-1)}C} C)_0^d = (p^{(n-2)}(A_0\times_{q^{(n-2)}C_0} C_0))^d = A_0^d = A_0^d\times_{q^{(n-2)}C_0^d} C_0^d,$$

where we used the fact that

$$q^{(n-2)}C_0^d = q^{(n-2)}\dots p^{(0)}\dots p^{(n-2)}C_0 = p^{(0)}\dots p^{(n-2)}C_0 = C_0^d \ .$$

Recalling that $q^{(n-2)}$ commutes with pullbacks over discrete objects, we obtain

$$\begin{aligned}
&(A\times_{q^{(n-1)}C} C)_1 \times_{(A\times_{q^{(n-1)}C} C)_0^d} (A\times_{q^{(n-1)}C} C)_1\\
&= (A_1\times_{q^{(n-2)}C_1} C_1) \times_{(A_0^d\times_{q^{(n-2)}C_0^d} C_0^d)} (A_1\times_{q^{(n-2)}C_1} C_1)\\
&= (A_1 \times_{A_0^d} A_1)\times_{q^{(n-2)}(C_1\times_{C_0^d} C_1)} (C_1 \times_{C_0^d} C_1)\ .
\end{aligned}$$

On the other hand

$$(A\times_{q^{(n-1)}C} C)_2 = A_2\times_{q^{(n-2)}C_2} C_2\ .$$

Hence we see that the map $\hat{\mu}_2$ is the induced map on pullbacks from the diagram in $\mathsf{Ta}_{\mathsf{wg}}^{\mathsf{n-1}}$

$$\begin{array}{ccccc}
A_2 & \longrightarrow & q^{(n-2)}C_2 & \longleftarrow & C_2\\
\downarrow & & \downarrow & & \downarrow\\
A_1\times_{A_0^d} A_1 & \longrightarrow & q^{(n-2)}(C_1\times_{C_0^d} C_1) & \longleftarrow & C_1\times_{C_0^d} C_1
\end{array} \tag{7.17}$$

In this diagram, the left and right vertical maps are $(n-1)$-equivalences since they are induced Segal maps of A and C respectively. The map

$$q^{(n-2)}C_2 \to q^{(n-2)}(C_1\times_{C_0^d} C_1) = q^{(n-2)}C_1\times_{(q^{(n-2)}C_0)^d} q^{(n-2)}C_1$$

is the induced Segal map for $q^{(n-1)}C \in \mathsf{Ta}_{\mathsf{wg}}^{\mathsf{n-1}}$, and is therefore an $(n-2)$-equivalence. The central vertical map in (7.17) is therefore in particular an $(n-1)$-equivalence. We can therefore apply induction hypothesis b) to the diagram (7.17) and conclude that the induced map on pullbacks is an $(n-1)$-equivalence. That is, $\hat{\mu}_2$ is an $(n-1)$-equivalence.

Similarly one shows that $\hat{\mu}_s$ is an $(n-1)$-equivalence for all $s \geq 2$. This concludes the proof that

$$A\times_{q^{(n-1)}C} C \in \mathsf{Ta}_{\mathsf{wg}}^{\mathsf{n}}$$

and therefore $p^{(n-1)}(A\times_{q^{(n-1)}C} C) \in \mathsf{Ta}_{\mathsf{wg}}^{\mathsf{n-1}}$. We observe that

$$p^{(n-1)}(A\times_{q^{(n-1)}C} C) = p^{(n-1)}A\times_{q^{(n-1)}C} p^{(n-1)}C\ . \tag{7.18}$$

In fact, for each $\underline{k} \in \Delta^{n-1^{op}}$, since p commutes with pullbacks over discrete objects,

$$\begin{aligned}(p^{(n-1)}(A\times_{q^{(n-1)}C} C))_{\underline{k}} &= p(A_{\underline{k}}\times_{qC_{\underline{k}}} C_{\underline{k}}) = pA_{\underline{k}}\times_{qC_{\underline{k}}} pC_{\underline{k}} \\ &= (p^{(n-1)}A)_{\underline{k}}\times_{(q^{(n-1)}C)_{\underline{k}}} (p^{(n-1)}A)_{\underline{k}}\ ,\end{aligned}$$

so (7.18) follows. We conclude that $p^{(n-1)}A\times_{q^{(n-1)}C} p^{(n-1)}A \in \mathsf{Ta}^{\mathsf{n-1}}_{\mathsf{wg}}$ so the hypotheses of Lemma 7.1.11 are satisfied and therefore (7.15) follows.

b) We now show that (a, c) is an n-equivalence. By Proposition 7.1.2 b), it is enough to show that it is a local $(n-1)$-equivalence and that $p^{(0)}(a, c)$ is an isomorphism.

Let

$$(x_1, y_1), (x_2, y_2) \in (A\times_{q^{(n-1)}C} C)^d_0 = A^d_0\times_{q^{(n-2)}C^d_0} C^d_0.$$

Then the map

$$(a, c)((x_1, y_1), (x_2, y_2))$$

is the induced map on pullbacks in the diagram

$$\begin{array}{ccccc} A(x_1, x_2) & \xrightarrow{f(x_1,x_2)} & q^{(n-2)}C(fx_1, fx_2) & \xleftarrow{\gamma^{(n-2)}_{C(y_1,y_2)}} & C(y_1, y_2) \\ \Big\downarrow{\scriptstyle a(x_1,x_2)} & & \Big\downarrow{\scriptstyle b(fx_1,fx_2)} & & \Big\downarrow{\scriptstyle c(y_1,y_2)} \\ D(ax_1, ax_2) & \xrightarrow[h(ax_1,ax_2)]{} & q^{(n-2)}F(hax_1, hax_2) & \xleftarrow[\gamma^{(n-2)}_{F(cy_1,cy_2)}]{} & F(cy_1, cy_2) \end{array} \tag{7.19}$$

The vertical maps in (7.19) are $(n-1)$-equivalences. By inductive hypothesis b), since a, b, c are n-equivalences we conclude that the induced map of pullbacks is an $(n-1)$-equivalence. That is, (a, c) is a local $(n-1)$-equivalence. By Lemma 7.1.11,

$$p^{(0)}(a, c) = (p^{(0)}a, p^{(0)}c)\ . \tag{7.20}$$

Applying the functor $p^{(0)}$ to the diagram (7.16) we obtain a commutative diagram in Set

$$\begin{array}{ccccc} p^{(0)}A & \longrightarrow & p^{(0)}q^{(n-1)}C & \longleftarrow & p^{(0)}p^{(n-1)}C \\ \Big\downarrow{\scriptstyle p^{(0)}a} & & \Big\downarrow{\scriptstyle p^{(0)}b} & & \Big\downarrow{\scriptstyle p^{(0)}c} \\ p^{(0)}D & \longrightarrow & p^{(0)}q^{(n-1)}F & \longleftarrow & p^{(0)}p^{(n-1)}F \end{array}$$

Since, by hypothesis, a, b and c are n-equivalences, by Proposition 7.1.2 a) the vertical maps are isomorphisms, hence by (7.20) so is $p^{(0)}(a, c)$. We conclude from Proposition 7.1.2 b) that (a, c) is an n-equivalence.

c) By a), we have a commutative diagram in $\mathsf{Ta}^{\mathsf{n}}_{\mathsf{wg}}$

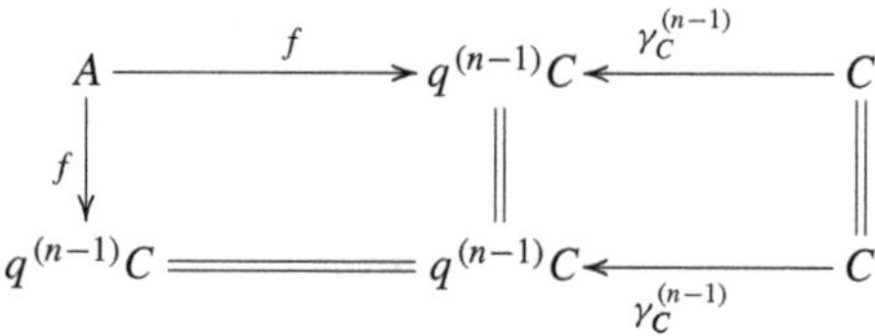

in which the vertical maps are n-equivalences. It follows by b) that the induced map of pullbacks

$$P = A \times_{q^{(n-1)}C} C \xrightarrow{w} q^{(n-1)}C \times_{q^{(n-1)}C} C = C$$

is an n-equivalence.

□

7.2 Properties of Weakly Globular n-Fold Categories

In this section we discuss the main properties of weakly globular n-fold categories, starting in Sect. 7.2.1 with some properties of n-equivalences in $\mathsf{Cat}^{\mathsf{n}}_{\mathsf{wg}}$.

The main result of Sect. 7.2.2 is Proposition 7.2.8 which has two parts. Part a) of Proposition 7.2.8 states that if $X \in \mathsf{Cat}^{\mathsf{n}}_{\mathsf{wg}}$, then $X^{\{2\}}_k \in \mathsf{Cat}^{\mathsf{n-1}}_{\mathsf{wg}}$ for all $k \geq 0$, where $X^{\{2\}}_k$ is the n-fold category X viewed as an internal category in $\mathsf{Cat}^{\mathsf{n-1}}$ along direction 2 (see Notational convention 2.5.4). Part b) of Proposition 7.2.8 gives a criterion for an n-fold category to be weakly globular. This criterion requires certain sub-structures in the n-fold category X to be homotopically discrete, as well as $p^{(n-1)}X \in \mathsf{Cat}^{\mathsf{n-1}}_{\mathsf{wg}}$. This criterion will be used crucially in the proof of Proposition 8.2.1 to characterize n-fold categories levelwise equivalent via a pseudo-natural transformation to Segalic pseudo-functors. This result leads to the main result, Theorem 8.2.3, on the strictification of Segalic pseudo-functors.

Using Proposition 7.2.8 we also deduce, in Proposition 7.2.12, that the functor $q^{(n-1)} : \mathsf{Ta}^{\mathsf{n}}_{\mathsf{wg}} \to \mathsf{Ta}^{\mathsf{n-1}}_{\mathsf{wg}}$ restricts to a functor $q^{(n-1)} : \mathsf{Cat}^{\mathsf{n}}_{\mathsf{wg}} \to \mathsf{Cat}^{\mathsf{n-1}}_{\mathsf{wg}}$.

In Sect. 7.2.3 we illustrate some consequences of Proposition 7.2.8 which will be used in Chap. 9, Sect. 9.1, in defining the category $\mathsf{LTa}^{\mathsf{n}}_{\mathsf{wg}}$.

7.2.1 Weakly Globular n-Fold Categories and n-Equivalences

In Proposition 7.2.1 we show that a weakly globular n-fold category n-equivalent to a homotopically discrete one is homotopically discrete. This result generalizes to higher dimensions the fact that a category equivalent to an equivalence relation is itself an equivalence relation. We deduce in Corollary 7.2.2 a criterion for a weakly globular n-fold category to be homotopically discrete.

Proposition 7.2.1 *Let $f : X \to Y$ be an n-equivalence in $\mathsf{Cat}^n_{\mathsf{wg}}$ with $Y \in \mathsf{Cat}^n_{\mathsf{hd}}$, then $X \in \mathsf{Cat}^n_{\mathsf{hd}}$.*

Proof By induction on n. It is clear for $n = 1$. Suppose it is true for $n-1$ and let f be as in the hypothesis. Then $p^{(n-1)}f : p^{(n-1)}X \to p^{(n-1)}Y$ is an $(n-1)$-equivalence with $p^{(n-1)}Y \in \mathsf{Cat}^{n-1}_{\mathsf{hd}}$ since $Y \in \mathsf{Cat}^n_{\mathsf{hd}}$. It follows by the induction hypothesis that $p^{(n-1)}X \in \mathsf{Cat}^{n-1}_{\mathsf{hd}}$. We have

$$X_1 = \coprod_{a,b\in X_0^d} X(a,b)\,. \tag{7.21}$$

Since f is an n-equivalence, there are $(n-1)$-equivalences

$$f(a,b) : X(a,b) \to Y(fa, fb)\,,$$

where $Y(fa, fb) \in \mathsf{Cat}^{n-1}_{\mathsf{hd}}$, since $Y \in \mathsf{Cat}^n_{\mathsf{hd}}$. By the induction hypothesis, it follows that $X(a,b) \in \mathsf{Cat}^{n-1}_{\mathsf{hd}}$. From (7.21) and the fact that $\mathsf{Cat}^{n-1}_{\mathsf{hd}}$ is closed under coproducts (see Corollary 5.2.2), we conclude that $X_1 \in \mathsf{Cat}^{n-1}_{\mathsf{hd}}$.

Since $X \in \mathsf{Cat}^n_{\mathsf{wg}}$, the induced Segal map for all $s \geq 2$

$$\hat{\mu}_s : X_s = X_1 \times_{X_0} \overset{s}{\cdots} \times_{X_0} X_1 \to X_1 \times_{X_0^d} \overset{s}{\cdots} \times_{X_0^d} X_1$$

is an $(n-1)$-equivalence. Since, from above, X_1 is homotopically discrete and X_0^d is discrete, by Corollary 5.2.2,

$$X_1 \times_{X_0^d} \overset{s}{\cdots} \times_{X_0^d} X_1 \in \mathsf{Cat}^{n-1}_{\mathsf{hd}}\,.$$

Thus by the induction hypothesis applied to the induced Segal map $\hat{\mu}_s$, we conclude that $X_s \in \mathsf{Cat}^{n-1}_{\mathsf{hd}}$ for all $s \geq 0$.

In summary, we showed that $X \in \mathsf{Cat}^n_{\mathsf{wg}}$ is such that $X_s \in \mathsf{Cat}^{n-1}_{\mathsf{hd}}$ for all $s \geq 0$ and $p^{(n-1)}X \in \mathsf{Cat}^{n-1}_{\mathsf{hd}}$. Therefore, by Lemma 5.1.6, $X \in \mathsf{Cat}^n_{\mathsf{hd}}$. □

Corollary 7.2.2 *Let $X \in \mathsf{Cat}^n_{\mathsf{wg}}$ be such that X_1 and $p^{(n-1)}X$ are in $\mathsf{Cat}^{n-1}_{\mathsf{hd}}$. Then $X \in \mathsf{Cat}^n_{\mathsf{hd}}$.*

Proof Since $X \in \mathsf{Cat}^n_{\mathsf{wg}}$, the induced Segal maps

$$\hat{\mu}_s : X_s \to X_1 \times_{X_0^d} \overset{s}{\cdots} \times_{X_0^d} X_1$$

are $(n-1)$-equivalences for all $s \geq 2$. Since by hypothesis $X_1 \in \mathsf{Cat}^{n-1}_{\mathsf{hd}}$ and X_0^d is discrete, by Corollary 5.2.2,

$$X_1 \times_{X_0^d} \overset{s}{\cdots} \times_{X_0^d} X_1 \in \mathsf{Cat}^{n-1}_{\mathsf{hd}}\,.$$

By Proposition 7.2.1 applied to $\hat{\mu}_s$ we conclude that $X_s \in \mathsf{Cat}^{n-1}_{\mathsf{hd}}$ for all $s \geq 2$. Therefore $X \in \mathsf{Cat}^n_{\mathsf{wg}}$ is such that $X_s \in \mathsf{Cat}^{n-1}_{\mathsf{hd}}$ for all $s \geq 0$ and $p^{(n-1)}X \in \mathsf{Cat}^{n-1}_{\mathsf{hd}}$. By Lemma 5.1.6, $X \in \mathsf{Cat}^n_{\mathsf{hd}}$. □

Corollary 7.2.3 *Let $X \in \mathsf{Cat}^n_{\mathsf{wg}}$, then $X \in \mathsf{Cat}^n_{\mathsf{hd}}$ if and only if there is an n-equivalence $\gamma : X \to Y$ with Y discrete.*

Proof If $X \in \mathsf{Cat}^n_{\mathsf{hd}}$ then by Corollary 5.2.7, $\gamma_{(n)} : X \to X^d$ is an n-equivalence. Conversely, suppose that there is an n-equivalence $\gamma : X \to Y$ with Y discrete, then in particular $Y \in \mathsf{Cat}^n_{\mathsf{hd}}$ so, by Proposition 7.2.1, $X \in \mathsf{Cat}^n_{\mathsf{hd}}$. □

Definition 7.2.4 Given $X \in \mathsf{Cat}^n$ and $k \geq 0$, let $X^{\{2\}} \in \mathsf{Cat}^n$ be as in Notational convention 2.5.4 so that for each $k \geq 0$, the $(n-1)$-fold category $X_k^{\{2\}}$ is given by

$$(X_k^{\{2\}})_s = \begin{cases} X_{0k}, & s = 0; \\ X_{1k}, & s = 1; \\ X_{sk} = X_{1k} \times_{X_{0k}} \overset{s}{\cdots} \times_{X_{0k}} X_{1k}, & s \geq 2. \end{cases}$$

7.2.2 A Criterion for an n-Fold Category to Be Weakly Globular

This section contains one of the main results of this chapter, Proposition 7.2.8, giving a very useful criterion for an n-fold category to be weakly globular.

The following lemmas are needed in the initial steps of the induction in the proof of Proposition 7.2.8.

Lemma 7.2.5 *Let $X \in \mathsf{Cat}^n_{\mathsf{wg}}$. Then for all $1 \leq r < n$ the $(n-1)$-fold category $X_0^{\{r\}}$ is a levelwise equivalence relation.*

Proof If $r = 1$, $X_0^{\{1\}} = X_0 \in \mathsf{Cat}^{n-1}_{\mathsf{hd}}$ so by Definition 5.1.2, X_0 is a levelwise equivalence relation. Suppose that $1 < r < n$. By Lemma 6.3.5 for each $\underline{k} = (k_1, \ldots, k_{r-1}) \in \Delta^{r-1^{op}}$, $X_{\underline{k}} \in \mathsf{Cat}^{n-r+1}_{\mathsf{wg}}$, therefore $X_{\underline{k}0} \in \mathsf{Cat}^{n-r}_{\mathsf{hd}}$. Thus by

Definition 5.1.2 $X_{\underline{k}0}$ is a levelwise equivalence relation. Since this holds for each $\underline{k} \in \Delta^{r-1^{op}}$, we conclude that $X_0^{\{r\}}$ is a levelwise equivalence relation. □

Lemma 7.2.6 *Let $X \in \mathsf{Cat}^2$ be such that*

i) $X_0 \in \mathsf{Cat}_{\mathsf{hd}}$,
ii) $p^{(1)}X \in \mathsf{Cat}$.

Then $X \in \mathsf{Cat}^2_{\mathsf{wg}}$.

Proof Since $X_0 \in \mathsf{Cat}_{\mathsf{hd}}$, $pX_0 = X_0^d$. By hypothesis, $pX_2 \cong pX_1 \times_{pX_0} pX_1$ and $X_2 \cong X_1 \times_{X_0} X_1$. Using the fact (Lemma 4.1.4) that p commutes with pullbacks over discrete objects, we obtain

$$p(X_1 \times_{X_0} X_1) \cong pX_2 \cong pX_1 \times_{pX_0} pX_1 = pX_1 \times_{pX_0^d} pX_1 = p(X_1 \times_{X_0^d} X_1) \,.$$

This shows that the map

$$\hat{\mu}_2 : X_1 \times_{X_0} X_1 \to X_1 \times_{X_0^d} X_1$$

is essentially surjective on objects. On the other hand, this map is also fully faithful. In fact, given $(a, b), (c, d) \in X_{10} \times_{X_{00}} X_{10}$, we have

$$\begin{aligned}(X_1 \times_{X_0} X_1)((a, b), (c, d)) &\cong X_1(a, c) \times_{X_0(\partial_0 a, \partial_0 c)} X_1(b, d) \\ &\cong X_1(a, c) \times X_1(b, d) \cong (X_1 \times_{X_0^d} X_1)(\hat{\mu}_2(a, b), \hat{\mu}_2(c, d)) \,,\end{aligned}$$

where we used the fact that $X_0(\partial_0 a, \partial_0 c)$ is the one-element set, since $X_0 \in \mathsf{Cat}_{\mathsf{hd}}$. We conclude that $\hat{\mu}_2$ is an equivalence of categories.

Similarly one shows that for all $k \geq 2$

$$\hat{\mu}_k : X_1 \times_{X_0} \overset{k}{\cdots} \times_{X_0} X_1 \to X_1 \times_{X_0^d} \overset{k}{\cdots} \times_{X_0^d} X_1$$

is an equivalence of categories. By definition (see also Example 6.3.10), this means that $X \in \mathsf{Cat}^2_{\mathsf{wg}}$. □

Lemma 7.2.7

a) For each $X \in \mathsf{Cat}^3_{\mathsf{wg}}$ *and* $k \geq 0$, $X_k^{\{2\}} \in \mathsf{Cat}^2_{\mathsf{wg}}$.
b) Let $X \in \mathsf{Cat}^3$ *be such that*

   i) $X_{0t} \in \mathsf{Cat}_{\mathsf{hd}}$, $X_{s0} \in \mathsf{Cat}_{\mathsf{hd}}$, *for all* $t, s \in \Delta^{op}$,
   ii) $p^{(2)}X \in \mathsf{Cat}^2_{\mathsf{wg}}$.

   Then $X \in \mathsf{Cat}^3_{\mathsf{wg}}$.

Proof

a) We show that, if $X \in \mathsf{Cat}^3_{\mathsf{wg}}$ and $k \geq 0$, $X_k^{\{2\}} \in \mathsf{Cat}^2$ satisfies the hypotheses of Lemma 7.2.6, so that $X_k^{\{2\}} \in \mathsf{Cat}^2_{\mathsf{wg}}$.

Since $X \in \mathsf{Cat}^3_{\mathsf{wg}}$, $X_0 \in \mathsf{Cat}^2_{\mathsf{hd}}$ and thus for each $k \geq 0$

$$X_{k0}^{\{2\}} = X_{0k} \in \mathsf{Cat}_{\mathsf{hd}} \ .$$

So hypothesis i) of Lemma 7.2.6 holds. Further,

$$p^{(1)} X_k^{\{2\}} = (p^{(2)} X)_k^{\{2\}}$$

is the nerve of a category since $p^{(2)} X \in \mathsf{Cat}^2_{\mathsf{wg}}$, so hypothesis ii) of Lemma 7.2.6 also holds. We conclude that $X_k^{\{2\}} \in \mathsf{Cat}^2_{\mathsf{wg}}$.

b) By hypothesis, $X_s \in \mathsf{Cat}^2$ is such that $X_{s0} \in \mathsf{Cat}_{\mathsf{hd}}$ and $p^{(1)} X_s$ is the nerve of a category. Thus by Lemma 7.2.6, $X_s \in \mathsf{Cat}^2_{\mathsf{wg}}$.

Also by hypothesis $p^{(2)} X \in \mathsf{Cat}^2_{\mathsf{wg}}$. To show that $X \in \mathsf{Cat}^3_{\mathsf{wg}}$ it remains to prove that for each $s \geq 2$ the induced Segal map

$$\hat{\mu}_s : X_1 \times_{X_0} \overset{s}{\cdots} \times_{X_0} X_1 \to X_1 \times_{X_0^d} \overset{s}{\cdots} \times_{X_0^d} X_1$$

is a 2-equivalence. We show this for $s = 2$, the case $s > 2$ being similar. We first show that it is a local equivalence. By part a) $X_1^{\{2\}} \in \mathsf{Cat}^2_{\mathsf{wg}}$. Thus there is an equivalence of categories

$$X_{11} \times_{X_{01}} X_{11} \to X_{11} \times_{X_{01}^d} X_{11} = X_{11} \times_{(p^{(1)} X_0)_1} X_{11} \ . \tag{7.22}$$

From hypothesis ii) by Remark 6.3.9 using the fact that $pp^{(1)} X_{s0} = X_{s0}^d$ we have

$$X_{20}^d = (X_{10} \times_{X_{00}} X_{10})^d \cong X_{10}^d \times_{X_{00}^d} X_{10}^d \ .$$

Let $(a, b), (c, d) \in X_{10}^d \times_{X_{00}^d} X_{10}^d$. By (7.22) there is an equivalence of categories

$$\begin{aligned} &(X_1 \times_{X_0} X_1)((a, b), (c, d)) \\ &= X_1(a, c) \times_{X_0(\partial_0 a, \partial_0 c)} X_1(b, d) \to X_1(a, c) \times_{(p^{(1)} X_0)(\tilde{\partial}_0 a, \tilde{\partial}_0 c)} X_1(b, d) \ . \end{aligned} \tag{7.23}$$

On the other hand, since $p^{(1)} X_0 \in \mathsf{Cat}_{\mathsf{hd}}$, $p^{(1)} X_0(\tilde{\partial}_0 a, \tilde{\partial}_0 c)$ is the one-element set. Therefore

$$\begin{aligned} &X_1(a, c) \times_{(p^{(1)} X_0)(\tilde{\partial}_0 a, \tilde{\partial}_0 c)} X_1(b, d) \\ &\cong X_1(a, c) \times X_1(b, d) \cong (X_1 \times_{X_0^d} X_1)(\hat{\mu}_2(a, b), \hat{\mu}_2(c, d)) \ . \end{aligned} \tag{7.24}$$

From (7.23) and (7.24) we conclude that $\hat{\mu}_2$ is a local equivalence. Further, by hypothesis ii), there is an equivalence of categories

$$p^{(1)}\hat{\mu}_2 : p^{(1)}(X_1 \times_{X_0} X_1) = p^{(1)}X_1 \times_{p^{(1)}X_0} p^{(1)}X_1$$

$$\xrightarrow{\sim} p^{(1)}X_1 \times_{(p^{(1)}X_0)^d} p^{(1)}X_1 = p^{(1)}(X_1 \times_{X_0^d} X_1)\ .$$

In conclusion, $\hat{\mu}_2$ is a 2-equivalence, as required.

□

The following proposition consists of two parts. Part a) stipulates that if $X \in \mathsf{Cat}^{\mathsf{n}}_{\mathsf{wg}}$ then $X_k^{\{2\}} \in \mathsf{Cat}^{\mathsf{n-1}}_{\mathsf{wg}}$ for all $k \geq 2$, where $X_k^{\{2\}}$ is as in Definition 7.2.4. This result will lead to the existence of the functor $q^{(n-1)} : \mathsf{Cat}^{\mathsf{n}}_{\mathsf{wg}} \to \mathsf{Cat}^{\mathsf{n-1}}_{\mathsf{wg}}$, which is used throughout this work. Part b) of Proposition 7.2.8 establishes an important criterion for an n-fold category to be weakly used crucially in Chap. 8.

The two parts in Proposition 7.2.8 may at first look unrelated, however a careful inspection of the proof shows that they are closely intertwined. Namely, the proof is by induction on n, the base of the induction being Lemmas 7.2.6 and 7.2.7. For the inductive step, the idea is to prove a) at step n by using b) at step $(n-1)$ and to prove b) at step n by using a) at step $(n-1)$.

More precisely, for a), given $X \in \mathsf{Cat}^{\mathsf{n}}_{\mathsf{wg}}$, we show that $X_k^{\{2\}} \in \mathsf{Cat}^{\mathsf{n-1}}$ satisfies hypotheses b) i) and b) ii), so by induction hypothesis b) $X_k^{\{2\}} \in \mathsf{Cat}^{\mathsf{n-1}}_{\mathsf{wg}}$, proving the inductive step for a). For b), given $X \in \mathsf{Cat}^{\mathsf{n}}$ such that b) i) and b) ii) hold, to show that the induced Segal maps of X are $(n-1)$-equivalences, we check that $X_k^{\{2\}} \in \mathsf{Cat}^{\mathsf{n-1}}$ satisfies inductive hypothesis b) so that $X_k^{\{2\}} \in \mathsf{Cat}^{\mathsf{n-1}}_{\mathsf{wg}}$. To check that hypothesis b) ii) holds for $X_k^{\{2\}}$ we note that

$$p^{(n-2)}X_k^{\{2\}} = (p^{(n-1)}X)_k^{\{2\}}\ . \tag{7.25}$$

Hence, by inductive hypothesis a) applied to $p^{(n-1)}X \in \mathsf{Cat}^{\mathsf{n-1}}_{\mathsf{wg}}$ we conclude from (7.25) that $p^{(n-2)}X_k^{\{2\}} \in \mathsf{Cat}^{\mathsf{n-2}}_{\mathsf{wg}}$, showing that $X_k^{\{2\}}$ satisfies inductive hypothesis b) ii).

Proposition 7.2.8

a) For each $X \in \mathsf{Cat}^{\mathsf{n}}_{\mathsf{wg}}$ and $k \geq 0$, $X_k^{\{2\}} \in \mathsf{Cat}^{\mathsf{n-1}}_{\mathsf{wg}}$.
b) Let $X \in \mathsf{Cat}^{\mathsf{n}}$ be such that

i) for all $1 \leq r < n$ the $(n-1)$-fold category $X_0^{\{r\}}$ is a levelwise equivalence relation,
ii) $p^{(n-1)}X \in \mathsf{Cat}^{\mathsf{n-1}}_{\mathsf{wg}}$.

Then $X \in \mathsf{Cat}^{\mathsf{n}}_{\mathsf{wg}}$.

Proof By induction on n. For $n = 2, 3$ see Lemmas 7.2.6 and 7.2.7. Suppose, inductively, that it holds for $(n-1)$.

a) Since $X \in \mathsf{Cat}^n$ $X_k^{\{2\}} \in \mathsf{Cat}^{n-1}$; we show that $X_k^{\{2\}}$ satisfies inductive hypothesis b) and thus conclude that $X_k^{\{2\}} \in \mathsf{Cat}^{n-1}_{\mathsf{wg}}$.

   Note that, for each $1 \le r < n-1$, $(X_k^{\{2\}})_0^{\{r\}} = (X_0^{\{r+1\}})_k^{\{2\}}$. Since, by Lemma 7.2.5, $X_0^{\{r+1\}}$ is a levelwise equivalence relation, the same holds for $(X_0^{\{r+1\}})_k^{\{2\}}$. Thus condition i) in inductive hypothesis b) holds for $X_k^{\{2\}}$. To show that condition ii) holds, note that

$$p^{(n-2)} X_k^{\{2\}} = (p^{(n-1)} X)_k^{\{2\}} \ . \tag{7.26}$$

   In fact, for all $(r_1, \ldots, r_{n-2}) \in \Delta^{n-2^{op}}$,

$$(p^{(n-1)} X_k^{\{2\}})_{r_1 \ldots r_{n-2}} = pX_{r_1 k r_2 \ldots r_{n-2}} = ((p^{(n-1)} X)_k^{\{2\}})_{r_1 \ldots r_{n-2}} \ .$$

   Since this holds for all $r_1, \ldots, r_{n-2}$, (7.26) follows.

   By induction hypothesis a) applied to $p^{(n-1)}X$, $(p^{(n-1)}X)_k^{\{2\}} \in \mathsf{Cat}^{n-2}_{\mathsf{wg}}$. Therefore (7.26) means that $X_k^{\{2\}} \in \mathsf{Cat}^{n-1}$ satisfies condition ii) in inductive hypothesis b). Thus we conclude that $X_k^{\{2\}} \in \mathsf{Cat}^{n-1}_{\mathsf{wg}}$, proving a).

b) Suppose, inductively, that the statement holds for $n-1$ and let X be as in the hypothesis. For each $s \ge 0$ consider $X_s \in \mathsf{Cat}^{n-1}$.

   Since, for each $1 \le r < n-1$, $(X_s)_0^{\{r\}} = (X_0^{\{r+1\}})_s$ and by hypothesis $X_0^{\{r+1\}}$ is a levelwise equivalence relation, the same holds for $(X_s)_0^{\{r\}}$; that is, $X_s \in \mathsf{Cat}^{n-1}$ satisfies inductive hypothesis b) i). Also, since X satisfies b) ii) and $(p^{(n-1)}X)_s = p^{(n-2)}X_s \in \mathsf{Cat}^{n-2}_{\mathsf{wg}}$, it follows that X_s satisfies inductive hypothesis b) ii). We conclude by induction that $X_s \in \mathsf{Cat}^{n-1}_{\mathsf{wg}}$.

   Since, by hypothesis b) ii), $p^{(n-1)}X \in \mathsf{Cat}^{n-1}_{\mathsf{wg}}$, to prove that $X \in \mathsf{Cat}^{n}_{\mathsf{wg}}$ it remains to prove that the induced Segal maps

$$\hat{\mu}_s : X_1 \times_{X_0} \overset{s}{\cdots} \times_{X_0} X_1 \to X_1 \times_{X_0^d} \overset{s}{\cdots} \times_{X_0^d} X_1$$

   are $(n-1)$-equivalences for all $s \ge 2$. We prove this for $s = 2$, the case $s > 2$ being similar. We claim that $X_k^{\{2\}} \in \mathsf{Cat}^{n-1}$ satisfies inductive hypothesis b).

   In fact, $(X_k^{\{2\}})_0^{\{r\}} = (X_0^{\{r+1\}})_k^{\{2\}}$ is a levelwise equivalence relation since X satisfies hypothesis b) i). Since $p^{(n-1)}X \in \mathsf{Cat}^{n-1}_{\mathsf{wg}}$, we obtain by a) applied to $p^{(n-1)}X$ that $p^{(n-2)}X_k^{\{2\}} = (p^{(n-1)}X)_k^{\{2\}} \in \mathsf{Cat}^{n-2}_{\mathsf{wg}}$. Thus $X_k^{\{2\}}$ also satisfies inductive hypothesis b) ii) and we conclude that $X_k^{\{2\}} \in \mathsf{Cat}^{n-1}_{\mathsf{wg}}$. It follows that the induced Segal map

$$X_{1k} \times_{X_{0k}} X_{1k} \to X_{1k} \times_{X_{0k}^d} X_{1k} = X_{1k} \times_{(p^{(1)}X_0)_k} X_{1k} \tag{7.27}$$

is an $(n-2)$-equivalence. Since $p^{(n-1)}X \in \mathsf{Cat}^{\mathsf{n-1}}_{\mathsf{wg}}$, using Remark 6.3.9 and the fact that $(p^{(n-1)}X)^d_{s0} = X^d_{s0}$ we obtain

$$(X_{10} \times_{X_{00}} X_{10})^d = X^d_{10} \times_{X^d_{00}} X^d_{10} \, .$$

Let $(a,b),(c,d) \in (X_{10} \times_{X_{00}} X_{10})^d = X^d_{10} \times_{X^d_{00}} X^d_{10}$. By (7.27) there is an $(n-2)$-equivalence

$$\begin{aligned}&(X_1 \times_{X_0} X_1)((a,b),(c,d)) = X_1(a,c) \times_{X_0(\partial_0 a, \partial_0 c)} X_1(b,d)\\ &\to X_1(a,c) \times_{(p^{(1)}X_0)(\tilde{\partial}_0 a, \tilde{\partial}_0 c)} X_1(b,d) \, .\end{aligned} \tag{7.28}$$

On the other hand, $p^{(1)}X_0 \in \mathsf{Cat}_{\mathsf{hd}}$ is an equivalence relation, therefore

$$p^{(1)}X_0(\tilde{\partial}_0 a, \tilde{\partial}_0 c)$$

is the one-element set. It follows that

$$\begin{aligned}&X_1(a,c) \times_{(p^{(1)}X_0)(\tilde{\partial}_0 a, \tilde{\partial}_0 c)} X_1(b,d)\\ &\cong X_1(a,c) \times X_1(b,d) \cong (X_1 \times_{X^d_0} X_1)(\hat{\mu}_2(a,b), \hat{\mu}_2(c,d)) \, .\end{aligned} \tag{7.29}$$

Thus (7.28) and (7.29) imply that $\hat{\mu}_2$ is a local $(n-2)$-equivalence.

To show that $\hat{\mu}_2$ is a $(n-1)$-equivalence it remains to prove that $p^{(n-2)}\hat{\mu}_2$ is an $(n-2)$-equivalence. Since from above, $p^{(n-1)}X \in \mathsf{Cat}^{\mathsf{n-1}}_{\mathsf{wg}}$, we have

$$\begin{aligned}&p^{(n-2)}\hat{\mu}_2 : p^{(n-2)}(X_1 \times_{X_0} X_1) \cong p^{(n-2)}X_1 \times_{p^{(n-2)}X_0} p^{(n-2)}X_1\\ &\to p^{(n-2)}X_1 \times_{(p^{(n-2)}X_0)^d} p^{(n-2)}X_1 = p^{(n-2)}(X_1 \times_{X^d_0} X_1)\end{aligned}$$

is an $(n-2)$-equivalence, as required.

□

In the rest of this chapter we prove some consequences of Proposition 7.2.8. The first one is a property that will be used in the proof of Corollary 9.1.6.

Corollary 7.2.9 *Let $f : X \to Y$ be a morphism in $\mathsf{Cat}^{\mathsf{n}}$ such that $f_{\underline{k}}$ is an equivalence of categories for all $\underline{k} \in \Delta^{n-1^{op}}$. Then*

a) If $X \in \mathsf{Cat}^{\mathsf{n}}_{\mathsf{wg}}$, then $Y \in \mathsf{Cat}^{\mathsf{n}}_{\mathsf{wg}}$. If $Y \in \mathsf{Cat}^{\mathsf{n}}_{\mathsf{wg}}$, then $X \in \mathsf{Cat}^{\mathsf{n}}_{\mathsf{wg}}$.
b) If $X \in \mathsf{Cat}^{\mathsf{n}}_{\mathsf{hd}}$, then $Y \in \mathsf{Cat}^{\mathsf{n}}_{\mathsf{hd}}$. If $Y \in \mathsf{Cat}^{\mathsf{n}}_{\mathsf{hd}}$, then $X \in \mathsf{Cat}^{\mathsf{n}}_{\mathsf{hd}}$.

Proof We proceed by induction on n.

When $n = 2$ let $X \in \mathsf{Cat}^2_{\mathsf{wg}}$. Since $X_0 \simeq Y_0$ and $X_0 \in \mathsf{Cat}_{\mathsf{hd}}$, also $Y_0 \in \mathsf{Cat}_{\mathsf{hd}}$ and $p^{(1)}Y \cong p^{(1)}X \in \mathsf{Cat}$. Thus by Lemma 7.2.6, $Y \in \mathsf{Cat}^2_{\mathsf{wg}}$. If $X \in \mathsf{Cat}^2_{\mathsf{hd}}$, in particular $X \in \mathsf{Cat}^2_{\mathsf{wg}}$, so by a) $Y \in \mathsf{Cat}^2_{\mathsf{wg}}$; but we also have $X_1 \simeq Y_1$ so, since

$X_1 \in \mathsf{Cat}_{\mathsf{hd}}$, $Y_1 \in \mathsf{Cat}_{\mathsf{hd}}$. Further, $p^{(1)}Y \cong p^{(1)}X \in \mathsf{Cat}_{\mathsf{hd}}$ (as $X \in \mathsf{Cat}^2_{\mathsf{hd}}$). Thus the hypotheses of Corollary 7.2.2 are satisfied for Y and we conclude that $Y \in \mathsf{Cat}^2_{\mathsf{hd}}$.

Suppose, inductively, that the corollary holds for $1 \leq k \leq n-1$.

a) We verify that $Y \in \mathsf{Cat}^{\mathsf{n}}$ satisfies the hypotheses of Proposition 7.2.8 b). Since $X \in \mathsf{Cat}^{\mathsf{n}}_{\mathsf{wg}}$, by Lemma 7.2.5, $X_0^{\{r\}}$ is a levelwise equivalence relation. Since $(X_0^{\{r\}})_{\underline{k}} \simeq (Y_0^{\{r\}})_{\underline{k}}$ for all $\underline{k} \in \Delta^{n-2^{op}}$, it follows that $Y_0^{\{r\}}$ is also a levelwise equivalence relation. So hypothesis b) i) of Proposition 7.2.8 is satisfied for Y.

   Hypothesis b) ii) of Proposition 7.2.8 also holds because, since $f_{\underline{k}}$ is an equivalence of categories for each $\underline{k}$, $pf_{\underline{k}}$ is an isomorphism, so that

$$p^{(n-1)}Y \cong p^{(n-1)}X \in \mathsf{Cat}^{\mathsf{n-1}}_{\mathsf{wg}} .$$

   We conclude by Proposition 7.2.8 b) that $Y \in \mathsf{Cat}^{\mathsf{n-1}}_{\mathsf{wg}}$. The other case is similar.

b) If $X \in \mathsf{Cat}^{\mathsf{n}}_{\mathsf{hd}}$, in particular $X \in \mathsf{Cat}^{\mathsf{n}}_{\mathsf{wg}}$, so by a) $Y \in \mathsf{Cat}^{\mathsf{n}}_{\mathsf{wg}}$. In addition, $X_1 \in \mathsf{Cat}^{\mathsf{n-1}}_{\mathsf{hd}}$ (since $X \in \mathsf{Cat}^{\mathsf{n}}_{\mathsf{hd}}$), so by inductive hypothesis b) applied to $f_1 : X_1 \to Y_1$, we conclude that $Y_1 \in \mathsf{Cat}^{\mathsf{n-1}}_{\mathsf{hd}}$. Since $f_{\underline{k}}$ is an equivalence of categories for all $\underline{k} \in \Delta^{n-1^{op}}$ we also have

$$p^{(n-1)}Y \cong p^{(n-1)}X \in \mathsf{Cat}^{\mathsf{n-1}}_{\mathsf{hd}} .$$

   Thus Y satisfies all the hypotheses of Corollary 7.2.2 and we conclude that $Y \in \mathsf{Cat}^{\mathsf{n}}_{\mathsf{hd}}$. The other case is similar.

□

We next use Proposition 7.2.8 to show the existence of the functor $q^{(n-1)} : \mathsf{Cat}^{\mathsf{n}}_{\mathsf{wg}} \to \mathsf{Cat}^{\mathsf{n-1}}_{\mathsf{wg}}$, which will be used throughout this work. We first need a preliminary lemma and one of its consequences.

Lemma 7.2.10 *Let* $X \in \mathsf{Ta}^{\mathsf{n}}_{\mathsf{wg}}$ *be such that*

a) $X_s \in \mathsf{Cat}^{\mathsf{n-1}}_{\mathsf{wg}}$ *for all* $s \geq 0$.
b) $X_s \cong X_1 \times_{X_0} \overset{s}{\cdots} \times_{X_0} X_1$ *for all* $s \geq 2$.
c) For all $s \geq 2$ *and* $0 \leq j \leq n-1$

$$p^{(j)}X_s \cong p^{(j)}(X_1 \times_{X_0} \overset{s}{\cdots} \times_{X_0} X_1) = p^{(j)}X_1 \times_{p^{(j)}X_0} \overset{s}{\cdots} \times_{p^{(j)}X_0} p^{(j)}X_1 \tag{7.30}$$

then $X \in \mathsf{Cat}^{\mathsf{n}}_{\mathsf{wg}}$.

Proof From a) and b), $X \in \mathsf{Cat}^{\mathsf{n}}$ and from c), $X \in \mathsf{Cat}^{\mathsf{n}}_{\mathsf{t}}$. Since also $X \in \mathsf{Ta}^{\mathsf{n}}_{\mathsf{wg}}$, by Definition 6.3.3, $X \in \mathsf{Cat}^{\mathsf{n}}_{\mathsf{wg}}$. □

We now deduce another useful criterion for a weakly globular Tamsamani n-category to be in $\mathsf{Cat}^n_{\mathsf{wg}}$.

Corollary 7.2.11 *Let $X \in \mathsf{Ta}^n_{\mathsf{wg}}$. Then $X \in \mathsf{Cat}^n_{\mathsf{wg}}$ if and only if*

a) $X_s \in \mathsf{Cat}^{n-1}_{\mathsf{wg}}$ for all $s \geq 0$.
b) $X^{\{2\}}_k \in \mathsf{Cat}^{n-1}_{\mathsf{wg}}$ for all $k \geq 1$.

Proof If $X \in \mathsf{Cat}^n_{\mathsf{wg}}$ then a) and b) hold by Lemma 6.3.5 and Proposition 7.2.8. Suppose conversely that $X \in \mathsf{Ta}^n_{\mathsf{wg}}$ satisfies a) and b). We show that X satisfies the hypotheses of Lemma 7.2.10 and therefore deduce that $X \in \mathsf{Cat}^n_{\mathsf{wg}}$. By hypothesis b), for all $k \geq 0$ and $s \geq 2$

$$X_{sk} \cong X_{1k} \times_{X_{0k}} \overset{s}{\cdots} \times_{X_{0k}} X_{1k} \; .$$

Therefore $X_s \cong X_1 \times_{X_0} \overset{s}{\cdots} \times_{X_0} X_1$ for $s \geq 2$ so hypothesis b) in Lemma 7.2.10 holds. Also by hypothesis b) and by Lemma 6.3.8 we have, for all $k \geq 0$, $s \geq 2$ and $0 \leq j \leq n$,

$$p^{(j)}(X_{1k} \times_{X_{0k}} \overset{s}{\cdots} \times_{X_{0k}} X_{1k}) \cong p^{(j)} X_{1k} \times_{p^{(j)} X_{0k}} \overset{s}{\cdots} \times_{p^{(j)} X_{0k}} p^{(j)} X_{1k} \; .$$

Since this holds for each k and $(p^{(j)} X_s)_k = p^{(j-1)} X_{sk}$, we deduce that, for each $0 \leq j \leq n-1$,

$$p^{(j)}(X_1 \times_{X_0} \overset{s}{\cdots} \times_{X_0} X_1) \cong p^{(j)} X_1 \times_{p^{(j)} X_0} \overset{s}{\cdots} \times_{p^{(j)} X_0} p^{(j)} X_1 \; ,$$

which is hypothesis c) in Lemma 7.2.10. We conclude that $X \in \mathsf{Cat}^n_{\mathsf{wg}}$, as required. □

Proposition 7.2.12 *The functor $q^{(n-1)} : \mathsf{Ta}^n_{\mathsf{wg}} \to \mathsf{Ta}^{n-1}_{\mathsf{wg}}$ restricts to a functor $q^{(n-1)} : \mathsf{Cat}^n_{\mathsf{wg}} \to \mathsf{Cat}^{n-1}_{\mathsf{wg}}$ and for each $X \in \mathsf{Cat}^n_{\mathsf{wg}}$ there is a map $\gamma^{(n-1)} : X \to q^{(n-1)} X$, natural in X.*

Proof By induction on n. When $n = 2$, $q^{(1)} X \in \mathsf{Cat}$. Suppose the statement holds for $(n-1)$ and let $X \in \mathsf{Cat}^n_{\mathsf{wg}}$. To prove that $q^{(n-1)} X \in \mathsf{Cat}^{n-1}_{\mathsf{wg}}$ we show that it satisfies the hypotheses of Corollary 7.2.11. For each $s \geq 0$, $(q^{(n-1)} X)_s = q^{(n-2)} X_s \in \mathsf{Cat}^{n-2}_{\mathsf{wg}}$, by the inductive hypothesis applied to X_s. Thus condition a) in Corollary 7.2.11 holds. Also, since $X^{\{2\}}_k \in \mathsf{Cat}^{n-1}_{\mathsf{wg}}$ (as $X \in \mathsf{Cat}^n_{\mathsf{wg}}$ using Proposition 7.2.8),

$$(q^{(n-1)} X)^{\{2\}}_k = q^{(n-2)} X^{\{2\}}_k \in \mathsf{Cat}^{n-2}_{\mathsf{wg}} \; .$$

Thus condition b) in Corollary 7.2.11 is satisfied and we conclude that $q^{(n-1)} X \in \mathsf{Cat}^{n-1}_{\mathsf{wg}}$, as required. The map $\gamma^{(n-1)}$ is as in Proposition 7.1.7. □

Remark 7.2.13 From Proposition 7.2.12, for each $0 \leq r < n$, the functor $q^{(r)}$: $\mathsf{Ta}^{\mathsf{n}}_{\mathsf{wg}} \to \mathsf{Ta}^{\mathsf{r}}_{\mathsf{wg}}$ restricts to a functor $q^{(r)} : \mathsf{Cat}^{\mathsf{n}}_{\mathsf{wg}} \to \mathsf{Cat}^{\mathsf{r}}_{\mathsf{wg}}$. It follows that for each $X \in \mathsf{Cat}^{\mathsf{n}}_{\mathsf{wg}}$ and $s \geq 2$ we have

$$q^{(r)}X_s \cong q^{(r)}(X_1 \times_{X_0} \overset{s}{\cdots} \times_{X_0} X_1) = q^{(r)}X_1 \times_{q^{(r)}X_0} \overset{s}{\cdots} \times_{q^{(r)}X_0} q^{(r)}X_1 \,. \tag{7.31}$$

We also have a map $\gamma^{(r)} : X \to q^{(r)}X$, natural in X.

7.2.3 A Geometric Interpretation

We now illustrate some consequences of Lemma 7.2.6 which will be used in Sect. 9.1 to define the category $\mathsf{LTa}^{\mathsf{n}}_{\mathsf{wg}}$. We also discuss a geometric interpretation implied by this. Although this geometric interpretation is interesting in its own right, it is not strictly needed for the rest of this work, so the second part of this section can be skipped at first reading.

Corollary 7.2.14 *Given* $X \in \mathsf{Cat}^{\mathsf{n}}_{\mathsf{wg}}$, $\underline{k} = (k_2, \dots, k_{n-1}) \in \Delta^{n-2^{op}}$ *and* $0 < r \leq n-1$, *let*

$$X^{\{r\}}(\text{-}, \underline{k}) : \Delta^{op} \to \mathsf{Cat}$$

be the functor associating to $k_1 \in \Delta^{op}$ *the category*

$$X^{\{r\}}(k_1, k_2, \dots, k_{n-1}) = X_{k_2,\dots,k_r,k_1,k_{r+1},\dots,k_{n-1}}$$

(see (2.3)*). Then* $X^{\{r\}}(\text{-}, \underline{k}) \in \mathsf{Cat}^2_{\mathsf{wg}}$.

Proof We show that $X^{\{r\}}(\text{-}, \underline{k})$ satisfies the hypotheses of Lemma 7.2.6. In fact, since $X \in \mathsf{Cat}^{\mathsf{n}}$, $X^{\{r\}}(\text{-}, \underline{k}) \in \mathsf{Cat}^2$. Also $X^{\{r\}}(0, \underline{k}) = X_{k_2,\dots,k_r,0,\dots,k_{n-1}} \in \mathsf{Cat}_{\mathsf{hd}}$, since $X_{k_2,\dots,k_r,0} \in \mathsf{Cat}^{\mathsf{n-r}}_{\mathsf{hd}}$. Since $p^{(n-1)}X \in \mathsf{Cat}^{\mathsf{n-1}}_{\mathsf{wg}}$, using Lemma 2.2.7 we obtain

$$p^{(1)}X^{\{r\}}(\text{-}, \underline{k}) = (p^{(n-1)}X^{\{r\}})(\text{-}, \underline{k}) = (p^{(n-1)}X)^{\{r\}}(\text{-}, \underline{k}) \in \mathsf{Cat} \,.$$

Therefore all the hypotheses of Lemma 7.2.6 are satisfied and we conclude that $X^{\{r\}}(\text{-}, \underline{k}) \in \mathsf{Cat}^{\mathsf{n-1}}_{\mathsf{wg}}$. □

Remark 7.2.15 Let $X \in \mathsf{Cat}^{\mathsf{n}}_{\mathsf{wg}}$, $\underline{k}$ and r as in Corollary 7.2.14. By definition of $\mathsf{Cat}^2_{\mathsf{wg}}$ it follows that the induced Segal maps of $X^{\{r\}}(\text{-}, \underline{k}) \in \mathsf{Cat}^2_{\mathsf{wg}}$ are equivalences of categories. That is, for each $k_1 \geq 2$ the maps

$$X^{\{r\}}(k_1, \underline{k}) \to X^{\{r\}}(1, \underline{k}) \times_{\{X^{\{r\}}(0,\underline{k})\}^d} \overset{k_1}{\cdots} \times_{\{X^{\{r\}}(0,\underline{k})\}^d} X^{\{r\}}(1, \underline{k}) \tag{7.32}$$

are equivalences of categories. Since

$$\{X^{\{r\}}(0,\underline{k})\}^d = p^{(0)}X^{\{r\}}(0,\underline{k}) = (p^{(n-2)}X_0^{\{r\}})(\underline{k})$$

and, for all $k_1 \in \Delta^{op}$

$$X^{\{r\}}(k_1,\underline{k}) = X_{k_1}^{\{r\}}(\underline{k}) \tag{7.33}$$

and (7.33) holds for each $\underline{k}$ we deduce that for each $k_1 \geq 2$ the induced Segal maps in $[\Delta^{n-2^{op}}, \mathsf{Cat}]$

$$X_{k_1}^{\{r\}} \to X_1^{\{r\}} \times_{p^{(n-2)}X_0^{\{r\}}} \overset{k_1}{\cdots} \times_{p^{(n-2)}X_0^{\{r\}}} X_1^{\{r\}}$$

are levelwise equivalences of categories. In Sect. 9.1.1 we will see how this condition motivates the definition of $\mathsf{LTa}^{\mathsf{n}}_{\mathsf{wg}}$.

In the rest of this section we discuss a geometric interpretation of Corollary 7.2.14. We first give some preliminary definitions, which hold in any n-fold category.

Definition 7.2.16 Let $\underline{s} \in \Delta^{n^{op}}$. We call $\underline{s}$ an orientation if $s_j \in \{0, 1\}$ for all $j = 1, \dots, n$ and we define $t(\underline{s}) = \sum_{j=1}^{n} s_j$.

Definition 7.2.17 Let $X \in \mathsf{Cat}^{\mathsf{n}}$ and let $\underline{s} \in \Delta^{n^{op}}$ be an orientation. An $(n, t(\underline{s}))$-hypercube in X with orientation $\underline{s}$ is an element of the set $X_{\underline{s}}$.

Definition 7.2.18 Let $0 \leq t \leq n$. We denote by Cube(n, t) the set of all (n, t)-hypercubes in X with orientations $\underline{s}$ such that $t(\underline{s}) = t$, that is,

$$\text{Cube}(n,t) = \bigcup_{\substack{\underline{s}\in\Delta^{n^{op}} \text{ orientation} \\ \sum_{j=1}^{n} s_j = t}} X_{\underline{s}}\ .$$

Definition 7.2.19 Let $X \in \mathsf{Cat}^{\mathsf{n}}$, $1 \leq i \leq n$ and let $\underline{s}$ be an orientation with $s_i = 1$ and $t(\underline{s}) = t$. Define

$$\underline{s}(0,i) = (s_1, \dots, s_{i-1}, 0, s_{i+1}, \dots, s_n)\ .$$

Note that $\underline{s}(0, i)$ is also an orientation, with $t(\underline{s}(0, i)) = t - 1$. Let $r \geq 2$. An (r, i)-string of (n, t)-hypercubes with orientations $\underline{s}$ is an element of

$$X_{\underline{s}} \times_{X_{\underline{s}(0,i)}} \overset{r}{\cdots} \times_{X_{\underline{s}(0,i)}} X_{\underline{s}} \cong X_{(s_1,\dots,s_{i-1},r,s_{i+1},\dots,s_n)}\ , \tag{7.34}$$

where the isomorphism in (7.34) follows by the Segal condition characterizing multinerves of n-fold categories (see Proposition 2.4.5).

Example 7.2.20 Let $X \in \mathsf{Cat}^2$. Then

$$\mathrm{Cube}(2, 0) = X_{00}$$

consists of the set of objects in the double category X. Also,

$$\mathrm{Cube}(2, 1) = X_{10} \cup X_{01}$$

consists of the set of arrows with orientations $(1, 0)$ and $(0, 1)$ which are respectively direction 1 and 2 in Figs. 2.2 and 2.3 on pages 37 and 38. Also

$$\mathrm{Cube}(2, 2) = X_{11}$$

is the set of squares.

If $r \geq 2$, an $(r, 1)$-string of $(2, 1)$-hypercubes is an element of

$$X_{10} \times_{X_{00}} \overset{r}{\cdots} \times_{X_{00}} X_{10} \cong X_{r0} \ ,$$

that is, a sequence of r composable arrows in direction 1. An $(r, 1)$-string of $(2, 2)$-hypercubes is an element of

$$X_{11} \times_{X_{01}} \overset{r}{\cdots} \times_{X_{01}} X_{11} \cong X_{r1} \ ,$$

that is, a sequence of r composable squares in direction 1.

Similarly, an $(r, 2)$-string of $(2, 1)$-hypercubes is an element of

$$X_{01} \times_{X_{00}} \overset{r}{\cdots} \times_{X_{00}} X_{01} \cong X_{0r}$$

and an $(r, 2)$-string of $(2, 2)$-hypercubes is an element of

$$X_{11} \times_{X_{10}} \overset{r}{\cdots} \times_{X_{10}} X_{11} \cong X_{1r} \ ,$$

with geometric interpretations similar to the above, but in direction 2 rather than direction 1.

Example 7.2.21 Let $X \in \mathsf{Cat}^3$. Then

$$\mathrm{Cube}(3, 0) = X_{000}$$

is the set of objects of the 3-fold category X. Also,

$$\mathrm{Cube}(3, 1) = X_{100} \cup X_{010} \cup X_{001}$$

is the set of edges in the three orientations (see Figs. 2.4 and 2.5 on pages 39 and 40). Similarly,

$$\text{Cube}(3, 2) = X_{110} \cup X_{101} \cup X_{011}$$

is the set of squares, and

$$\text{Cube}(3, 3) = X_{111}$$

is the set of cubes.

Given $r \geq 2$, an $(r, 1)$-string of $(3, 1)$-hypercubes is an element of

$$X_{100} \times_{X_{000}} \overset{r}{\cdots} \times_{X_{000}} X_{100} \cong X_{r00} ,$$

that is, a sequence of r-composable edges in direction 1. Similarly for $(r, 2)$-strings and $(r, 3)$-strings of $(3, 1)$-hypercubes.

An $(r, 1)$-string of $(3, 2)$-hypercubes is an element of

$$X_{110} \times_{X_{010}} \overset{r}{\cdots} \times_{X_{010}} X_{110} \cong X_{r10} ,$$

that is, a sequence of r composable squares in direction 1. Similarly for $(r, 2)$-strings and $(r, 3)$-strings of $(3, 2)$-hypercubes.

An $(r, 1)$-string of $(3, 3)$-hypercubes is an element of

$$X_{111} \times_{X_{011}} \overset{r}{\cdots} \times_{X_{011}} X_{111} \cong X_{r11} ,$$

that is, a sequence of r composable cubes in direction 1. Similarly for $(r, 2)$-strings and $(r, 3)$-strings of $(3, 3)$-hypercubes.

Let $X \in \mathsf{Cat}^n$, $\underline{s} \in \Delta^{n^{op}}$ be an orientation and $1 \leq i \leq n-1$. Let

$$s^{(i,n)}(t, q) = (s_1, \ldots, s_{i-1}, t, s_{i+1}, \ldots, s_{n-1}, q) \in \Delta^{n^{op}} .$$

Clearly $s^{(i,n)}(0, 0)$, $s^{(i,n)}(0, 1)$, $s^{(i,n)}(1, 0)$, $s^{(i,n)}(1, 1)$ are orientations and if we define

$$t = 1 + \sum_{\substack{j=1 \\ j \neq i,n}}^{n} s_j$$

we have

$$\begin{aligned}
t(s^{(i,n)}(0, 0)) &= t - 1 , \\
t(s^{(i,n)}(1, 0)) &= t(s^{(i,n)}(0, 1)) = t , \\
t(s^{(i,n)}(1, 1)) &= t + 1 .
\end{aligned}$$

Let $X \in \mathsf{Cat}^n$ and let $X(\underline{s}, i, n) \in [\Delta^{2^{op}}, \mathsf{Set}]$ with

$$X(\underline{s}, i, n)(v, w) = X_{(s_1,\dots,s_{i-1},v,s_{i+1},\dots,s_{n-1},w)} \; ,$$

then $X(\underline{s}, i, n)$ is the double nerve of a double category, which we still denote by $X(\underline{s}, i, n)$. The objects of this double category are

$$\{X(\underline{s}, i, n)\}_{00} = X_{\underline{s}^{(i,n)}(0,0)} = \{(n, t-1)\text{-hypercubes in } X \text{ with orientation } \underline{s}^{(i,n)}(0,0)\}.$$

The arrows in the double category are

$$\{X(\underline{s}, i, n)\}_{10} = X_{\underline{s}^{(i,n)}(1,0)} = \{(n, t)\text{-hypercubes in } X \text{ with orientation } \underline{s}^{(i,n)}(1,0)\},$$

and

$$\{X(\underline{s}, i, n)\}_{01} = X_{\underline{s}^{(i,n)}(0,1)} = \{(n, t)\text{-hypercubes in } X \text{ with orientation } \underline{s}^{(i,n)}(0,1)\}.$$

The squares in the double category are

$$\{X(\underline{s}, i, n)\}_{11} = X_{\underline{s}^{(i,n)}(1,1)} = \{(n, t+1)\text{-hypercubes in } X \text{ with orientation } \underline{s}^{(i,n)}(1,1)\} \; .$$

If $r \geq 2$, a sequence of r composable arrows in the double category in direction 1 is an element of

$$X_{\underline{s}^{(i,n)}(1,0)} \times_{X_{\underline{s}^{(i,n)}(0,0)}} \overset{r}{\cdots} \times_{X_{\underline{s}^{(i,n)}(0,0)}} X_{\underline{s}^{(i,n)}(1,0)} \; .$$

By the above geometric interpretation of arrows of $X(\underline{s}, i, n)$ it follows that this is an (r, i)-string of (n, t)-hypercubes in X with orientation $\underline{s}^{(i,n)}(1, 0)$ in the sense of Definition 7.2.19.

Suppose now that $X \in \mathsf{Cat}^n_{\mathsf{wg}}$. If we let $\underline{k} = (s_1, \dots, s_{n-1}) \in \Delta^{n-1^{op}}$ then

$$\begin{aligned}
(\underline{k}(0, i), 0) &= \underline{s}^{(i,n)}(0, 0) \; , \\
(\underline{k}(1, i), 0) &= \underline{s}^{(i,n)}(1, 0) \; , \\
(\underline{k}(0, i), 1) &= \underline{s}^{(i,n)}(0, 1) \; , \\
(\underline{k}(1, i), 1) &= \underline{s}^{(i,n)}(1, 1) \; .
\end{aligned}$$

Thus

$$\begin{aligned} X_{\underline{s}^{(i,n)}(0,0)} &= obj\, X_{\underline{k}(0,i)}\ , \\ X_{\underline{s}^{(i,n)}(1,0)} &= obj\, X_{\underline{k}(1,i)}\ , \\ X_{\underline{s}^{(i,n)}(0,1)} &= mor\, X_{\underline{k}(0,i)}\ , \\ X_{\underline{s}^{(i,n)}(1,1)} &= mor\, X_{\underline{k}(1,i)}. \end{aligned} \tag{7.35}$$

By Proposition 9.1.5, $X \in \mathsf{LTa}^n_{\mathsf{wg}}$, so by Definition 9.1.1 and Remark 9.1.2 there is an equivalence of categories for each $r \geq 2$

$$X_{\underline{k}(1,i)} \times_{X_{\underline{k}(0,i)}} \overset{r}{\cdots} \times_{X_{\underline{k}(0,i)}} X_{\underline{k}(1,i)} \to X_{\underline{k}(1,i)} \times_{X^d_{\underline{k}(0,i)}} \overset{r}{\cdots} \times_{X^d_{\underline{k}(0,i)}} X_{\underline{k}(1,i)}\ .$$

By (7.35) and the above this means that the double category $X(\underline{s}, i, n)$ is weakly globular. As discussed in Example 6.3.10 it follows that every staircase of length r of horizontal arrows in $X(\underline{s}, i, n)$ can be lifted to a string of r composable horizontal arrows by strings of vertically invertible squares. To give a geometric interpretation of this lifting condition for the weakly globular double category $X(\underline{s}, i, n)$ we introduce the following definition:

Definition 7.2.22 A staircase of length r of horizontal morphisms in $X(\underline{s}, i, n)$ is called an (r, i)-staircase of (n, t)-hypercubes in X with orientations $\underline{s}^{(i,n)}(1, 0)$.

Definition 7.2.22 means that the (n, t)-hypercubes are not composable in direction i, but are joined by strings of (n, t)-hypercubes with orientations $\underline{s}^{(i,n)}(0, 1)$.

From the geometric interpretation of arrows and squares in the double category $X(\underline{s}, i, n)$ given above, we deduce that the lifting condition in the weakly globular double category $X(\underline{s}, i, n)$ implies the following:

Proposition 7.2.23 *Given $X \in \mathsf{Cat}^n_{\mathsf{wg}}$, every (r, i)-staircase of (n, t)-hypercubes in X with orientations $\underline{s}^{(i,n)}(1, 0)$ can be lifted to an (r, i)-string of (n, t)-hypercubes in X via strings of $(n, t+1)$-hypercubes with orientations $\underline{s}^{(i,n)}(1, 1)$ which are invertible in direction n.*

We illustrate below the case $n = 3$.

Example 7.2.24 Let $X \in \mathsf{Cat}^3_{\mathsf{wg}}$, $r \geq 2$. An $(r, 1)$-staircase of $(3, 1)$-hypercubes with orientation (100) is an element of

$$X_{100} \times_{X^d_{000}} \overset{r}{\cdots} \times_{X^d_{000}} X_{100}.$$

This can be represented as a staircase
with edges in direction 1 and vertical sides in direction 3. The equivalence of categories

$$X_{100} \times_{X_{000}} \overset{r}{\cdots} \times_{X_{000}} X_{100} \simeq X_{100} \times_{X^d_{000}} \overset{r}{\cdots} \times_{X^d_{000}} X_{100}$$

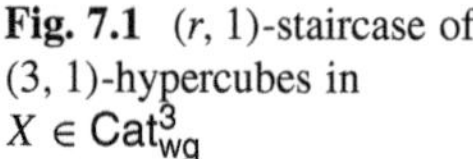

Fig. 7.1 $(r, 1)$-staircase of $(3, 1)$-hypercubes in $X \in \mathsf{Cat}^3_{\mathsf{wg}}$

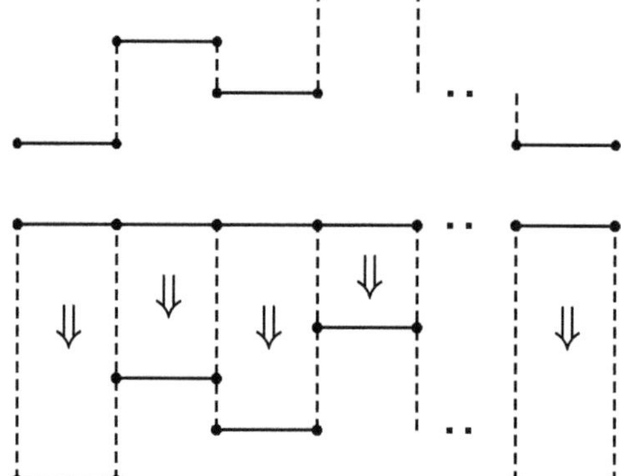

Fig. 7.2 Lifting condition for the staircase in Fig. 7.1

means that this staircase can be lifted to a string of r composable arrows in direction 1 through strings of vertically invertible squares.

An $(r, 2)$-staircase of $(3, 1)$-hypercubes with orientation (010) is an element of

$$X_{010} \times_{X^d_{000}} \overset{r}{\cdots} \times_{X^d_{000}} X_{010}\ .$$

This can be pictured as a staircase like the one in Fig. 7.1 but with horizontal edges in direction 2 and vertical edges in direction 3. The lifting condition is similar to Fig. 7.2, but now the vertical squares have orientation in directions 2 and 3 rather than 1 and 3.

An $(r, 1)$-staircase of $(3, 2)$-hypercubes with orientation (110) is an element of

$$X_{110} \times_{X^d_{010}} \overset{r}{\cdots} \times_{X^d_{010}} X_{110}\ .$$

This can be pictured as in Fig. 7.3 on page 161 (where for simplicity we choose $r = 3$).

The equivalence of categories

$$X_{110} \times_{X_{010}} \overset{r}{\cdots} \times_{X_{010}} X_{110} \simeq X_{110} \times_{X^d_{010}} \overset{r}{\cdots} \times_{X^d_{010}} X_{110}$$

means that this staircase can be lifted to a sequence of r composable squares in direction 1 via strings of cubes which are vertically invertible. See Fig. 7.4 on page 161 (when $r = 3$).

An $(r, 2)$-staircase of $(3, 2)$-hypercubes with orientation (110) is an element of

$$X_{110} \times_{X^d_{100}} \overset{r}{\cdots} \times_{X^d_{100}} X_{110}\ .$$

The geometric picture is obtained by rotating Fig. 7.3 by 90° around the vertical axis, and the lifting condition is similar to Fig. 7.4 in the rotated picture.

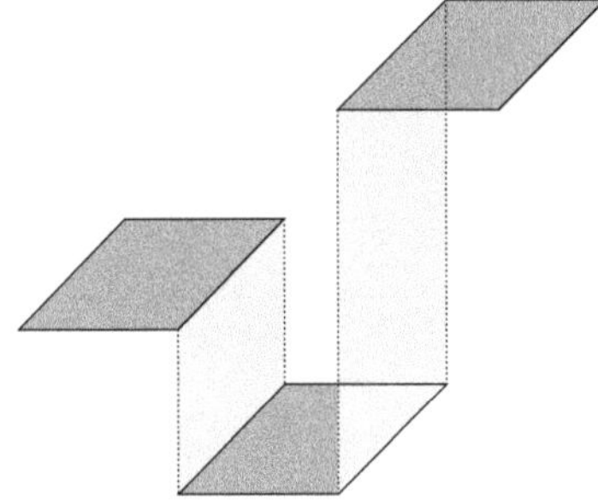

Fig. 7.3 A (3, 1)-staircase of (3, 2)-hypercubes in $X \in \mathsf{Cat}^3_{\mathsf{wg}}$ with orientation (1, 1, 0)

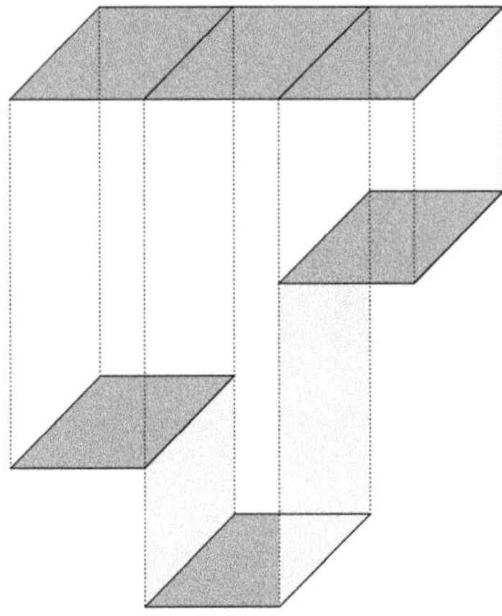

Fig. 7.4 Lifting condition for the staircase in Fig. 7.3

Chapter 8
Pseudo-Functors Modelling Higher Structures

Abstract In this chapter we introduce another main structure of this work: the category $\mathsf{SegPs}[\Delta^{n-1^{op}}, \mathsf{Cat}]$ of Segalic pseudo-functors. This category is a full subcategory of the category $\mathsf{Ps}[\Delta^{n-1^{op}}, \mathsf{Cat}]$ of pseudo-functors. The main result of this chapter is that the classical strictification functor form pseudo-functors $\mathsf{Ps}[\Delta^{n-1^{op}}, \mathsf{Cat}]$ to strict functors $[\Delta^{n-1^{op}}, \mathsf{Cat}]$ restrict to a functor from Segalic pseudo-functors to weakly globular n-fold categories. This result is crucially used in Chap. 10 to construct the rigidification functor Q_n from weakly globular Tamsamani n-categories to weakly globular n-fold categories.

In this chapter we connect the category $\mathsf{Cat}^{\mathsf{n}}_{\mathsf{wg}}$ of weakly globular n-fold categories introduced in Chap. 6 to the notion of pseudo-functor. Our main result in this chapter is that weakly globular n-fold categories arise as the strictification of a special class of pseudo-functors, which we call Segalic pseudo-functors. This result is used in Chaps. 10 and 12 to build the comparison functors between $\mathsf{Cat}^{\mathsf{n}}_{\mathsf{wg}}$ and Tamsamani n-categories and show that they are suitably equivalent after localization.

The classical theory of strictification of pseudo-algebras [81, 106] affords the strictification functor

$$St : \mathsf{Ps}[\Delta^{n-1^{op}}, \mathsf{Cat}] \to [\Delta^{n-1^{op}}, \mathsf{Cat}] \tag{8.1}$$

left adjoint to the inclusion.

The coherence axioms in a pseudo-functor are reminiscent of the coherence data for the compositions of higher cells in a weak higher category. So it is natural to ask if a subcategory of pseudo-functors can model, in a suitable sense, higher structures. In this chapter we positively answer this question by introducing a subcategory

$$\mathsf{SegPs}[\Delta^{n-1^{op}}, \mathsf{Cat}] \subset \mathsf{Ps}[\Delta^{n-1^{op}}, \mathsf{Cat}]$$

S. Paoli, *Simplicial Methods for Higher Categories*, Algebra and Applications 26,
https://doi.org/10.1007/978-3-030-05674-2_8

of *Segalic pseudo-functors*. Our main result, Theorem 8.2.3, is that the strictification functor (8.1) restricts to a functor

$$St \ : \mathsf{SegPs}[\Delta^{n-1^{op}}, \mathsf{Cat}\,] \to \mathsf{Cat}^{\mathsf{n}}_{\mathsf{wg}}\ .$$

In Chap. 9 we will functorially associate to a weakly globular Tamsamani n-category a Segalic pseudo-functor and build the rigidification functor Q_n from weakly globular Tamsamani n-categories to weakly globular n-fold categories as a composite

$$Q_n : \mathsf{Ta}^{\mathsf{n}}_{\mathsf{wg}} \to \mathsf{SegPs}[\Delta^{n-1^{op}}, \mathsf{Cat}\,] \xrightarrow{St} \mathsf{Cat}^{\mathsf{n}}_{\mathsf{wg}}\ .$$

This chapter is organized as follows. In Sect. 8.1 we define Segalic pseudo-functors, in Sect. 8.2 we discuss their strictification. We show in Proposition 8.2.1 that an n-fold category that is levelwise equivalent (via a pseudo-natural transformation) to a Segalic pseudo-functor satisfies the hypothesis of Proposition 7.2.8 b) and is therefore weakly globular.

In Theorem 8.2.3, using the properties of the monad corresponding to Segalic pseudo-functors proved in Lemma 8.2.2, we show that the strictification of a Segalic pseudo-functor is an n-fold category and that it satisfies the hypotheses of Proposition 8.2.1. We therefore conclude that the strictification functor restricts to the functor St from Segalic pseudo-functors to weakly globular n-fold categories.

8.1 The Definition of a Segalic Pseudo-Functor

In this section we give the definition of the category of Segalic pseudo-functors as a full subcategory of the category

$$\mathsf{Ps}[\Delta^{n^{op}}, \mathsf{Cat}\,]$$

of pseudo-functors and pseudo-natural transformations [32], recalled in Sect. 4.2.2.

8.1.1 Notational Conventions for Segalic Pseudo-Functors

We use the following notational conventions for $\mathsf{Ps}[\Delta^{n^{op}}, \mathsf{Cat}\,]$, similar to the ones established in Chap. 2 for $[\Delta^{n^{op}}, \mathsf{Set}]$.

Notational Convention 8.1.1

i) We identify $[\Delta^{n^{op}}, \mathsf{Set}]$ with the corresponding full subcategory of $\mathsf{Ps}[\Delta^{n^{op}}, \mathsf{Cat}\,]$ of levelwise discrete objects.

ii) We denote by

$$p^{(n)}: \mathsf{Ps}[\Delta^{n^{op}}, \mathsf{Cat}] \to [\Delta^{n^{op}}, \mathsf{Set}]$$

the functor $(p^{(n)}X)_{\underline{k}} = pX_{\underline{k}}$ for $X \in \mathsf{Ps}[\Delta^{n^{op}}, \mathsf{Cat}]$, $\underline{k} \in \Delta^{n^{op}}$.

iii) Given $X \in \mathsf{Ps}[\Delta^{n^{op}}, \mathsf{Cat}]$ and $(k_1, \dots, k_r) \in \Delta^{r^{op}}$, we denote by $X_{k_1 \dots k_r} \in \mathsf{Ps}[\Delta^{n-r^{op}}, \mathsf{Cat}]$ the pseudo-functor obtained by fixing the first r indices.

iv) Given $\alpha \in \Sigma_n$, let

$$(\text{-})^{\alpha}: \mathsf{Ps}[\Delta^{n^{op}}, \mathsf{Cat}] \to \mathsf{Ps}[\Delta^{n^{op}}, \mathsf{Cat}]$$

be given by

$$(X^{\alpha})_{\underline{k}} = X_{\alpha(\underline{k})} \text{ for } X \in \mathsf{Ps}[\Delta^{n^{op}}, \mathsf{Cat}], \quad \underline{k} \in \Delta^{n^{op}}, \quad \alpha(\underline{k}) = (k_{\alpha(1)}, \dots, k_{\alpha(n)}) \,.$$

In particular, for each $1 \le r \le n$,

$$X^{\{r\}}_{k_1 \dots k_n} = \begin{cases} X_{k_2 k_3 \dots k_r k_1 k_{r+1} \dots k_n}, & \text{if } \quad 1 \le r < n \,, \\ X_{k_2 k_3 \dots k_{n-1} k_1}, & \text{if } \quad r = n \,. \end{cases}$$

Note that $X^{\{1\}} = X$.

8.1.2 The Idea of a Segalic Pseudo-Functor

A topological intuition about an object of $\mathsf{Ps}[\Delta^{n^{op}}, \mathsf{Cat}]$ is that it consists of categories $X_{\underline{k}}$ for each object $\underline{k}$ of $\Delta^{n^{op}}$ together with multi-simplicial face and degeneracy maps satisfying the multi-simplicial identities not as equalities but as isomorphisms, and these isomorphisms satisfy coherence axioms. Guided by this intuition, we generalize to certain pseudo-functors the multi-simplicial notion of a Segal map.

For this purpose, consider a functor $H \in [\Delta^{n^{op}}, \mathsf{Cat}]$. By Remark 2.1.8, for each $1 \le i \le n$ there are Segal maps in $[\Delta^{n-1^{op}}, \mathsf{Cat}]$ for all $r \ge 2$

$$H^{\{i\}}_r \to H^{\{i\}}_1 \times_{H^{\{i\}}_0} \overset{r}{\cdots} \times_{H^{\{i\}}_0} H^{\{i\}}_1$$

identified by the commuting diagram

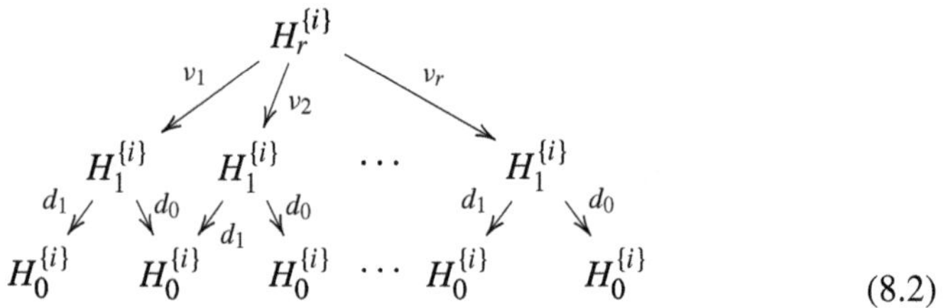

(8.2)

If H is not a functor but a pseudo-functor $H \in \mathsf{Ps}[\Delta^{n^{op}}, \mathsf{Cat}]$, diagram (8.2) no longer commutes but pseudo-commutes and thus we can no longer define Segal maps. However, if $H_0^{\{i\}}$ is levelwise discrete (or equivalently an object of $[\Delta^{n^{op}}, \mathsf{Set}] \subset \mathsf{Ps}[\Delta^{n^{op}}, \mathsf{Cat}]$), then diagram (8.2) commutes and therefore we can define Segal maps for H.

In the definition of a Segalic pseudo-functor we require the above discreteness conditions to be satisfied to be able to define Segal maps and then we require all Segal maps to be isomorphisms.

The last condition in the definition of a Segalic pseudo-functor is the existence of a truncation functor. We require the functor

$$p^{(n)} : \mathsf{Ps}[\Delta^{n^{op}}, \mathsf{Cat}] \to [\Delta^{n^{op}}, \mathsf{Set}]$$

to produce a weakly globular n-fold category, that is,

$$p^{(n)} : \mathsf{SegPs}[\Delta^{n^{op}}, \mathsf{Cat}] \to \mathsf{Cat}^{\mathsf{n}}_{\mathsf{wg}} .$$

The notion of a weakly globular n-fold category was discussed informally in Sect. 6.3.1. The category of Segalic pseudo-functors is essential in building the functor Tr_n and thus the rigidification functor Q_n, discussed informally in Sects. 10.1.1 and 10.2.1 respectively.

8.1.3 The Formal Definition of a Segalic Pseudo-Functor

Definition 8.1.2 We define the subcategory $\mathsf{SegPs}[\Delta^{n^{op}}, \mathsf{Cat}]$ of $\mathsf{Ps}[\Delta^{n^{op}}, \mathsf{Cat}]$ as follows:

For $n = 1$, $H \in \mathsf{SegPs}[\Delta^{^{op}}, \mathsf{Cat}]$ if H_0 is discrete and the Segal maps are isomorphisms: that is, for all $r \geq 2$

$$H_r \cong H_1 \times_{H_0} \overset{r}{\cdots} \times_{H_0} H_1 .$$

Note that, since p commutes with pullbacks over discrete objects, there is a functor

$$p^{(1)} : \mathsf{SegPs}[\Delta^{op}, \mathsf{Cat}\,] \to \mathsf{Cat}\ ,$$
$$(p^{(1)}X)_k = pX_k\ .$$

When $n > 1$, $\mathsf{SegPs}[\Delta^{n^{op}}, \mathsf{Cat}\,]$ is the full subcategory of $\mathsf{Ps}[\Delta^{n^{op}}, \mathsf{Cat}\,]$ whose objects H satisfy the following:

a) Discreteness condition: for each $1 \leq i \leq n$ the pseudo-functor $H_0^{\{i\}} \in \mathsf{Ps}[\Delta^{n-1^{op}}, \mathsf{Cat}\,]$ is levelwise discrete or equivalently it is an object of $[\Delta^{n-1^{op}}, \mathsf{Set}]$.
b) Segal maps condition: for each $1 \leq i \leq n$ and $r \geq 2$ the Segal maps

$$H_r^{\{i\}} \to H_1^{\{i\}} \times_{H_0^{\{i\}}} \overset{r}{\cdots} \times_{H_0^{\{i\}}} H_1^{\{i\}}$$

are isomorphisms for all $r \geq 2$.
c) The functor $p^{(n)} : \mathsf{Ps}[\Delta^{n^{op}}, \mathsf{Cat}\,] \to [\Delta^{n^{op}}, \mathsf{Set}]$ restricts to a functor

$$p^{(n)} : \mathsf{SegPs}[\Delta^{n^{op}}, \mathsf{Cat}\,] \to \mathsf{Cat}^{\mathsf{n}}_{\mathsf{wg}}\ .$$

Remark 8.1.3 Given $\underline{k} = (k_1, \ldots, k_n) \in \Delta^{n^{op}}$ and $1 \leq i \leq n$ we define

$$\begin{aligned} \underline{k}(1,i) &= (k_1, \ldots, k_{i-1}, 1, k_{i+1}, \ldots, k_n) \in \Delta^{n^{op}}\ , \\ \underline{k}(0,i) &= (k_1, \ldots, k_{i-1}, 0, k_{i+1}, \ldots, k_n) \in \Delta^{n^{op}}\ . \end{aligned} \tag{8.3}$$

Then condition b) in the definition of a pseudo-functor H is equivalent to asking that for each $\underline{k} = (k_1, \ldots, k_n) \in \Delta^{n^{op}}$ with $k_i \geq 2$, $1 \leq i \leq n$, the Segal maps

$$H_{\underline{k}} \to H_{\underline{k}(1,i)} \times_{H_{\underline{k}(0,i)}} \overset{k_i}{\cdots} \times_{H_{\underline{k}(0,i)}} H_{\underline{k}(1,i)}$$

are isomorphisms.

Lemma 8.1.4 *Let $X \in \mathsf{SegPs}[\Delta^{n^{op}}, \mathsf{Cat}\,]$, $n \geq 2$. Then for each $j \geq 0$*

$$X_j \in \mathsf{SegPs}[\Delta^{n-1^{op}}, \mathsf{Cat}\,].$$

Proof For each $1 \leq i \leq n-2$, $j \in \Delta^{op}$

$$X_0^{\{i+1\}} = (X_j)_0^{\{i\}}$$

is discrete since $X \in \mathsf{SegPs}[\Delta^{n^{op}}, \mathsf{Cat}]$; further, by hypothesis there are isomorphisms:

$$\begin{aligned}(X_j)_r^{\{i\}} &\cong X_r^{\{i+1\}} \cong X_1^{\{i+1\}} \times_{X_0^{\{i+1\}}} \overset{r}{\cdots} \times_{X_0^{\{i+1\}}} X_1^{\{i+1\}} \\ &\cong (X_j)_1^{\{i\}} \times_{(X_j)_0^{\{i\}}} \overset{r}{\cdots} \times_{(X_j)_0^{\{i\}}} (X_j)_1^{\{i\}} .\end{aligned}$$

To show that $X_j \in \mathsf{SegPs}[\Delta^{n-1^{op}}, \mathsf{Cat}]$ it remains to show that $p^{(n-1)}X_j \in \mathsf{Cat}^{n-1}_{\mathsf{wg}}$, where

$$(p^{(n-1)}X_j)_{\underline{r}} = pX_{j\underline{r}}$$

for each $\underline{r} \in \Delta^{n-1^{op}}$.

Since $X \in \mathsf{SegPs}[\Delta^{n^{op}}, \mathsf{Cat}]$, by definition $p^{(n)}X \in \mathsf{Cat}^{n}_{\mathsf{wg}}$, where $(p^{(n)}X)_{\underline{k}} = pX_{\underline{k}}$ for all $\underline{k} \in \Delta^{n^{op}}$. We also observe that, for each $j \geq 0$,

$$(p^{(n)}X)_j = p^{(n-1)}X_j \tag{8.4}$$

since, for each $\underline{r} \in \Delta^{n-1^{op}}$,

$$(p^{(n)}X)_{j\underline{r}} = pX_{j\underline{r}} = (p^{(n-1)}X_j)_{\underline{r}} .$$

Since $p^{(n)}X \in \mathsf{Cat}^{n}_{\mathsf{wg}}$, $(p^{(n)}X)_j \in \mathsf{Cat}^{n-1}_{\mathsf{wg}}$ so by (8.4) we conclude that $p^{(n-1)}X_j \in \mathsf{Cat}^{n-1}_{\mathsf{wg}}$, as required. □

Example 8.1.5 Let $X \in \mathsf{SegPs}[\Delta^{2^{op}}, \mathsf{Cat}]$. Then $X \in \mathsf{Ps}[\Delta^{2^{op}}, \mathsf{Cat}]$ with X_{k0}, X_{0s} discrete categories for each $k, s \in \Delta^{op}$ and $p^{(2)}X \in \mathsf{Cat}^{2}_{\mathsf{wg}}$. In Fig. 8.1 on page 168 we have depicted the corner of a pseudo-functor $X \in \mathsf{SegPs}[\Delta^{2^{op}}, \mathsf{Cat}]$ with pseudo-commuting squares containing the symbol $\cong$ since the simplicial relations hold only up to isomorphisms; the structures in red are discrete categories;

$$\begin{array}{ccccc}
(X_{11}\times_{X_{01}} X_{11})\times_{(X_{10}\times_{X_{00}} X_{10})}(X_{11}\times_{X_{01}} X_{11}) & \rightrightarrows & X_{11}\times_{X_{10}} X & \rightrightarrows & X_{01}\times_{X_{00}} X_{01} \\
\downarrow\downarrow\downarrow & \cong & \downarrow\downarrow\downarrow & \cong & \downarrow\downarrow\downarrow \\
\cdots \quad X_{11}\times_{X_{01}} X_{11} & \rightrightarrows & X_{11} & \rightrightarrows & X_{01} \\
\downarrow\downarrow & \cong & \downarrow\downarrow & \cong & \downarrow\downarrow \\
\cdots \quad X_{10}\times_{X_{00}} X_{10} & \rightrightarrows & X_{10} & \rightrightarrows & X_{00}
\end{array}$$

Fig. 8.1 Picture of the corner of $X \in \mathsf{SegPs}[\Delta^{2^{op}}, \mathsf{Cat}]$

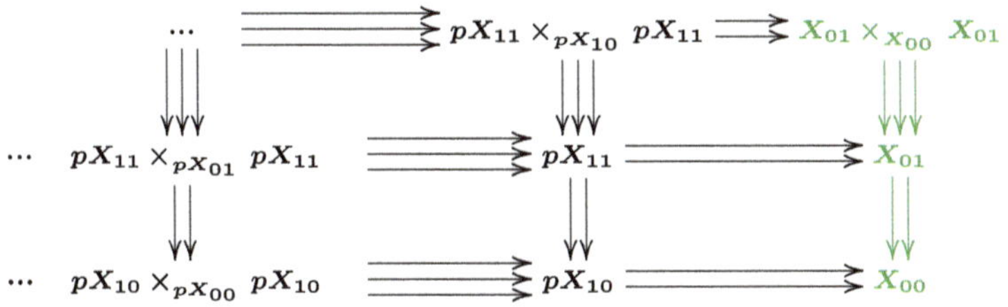

Fig. 8.2 Picture of the corner of $p^{(2)}X$, for $X \in \mathsf{SegPs}[\Delta^{2^{op}}, \mathsf{Cat}]$

Fig. 8.2 on page 169 depicts the corner of $p^{(2)}X$, which is a bisimplicial set, the double nerve of a weakly globular double category. The structure in green is homotopically discrete.

8.2 Strictification of Segalic Pseudo-Functors

In this section we prove the main result of this chapter, Theorem 8.2.3, that the strictification functor applied to the category of Segalic pseudo-functors gives a weakly globular n-fold category; that is, there is a functor

$$St : \mathsf{SegPs}[\Delta^{n-1^{op}}, \mathsf{Cat}] \to \mathsf{Cat}^{\mathsf{n}}_{\mathsf{wg}}.$$

The strategy to prove this result is based of the following main steps:

a) In Proposition 8.2.1 we show that if an n-fold category is levelwise equivalent through a pseudo-natural transformation to a Segalic pseudo-functor, then it is a weakly globular n-fold category. The proof of this result makes critical use of the criterion for an n-fold category to be weakly globular given in Proposition 7.2.8 b).
b) We show in the proof of Theorem 8.2.3 that the strictification of a Segalic pseudo-functor is an n-fold category and that it satisfies the hypotheses of Proposition 8.2.1. The proof depends on some properties of the monad corresponding to Segalic pseudo-functors which we establish in Lemma 8.2.2.
c) We immediately deduce from a) and b) that the strictification of a Segalic pseudo-functor is a weakly globular n-fold category.

Proposition 8.2.1 *Let* $H \in \mathsf{SegPs}[\Delta^{n-1^{op}}, \mathsf{Cat}]$, $L \in \mathsf{Cat}^{\mathsf{n}} \hookrightarrow \mathsf{Ps}[\Delta^{n-1^{op}}, \mathsf{Cat}]$ *and* $\phi : L \to H$ *be a pseudo-natural transformation in* $\mathsf{Ps}[\Delta^{n-1^{op}}, \mathsf{Cat}]$ *such that* $\phi_{\underline{k}}$ *is an equivalence of categories for all* $\underline{k} \in \Delta^{n-1^{op}}$*, then*

a) $L \in \mathsf{Cat}^{\mathsf{n}}_{\mathsf{wg}}$.
b) If, further, $H_{\underline{k}} \in \mathsf{Cat}_{\mathsf{hd}}$ *for all* $\underline{k} \in \Delta^{n-1^{op}}$ *and* $p^{(n-1)}H \in \mathsf{Cat}^{\mathsf{n-1}}_{\mathsf{hd}}$*, then* $L \in \mathsf{Cat}^{\mathsf{n}}_{\mathsf{hd}}$.

Proof The proof of a) is based on showing that $X \in \mathsf{Cat}^n$ satisfies the hypotheses of Proposition 7.2.8 b), which then implies that $X \in \mathsf{Cat}^n_{wg}$. The proof of b) is based on showing that $X \in \mathsf{Cat}^n_{wg}$ satisfies the hypotheses of Corollary 7.2.2, which then implies $X \in \mathsf{Cat}^n_{hd}$.

We proceed by induction on n. For $n = 2$, if $H \in \mathsf{SegPs}[\Delta^{op}, \mathsf{Cat}]$, then by definition H_0 is discrete; thus, since by hypothesis there is an equivalence of categories $L_0 \simeq H_0$, $L_0 \in \mathsf{Cat}_{hd}$. By hypothesis $L_k \simeq H_k$ for all $k \in \Delta^{op}$. Since ϕ is pseudo-natural, $p^{(1)}\phi : p^{(1)}L \to p^{(1)}H$ is natural, hence there is a commuting diagram for $k \geq 2$

$$\begin{array}{ccc} pL_k & \longrightarrow & pL_1 \times_{pL_0} \overset{k}{\cdots} \times_{pL_0} pL_1 \\ \wr\Vert & & \wr\Vert \\ pH_k & \xrightarrow{\cong} & pH_1 \times_{pH_0} \overset{k}{\cdots} \times_{pH_0} pH_1 \end{array}$$

where the bottom map is an isomorphism since $H \in \mathsf{SegPs}[\Delta^{op}, \mathsf{Cat}]$. It follows that the top map is also an isomorphism, therefore $p^{(1)}L \in \mathsf{Cat}$. So L satisfies the hypotheses of Lemma 7.2.6 and we conclude that $L \in \mathsf{Cat}^2_{wg}$.

If, further, $H_k \in \mathsf{Cat}_{hd}$ for all k and $p^{(1)}H \in \mathsf{Cat}_{hd}$, then $L_1 \in \mathsf{Cat}_{hd}$ (since $L_1 \sim H_1$) and $p^{(1)}L \cong p^{(1)}H \in \mathsf{Cat}_{hd}$. Therefore, by Corollary 7.2.2, $L \in \mathsf{Cat}^2_{hd}$.

Suppose, inductively, that the lemma holds for $(n-1)$.

a) Let L and H be as in the hypothesis a). We are going to show that $L \in \mathsf{Cat}^n$ satisfies the hypotheses of Proposition 7.2.8 b) which implies that $L \in \mathsf{Cat}^n_{wg}$.

Since, by hypothesis, $\phi_{\underline{k}}$ is an equivalence of categories for all $\underline{k} \in \Delta^{n-1^{op}}$, for each $1 \leq r < n$ there is an equivalence of categories for all $\underline{s} \in \Delta^{n-2^{op}}$

$$(L_0^{\{r\}})_{\underline{s}} \simeq (H_0^{\{r\}})_{\underline{s}}\,. \tag{8.5}$$

Since $H \in \mathsf{SegPs}[\Delta^{n-1^{op}}, \mathsf{Cat}]$, by definition $(H_0^{\{r\}})_{\underline{s}}$ is discrete; therefore from (8.5), $(L_0^{\{r\}})_{\underline{s}}$ is an equivalence relation. That is, L satisfies the hypothesis b) i) of Proposition 7.2.8. As for hypothesis b) ii) of Proposition 7.2.8, note that for each $\underline{k} \in \Delta^{n-2^{op}}$,

$$(p^{(n-1)}L)_{\underline{k}} = pL_{\underline{k}} \cong pH_{\underline{k}} = (p^{(n-1)}H)_{\underline{k}}\,.$$

Therefore

$$(p^{(n-1)}L)_{\underline{k}} \cong (p^{(n-1)}H)_{\underline{k}}\,. \tag{8.6}$$

Since $p^{(n-1)}\phi$ is natural (as ϕ is pseudo-natural) and all Segal maps of $p^{(n-1)}H$ are isomorphisms (as $p^{(n-1)}H \in \mathsf{Cat}^{n-1}$) we conclude from (8.6) that all Segal maps of $p^{(n-1)}L$ are also isomorphisms, so $p^{(n-1)}L \in \mathsf{Cat}^{\mathsf{n-1}}$. Thus $p^{(n-1)}\phi$ is a morphism in $\mathsf{Cat}^{\mathsf{n-1}}$ which is a levelwise equivalence of categories (by (8.6)) with $p^{(n-1)}H \in \mathsf{Cat}^{\mathsf{n-1}}_{\mathsf{wg}}$, so by Corollary 7.2.9, $p^{(n-1)}L \in \mathsf{Cat}^{\mathsf{n-1}}_{\mathsf{wg}}$. By Proposition 7.2.8 b), we conclude that $L \in \mathsf{Cat}^{\mathsf{n}}_{\mathsf{wg}}$, proving a) at step n.

b) Suppose that H is as in b). By Corollary 7.2.2, to show that $L \in \mathsf{Cat}^{\mathsf{n}}_{\mathsf{hd}}$, it is enough to show that $L_1 \in \mathsf{Cat}^{\mathsf{n-1}}_{\mathsf{hd}}$ and $p^{(n-1)}L \in \mathsf{Cat}^{\mathsf{n-1}}_{\mathsf{hd}}$. For all $\underline{k} \in \Delta^{n-2^{op}}$ there is an equivalence of categories

$$(L_1)_{\underline{k}} \simeq (H_1)_{\underline{k}} \,. \tag{8.7}$$

Since by hypothesis $(H_1)_{\underline{k}} \in \mathsf{Cat}_{\mathsf{hd}}$ we conclude from (8.7) that $(L_1)_{\underline{k}} \in \mathsf{Cat}_{\mathsf{hd}}$. Since $L \in \mathsf{Cat}^{\mathsf{n}}$, $L_i \in \mathsf{Cat}^{\mathsf{n-1}}$ and since $H \in \mathsf{SegPs}[\Delta^{n^{op}}, \mathsf{Cat}]$, by Lemma 7.1.3, $H_i \in \mathsf{SegPs}[\Delta^{n-1^{op}}, \mathsf{Cat}]$. Further, since $p^{(n-1)}H \in \mathsf{Cat}^{\mathsf{n-1}}_{\mathsf{hd}}$, we have

$$p^{(n-2)}H_1 = (p^{(n-1)}H)_1 \in \mathsf{Cat}^{\mathsf{n-2}}_{\mathsf{hd}} \,.$$

Thus $\phi_1 : L_1 \to H_1$ satisfies the induction hypothesis and we conclude that $L_1 \in \mathsf{Cat}^{\mathsf{n-1}}_{\mathsf{hd}}$. Finally,

$$p^{(n-1)}L \cong p^{(n-1)}H \in \mathsf{Cat}^{\mathsf{n-1}}_{\mathsf{hd}} \,.$$

Thus by Corollary 7.2.2 we conclude that $L \in \mathsf{Cat}^{\mathsf{n}}_{\mathsf{hd}}$. □

In the next lemma we show some properties of the monad corresponding to Segalic pseudo-functors. These properties will be used in the proof of the main result of this chapter, Theorem 8.2.3 on the strictification of Segalic pseudo-functors. We refer to Sect. 4.2 in Chap. 4 for background regarding the monad for pseudo-functors and about strictification of pseudo-functors.

In the following lemma, $\underline{k}(1,i)$ and $\underline{k}(0,i)$ are as in Remark 8.1.3. We refer to Sect. 4.2.4 for background about the strictification of pseudo-functors.

Lemma 8.2.2 *Let T be the monad corresponding to the adjunction given by the forgetful functor*

$$U : [\Delta^{n^{op}}, \mathsf{Cat}] \to [ob(\Delta^{n^{op}}), \mathsf{Cat}]$$

$(UH)_{\underline{k}} = H_{\underline{k}}$ for all $\underline{k} \in \Delta^{n^{op}}$ and its left adjoint F. Let $H \in \mathsf{SegPs}[\Delta^{n^{op}}, \mathsf{Cat}]$, then

a) There are functors for each $\underline{k} \in \Delta^n$, $0 \le i \le n$,

$$\partial_{i1}, \partial_{i0} : (TUH)_{\underline{k}(1,i)} \rightrightarrows (TUH)_{\underline{k}(0,i)}$$

such that the following diagram commutes

$$\begin{array}{ccc} (TUH)_{\underline{k}(1,i)} & \xrightarrow{h_{\underline{k}(1,i)}} & (UH)_{\underline{k}(1,i)} \\ \partial_{i0} \Big\downarrow \Big\downarrow \partial_{i1} & & d_{i0} \Big\downarrow \Big\downarrow d_{i1} \\ (TUH)_{\underline{k}(0,i)} & \xrightarrow[h_{\underline{k}(0,i)}]{} & (UH)_{\underline{k}(0,i)} \end{array} \tag{8.8}$$

b) *For all* $\underline{k} = (k_1, \dots, k_n) \in \Delta^n$ *with* $k_i \geq 2$ *and* $0 \leq i \leq n$ *there are isomorphisms*

$$(TUH)_{\underline{k}} \cong (TUH)_{\underline{k}(1,i)} \times_{(TUH)_{\underline{k}(0,i)}} \overset{k_i}{\cdots} \times_{(TUH)_{\underline{k}(0,i)}} (TUH)_{\underline{k}(1,i)} \,.$$

c) *The morphism* $h_{\underline{k}} : (TUH)_{\underline{k}} \to (UH)_{\underline{k}}$ *is given by*

$$h_{\underline{k}} = (h_{\underline{k}(1,i)}, \dots, h_{\underline{k}(1,i)}).$$

Proof

a) For each $j = 0, 1$ let $\nu_j : [0] \to [1]$, $\nu_0(0) = 0$, $\nu_1(0) = 1$ and let $\delta_{ij} : \underline{k}(0,i) \to \underline{k}(1,i)$ be given by

$$\delta_{ij}(k_s) = \begin{cases} k_s, & s \neq i; \\ \nu_j(k_i), & s = i. \end{cases}$$

Recall from Sect. 4.2.4 that

$$(TUH)_{\underline{k}} = \coprod_{\underline{r} \in \Delta^n} \Delta^n(\underline{k}, \underline{r}) \times H_{\underline{r}} = \coprod_{\underline{r} \in \Delta^n} \coprod_{\Delta^n(\underline{k}, \underline{r})} H_{\underline{r}} \,.$$

If $f \in \Delta^n(\underline{k}, \underline{r})$, let

$$i_{\underline{r}} = \coprod_{\Delta^n(\underline{k}, \underline{r})} H_{\underline{r}} \to \coprod_{\underline{r} \in \Delta^n} \coprod_{\Delta^n(\underline{k}, \underline{r})} H_{\underline{r}} = (TUH)_{\underline{r}} \,,$$

$$j_f : H_{\underline{r}} \to \coprod_{\Delta^n(\underline{k}, \underline{r})} H_{\underline{r}}$$

be the coproduct inclusions. Let

$$\partial_{ij} : (TUH)_{\underline{k}(1,i)} \to (TUH)_{\underline{k}(0,i)}$$

be the functors determined by

$$\partial_{ij} i_{\underline{r}} j_f = i_{\underline{r}} j_{f\delta_{ij}} \ ,$$

where $f \in \Delta^n(\underline{k}(1,i), \underline{r})$. From Sect. 4.2.4, since $k_{\underline{k}} i_{\underline{r}} j_f = H(f)$, denoting $d_{ij} = H(\delta_{ij})$, we obtain

$$\begin{aligned} h_{\underline{k}(0,i)}\, \partial_{ij}\, i_{\underline{r}}\, j_f &= h_{\underline{k}(0,i)}\, i_{\underline{r}}\, j_{f\delta_{ij}} = H(f\delta_{ij}) \ , \\ d_{ij}\, h_{\underline{k}(1,i)}\, i_{\underline{r}}\, j_f &= H(\delta_{ij})\, H(f) \ . \end{aligned}$$

Since $H \in \mathsf{Ps}[\Delta^{n^{op}}, \mathsf{Cat}\,]$ and $H_{\underline{k}(0,i)}$ is discrete, we have

$$H(f\delta_{ij}) = H(\delta_{ij})H(f)$$

so that, from above,

$$h_{\underline{k}(0,i)}\partial_{ij} i_{\underline{r}} j_f = d_{ij} h_{\underline{k}(1,i)} i_{\underline{r}} j_f$$

for each $\underline{r}$, f. We conclude that

$$h_{\underline{k}(0,i)}\partial_{ij} = d_{ij} h_{\underline{k}(1,i)} \ .$$

That is, diagram (8.8) commutes.

b) Since, for each $k_i \geq 2$

$$[k_i] = [1] \coprod_{[0]} \overset{k_i}{\cdots} \coprod_{[0]} [1]$$

we have, for each $\underline{k} = (k_1, \ldots, k_n) \in \Delta^n$ with $k_i \geq 2$ and $1 \leq i \leq n$,

$$\underline{k} = \underline{k}(1,i) \coprod_{\underline{k}(0,i)} \overset{k_i}{\cdots} \coprod_{\underline{k}(0,i)} \underline{k}(1,i) \ .$$

Therefore there is a bijection

$$\Delta^n(\underline{k}, \underline{r}) = \Delta^n\big(\underline{k}(1,i), \underline{r}\big) \times_{\Delta^n(\underline{k}(0,i),\underline{r})} \overset{k_i}{\cdots} \times_{\Delta^n(\underline{k}(0,i),\underline{r})} \Delta^n\big(\underline{k}(1,i), \underline{r}\big) \ .$$

From the proof of a), the functors

$$\partial_{ij} : (TUH)_{\underline{k}(1,i)} \rightarrow (TUH)_{\underline{k}(0,i)}$$

for $j = 0, 1$ are determined by the functors

$$(\overline{\delta_{ij}}, id) : \Delta^n\big(\underline{k}(1,i), \underline{r}\big) \times H_{\underline{r}} \rightarrow \Delta^n\big(\underline{k}(0,i), \underline{r}\big) \times H_{\underline{r}} \ ,$$

where $\overline{\delta_{ij}}(g) = g\delta_{ij}$ for $g \in \Delta^n\big(\underline{k}(1,i),\underline{r}\big)$ and

$$(TUH)_{\underline{k}(1,i)} = \coprod_{\underline{r}\in\Delta^n} \Delta^n\big(\underline{k}(1,i),\underline{r}\big) \times H_{\underline{r}}\ ,$$

$$(TUH)_{\underline{k}(0,i)} = \coprod_{\underline{r}\in\Delta^n} \Delta^n\big(\underline{k}(0,i),\underline{r}\big) \times H_{\underline{r}}\ .$$

It follows that

$$\begin{aligned}
&(TUH)_{\underline{k}(1,i)}\times_{(TUH)_{\underline{k}(0,i)}} \overset{k_i}{\cdots}\times_{(TUH)_{\underline{k}(0,i)}} (TUH)_{\underline{k}(1,i)}\\
&= \coprod_{\underline{r}\in\Delta^n}\{\Delta^n\big(\underline{k}(1,i),\underline{r}\big)\times_{\Delta^n(\underline{k}(0,i),\underline{r})} \overset{k_i}{\cdots}\times_{\Delta^n(\underline{k}(0,i),\underline{r})} \Delta^n\big(\underline{k}(1,i),\underline{r}\big)\}\times H_{\underline{r}}\\
&= \coprod_{\underline{r}\in\Delta^n}\Delta^n(\underline{k},\underline{r})\times H_{\underline{r}} = (TUH)_{\underline{k}}\ .
\end{aligned}$$

This proves b).

c) From Sect. 4.2.4,

$$h_k\, i_r\, j_f = H(f)$$

for $f \in \Delta^n(\underline{k},\underline{r})$. Let f correspond to $(\delta_1,\ldots,\delta_{k_i})$ in the isomorphism

$$\Delta^n(\underline{k},\underline{r}) = \Delta^n\big(\underline{k}(1,i),\underline{r}\big)\times_{\Delta^n(\underline{k}(0,i),\underline{r})} \overset{k_i}{\cdots}\times_{\Delta^n(\underline{k}(0,i),\underline{r})} \Delta^n\big(\underline{k}(1,i),\underline{r}\big)\ .$$

Then $j_f = (j_{\delta_1},\ldots,j_{\delta_{k_i}})$. Since

$$H_{\underline{k}} \cong H_{\underline{k}(1,i)}\times_{H_{\underline{k}(0,i)}} \overset{k_i}{\cdots}\times_{H_{\underline{k}(0,i)}} H_{\underline{k}(1,i)}\ ,$$

$H(f)$ corresponds to $(H(\delta_1),\ldots,H(\delta_{k_i}))$ with $p_i H(f) = H(\delta_i)$. Then for all f we have

$$\begin{aligned}
h_{\underline{k}}\, i_{\underline{r}}\, j_f &= (H(\delta_1),\ldots,H(\delta_{k_i})) = (h_{\underline{k}(1,i)}\, i_{\underline{r}}\, j_{\delta_1},\ldots,h_{\underline{k}(1,i)}\, i_{\underline{r}}\, j_{\delta_{k_i}})\\
&= (h_{\underline{k}(1,i)},\ldots,h_{\underline{k}(1,i)})\, i_{\underline{r}}(j_{\delta_1},\ldots,j_{\delta_{k_i}}) = (h_{\underline{k}(1,i)},\ldots,h_{\underline{k}(1,i)})\, i_{\underline{r}}\, j_f\ .
\end{aligned}$$

It follows that $h_{\underline{k}} = (h_{\underline{k}(1,i)},\ldots,h_{\underline{k}(1,i)})$. □

We now show the main result of this chapter that the strictification of a Segalic pseudo-functor is a weakly globular n-fold category. This result will be used in Chap. 9 in the construction of the rigidification functor from $\mathsf{Ta}^{\mathsf{n}}_{\mathsf{wg}}$ to $\mathsf{Cat}^{\mathsf{n}}_{\mathsf{wg}}$.

Theorem 8.2.3 *The strictification functor*

$$St\ : \mathsf{Ps}[\Delta^{n-1^{op}},\mathsf{Cat}] \to [\Delta^{n-1^{op}},\mathsf{Cat}]$$

restricts to a functor

$$St : \mathsf{SegPs}[\Delta^{n-1^{op}}, \mathsf{Cat}] \to \mathsf{Cat}^{\mathsf{n}}_{\mathsf{wg}} .$$

Further, for each $H \in \mathsf{SegPs}[\Delta^{n-1^{op}}, \mathsf{Cat}]$ *and* $\underline{k} \in \Delta^{n-1^{op}}$, *the map*

$$(St\, H)_{\underline{k}} \to H_{\underline{k}}$$

is an equivalence of categories.

Proof Let $h : TUH \to UH$ be as in Sect. 4.2.4. As recalled there, to construct the strictification $L = St\, H$ of a pseudo-functor H we need to factorize $h = gv$ in such a way that for each $\underline{k} \in \Delta^{n-1^{op}}$, $h_{\underline{k}}$ factorizes as

$$(TUH)_{\underline{k}} \overset{v_{\underline{k}}}{\to} L_{\underline{k}} \overset{g_{\underline{k}}}{\to} (UH)_{\underline{k}} = H_{\underline{k}} ,$$

with $v_{\underline{k}}$ bijective on objects and $g_{\underline{k}}$ fully faithful. As explained in [106], $g_{\underline{k}}$ is in fact an equivalence of categories.

Since the bijective on objects functors and the fully faithful functors form a factorization system in Cat, the commutativity of (8.8) implies that there are functors

$$\tilde{d}_{ij} : L_{\underline{k}(1,i)} \rightrightarrows L_{\underline{k}(0,i)} \qquad j = 0, 1, \quad 0 \le i \le n$$

such that the following diagram commutes:

$$\begin{array}{ccccc}
(TUH)_{\underline{k}(1,i)} & \xrightarrow{v_{\underline{k}(1,i)}} & L_{\underline{k}(1,i)} & \xrightarrow{g_{\underline{k}(1,i)}} & H_{\underline{k}(1,i)} \\
{\scriptstyle \partial_{i0}} \downarrow\downarrow {\scriptstyle \partial_{i1}} & & {\scriptstyle \tilde{d}_{i0}} \downarrow\downarrow {\scriptstyle \tilde{d}_{i1}} & & {\scriptstyle d_{i0}} \downarrow\downarrow {\scriptstyle d_{i1}} \\
(TUH)_{\underline{k}(0,i)} & \xrightarrow[v_{\underline{k}(0,i)}]{} & L_{\underline{k}(0,i)} & \xrightarrow[g_{\underline{k}(0,i)}]{} & H_{\underline{k}(0,i)} .
\end{array}$$

By Lemma 8.2.2, $h_{\underline{k}}$ factorizes as

$$\begin{aligned}
(TUH)_{\underline{k}} &\cong (TUH)_{\underline{k}(1,i)} \times_{(TUH)_{\underline{k}(0,i)}} \overset{k_i}{\cdots} \times_{(TUH)_{\underline{k}(0,i)}} (TUH)_{\underline{k}(1,i)} \\
&\xrightarrow{(v_{\underline{k}(1,i)},\dots,v_{\underline{k}(1,i)})} L_{\underline{k}(1,i)} \times_{L_{\underline{k}(0,i)}} \overset{k_i}{\cdots} \times_{L_{\underline{k}(0,i)}} L_{\underline{k}(1,i)} \\
&\xrightarrow{(g_{\underline{k}(1,i)},\dots,g_{\underline{k}(1,i)})} H_{\underline{k}(1,i)} \times_{H_{\underline{k}(0,i)}} \overset{k_i}{\cdots} \times_{H_{\underline{k}(0,i)}} H_{\underline{k}(1,i)} \cong H_{\underline{k}} .
\end{aligned}$$

Since $v_{\underline{k}(1,i)}$ and $v_{\underline{k}(0,i)}$ are bijective on objects, so is $(v_{\underline{k}(1,i)}, \dots, v_{\underline{k}(1,i)})$. Since $g_{\underline{k}(1,i)}, g_{\underline{k}(0,i)}$ are fully faithful, so is $(g_{\underline{k}(1,i)}, \dots, g_{\underline{k}(1,i)})$. Therefore the above is the required factorization of $h_{\underline{k}}$ and we conclude that

$$L_{\underline{k}} \cong L_{\underline{k}(1,i)} \times_{L_{\underline{k}(0,i)}} \overset{k_i}{\cdots} \times_{L_{\underline{k}(0,i)}} L_{\underline{k}(1,i)} \ .$$

Since this holds for each $\underline{k} \in \Delta^{{n-1}^{op}}$, it is the same as

$$L_{k_i}^{\{i\}} \cong L_1^{\{i\}} \times_{L_0^{\{i\}}} \overset{k_i}{\cdots} \times_{L_0^{\{i\}}} L_1^{\{i\}}$$

for each $k_i \geq 2$ and $0 \leq i \leq n$. Since $L \in [\Delta^{{n-1}^{op}}, \mathsf{Cat}]$, by Proposition 2.4.5 this implies that $L \in \mathsf{Cat}^n$. As recalled in Sect. 4.2.4, by Power [106] $g : L \to H$ is a pseudo-natural transformation with $g_{\underline{k}} : L_{\underline{k}} \to H_{\underline{k}}$ an equivalence of categories for all $\underline{k} \in \Delta^{{n-1}^{op}}$; thus the hypotheses of Proposition 8.2.1 are satisfied and we conclude that $L \in \mathsf{Cat}^n_{\mathsf{wg}}$. □

Part III
Rigidification of Weakly Globular Tamsamani n-Categories

This Part is devoted to the rigidification of weakly globular Tamsamani n-categories. The main result of this Part is Theorem 10.2.1, which constructs the rigidification functor

$$Q_n : \mathsf{Ta}^{\mathsf{n}}_{\mathsf{wg}} \to \mathsf{Cat}^{\mathsf{n}}_{\mathsf{wg}}$$

and, for each $X \in \mathsf{Ta}^{\mathsf{n}}_{\mathsf{wg}}$, an n-equivalence $s_n(X) : Q_n X \to X$ natural in X. A schematic summary of the main results of this Part is given in Fig. 8.3.

The construction of the rigidification functor when $n = 2$ is quite straightforward and was already done in [103]. The construction of Q_n when $n > 2$ is much more complex and is new to this work: it needs in particular the subcategory $\mathsf{LTa}^{\mathsf{n}}_{\mathsf{wg}}$ of $\mathsf{Ta}^{\mathsf{n}}_{\mathsf{wg}}$. The idea of the category $\mathsf{LTa}^{\mathsf{n}}_{\mathsf{wg}}$ is to produce a generalization of the category $\mathsf{Cat}^{\mathsf{n}}_{\mathsf{wg}}$ and there are embeddings $\mathsf{Cat}^{\mathsf{n}}_{\mathsf{wg}} \subset \mathsf{LTa}^{\mathsf{n}}_{\mathsf{wg}} \subset \mathsf{Ta}^{\mathsf{n}}_{\mathsf{wg}}$. We introduce the idea of this subcategory in Sect. 9.1.1, before the formal definition.

In Chap. 9 we prove two important results involving this subcategory which are used in the construction of the rigidification functor: Theorems 9.2.4 and 10.1.1. Theorem 9.2.4 establishes a procedure to approximate, up to n-equivalence, an object of $\mathsf{Ta}^{\mathsf{n}}_{\mathsf{wg}}$ with an object of $\mathsf{LTa}^{\mathsf{n}}_{\mathsf{wg}}$: its proof is based on the properties of the pullback constructions of Sect. 7.1.3 as well as on the criterion given in Proposition 7.1.5 for an n-equivalence in $\mathsf{Ta}^{\mathsf{n}}_{\mathsf{wg}}$ to be a levelwise equivalence of categories. The main steps needed in these constructions are explained informally in Sect. 9.2.1.

In Theorem 10.1.1 we construct the functor

$$Tr_n : \mathsf{LTa}^{\mathsf{n}}_{\mathsf{wg}} \to \mathsf{SegPs}[\Delta^{n-1^{op}}, \mathsf{Cat}]\,.$$

The idea of the functor Tr_n is explained in Sect. 10.1.1. The proof of Theorem 10.1.1 relies on a technique to produce pseudo-functors that is an instance of 'transport of structure along an adjunction' (recalled in Sect. 4.3), as well as on the definition of the category $\mathsf{LTa}^{\mathsf{n}}_{\mathsf{wg}}$.

We finally construct the rigidification functor $Q_n : \mathsf{Ta}^{\mathsf{n}}_{\mathsf{wg}} \rightarrow \mathsf{Cat}^{\mathsf{n}}_{\mathsf{wg}}$. The idea of the construction of Q_n is explained in Sect. 10.2.1. In the case $n = 2$, the rigidification functor Q_2 is the composite

$$Q_2 : \mathsf{Ta}^{2}_{\mathsf{wg}} \xrightarrow{Tr_2} \mathsf{SegPs}[\Delta^{^{op}}, \mathsf{Cat}] \xrightarrow{St} \mathsf{Cat}^{2}_{\mathsf{wg}} ,$$

where Tr_2 is as in Theorem 10.1.1. When $n > 2$ the functor Q_n is given as a composite

$$\mathsf{Ta}^{\mathsf{n}}_{\mathsf{wg}} \xrightarrow{P_n} \mathsf{LTa}^{\mathsf{n}}_{\mathsf{wg}} \xrightarrow{Tr_n} \mathsf{SegPs}[\Delta^{n-1^{op}}, \mathsf{Cat}] \xrightarrow{St} \mathsf{Cat}^{\mathsf{n}}_{\mathsf{wg}} .$$

This relies on Theorem 9.2.4 (for the construction of the functor P_n), Theorem 10.1.1 (for the functor Tr_n) and Theorem 8.2.3 (for the functor St).

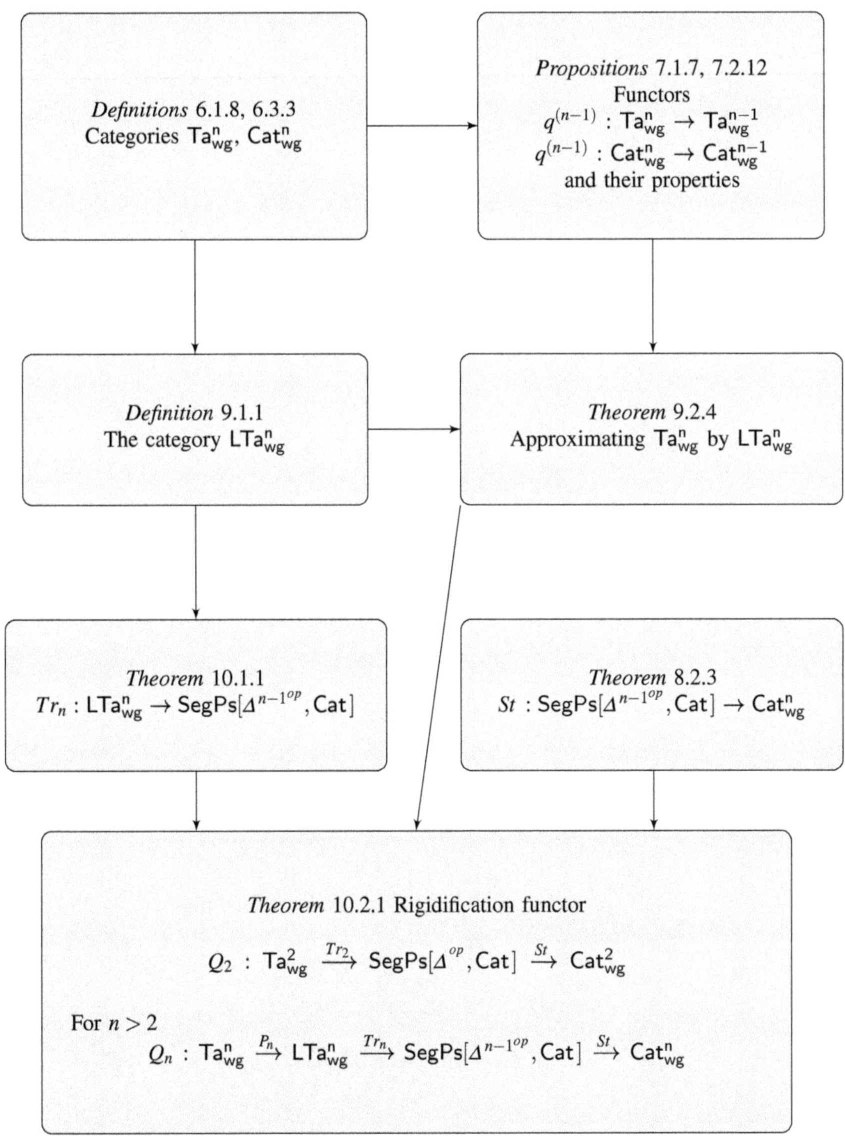

Fig. 8.3 The construction of the rigidification functor Q_n

Chapter 9
Approximating Weakly Globular Tamsamani n-Categories by Simpler Ones

Abstract In this chapter we introduce a new subcategory $\mathsf{LTa}^{\mathsf{n}}_{\mathsf{wg}}$ of the category $\mathsf{Ta}^{\mathsf{n}}_{\mathsf{wg}}$ and we establish its main properties. In the next chapter we functorially build Segalic pseudo-functors from the category $\mathsf{LTa}^{\mathsf{n}}_{\mathsf{wg}}$. The main result of this chapter is that, if $X \in \mathsf{Ta}^{\mathsf{n}}_{\mathsf{wg}}$ is such that $q^{(n-1)}X$ can be approximated up to $(n-1)$-equivalence with an object of $\mathsf{Cat}^{\mathsf{n-1}}_{\mathsf{wg}}$, then X can be approximated up to n-equivalence with an object of $\mathsf{LTa}^{\mathsf{n}}_{\mathsf{wg}}$. This will be used in Chap. 10 in building the rigidification functor Q_n from weakly globular Tamsamani n-categories to weakly globular n-fold categories.

In this chapter we continue the study of the category $\mathsf{Ta}^{\mathsf{n}}_{\mathsf{wg}}$ of weakly globular Tamsamani n-categories introduced in Chap. 6. We define a full subcategory $\mathsf{LTa}^{\mathsf{n}}_{\mathsf{wg}}$ of $\mathsf{Ta}^{\mathsf{n}}_{\mathsf{wg}}$ containing $\mathsf{Cat}^{\mathsf{n}}_{\mathsf{wg}}$, so that there are embeddings

$$\mathsf{Cat}^{\mathsf{n}}_{\mathsf{wg}} \subset \mathsf{LTa}^{\mathsf{n}}_{\mathsf{wg}} \subset \mathsf{Ta}^{\mathsf{n}}_{\mathsf{wg}} .$$

Objects X of $\mathsf{LTa}^{\mathsf{n}}_{\mathsf{wg}}$ are simpler than those of $\mathsf{Ta}^{\mathsf{n}}_{\mathsf{wg}}$, not only because $p^{(n-1)}X \in \mathsf{Cat}^{\mathsf{n-1}}_{\mathsf{wg}}$ but also because certain induced Segal maps are levelwise equivalences of categories (see Definition 9.1.1 for more details). In Chap. 10 we will see how these properties of $\mathsf{LTa}^{\mathsf{n}}_{\mathsf{wg}}$ allow us to build a functor Tr_n from $\mathsf{LTa}^{\mathsf{n}}_{\mathsf{wg}}$ to the category of Segalic pseudo-functors of Chap. 8; using Theorem 8.2.3 this functor will lead to a composite functor

$$\mathsf{LTa}^{\mathsf{n}}_{\mathsf{wg}} \xrightarrow{Tr_n} \mathsf{SegPs}[\Delta^{n-1^{op}}, \mathsf{Cat}\,] \xrightarrow{St} \mathsf{Cat}^{\mathsf{n}}_{\mathsf{wg}}$$

which rigidifies objects of $\mathsf{LTa}^{\mathsf{n}}_{\mathsf{wg}}$ to n-equivalent objects of $\mathsf{Cat}^{\mathsf{n}}_{\mathsf{wg}}$. For this rigidification process to work for the whole category $\mathsf{Ta}^{\mathsf{n}}_{\mathsf{wg}}$ we will build in Chap. 10 a functor

$$P_n : \mathsf{Ta}^{\mathsf{n}}_{\mathsf{wg}} \to \mathsf{LTa}^{\mathsf{n}}_{\mathsf{wg}}$$

which produces from $X \in \mathsf{Ta}^{\mathsf{n}}_{\mathsf{wg}}$ an n-equivalent $P_nX \in \mathsf{LTa}^{\mathsf{n}}_{\mathsf{wg}}$.

S. Paoli, *Simplicial Methods for Higher Categories*, Algebra and Applications 26,
https://doi.org/10.1007/978-3-030-05674-2_9

The construction of P_n is based on the main result of this chapter, Theorem 9.2.4, which states that if $X \in \mathsf{Ta}^{\mathsf{n}}_{\mathsf{wg}}$ is such that $q^{(n-1)}X$ can be approximated up to an $(n-1)$-equivalence by an object of $\mathsf{Cat}^{\mathsf{n-1}}_{\mathsf{wg}}$, then X can be approximated up to an n-equivalence by an object of $\mathsf{LTa}^{\mathsf{n}}_{\mathsf{wg}}$.

This chapter is organized as follows: In Sect. 9.1 we introduce the subcategory $\mathsf{LTa}^{\mathsf{n}}_{\mathsf{wg}} \subset \mathsf{Ta}^{\mathsf{n}}_{\mathsf{wg}}$ and establish its properties. In Sect. 9.2, using a pullback construction and the properties established in Sect. 7.1.3, we show in Theorem 9.2.4 how to approximate up to n-equivalence objects of $\mathsf{Ta}^{\mathsf{n}}_{\mathsf{wg}}$ by objects of $\mathsf{LTa}^{\mathsf{n}}_{\mathsf{wg}}$.

9.1 The Category $\mathsf{LTa}^{\mathsf{n}}_{\mathsf{wg}}$

In this section we introduce the subcategory $\mathsf{LTa}^{\mathsf{n}}_{\mathsf{wg}}$ of $\mathsf{Ta}^{\mathsf{n}}_{\mathsf{wg}}$. We will show in Sect. 10.2 how to build a functor from this category to the category of Segalic pseudo-functors, which in turn will lead to the construction of the rigidification functor Q_n.

9.1.1 The Idea of the Category $\mathsf{LTa}^{\mathsf{n}}_{\mathsf{wg}}$

Recall from Remark 7.2.15 that any $X \in \mathsf{Cat}^{\mathsf{n}}_{\mathsf{wg}}$ satisfies the following properties:

(a) $p^{(n-1)}X \in \mathsf{Cat}^{\mathsf{n-1}}_{\mathsf{wg}}$.
(b) For all $1 \leq r < n$ the induced Segal map in $[\Delta^{n-2^{op}}, \mathsf{Cat}]$ for $k \geq 2$

$$v_k^{\{r\}} : X_k^{\{r\}} \to X_1^{\{r\}} \times_{p^{(n-2)}X_0^{\{r\}}} \overset{k}{\cdots} \times_{p^{(n-2)}X_0^{\{r\}}} X_1^{\{r\}}$$

is a levelwise equivalence of categories.

The idea of the category $\mathsf{LTa}^{\mathsf{n}}_{\mathsf{wg}}$ is to produce a generalization of the category $\mathsf{Cat}^{\mathsf{n}}_{\mathsf{wg}}$ so that (a) and (b) hold, and there are embeddings

$$\mathsf{Cat}^{\mathsf{n}}_{\mathsf{wg}} \subset \mathsf{LTa}^{\mathsf{n}}_{\mathsf{wg}} \subset \mathsf{Ta}^{\mathsf{n}}_{\mathsf{wg}}\ .$$

In Chap. 10 we will see that the properties (a) and (b) are the key to constructing Segalic pseudo-functors from the category $\mathsf{LTa}^{\mathsf{n}}_{\mathsf{wg}}$.

It is useful to note that, given $X \in \mathsf{Cat}^{\mathsf{n}}_{\mathsf{wg}}$, the fact that X satisfies (a) implies that (b) holds, but this is no longer the case for any $X \in \mathsf{Ta}^{\mathsf{n}}_{\mathsf{wg}}$. In fact, given $X \in \mathsf{Ta}^{\mathsf{n}}_{\mathsf{wg}}$, $\underline{k} = (k_2, \ldots, k_{n-1}) \in \Delta^{n-2^{op}}$ and $1 \leq r < n$, as in Corollary 7.2.14, let

$$X^{\{r\}}(\text{-}, \underline{k}) : \Delta^{op} \to \mathsf{Cat}$$

be the functor associating to $k_1 \in \Delta^{op}$ the category

$$X^{\{r\}}(k_1, k_2, \ldots, k_{n-1}) = X_{k_2,\ldots,k_r,k_1,\ldots,k_{n-1}}.$$

Then $X^{\{r\}}(0, \underline{k}) \in \mathsf{Cat}_{\mathsf{hd}}$ and $p^{(1)} X^{\{r\}}(\text{-}, \underline{k}) \in \mathsf{Cat}$. However, this property is not sufficient to conclude that the induced Segal maps of $X^{\{r\}}(\text{-}, \underline{k})$

$$X^{\{r\}}(k_1, \underline{k}) \to X^{\{r\}}(1, \underline{k}) \times_{(X^{\{r\}}(0,\underline{k}))^d} \overset{k_1}{\cdots} \times_{(X^{\{r\}}(0,\underline{k}))^d} X^{\{r\}}(1, \underline{k})$$

are equivalences of categories for all $\underline{k} \in \Delta^{n-2^{op}}$. If this was true, it would imply (b), since

$$(p^{(n-2)} X_0^{\{r\}})_{\underline{k}} = p(X_0^{\{r\}}(\underline{k})) = pX^{\{r\}}(0, \underline{k}) = (X^{\{r\}}(0, \underline{k}))^d .$$

However, in this case $p^{(1)} X^{\{r\}}(\text{-}, \underline{k}) \in \mathsf{Cat}$ only implies that the induced Segal maps of $X^{\{r\}}(\text{-}, \underline{k})$ are essentially surjective on objects (as they become isomorphisms after applying p). But in general these induced Segal maps are not fully faithful. Instead, when $X \in \mathsf{Cat}^{\mathsf{n}}_{\mathsf{wg}}$, they are fully faithful because for each $k_1 \geq 2$

$$X^{\{r\}}(k_1, \underline{k}) \cong X^{\{r\}}(1, \underline{k}) \times_{X^{\{r\}}(0,\underline{k})} \overset{k_1}{\cdots} \times_{X^{\{r\}}(0,\underline{k})} X^{\{r\}}(1, \underline{k})$$

and $X^{\{r\}}(0, \underline{k}) \in \mathsf{Cat}_{\mathsf{hd}}$.

The category $\mathsf{LTa}^{\mathsf{n}}_{\mathsf{wg}}$ is essential in building the functors Tr_n and thus the rigidification functor Q_n, discussed informally in Sects. 10.1.1 and 10.2.1 respectively.

9.1.2 The Formal Definition of the Category $\mathsf{LTa}^{\mathsf{n}}_{\mathsf{wg}}$

Given $X \in \mathsf{Ta}^{\mathsf{n}}_{\mathsf{wg}}$ and $1 \leq r < n$, $X_k^{\{r\}} \in [\Delta^{n-2^{op}}, \mathsf{Cat}]$. There is a map $X_0^{\{r\}} \to p^{(n-2)} X_0^{\{r\}}$ in $[\Delta^{n-2^{op}}, \mathsf{Cat}]$ and therefore a corresponding induced Segal map in $[\Delta^{n-2^{op}}, \mathsf{Cat}]$ for $k \geq 2$

$$v_k^{\{r\}} : X_k^{\{r\}} \to X_1^{\{r\}} \times_{p^{(n-2)} X_0^{\{r\}}} \overset{k}{\cdots} \times_{p^{(n-2)} X_0^{\{r\}}} X_1^{\{r\}} . \tag{9.1}$$

We next define the subcategory $\mathsf{LTa}^{\mathsf{n}}_{\mathsf{wg}}$ of $\mathsf{Ta}^{\mathsf{n}}_{\mathsf{wg}}$.

Definition 9.1.1 $\mathsf{LTa}^{\mathsf{n}}_{\mathsf{wg}}$ is the full subcategory of $\mathsf{Ta}^{\mathsf{n}}_{\mathsf{wg}}$ whose objects X are such that

(a) $p^{(n-1)}X \in \mathsf{Cat}^{\mathsf{n-1}}_{\mathsf{wg}}$.
(b) For all $1 \le r < n$ the induced Segal map in $[\Delta^{n-2^{op}}, \mathsf{Cat}\,]$ for $k \geq 2$

$$v_k^{\{r\}} : X_k^{\{r\}} \to X_1^{\{r\}} \times_{p^{(n-2)}X_0^{\{r\}}} \overset{k}{\cdots} \times_{p^{(n-2)}X_0^{\{r\}}} X_1^{\{r\}}$$

is a levelwise equivalence of categories.

Remark 9.1.2 From Definition 9.1.1, $\mathsf{LTa}^{\mathsf{n}}_{\mathsf{wg}} = \mathsf{Ta}^{\mathsf{n}}_{\mathsf{wg}}$ for $n = 0, 1, 2$. Also note that, using notation (8.3), condition (b) in Definition 9.1.1 can be re-written as stating that, for all $\underline{k} = (k_1, \dots, k_{n-1}) \in \Delta^{n-1^{op}}$, $1 \le r \le n-1$, the induced Segal maps in Cat for $k_r \geq 2$

$$X_{k_r} \to X_{\underline{k}(1,r)} \times_{X^d_{\underline{k}(0,r)}} \overset{k_r}{\cdots} \times_{X^d_{\underline{k}(0,r)}} X_{\underline{k}(1,r)}$$

are equivalences of categories.

9.1.3 Properties of the Category $\mathsf{LTa}^{\mathsf{n}}_{\mathsf{wg}}$

In the following proposition we give an inductive characterization of the category $\mathsf{LTa}^{\mathsf{n}}_{\mathsf{wg}}$.

Proposition 9.1.3 *Let $X \in \mathsf{Ta}^{\mathsf{n}}_{\mathsf{wg}}$. Then $X \in \mathsf{LTa}^{\mathsf{n}}_{\mathsf{wg}}$ if and only if*

(a) $p^{(n-1)}X \in \mathsf{Cat}^{\mathsf{n-1}}_{\mathsf{wg}}$.
(b) (i) For each $k \geq 2$ the induced Segal map in $[\Delta^{n-2^{op}}, \mathsf{Cat}\,]$

$$v_k : X_k \to X_1 \times_{p^{(n-2)}X_0} \overset{k}{\cdots} \times_{p^{(n-2)}X_0} X_1$$

is a levelwise equivalence of categories.
(ii) For all $k \in \Delta^{op}$, $X_k \in \mathsf{LTa}^{\mathsf{n-1}}_{\mathsf{wg}}$.

Proof Let $X \in \mathsf{LTa}^{\mathsf{n}}_{\mathsf{wg}}$. Then, by Definition 9.1.1, (a) and (b) (i) hold. As for (b) (ii), since $p^{(n-1)}X \in \mathsf{Cat}^{\mathsf{n-1}}_{\mathsf{wg}}$, $(p^{(n-1)}X)_k = p^{(n-2)}X_k \in \mathsf{Cat}^{\mathsf{n-2}}_{\mathsf{wg}}$. Thus to show that $X_k \in \mathsf{LTa}^{\mathsf{n-1}}_{\mathsf{wg}}$ it remains to prove that, for each $1 \le r < n-1$, the induced Segal map in $[\Delta^{n-3^{op}}, \mathsf{Cat}\,]$ for $s \geq 2$

$$(X_k)_s^{\{r\}} \to (X_k)_1^{\{r\}} \times_{p^{(n-3)}(X_k)_0^{\{r\}}} \overset{k}{\cdots} \times_{p^{(n-3)}(X_k)_0^{\{r\}}} (X_k)_1^{\{r\}}$$

is a levelwise equivalence of categories; that is, for each $(k_1 \dots k_{n-3}) \in \Delta^{n-3^{op}}$ the map in Cat

$$X_{k\,k_1\dots k_r\,s\,k_{r+1}\dots k_{n-3}} \\ \to X_{k\,k_1\dots k_r\,1\,k_{r+1}\dots k_{n-3}} \times_{pX_{k\,k_1\dots k_r\,0\,k_{r+1}\dots k_{n-3}}} \overset{s}{\cdots} \times_{pX_{k\,k_1\dots k_r\,0\,k_{r+1}\dots k_{n-3}}} X_{k\,k_1\dots k_r\,1\,k_{r+1}\dots k_{n-3}} \tag{9.2}$$

is an equivalence of categories.

Since $X \in \mathsf{LTa}^{\mathsf{n}}_{\mathsf{wg}}$, for each $s \geq 2$ the map in $[\Delta^{n-2^{op}}, \mathsf{Cat}]$

$$X^{\{r+1\}}_s \to X^{\{r+1\}}_1 \times_{X^{\{r+1\}}_0} \overset{s}{\cdots} \times_{X^{\{r+1\}}_0} X^{\{r+1\}}_1$$

is a levelwise equivalence of categories. Noting that, for each $s \geq 0$,

$$(X^{\{r+1\}}_s)_{k\,k_1\dots k_{n-3}} = X_{k\,k_1\dots k_r\,s\,k_{r+1}\dots k_{n-3}} \tag{9.3}$$

it follows that (9.2) is an equivalence of categories, as required. Conversely, suppose that $X \in \mathsf{Ta}^{\mathsf{n}}_{\mathsf{wg}}$ satisfies (a) and (b). Since by hypothesis $p^{(n-1)}X \in \mathsf{Cat}^{\mathsf{n-1}}_{\mathsf{wg}}$, by Definition 9.1.1 to prove that $X \in \mathsf{LTa}^{\mathsf{n}}_{\mathsf{wg}}$ we need to show that, for each $1 \leq r < n$, the induced Segal maps in $[\Delta^{n-2^{op}}, \mathsf{Cat}]$, for $s \geq 2$,

$$X^{\{r\}}_s \to X^{\{r\}}_1 \times_{p^{(n-2)}X^{\{r\}}_0} \overset{s}{\cdots} \times_{p^{(n-2)}X^{\{r\}}_0} X^{\{r\}}_1 \tag{9.4}$$

are levelwise equivalences of categories. When $r = 1$ this holds by hypothesis (b) (i). Let $r > 1$, then for each $k_1, \dots, k_{n-2}$, s

$$(X^{\{r\}}_s)(k_1 \dots k_{n-2}) = (X_{k_1})^{\{r-1\}}_s (k_2 \dots k_{n-2})\ .$$

Thus when $r > 1$, the map (9.4) is a levelwise equivalence of categories if and only if, for all $k_1 \in \Delta^{op}$,

$$(X_{k_1})^{\{r-1\}}_s \to (X_{k_1})^{\{r-1\}}_1 \times_{p^{(n-3)}(X_{k_1})^{\{r-1\}}_0} \overset{s}{\cdots} \times_{p^{(n-3)}(X_{k_1})^{\{r-1\}}_0} (X_{k_1})^{\{r-1\}}_1$$

is a levelwise equivalence of categories. But this holds because, by hypothesis, $X_{k_1} \in \mathsf{LTa}^{\mathsf{n-1}}_{\mathsf{wg}}$. □

The following criterion will be used in the proof of the central result of this chapter, Theorem 9.2.4.

Lemma 9.1.4 *Let $X \in \mathsf{Ta}^{\mathsf{n}}_{\mathsf{wg}}$. Then $X \in \mathsf{LTa}^{\mathsf{n}}_{\mathsf{wg}}$ if and only if* *(a)* $p^{(n-1)}X \in \mathsf{Cat}^{n-1}_{\mathsf{wg}}$.

(b) For each $k \geq 2$ *the induced Segal map in* $[\Delta^{n-2^{op}}, \mathsf{Cat}]$

$$v_k : X_k \to X_1 \times_{p^{(n-2)}X_0} \overset{k}{\cdots} \times_{p^{(n-2)}X_0} X_1 \tag{9.5}$$

is a levelwise equivalence of categories.

(c) For each $1 < r \leq n-1$, $k_1, \dots, k_{n-r} \in \Delta^{op}$, $s \geq 2$ *the induced Segal map in* $[\Delta^{r-1^{op}}, \mathsf{Cat}]$

$$v_{k_1 \dots k_{n-r}} : X_{k_1 \dots k_{n-r}\, s} \to X_{k_1 \dots k_{n-r}\, 1} \times_{p^{(r-2)}X_{k_1 \dots k_{n-r}\, 0}} \overset{s}{\cdots} \times_{p^{(r-2)}X_{k_1 \dots k_{n-r}\, 0}} X_{k_1 \dots k_{n-r}\, 1} \tag{9.6}$$

is a levelwise equivalence of categories.

Proof Let $X \in \mathsf{LTa}^{n}_{\mathsf{wg}}$. Then (a) and (b) hold by Proposition 9.1.3(a) and (b) (i). Also, by Proposition 9.1.3(b) (ii), $X_{k_1 \dots k_{n-r}} \in \mathsf{LTa}^{r}_{\mathsf{wg}}$. Thus (c) holds by Proposition 9.1.3(b) (i) applied to $X_{k_1 \dots k_{n-r}}$.

Conversely, let $X \in \mathsf{LTa}^{n}_{\mathsf{wg}}$ satisfy (a), (b), (c). We show that $X \in \mathsf{LTa}^{n}_{\mathsf{wg}}$ by induction on n. It holds for $n = 2$ since $\mathsf{LTa}^{2}_{\mathsf{wg}} = \mathsf{Ta}^{2}_{\mathsf{wg}}$. Suppose, conversely, that the statement holds for $(n-1)$. By Proposition 9.1.3 to prove that $X \in \mathsf{LTa}^{n}_{\mathsf{wg}}$ it is enough to show that $X_k \in \mathsf{LTa}^{n-1}_{\mathsf{wg}}$ for all $k \in \Delta^{op}$.

Since by hypothesis $p^{(n-1)}X \in \mathsf{Cat}^{n-1}_{\mathsf{wg}}$, $(p^{(n-1)}X)_k = p^{(n-2)}X_k \in \mathsf{Cat}^{n-2}_{\mathsf{wg}}$, that is, $X_k \in \mathsf{Ta}^{n-1}_{\mathsf{wg}}$ satisfies inductive hypothesis (a). By hypothesis (c), for each $s \geq 2$ the map

$$X_{ks} \to X_{k1} \times_{p^{(n-3)}X_{k0}} \overset{s}{\cdots} \times_{p^{(n-3)}X_{k0}} X_{k1}$$

is a levelwise equivalence of categories; that is, X_k satisfies inductive hypothesis (b).

Also by hypothesis (c), for each $1 < r \leq n-2$, $s \geq 2$,

$$\begin{aligned}&(X_k)_{k_2 \dots k_{n-r}\, s}\\ &\to (X_k)_{k_2 \dots k_{n-r}\, 1} \times_{p^{(r-2)}(X_k)_{k_2 \dots k_{n-r}\, 0}} \overset{s}{\cdots} \times_{p^{(r-2)}(X_k)_{k_2 \dots k_{n-r}\, 0}} (X_k)_{k_2 \dots k_{n-r}\, 1}\end{aligned}$$

is a levelwise equivalence of categories. So X_k satisfies inductive hypothesis (c). In conclusion, X_k satisfies the inductive hypothesis so we conclude that $X_k \in \mathsf{LTa}^{n-1}_{\mathsf{wg}}$, as required. □

9.1.4 $\mathsf{Cat}^{n}_{\mathsf{wg}}$ *and the Category* $\mathsf{LTa}^{n}_{\mathsf{wg}}$

Proposition 9.1.5 *Let* $X \in \mathsf{Cat}^{n}_{\mathsf{wg}}$. *Then* $X \in \mathsf{LTa}^{n}_{\mathsf{wg}}$.

Proof By Definition 9.1.1 of $\mathsf{LTa}^{n}_{\mathsf{wg}}$, the proposition follows by Corollary 7.2.14 and Remark 7.2.15. □

The following corollary will be used in the proof of the main result of this chapter, Theorem 9.2.4.

Corollary 9.1.6 *Let* $X \in \mathsf{Cat}^{\mathsf{n}}_{\mathsf{wg}}$, $1 \leq j \leq n-1$, $k \geq 2$. *Define*

$$Y = X_1 \times_{p^{(j-1)}X_0} \overset{k}{\cdots} \times_{p^{(j-1)}X_0} X_1 \in \mathsf{Cat}^{\mathsf{n-1}}.$$

Then

(a) $Y \in \mathsf{Cat}^{\mathsf{n-1}}_{\mathsf{wg}}$.
(b) If $X_t \in \mathsf{Cat}^{\mathsf{n-1}}_{\mathsf{hd}}$ *for all* $t \geq 0$, *then* $Y \in \mathsf{Cat}^{\mathsf{n-1}}_{\mathsf{hd}}$.

Proof We prove the corollary for $k = 2$, the case $k > 2$ being similar.

When $j = 1$, $p^{(j-1)}X_0 = X^d_0$, so (a) holds, since $X \in \mathsf{Cat}^{\mathsf{n}}_{\mathsf{wg}}$, while (b) follows by Corollary 5.2.2(b). When $j = n-1$,

$$Y = X_1 \times_{p^{(n-2)}X_0} X_1 \ .$$

By Proposition 9.1.5 the induced Segal map $X_2 \to Y$ in $\mathsf{Cat}^{\mathsf{n-1}}_{\mathsf{wg}}$ is a levelwise equivalence of categories and $X_2 \in \mathsf{Cat}^{\mathsf{n-1}}_{\mathsf{wg}}$. Thus by Corollary 7.2.9, $Y \in \mathsf{Cat}^{\mathsf{n-1}}_{\mathsf{wg}}$, proving (a). As for (b), if $X_2 \in \mathsf{Cat}^{\mathsf{n-1}}_{\mathsf{hd}}$ by Corollary 7.2.9 then also $Y \in \mathsf{Cat}^{\mathsf{n-1}}_{\mathsf{hd}}$.

Let $1 < j < n-1$. We proceed by induction on n. For $n = 2$ it holds since $Y = X_1 \times_{X^d_0} X_1$. Suppose, inductively, that it holds for each $2 \leq r \leq (n-1)$.

(a) We verify that $Y \in \mathsf{Cat}^{\mathsf{n-1}}$ satisfies the hypotheses of Proposition 7.2.8(b). In fact, for each $1 \leq r < n$,

$$Y^{\{r\}}_0 = (X_1)^{\{r\}}_0 \times_{(p^{(j-1)}X_0)^{\{r\}}_0} (X_1)^{\{r\}}_0 \ .$$

Since $X \in \mathsf{Cat}^{\mathsf{n}}_{\mathsf{wg}}$, $X_1 \in \mathsf{Cat}^{\mathsf{n-1}}_{\mathsf{wg}}$, so by Lemma 7.2.5, $(X_1)^{\{r\}}_0$ is a levelwise equivalence relation. Since $(p^{(j-1)}X_0)^{\{r\}}_0$ is levelwise discrete, it follows by Corollary 5.2.2 that $Y^{\{r\}}_0$ is a levelwise equivalence relation. Thus Y satisfies hypothesis (i) in Proposition 7.2.8(b).

By inductive hypothesis (a) applied to $p^{(n-1)}X \in \mathsf{Cat}^{\mathsf{n-1}}_{\mathsf{wg}}$, using the fact that $j \leq n-2$, we obtain

$$p^{(n-2)}Y = (p^{(n-1)}X)_1 \times_{p^{(j-1)}(p^{(n-1)}X)_0} (p^{(n-1)}X)_1 \in \mathsf{Cat}^{\mathsf{n-2}}_{\mathsf{wg}} \ . \tag{9.7}$$

This shows that Y satisfies hypothesis (ii) of Proposition 7.2.8(b). We conclude from Proposition 7.2.8 that $Y \in \mathsf{Cat}^{\mathsf{n-1}}_{\mathsf{wg}}$.

b) Since $X_1 \in \mathsf{Cat}^{\mathsf{n-1}}_{\mathsf{hd}}$, X_1 is a levelwise equivalence relation and thus by Corollary 5.2.2 so is Y. Further, since $X_t \in \mathsf{Cat}^{\mathsf{n-1}}_{\mathsf{hd}}$, $(p^{(n-1)}X)_t = p^{(n-2)}X_t \in \mathsf{Cat}^{\mathsf{n-2}}_{\mathsf{hd}}$, so by (9.7) and inductive hypothesis (b) applied to $p^{(n-1)}X$ we conclude that $p^{(n-2)}Y \in \mathsf{Cat}^{\mathsf{n-2}}_{\mathsf{hd}}$. By definition, this means that $Y \in \mathsf{Cat}^{\mathsf{n-1}}_{\mathsf{hd}}$.

□

9.2 Approximating $\mathsf{Ta}^{\mathsf{n}}_{\mathsf{wg}}$ by $\mathsf{LTa}^{\mathsf{n}}_{\mathsf{wg}}$

The main result of this section, Theorem 9.2.4, states that if $X \in \mathsf{Ta}^{\mathsf{n}}_{\mathsf{wg}}$ is such that $q^{(n-1)}X$ can be approximated up to $(n-1)$-equivalence by an object of $\mathsf{Cat}^{(\mathsf{n}-1)}_{\mathsf{wg}}$, then X can be approximated up to an n-equivalence by an object of $\mathsf{LTa}^{\mathsf{n}}_{\mathsf{wg}}$. In the next section this property is used in the proof of Theorem 10.2.1 to construct the functor

$$P_n : \mathsf{Ta}^{\mathsf{n}}_{\mathsf{wg}} \to \mathsf{LTa}^{\mathsf{n}}_{\mathsf{wg}}$$

from which the rigidification functor Q_n will be built.

9.2.1 *The Main Steps in Approximating* $\mathsf{Ta}^{\mathsf{n}}_{\mathsf{wg}}$ *by* $\mathsf{LTa}^{\mathsf{n}}_{\mathsf{wg}}$

We now discuss informally the main steps needed in approximating $\mathsf{Ta}^{\mathsf{n}}_{\mathsf{wg}}$ by $\mathsf{LTa}^{\mathsf{n}}_{\mathsf{wg}}$, before the formal proofs in the next section. This approximation is crucial in building the rigidification functor Q_n. The ideas of the categories $\mathsf{Ta}^{\mathsf{n}}_{\mathsf{wg}}$ and $\mathsf{LTa}^{\mathsf{n}}_{\mathsf{wg}}$ were discussed in Sects. 6.1.1 and 6.3.1 respectively.

The basic construction is the pullback in $[\Delta^{n-1^{op}}, \mathsf{Cat}]$

$$\begin{array}{ccc} P & \xrightarrow{\quad w \quad} & X \\ \big\downarrow & & \big\downarrow \gamma^{(n-1)} \\ Z & \xrightarrow[\quad r \quad]{} & q^{(n-1)}X \end{array} \tag{9.8}$$

with $X \in \mathsf{Ta}^{\mathsf{n}}_{\mathsf{wg}}$, $Z \in \mathsf{Cat}^{\mathsf{n}-1}_{\mathsf{wg}}$ and $r : Z \to q^{(n-1)}X$ an $(n-1)$-equivalence in $\mathsf{Ta}^{\mathsf{n}-1}_{\mathsf{wg}}$. We show in Theorem 9.2.4 that $P \in \mathsf{LTa}^{\mathsf{n}}_{\mathsf{wg}}$ and w is an n-equivalence.

This construction is crucial to proving the existence of the rigidification functor Q_n: in the proof of Theorem 10.2.1 we will use the above construction with $Z = Q_{n-1}q^{(n-1)}X$ for

$$Q_{n-1} : \mathsf{Ta}^{\mathsf{n}-1}_{\mathsf{wg}} \to \mathsf{Cat}^{\mathsf{n}-1}_{\mathsf{wg}}$$

inductively defined and the $(n-1)$-equivalence r also given by the inductive hypothesis.

By Proposition 7.1.12, $P \in \mathsf{Ta}^{\mathsf{n}}_{\mathsf{wg}}$. To prove that $P \in \mathsf{LTa}^{\mathsf{n}}_{\mathsf{wg}}$ it is enough to show that P satisfies the hypotheses of Lemma 9.1.4.

To verify hypothesis (b) of Lemma 9.1.4 we apply the criterion of Proposition 7.1.5 (for an n-equivalence in $\mathsf{Ta}^{\mathsf{n}}_{\mathsf{wg}}$ to be a levelwise equivalence of categories) to the map

$$P_k \to P_1 \times_{p^{(n-2)}P_0} \overset{k}{\cdots} \times_{p^{(n-2)}P_0} P_1 \;, \tag{9.9}$$

for each $k \geq 2$. That is, we want to show this is an $(n-1)$-equivalence in $\mathsf{Ta}^{\mathsf{n-1}}_{\mathsf{wg}}$ satisfying the additional conditions in hypotheses (b) and (c) of Proposition 7.1.5: the idea is that these additional conditions are forced by the fact that $Z \in \mathsf{Cat}^{\mathsf{n-1}}_{\mathsf{wg}}$.

The intuition is as follows. Since (9.8) is a pullback in $[\Delta^{n-1^{op}}, \mathsf{Cat}]$, it is computed levelwise, so for each $k \geq 0$ there is a pullback in $[\Delta^{n-3^{op}}, \mathsf{Cat}]$

$$\begin{array}{ccc} P_{k0} & \longrightarrow & X_{k0} \\ \big\downarrow & & \big\downarrow {\scriptstyle (\gamma^{(n-1)})_{k0}} \\ Z_{k0} & \longrightarrow & (q^{(n-1)}X)_{k0} = q^{(n-3)}X_{k0} \end{array} \tag{9.10}$$

As X_{k0} is homotopically discrete,

$$(q^{(n-1)}X)_{k0} = q^{(n-3)}X_{k0} = p^{(n-3)}X_{k0} \;.$$

Applying p levelwise to (9.10) and using the fact that p commutes with pullbacks over discrete objects, we deduce that

$$p^{(n-3)}P_{k0} = Z_{k0} \;.$$

Thus, for instance,

$$\begin{aligned} & p^{(n-3)}(P_1 \times_{p^{(n-2)}P_0} P_1)_0 = p^{(n-3)}(P_{10} \times_{p^{(n-3)}P_{00}} P_{10}) \\ = {} & p^{(n-3)}P_{10} \times_{p^{(n-3)}P_{00}} p^{(n-3)}P_{10} = Z_{10} \times_{Z_{00}} Z_{10} \;. \end{aligned}$$

Since $Z \in \mathsf{Cat}^{\mathsf{n-1}}_{\mathsf{wg}}$, $Z_{20} \cong Z_{10} \times_{Z_{00}} Z_{10}$ so from above

$$p^{(n-3)}P_{20} \cong p^{(n-3)}(P_1 \times_{p^{(n-2)}P_0} P_1)_0 \;.$$

Thus condition (b) in the hypotheses of Proposition 7.1.5 holds for the map

$$P_2 \to P_1 \times_{p^{(n-2)}P_0} P_1 \;.$$

Similarly for the maps $P_k \to P_1 \times_{p^{(n-2)}P_0} \overset{k}{\cdots} \times_{p^{(n-2)}P_0} P_1$ for each $k \geq 2$.

The main steps in the formal proof that $P \in \mathsf{LTa}^n_{wg}$ are as follows:

(a) We apply Proposition 7.1.12 to deduce that $P \in \mathsf{Ta}^n_{wg}$.
(b) We show in Proposition 9.2.3 that the map (9.9) is a levelwise equivalence of categories. The proof of this fact is based on Lemma 9.2.1 below applied to the induced Segal maps of P which are shown to factor as a composite of maps in Ta^{n-1}_{wg}

$$\hat{\mu}_k : P_k \xrightarrow{\alpha} P_1 \times_{p^{(n-2)} P_0} \overset{k}{\cdots} \times_{p^{(n-2)} P_0} P_1 \xrightarrow{\beta} P_1 \times_{P_0^d} \overset{k}{\cdots} \times_{P_0^d} P_1 \; .$$

The fact that $Z \in \mathsf{Cat}^{n-1}_{wg}$ forces additional conditions (hypotheses (i) and (ii) of Lemma 9.2.1) which allow us to apply the criterion of Proposition 7.1.5 to the map α and deduce that it is a levelwise equivalence of categories.
(c) In Theorem 9.2.4 we use Proposition 9.2.3 and Lemma 7.2.10 to show that $p^{(n-1)}P$ satisfies the hypotheses of Lemma 7.2.10 and therefore $p^{(n-1)}P \in \mathsf{Cat}^{n-1}_{wg}$. Thus P satisfies hypothesis (a) of Lemma 9.1.4. Working inductively we then easily establish that hypothesis (c) of Lemma 9.1.4 also holds. From point (b) above it follows, by Lemma 9.1.4, that $P \in \mathsf{LTa}^n_{wg}$.

9.2.2 *Approximating* Ta^n_{wg} *by* LTa^n_{wg}*: The Formal Proofs*

The following lemma and its corollary are used in the proof of Theorem 9.2.4. Their proofs use the criterion given in Proposition 7.1.5 for an n-equivalence in Ta^n_{wg} to be a levelwise equivalence of categories and the properties of pullbacks along the map $\gamma^{(n-1)}$ established in Sect. 7.1.3.

Lemma 9.2.1 *Let*

$$B \overset{\partial_0}{\longrightarrow} X \overset{\partial_1}{\longleftarrow} B$$

be a diagram in Ta^n_{wg} *with* $X \in \mathsf{Cat}^n_{hd}$, $B \times_X B \in \mathsf{Ta}^n_{wg}$ *and let*

$$A \xrightarrow{\alpha} B \times_X B \xrightarrow{\beta} B \times_{X^d} B$$

be maps Ta^n_{wg} *(where β is induced by the map $\gamma : X \to X^d$) such that*

(i) $p^{(n-2)}\alpha_0$, *and* $p^{(n-r-2)}\alpha_{k_1 \ldots k_r 0}$ *are isomorphisms for all* $1 \le r < n-1$.
(ii) $(B \times_X B)^d_0 \cong B^d_0 \times_{X^d_0} B^d_0$,
$(B \times_X B)^d_{k_1 \ldots k_r 0} \cong B^d_{k_1 \ldots k_r 0} \times_{X^d_{k_1 \ldots k_r 0}} B^d_{k_1 \ldots k_r 0}$ *for all* $1 \le r < n-1$.
(iii) $\beta\alpha$ *is an n-equivalence.*

Then α is a levelwise equivalence of categories.

Proof Let $x, x' \in A^d_0$. By hypothesis (i) and (ii),

$$A^d_0 \cong (B \times_X B)^d_0 = B^d_0 \times_{X^d_0} B^d_0 \subset B^d_0 \times_{X^d} B^d_0 \ ,$$

where the last inclusion holds because the map $\gamma_0 : X_0 \to (X^d)_0 = X^d$ factors through X^d_0. Let

$$\alpha x = (a, b) \in B^d_0 \times_{X^d_0} B^d_0, \qquad \alpha x' = (a', b') \in B^d_0 \times_{X^d_0} B^d_0.$$

We claim that the composite map

$$A(x, x') \xrightarrow{\alpha(x,x')} B(a, a') \times_{X(\partial a, \partial a')} B(b, b') \xrightarrow{s} B(a, a') \times_{X(\partial a, \partial a')^d} B(b, b') \tag{9.11}$$

is an $(n-1)$-equivalence. We have

$$(B \times_X B)(\alpha x, \beta x') = B(a, a') \times_{X(\partial_0 a, \partial_0 a')} B(b, b') \ ,$$
$$(B \times_{X^d} B)(\beta\alpha x, \beta\alpha x') = B(a, a') \times_{X^d(\gamma a, \gamma a')} B(b, b') = B(a, a') \times B(b, b') \ ,$$

where in the last equality we used the fact that $\gamma a = \gamma a'$, so that $X^d(\gamma a, \gamma a') = \{\cdot\}$, since X^d is discrete.

The map $\gamma : X \to X^d$ factors as

$$X \to p^{(1)} \dots p^{(n-1)} X \to p \dots p^{(n-1)} X = X^d$$

and we have

$$(p^{(1)} \dots p^{(n-1)} X)(p\partial_0 a, p\partial_0 a') = X(\partial_0 a, \partial_0 a')^d \ .$$

Thus the map $\beta(x, x')$ factors as

$$B(a, a') \times_{X(\partial_0 a, \partial_0 a')} B(b, b') \xrightarrow{s} B(a, a') \times_{X(\partial_0 a, \partial_0 a')^d} B(b, b') \xrightarrow{t} B(a, a') \times B(b, b') \ .$$

On the other hand, since $p^{(1)} X \in \mathsf{Cat}_{\mathsf{hd}}$, the set

$$(p^{(1)} X)(p\partial_0 a, p\partial_0 a')$$

contains only one element. Thus $X(\partial_0 a, \partial_0 a')^d$ is the terminal object, and $t = \mathrm{Id}$, so that

$$(\beta\alpha)(x, x') = s\alpha(x, x'). \tag{9.12}$$

Since, by hypothesis, $\beta\alpha$ is an n-equivalence $\beta\alpha(x, x')$ is an $(n-1)$-equivalence, so by (9.12) the composite (9.11) is an $(n-1)$-equivalence. This proves the claim. We now proceed to the rest of the proof by induction on n. The strategy is to show that α satisfies the hypotheses of Proposition 7.1.5.

When $n = 2$, since $X(\partial_0 a, \partial_0 a') \in \mathsf{Cat}_{\mathsf{hd}}$, the map s in (9.11) is fully faithful. Since $s\alpha(x, y)$ is an equivalence of categories, it is essentially surjective on objects, and therefore s is essentially surjective on objects. It follows that s is an equivalence of categories, and therefore so is $\alpha(x, y)$. Since by hypothesis $p\alpha_0$ is a bijection, the map $p^{(1)}\alpha$ is bijective on objects, thus $pp^{(1)}\alpha$ is surjective. From Proposition 7.1.2 we deduce that α is a 2-equivalence. Thus α satisfies the hypotheses of Proposition 7.1.5 and we conclude that it is a levelwise equivalence of categories.

Suppose, inductively, that the lemma holds for $n-1$.

We show that the maps (9.11) satisfy the inductive hypothesis. Since, as proved above, $s\alpha(x, y)$ is an $(n-1)$-equivalence, inductive hypothesis (iii) holds. Since by hypothesis $p^{(n-2)}\alpha_0$ is an isomorphism, so is

$$p^{(n-3)}\alpha_0(x, x') = (p^{(n-2)}\alpha_0)(x, x')$$

as well as

$$p^{(n-r-3)}\{\alpha_{k_1\dots k_r 0}(x, x')\} = (p^{(n-r-2)}\alpha_{k_1\dots k_r 0})(x, x') .$$

Thus inductive hypothesis (i) holds for the maps (9.11). Further, using hypothesis (ii) we compute

$$\begin{aligned} &((B\times_X B)(\alpha x, \alpha x'))^d_0 = (B_1 \times_{X_1} B_1)^d_0(\alpha x, \alpha x') \\ &= (B^d_{10} \times_{X^d_{10}} B^d_{10})(\alpha x, \alpha x') = B(\alpha x, \alpha x')^d_0 \times_{X(\alpha x, \alpha x')^d_0} B(\alpha x, \alpha x')^d_0 \end{aligned}$$

and similarly

$$\begin{aligned} &((B\times_X B)(\alpha x, \alpha x'))^d_{k_1\dots k_r 0} = (B\times_X B)^d_{1k_1\dots k_r 0}(\alpha x, \alpha x') \\ &\cong B^d_{1k_1\dots k_r 0}(\alpha x, \alpha x') \times_{X^d_{1k_1\dots k_r 0}(\alpha x, \alpha x')} B^d_{1k_1\dots k_r 0}(\alpha x, \alpha x') \\ &= B_{k_1\dots k_r 0}(\alpha x, \alpha x')^d \times_{X_{k_1\dots k_r 0}(\alpha x, \alpha x')^d} B_{k_1\dots k_r 0}(\alpha x, \alpha x')^d . \end{aligned}$$

Thus inductive hypothesis (ii) holds for the maps (9.11). We conclude by induction that $\alpha(x, x')$ is a levelwise equivalence of categories. It follows by Remark 7.1.4 that $\alpha(x, x')$ is an $(n-1)$-equivalence. That is, α is a local $(n-1)$-equivalence.

Since by hypothesis $p^{(n-2)}\alpha_0$ is an isomorphism, so is

$$(p^{(1)}\alpha)_0 = p\alpha_0 ,$$

so that $p\alpha$ is surjective. Since, from above, α is a local $(n-1)$-equivalence, from Proposition 7.1.2 we conclude that α is an n-equivalence. Together with hypothesis (i) this shows that α satisfies the hypotheses of Proposition 7.1.5 and we conclude that α is a levelwise equivalence of categories. □

Remark 9.2.2 Lemma 9.2.1 also generalizes as follows, with a completely analogous proof. If B and X are as in Lemma 9.2.1, $k \geq 2$ and

$$A \xrightarrow{\alpha} B\times_X \overset{k}{\cdots}\times_X B \xrightarrow{\beta} B\times_{X^d} \overset{k}{\cdots}\times_{X^d} B$$

are maps in $\mathsf{Ta}^{\mathsf{n}}_{\mathsf{wg}}$ (where β is induced by the map $\gamma : X \to X^d$) such that

(i) $p^{(n-2)}\alpha_0$ and $p^{(n-r-2)}\alpha_{k_1\dots k_r 0}$ are isomorphisms for all $1 \leq r \leq n-1$.

(ii) $(B\times_X \overset{k}{\cdots}\times_X B)^d_0 \cong B^d_0\times_{X^d} \overset{k}{\cdots}\times_{X^d} B^d_0$,
$(B\times_X \overset{k}{\cdots}\times_X B)^d_{k_1\dots k_r 0}$
$\cong B^d_{k_1\dots k_r 0}\times_{X^d_{k_1\dots k_r 0}} \overset{k}{\cdots}\times_{X^d_{k_1\dots k_r 0}} B^d_{k_1\dots k_r 0}$ for all $1 \leq r \leq n-1$.

(iii) $\beta\alpha$ is an n-equivalence.

Then α is a levelwise equivalence of categories.

Using the lemma and remark above, we now prove the following proposition, which will be used in the proof of Theorem 9.2.4. In the proof of this result we use the properties of pullbacks along the map $\gamma^{(n-1)}$ established in Sect. 7.1.3.

Proposition 9.2.3 *Let $X \in \mathsf{Ta}^{\mathsf{n}}_{\mathsf{wg}}$, and let*

$$r : Z \to q^{(n-1)}X$$

be a map in $\mathsf{Ta}^{\mathsf{n-1}}_{\mathsf{wg}}$ with $Z \in \mathsf{Cat}^{\mathsf{n-1}}_{\mathsf{wg}}$ and consider the pullback in $[\Delta^{n-1^{op}}, \mathsf{Cat}]$

$$\begin{array}{ccc} P & \xrightarrow{w} & X \\ \downarrow & & \downarrow{\scriptstyle \gamma^{(n-1)}} \\ Z & \xrightarrow[r]{} & q^{(n-1)}X \end{array}$$

Then $P \in \mathsf{Ta}^{\mathsf{n}}_{\mathsf{wg}}$, $P_1\times_{p^{(n-1)}P_0} \overset{k}{\cdots}\times_{p^{(n-1)}P_0} P_1 \in \mathsf{Ta}^{\mathsf{n-1}}_{\mathsf{wg}}$ and for all $k \geq 2$ the map

$$v_k : P_k \to P_1\times_{p^{(n-2)}P_0} \overset{k}{\cdots}\times_{p^{(n-2)}P_0} P_1$$

is a levelwise equivalence of categories.

Proof Throughout this proof we will, for simplicity, denote the map v_k by α. By Proposition 7.1.12, $P \in \mathsf{Ta}^{\mathsf{n}}_{\mathsf{wg}}$, therefore its induced Segal maps $\hat{\mu}_k$ are $(n-1)$-equivalences. The strategy of the proof is to show that for each $k \geq 2$ $\hat{\mu}_k$ factors in $\mathsf{Ta}^{\mathsf{n-1}}_{\mathsf{wg}}$ as

$$\hat{\mu}_k : P_k \xrightarrow{\alpha} P_1 \times_{p^{(n-2)}P_0} \overset{k}{\cdots} \times_{p^{(n-2)}P_0} P_1 \xrightarrow{\beta} P_1 \times_{P_0^d} \overset{k}{\cdots} \times_{P_0^d} P_1$$

and this satisfies the hypotheses of Lemma 9.2.1 (see also Remark 9.2.2): the hypothesis (iii) of Lemma 9.2.1 holds since $P \in \mathsf{Ta}^{\mathsf{n}}_{\mathsf{wg}}$ while hypotheses (i) and (ii) will be a direct consequence of the fact that $Z \in \mathsf{Cat}^{\mathsf{n-1}}_{\mathsf{wg}}$, as illustrated below.

We first show that for each $k \geq 2$

$$P_1 \times_{p^{(n-2)}P_0} \overset{k}{\cdots} \times_{p^{(n-2)}P_0} P_1 \in \mathsf{Ta}^{\mathsf{n-1}}_{\mathsf{wg}} .$$

We illustrate the proof of this fact for $k = 2$, the case $k > 2$ being similar. Since p commutes with pullbacks over discrete objects we have

$$p^{(n-1)}P = Z \times_{q^{(n-1)}X} p^{(n-1)}X$$

and, (since $q^{(n-2)}X_0 = p^{(n-2)}X_0$ as $X_0 \in \mathsf{Cat}^{\mathsf{n-1}}_{\mathsf{hd}}$)

$$p^{(n-2)}P_0 = Z_0 \times_{q^{(n-2)}X_0} p^{(n-2)}X_0 = Z_0 .$$

Also, $P_1 = Z_1 \times_{q^{(n-2)}X_1} X_1$. Therefore

$$\begin{aligned} & P_1 \times_{p^{(n-2)}P_0} P_1 \\ &= (Z_1 \times_{q^{(n-2)}X_1} X_1) \times_{Z_0} (Z_1 \times_{q^{(n-2)}X_1} X_1) \\ &= (Z_1 \times_{Z_0} Z_1) \times_{q^{(n-2)}(X_1 \times X_1)} (X_1 \times X_1) . \end{aligned}$$

By Proposition 7.1.12, this is an object of $\mathsf{Ta}^{\mathsf{n-1}}_{\mathsf{wg}}$.

The induced Segal map $\hat{\mu}_2$ for P can therefore be written as a composite of maps in $\mathsf{Ta}^{\mathsf{n-1}}_{\mathsf{wg}}$

$$P_2 \xrightarrow{\alpha} P_1 \times_{p^{(n-2)}P_0} P_1 \xrightarrow{\beta} P_1 \times_{P_0^d} P_1 . \tag{9.13}$$

We show that the maps (9.13) satisfy the hypotheses of Lemma 9.2.1.

Since $P \in \mathsf{Ta}^{\mathsf{n}}_{\mathsf{wg}}$, $\hat{\mu}_2 = \beta\alpha$ is an $(n-1)$-equivalence, so hypothesis (iii) of Lemma 9.2.1 holds for the maps (9.13).

To check hypothesis (i), note that

$$\begin{aligned} & p^{(n-3)}P_{20} = p^{(n-3)}(Z_{20}\times_{q^{(n-3)}X_{10}} p^{(n-3)}X_{10}) = Z_{20}\,, \\ & p^{(n-3)}(P_1 \times_{p^{(n-2)}P_0} P_1)_0 \\ = {} & p^{(n-3)}P_{10}\times_{p^{(n-3)}P_{00}} p^{(n-3)}P_{10} = Z_{10}\times_{Z_{00}} Z_{10} \cong Z_{20}\,, \end{aligned} \tag{9.14}$$

where the last isomorphism holds since $Z \in \mathsf{Cat}^{\mathsf{n-1}}_{\mathsf{wg}}$. Hence $p^{(n-2)}\alpha_0$ is an isomorphism. Similarly

$$P_{s_1\dots s_r} = Z_{s_1\dots s_r}\times_{q^{(n-r-2)}X_{s_1\dots s_r}} X_{s_1\dots s_r}\,,$$

$$p^{(n-r-2)}P_{s_1\dots s_r} = p^{(n-r-2)}Z_{s_1\dots s_r}\times_{q^{(n-r-2)}X_{s_1\dots s_r}} p^{(n-r-2)}X_{s_1\dots s_r}\,.$$

Thus

$$\begin{aligned} & p^{(n-r-3)}P_{2k_1\dots k_r0} = Z_{2k_1\dots k_r0}\,, \\ & p^{(n-r-3)}(P_1\times_{p^{(n-2)}P_0} P_1)_{k_1\dots k_r0} \\ = {} & p^{(n-r-3)}(P_{1k_1\dots k_r0}\times_{p^{(n-r-3)}P_{0k_1\dots k_r0}} P_{1k_1\dots k_r0}) \\ = {} & p^{(n-r-3)}P_{1k_1\dots k_r0}\times_{p^{(n-r-3)}P_{0k_1\dots k_r0}} p^{(n-r-3)}P_{1k_1\dots k_r0} \\ = {} & Z_{1k_1\dots k_r0}\times_{Z_{0k_1\dots k_r0}} Z_{1k_1\dots k_r0}\cong Z_{2k_1\dots k_r0}\,, \end{aligned}$$

where the last isomorphism holds since $Z \in \mathsf{Cat}^{\mathsf{n-1}}_{\mathsf{wg}}$. This shows that $p^{(n-r-2)}\alpha_{k_1\dots k_r0}$ is an isomorphism, proving hypothesis (i) of Lemma 9.2.1 for the maps (9.13).

To check hypothesis (ii) of Lemma 9.2.1 note that by (9.14) $P^d_{20} = (p^{(n-3)}P_{20})^d = Z^d_{20}$ while

$$\begin{aligned} & (P_1\times_{p^{(n-2)}P_0} P_1)^d_0 = (p^{(n-3)}(P_1\times_{p^{(n-2)}P_0} P_1)_0)^d_0 \\ = {} & (Z_{10}\times_{Z_{00}} Z_{10})^d = Z^d_{10}\times_{Z^d_{00}} Z^d_{10}\cong Z^d_{20}\,, \end{aligned}$$

where the last isomorphism holds because $Z \in \mathsf{Cat}^{\mathsf{n-1}}_{\mathsf{wg}}$ (apply Remark 6.3.9 to $Z^{\{2\}}_0$, which is an object of $\mathsf{Cat}^{\mathsf{n-2}}_{\mathsf{wg}}$ by Proposition 7.2.8). Similarly

$$\begin{aligned} & (P_2)^d_{k_1\dots k_r0} = Z^d_{2k_1\dots k_r0}\,, \\ & (P_1\times_{p^{(n-2)}P_0} P_1)^d_{k_1\dots k_r0} \\ = {} & (Z_{1k_1\dots k_r0}\times_{p^{(n-2)}Z_{0k_1\dots k_r0}} Z_{1k_1\dots k_r0})^d \\ = {} & Z^d_{1k_1\dots k_r0}\times_{Z^d_{0k_1\dots k_r0}} Z^d_{1k_1\dots k_r0} = Z^d_{2k_1\dots k_r0}\,, \end{aligned}$$

where the last equality holds because $Z \in \mathsf{Cat}^{n-1}_{\mathsf{wg}}$ and therefore $Z_{k_0\dots k_r} \in \mathsf{Cat}^{n-r-1}_{\mathsf{wg}}$ by applying Remark 6.3.9 to $(Z_{k_0\dots k_r})^{\{2\}}_0$, which is an object of $\mathsf{Cat}^{n-r-1}_{\mathsf{wg}}$ by Proposition 7.2.8. This proves that hypothesis (ii) of Lemma 9.2.1 holds for (9.13).

So all hypotheses of Lemma 9.2.1 hold for the maps (9.13) and we conclude that α is a levelwise equivalence of categories. □

Theorem 9.2.4 *Let $X \in \mathsf{Ta}^{n}_{\mathsf{wg}}$, and let*

$$r : Z \to q^{(n-1)}X$$

be a map in $\mathsf{Ta}^{n-1}_{\mathsf{wg}}$ with $Z \in \mathsf{Cat}^{n-1}_{\mathsf{wg}}$ and consider the pullback in $[\Delta^{n-1^{op}}, \mathsf{Cat}]$

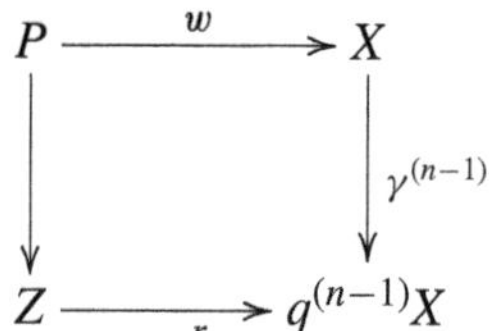

Then

(a) $q^{(n-1)}P$ and $p^{(n-1)}P$ are in $\mathsf{Cat}^{n-1}_{\mathsf{wg}}$.
(b) $P \in \mathsf{Ta}^{n}_{\mathsf{wg}}$ satisfies hypotheses (a) and (b) of Lemma 9.1.4.
(c) If r is an $(n-1)$-equivalence then w is an n-equivalence.
(d) $P \in \mathsf{LTa}^{n}_{\mathsf{wg}}$.

Proof By induction in n. When $n = 2$, we know by Proposition 7.1.12 that $P \in \mathsf{Ta}^{2}_{\mathsf{wg}} = \mathsf{LTa}^{2}_{\mathsf{wg}}$, and that (c) holds. Part (a) is trivial since $p^{(1)}P$ and $q^{(1)}P$ are in Cat. Part (d) holds since $\mathsf{Ta}^{2}_{\mathsf{wg}} = \mathsf{LTa}^{2}_{\mathsf{wg}}$.

Suppose, inductively, that the theorem holds for $n-1$.

(a) We have $q^{(n-1)}P = Z \in \mathsf{Cat}^{n-1}_{\mathsf{wg}}$, $p^{(n-1)}P \in \mathsf{Ta}^{n-1}_{\mathsf{wg}}$ since $P \in \mathsf{Ta}^{n}_{\mathsf{wg}}$ by Proposition 7.1.12. We now show that $p^{(n-1)}P$ satisfies the hypotheses of Lemma 7.2.10, which then shows that $p^{(n-1)}P \in \mathsf{Cat}^{n-1}_{\mathsf{wg}}$.

We have the pullback in $[\Delta^{n-2^{op}}, \mathsf{Cat}]$ for each $k \geq 0$,

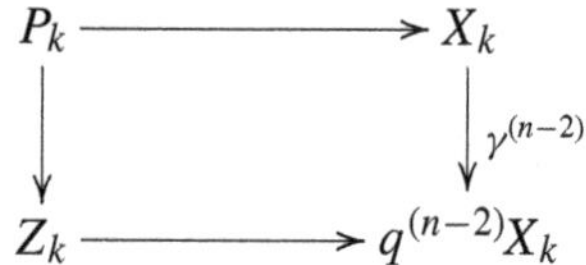

which satisfies the induction hypothesis. Therefore, by induction, $p^{(n-2)}P_k \in \mathsf{Cat}^{n-2}_{\mathsf{wg}}$ which is hypothesis (a) of Lemma 7.2.10. Since, by Proposition 9.2.3, the map

$$v_2 : P_2 \to P_1 \times_{p^{(n-2)}P_0} P_1$$

is a levelwise equivalence of categories, it induces an isomorphism

$$p^{(n-2)}P_2 \cong p^{(n-2)}(P_1 \times_{p^{(n-2)}P_0} P_1) \cong p^{(n-2)}P_1 \times_{p^{(n-2)}P_0} p^{(n-2)}P_1$$

and similarly all the other Segal maps of $p^{(n-1)}P$ are isomorphisms. This proves hypothesis (b) of Lemma 7.2.10 for $p^{(n-1)}X$.

To prove hypothesis (c) of Lemma 7.2.10, we first note that

$$P_1 \times_{p^{(j-1)}P_0} P_1 \in \mathsf{Ta}^{n-1}_{wg} \,. \tag{9.15}$$

In fact, since $p^{(n-1)}P_0 = Z_0$ we have

$$\begin{aligned} & P_1 \times_{p^{(j-1)}P_0} P_1 \\ &= (Z_1 \times_{q^{(n-2)}X_1} X_1) \times_{p^{(j-1)}Z_0} (Z_1 \times_{q^{(n-2)}X_1} X_1) \\ &= (Z_1 \times_{p^{(j-1)}Z_0} Z_1) \times_{q^{(n-2)}(X_1 \times X_1)} (X_1 \times X_1) \\ &= (Z_1 \times_{p^{(j-1)}Z_0} Z_1) \times_{q^{(n-2)}(X_1 \times X_1)} (X_1 \times X_1) \,. \end{aligned} \tag{9.16}$$

By Corollary 9.1.6, $Z_1 \times_{p^{(j-1)}Z_0} Z_1 \in \mathsf{Cat}^{n-2}_{wg}$; by Proposition 7.1.12 and (9.16) we conclude that (9.15) holds. It follows that

$$p^{(j)}(P_1 \times_{p^{(j-1)}P_0} P_1) = p^{(j)}P_1 \times_{p^{(j-1)}P_0} p^{(j)}P_1 \in \mathsf{Ta}^j_{wg} \,.$$

Since $P \in \mathsf{Ta}^n_{wg}$, $p^{(j+1)}P \in \mathsf{Ta}^{j+1}_{wg}$, so that the induced Segal map

$$p^{(j)}P_2 \to p^{(j)}P_1 \times_{(p^{(j)}P_0)^d} p^{(j)}P_1$$

is a j-equivalence in Ta^j_{wg}. From above, this map factorizes as a composite of maps in Ta^j_{wg}

$$p^{(j)}P_2 \xrightarrow{\alpha} p^{(j)}P_1 \times_{p^{(j-1)}P_0} p^{(j)}P_1 \xrightarrow{\beta} p^{(j)}P_1 \times_{(p^{(j)}P_0)^d} p^{(j)}P_1 \,. \tag{9.17}$$

We check that the maps (9.17) satisfy the hypotheses of Lemma 9.2.1. In fact since (as shown in the proof of Proposition 9.2.3) $p^{(n-3)}P_{k0} = Z_{k0}$, from Lemma 6.3.8 applied to $Z^{\{2\}}_0 \in \mathsf{Cat}^{n-2}_{wg}$ we obtain

$$\begin{aligned} & p^{(j-2)}(p^{(j)}P_2)_0 = p^{(j-2)}P_{20} = p^{(j-2)}Z_{20} \\ &\cong p^{(j-2)}Z_{10} \times_{p^{(j-2)}Z_{00}} p^{(j-2)}Z_{10} \\ &= p^{(j-2)}(p^{(j-1)}Z_{10} \times_{p^{(j-1)}Z_{00}} p^{(j-1)}Z_{10}) \\ &= p^{(j-2)}(p^{(j)}P_1 \times_{p^{(j)}P_0} p^{(j)}P_1)_0 \,. \end{aligned}$$

That is, $p^{(j-2)}\alpha_0$ is an isomorphism. Similarly one shows that $p^{(n-r-2)}\,\alpha_{k_1\dots k_r 0}$ is an isomorphism; thus hypothesis (i) of Lemma 9.2.1 holds. As for hypothesis (ii), since (as shown in the proof of Proposition 9.2.3) $P^d_{k0} \cong Z^d_{k0}$, we have

$$\begin{aligned}
&(p^{(j)}P_1 \times_{p^{(j)}P_0} p^{(j)}P_1)^d_0 \\
&\cong (p^{(j-1)}Z_{10} \times_{p^{(j-1)}Z_{00}} p^{(j-1)}Z_{10})^d \\
&\cong (p^{(j-1)}Z_{20})^d \cong Z^d_{20} \cong Z^d_{10} \times_{Z^d_{00}} Z^d_{10} \cong P^d_{10} \times_{P^d_{00}} P^d_{10} \\
&\cong (p^{(j)}P_1)^d_0 \times_{(p^{(j)}P_0)^d_0} (p^{(j)}P_1)^d_0 \,.
\end{aligned}$$

The rest of hypothesis (ii) of Lemma 9.2.1 is checked similarly, while hypothesis (iii) of Lemma 9.2.1 holds from above.

We can therefore apply Lemma 9.2.1 to (9.17) and conclude that α is a levelwise equivalence of categories. Therefore $p^{(j-1)}\alpha$ is an isomorphism, that is,

$$\begin{aligned}
p^{(j-1)}P_2 &\cong p^{(j-1)}(p^{(j)}P_1 \times_{p^{(j-1)}P_0} p^{(j)}P_1) \\
&= p^{(j-1)}P_1 \times_{p^{(j-1)}P_0} p^{(j-1)}P_1 \,.
\end{aligned}$$

Similarly one shows that all the other Segal maps for $p^{(j)}P$ are isomorphisms, which proves condition (c) in Lemma 7.2.10 for $p^{(n-1)}P$. Thus by Lemma 7.2.10 we conclude that $p^{(n-1)}P \in \mathsf{Cat}^{n-1}_{\mathsf{wg}}$.

(b) This follows from Proposition 9.2.3 and (a).

(c) Consider the commuting diagram in $\mathsf{Ta}^{n}_{\mathsf{wg}}$

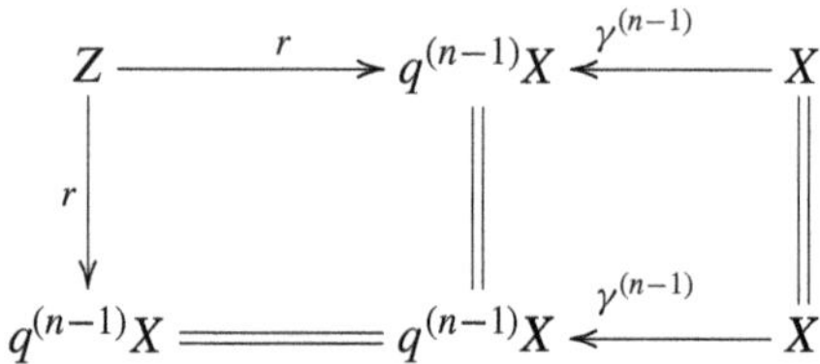

By hypothesis, r is an n-equivalence. Thus by Proposition 7.1.12 the induced map of pullbacks

$$P = Z \times_{q^{(n-1)}X} X \xrightarrow{w} q^{(n-1)}X \times_{q^{(n-1)}X} X = X$$

is an n-equivalence, as required.

(d) By (b), to show that $P \in \mathsf{LTa}^{n}_{\mathsf{wg}}$ it is enough to show that P satisfies hypothesis (c) of Lemma 9.1.4. For each $k_1, \dots, k_{n-r} \in \Delta^{op}$, $1 < r \leq n-1$, we have a pullback in $[\Delta^{r-1^{op}}, \mathsf{Cat}]$

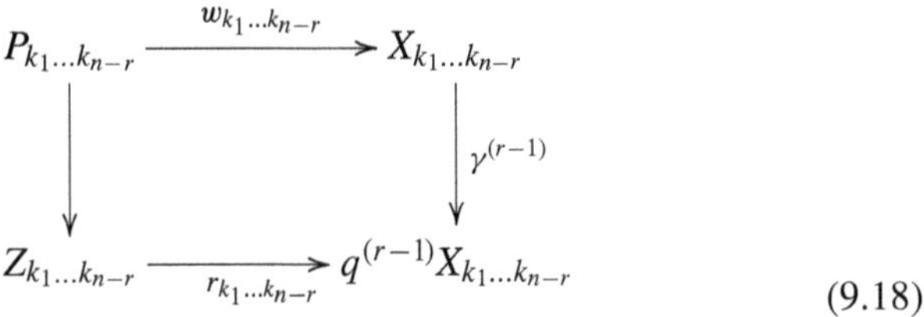

(9.18)

where $X_{k_1\dots k_{n-r}} \in \mathsf{Ta}^{\mathsf{r}}_{\mathsf{wg}}$ (since $X \in \mathsf{Ta}^{\mathsf{n}}_{\mathsf{wg}}$) and $Z_{k_1\dots k_{n-r}} \in \mathsf{Cat}^{\mathsf{r-1}}_{\mathsf{wg}}$ (since $Z \in \mathsf{Cat}^{\mathsf{n-1}}_{\mathsf{wg}}$).

Thus (9.18) satisfies the hypotheses of the theorem and we conclude from (b) that the maps

$$v_{k_1\dots k_{n-r}} : P_{k_1\dots k_{n-r}\,s} \to P_{k_1\dots k_{n-r}\,1} \times_{p^{(r-2)} P_{k_1\dots k_{n-r}\,0}} \overset{s}{\cdots} \times_{p^{(r-2)} P_{k_1\dots k_{n-r}\,0}} P_{k_1\dots k_{n-r}\,1}$$

are levelwise equivalences of categories. That is, P satisfies hypothesis (c) of Lemma 9.1.4, as required.

□

Corollary 9.2.5 *Let X, Z be as in the hypothesis of Theorem 9.2.4 and assume, further, that $X \in \mathsf{Cat}^{\mathsf{n}}_{\mathsf{wg}}$. Then $P \in \mathsf{Cat}^{\mathsf{n}}_{\mathsf{wg}}$.*

Proof By induction on n. When $n = 2$, $P \in \mathsf{Cat}^2$ with $p^{(1)}P \in \mathsf{Cat}$ (by Theorem 9.2.4). Therefore, by Lemma 7.2.10, $P \in \mathsf{Cat}^2_{\mathsf{wg}}$. Suppose, inductively, that the statement holds for $n-1$. We show that P satisfies the hypotheses of Lemma 7.2.10. For each $s \geq 0$ we have a pullback in $[\Delta^{n-2^{op}}, \mathsf{Cat}]$

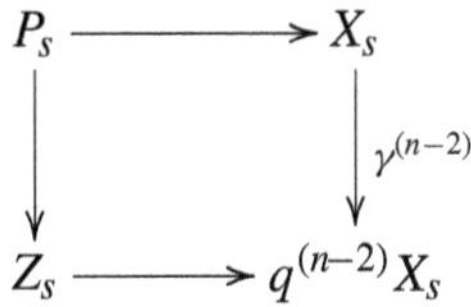

with $X_s \in \mathsf{Cat}^{\mathsf{n-1}}_{\mathsf{wg}}$ (since $X \in \mathsf{Cat}^{\mathsf{n}}_{\mathsf{wg}}$). So by the induction hypothesis, $P_s \in \mathsf{Cat}^{\mathsf{n-1}}_{\mathsf{wg}}$. Thus hypothesis (a) in Lemma 7.2.10 holds for P.

Hypothesis (b) also holds since $X \in \mathsf{Cat}^{\mathsf{n}}$, $q^{(n-1)}X \in \mathsf{Cat}^{\mathsf{n-1}}$, $Z \in \mathsf{Cat}^{\mathsf{n-1}}$, so $P \in \mathsf{Cat}^{\mathsf{n}}$. Finally, hypothesis (c) is satisfied because, by Theorem 9.2.4, $p^{(n-1)}P \in \mathsf{Cat}^{\mathsf{n-1}}_{\mathsf{wg}}$. We conclude by Lemma 7.2.10 that $P \in \mathsf{Cat}^{\mathsf{n}}_{\mathsf{wg}}$. □

Chapter 10
Rigidifying Weakly Globular Tamsamani n-Categories

Abstract This chapter contains one of the central results of this work: the existence of a rigidification functor Q_n from weakly globular Tamsamani n-categories to weakly globular n-fold categories, as well as n-equivalences $s_n(X) : Q_n X \to X$, natural in $X \in \mathsf{Ta}^{\mathsf{n}}_{\mathsf{wg}}$. We build a functor Tr_n from the category $\mathsf{LTa}^{\mathsf{n}}_{\mathsf{wg}}$ of Chap. 9 to the category $\mathsf{SegPs}[\Delta^{n-1^{op}}, \mathsf{Cat}]$ of Segalic pseudo-functors. The approximation results of Chap. 9 afford a functor P_n from $\mathsf{Ta}^{\mathsf{n}}_{\mathsf{wg}}$ to $\mathsf{LTa}^{\mathsf{n}}_{\mathsf{wg}}$. By pre-composing Tr_n with P_n and post-composing it with the strictification functor St (and using the main result of Chap. 8) we obtain the rigidification functor Q_n.

The main result of this chapter, Theorem 10.2.1, asserts the existence of a *rigidification functor*

$$Q_n : \mathsf{Ta}^{\mathsf{n}}_{\mathsf{wg}} \to \mathsf{Cat}^{\mathsf{n}}_{\mathsf{wg}}$$

such that for each $X \in \mathsf{Ta}^{\mathsf{n}}_{\mathsf{wg}}$ there is an n-equivalence, natural in X,

$$Q_n X \to X.$$

This result means that X can be approximated up to n-equivalence by the more rigid and therefore simpler structure $Q_n X$. In particular, this property implies (see Corollary 10.2.3) that the two categories $\mathsf{Cat}^{\mathsf{n}}_{\mathsf{wg}}$ and $\mathsf{Ta}^{\mathsf{n}}_{\mathsf{wg}}$ are equivalent after localization with respect to the n-equivalences.

The functor Q_n restricts in particular to a functor

$$Q_n : \mathsf{Ta}^{\mathsf{n}} \to \mathsf{Cat}^{\mathsf{n}}_{\mathsf{wg}}$$

from Tamsamani n-categories to weakly globular n-fold categories. In Chap. 12 we will show that this functor leads to an equivalence after localization between $\mathsf{Cat}^{\mathsf{n}}_{\mathsf{wg}}$ and Ta^{n}, exhibiting $\mathsf{Cat}^{\mathsf{n}}_{\mathsf{wg}}$ as a new model of weak n-categories satisfying, in particular, the homotopy hypothesis.

S. Paoli, *Simplicial Methods for Higher Categories*, Algebra and Applications 26,
https://doi.org/10.1007/978-3-030-05674-2_10

The rigidification functor factors through the category

$$\mathsf{SegPs}[\Delta^{n-1^{op}}, \mathsf{Cat}]$$

of Segalic pseudo-functors introduced in Chap. 8. More precisely, Q_n is the composite

$$Q_n : \mathsf{Ta^n_{wg}} \to \mathsf{SegPs}[\Delta^{n-1^{op}}, \mathsf{Cat}] \xrightarrow{St} \mathsf{Cat^n_{wg}} \subset [\Delta^{n-1^{op}}, \mathsf{Cat}]\ .$$

In the case $n = 2$, it is easy to build pseudo-functors from $\mathsf{Ta^2_{wg}}$, and was already done by Pronk and the author in [103]. More precisely, given $X \in \mathsf{Ta^2_{wg}}$, define $Tr_2X \in [ob(\Delta^{op}), \mathsf{Cat}]$ by

$$(Tr_2X)_k = \begin{cases} X_0^d & k = 0 \\ X_1 & k = 1 \\ X_1 \times_{X_0^d} \overset{k}{\cdots} \times_{X_0^d} X_1 & k > 1\ . \end{cases} \tag{10.1}$$

Since $X \in \mathsf{Ta^2_{wg}}$, $X_0 \in \mathsf{Cat_{hd}}$, so there are equivalences of categories

$$X_0 \simeq X_0^d\ ,$$
$$X_k \simeq X_1 \times_{X_0^d} \overset{k}{\cdots} \times_{X_0^d} X_1 \quad \text{for } k > 1.$$

Thus, for all $k \geq 0$ there is an equivalence of categories

$$(Tr_2X)_k \simeq X_k\ .$$

By using transport of structure (more precisely Lemma 4.3.2 with $\mathscr{C} = \Delta^{op}$) we can lift Tr_2X to a pseudo-functor

$$Tr_2X \in \mathsf{Ps}[\Delta^{op}, \mathsf{Cat}]$$

and by construction $Tr_2X \in \mathsf{SegPs}[\Delta^{op}, \mathsf{Cat}]$.

Building pseudo-functors from $\mathsf{Ta^n_{wg}}$ when $n > 2$ is much more complex, and is new to this work. The above approach cannot be applied directly because the induced Segal maps of $X \in \mathsf{Ta^n_{wg}}$, when $n > 2$, are $(n-1)$-equivalences but not in general levelwise equivalence of categories. For this reason we use the intermediate category $\mathsf{LTa^n_{wg}}$ introduced in Chap. 9, from which it is possible to build pseudo-functors using transport of structure. When $n > 2$, the functor from $\mathsf{Ta^n_{wg}}$ to Segalic pseudo-functors factorizes as

$$\mathsf{Ta^n_{wg}} \xrightarrow{P_n} \mathsf{LTa^n_{wg}} \xrightarrow{Tr_n} \mathsf{SegPs}[\Delta^{n-1^{op}}, \mathsf{Cat}]\ .$$

The functor P_n produces a functorial approximation (up to n-equivalence) of objects of $\mathsf{Ta^n_{wg}}$ with objects of $\mathsf{LTa^n_{wg}}$, while Tr_n is built using transport of structure.

This chapter is organized as follows. In Sect. 10.1 we show how to construct a pseudo-functor from the category $\mathsf{LTa^n_{wg}}$ introduced in Chap. 9, proving in Theorem 10.1.1 the existence of the functor

$$Tr_n : \mathsf{LTa^n_{wg}} \to \mathsf{SegPs}[\Delta^{n-1^{op}}, \mathsf{Cat}\,] \,.$$

In Sect. 10.2 we construct the rigidification functor $Q_n : \mathsf{Ta^n_{wg}} \to \mathsf{Cat^n_{wg}}$.

In the proof of Theorem 10.2.1 we define the functor $P_n : \mathsf{Ta^n_{wg}} \to \mathsf{LTa^n_{wg}}$ and then construct the rigidification functor as the composite

$$Q_n : \mathsf{Ta^n_{wg}} \xrightarrow{P_n} \mathsf{LTa^n_{wg}} \xrightarrow{Tr_n} \mathsf{SegPs}[\Delta^{n-1^{op}}, \mathsf{Cat}\,] \xrightarrow{St} \mathsf{Cat^n_{wg}} \,.$$

10.1 From $\mathsf{LTa^n_{wg}}$ to Pseudo-Functors

In this section we show that we can associate functorially to each object of $\mathsf{LTa^n_{wg}}$ a pseudo-functor which is Segalic. We build in Theorem 10.1.1 a functor

$$Tr_n : \mathsf{LTa^n_{wg}} \to \mathsf{SegPs}[\Delta^{n-1^{op}}, \mathsf{Cat}\,] \,,$$

together with a pseudo-natural transformation

$$t_n(X) : Tr_n X \to X$$

for each $X \in \mathsf{LTa^n_{wg}}$ which is a levelwise equivalence of categories. The functor Tr_n will be used in Sect. 10.2 to build the rigidification functor Q_n.

10.1.1 The Idea of the Functor Tr_n

The functor Tr_n involves the categories $\mathsf{LTa^n_{wg}}$ and $\mathsf{SegPs}[\Delta^{n-1^{op}}, \mathsf{Cat}\,]$, whose idea was discussed in Sects. 9.1.1 and 8.1.2, respectively. The construction of the functor Tr_n in Theorem 10.1.1 uses the defining property of $X \in \mathsf{LTa^n_{wg}}$ that for each $1 \le r < n$ and $k \ge 2$ the induced Segal map in $[\Delta^{n-2^{op}}, \mathsf{Cat}\,]$

$$v_k^{\{r\}} : X_k^{\{r\}} \to X_1^{\{r\}} \times_{p^{(n-2)} X_0^{\{r\}}} \overset{k}{\cdots} \times_{p^{(n-2)} X_0^{\{r\}}} X_1^{\{r\}}$$

is a levelwise equivalence of categories. Using induction we then build a diagram

$$Tr_n X \in [ob(\Delta^{n-1^{op}}), \mathsf{Cat}\,]$$

in which

(i) For all $\underline{k} \in \Delta^{n-1^{op}}$, $1 \leq i \leq n-1$

$$(Tr_n X)_0^{\{i\}} = (X_0^{\{i\}})^d$$

is discrete.

(ii) $(Tr_n X)_{\underset{1\dots 1}{n-1}} = X_{\underset{1\dots 1}{n-1}}$.

(iii) For $k_1 \geq 2$, $\underline{s} = (k_2, \dots, k_{n-1})$, $\underline{k} = (k_1, \underline{s})$,

$$(Tr_n X)_{\underline{k}} = (Tr_{n-1} X_1)_{\underline{s}} \times_{X^d_{\underline{k}(0,1)}} \overset{k_1}{\cdots} \times_{X^d_{\underline{k}(0,1)}} (Tr_{n-1} X_1)_{\underline{s}}$$

where $\underline{k}(0, i)$ is as in Remark 8.1.3.

For instance, when $n = 2$, we set

$$(Tr_2 X)_k = \begin{cases} X_0^d, & k = 0 \\ X_1, & k = 1 \\ X_1 \times_{X_0^d} \overset{k}{\cdots} \times_{X_0^d} X_1, & k > 1 . \end{cases}$$

When $n = 3$ we set

$$(Tr_3 X)_{k_1 k_2} = X^d_{k_1 k_2} \quad \text{if } k_1 = 0 \text{ or } k_2 = 0 ,$$
$$(Tr_3 X)_{11} = X_{11} ,$$
$$(Tr_3 X)_{k_1 1} = X_{11} \times_{X^d_{01}} \overset{k_1}{\cdots} \times_{X^d_{01}} X_{11} \quad \text{if } k_1 \geq 2 ,$$
$$(Tr_3 X)_{1 k_2} = X_{11} \times_{X^d_{10}} \overset{k_2}{\cdots} \times_{X^d_{10}} X_{11} \quad \text{if } k_2 \geq 2 .$$

If both $k_1 \geq 2$ and $k_2 \geq 2$, we set

$$\begin{aligned} (Tr_3 X)_{k_1 k_2} &= (Tr_2 X_1)_{1 k_2} \times_{X^d_{0k_2}} \overset{k_1}{\cdots} \times_{X^d_{0k_2}} (Tr_2 X_1)_{1 k_2} \\ &= (X_{11} \times_{X^d_{10}} \overset{k_2}{\cdots} \times_{X^d_{10}} X_{11}) \times_{(X^d_{01} \times_{X^d_{00}} \overset{k_2}{\cdots} \times_{X^d_{00}} X^d_{01})} \overset{k_1}{\cdots} \\ &\quad \overset{k_1}{\cdots} \times_{(X^d_{01} \times_{X^d_{00}} \overset{k_2}{\cdots} \times_{X^d_{00}} X^d_{01})} (X_{11} \times_{X^d_{10}} \overset{k_2}{\cdots} \times_{X^d_{10}} X_{11}) , \end{aligned}$$

where we used the fact that, since $X_0 \in \mathsf{Cat}^2_{\mathsf{hd}}$,

$$X^d_{0k_2} \cong X^d_{01} \times_{X^d_{00}} \overset{k_2}{\cdots} \times_{X^d_{00}} X^d_{01} .$$

Note that

$$X^d_{k_1 0} \cong X^d_{10} \times_{X^d_{00}} \overset{k_1}{\cdots} \times_{X^d_{00}} X^d_{10} \,.$$

By the commutation of pullbacks we obtain from the above

$$\begin{aligned}
&(Tr_3 X)_{k_1 k_2} \\
&= (X_{11} \times_{X^d_{01}} \overset{k_1}{\cdots} \times_{X^d_{01}} X_{11}) \times_{(X^d_{10} \times_{X^d_{00}} \overset{k_1}{\cdots} \times_{X^d_{00}} X^d_{10})} \overset{k_2}{\cdots} \\
&\overset{k_2}{\cdots} \times_{(X^d_{10} \times_{X^d_{00}} \overset{k_1}{\cdots} \times_{X^d_{00}} X^d_{10})} (X_{11} \times_{X^d_{01}} \overset{k_1}{\cdots} \times_{X^d_{01}} X_{11}) \\
&= (Tr_3 X)_{k_1 1} \times_{X^d_{k_1 0}} \overset{k_2}{\cdots} \times_{X^d_{k_1 0}} (Tr_3 X)_{k_1 1} \,.
\end{aligned}$$

After defining $Tr_n X \in [ob(\Delta^{n-1^{op}}), \mathsf{Cat}]$ we show that, for each $\underline{k} \in \Delta^{n-1^{op}}$, there is an equivalence of categories

$$(Tr_n X)_{\underline{k}} \simeq X_{\underline{k}} \,. \tag{10.2}$$

Using the 'transport of structure' technique of Lemma 4.3.2, we then lift $Tr_n X$ to a pseudo-functor

$$Tr_n X \in \mathsf{Ps}[\Delta^{n-1^{op}}, \mathsf{Cat}]$$

and we show that this is in fact a Segalic pseudo-functor. Conditions (a) and (b) in the definition of Segalic pseudo-functor depend on the conditions (i), (ii), (iii) in the definition of $(Tr_n X)_{\underline{k}}$, while condition (c) is a straightforward consequence of the equivalence of categories (10.2) and the fact that, since $X \in \mathsf{LTa}^{\mathsf{n}}_{\mathsf{wg}}$, $p^{(n-1)} X \in \mathsf{Cat}^{\mathsf{n-1}}_{\mathsf{wg}}$.

10.1.2 The Formal Construction of the Functor Tr_n

Theorem 10.1.1 *There is a functor*

$$Tr_n : \mathsf{LTa}^{\mathsf{n}}_{\mathsf{wg}} \to \mathsf{SegPs}[\Delta^{n-1^{op}}, \mathsf{Cat}]$$

together with a pseudo-natural transformation

$$t_n(X) : Tr_n X \to X$$

for each $X \in \mathsf{LTa}^{\mathsf{n}}_{\mathsf{wg}}$ which is a levelwise equivalence of categories.

Proof By induction on n. For $n = 2$, let $X \in \mathsf{LTa}^2_{\mathsf{wg}} = \mathsf{Ta}^2_{\mathsf{wg}}$. Define $Tr_2 X \in [ob(\Delta^{op}), \mathsf{Cat}]$

$$(Tr_2 X)_k = \begin{cases} X_0^d & k = 0 \\ X_1 & k = 1 \\ X_1 \times_{X_0^d} \overset{k}{\cdots} \times_{X_0^d} X_1 & k > 1\,. \end{cases} \tag{10.3}$$

Since $X \in \mathsf{Ta}^2_{\mathsf{wg}}$, $X_0 \in \mathsf{Cat}_{\mathsf{hd}}$, so there are equivalences of categories

$$X_0 \simeq X_0^d\,,$$
$$X_k \simeq X_1 \times_{X_0^d} \overset{k}{\cdots} \times_{X_0^d} X_1 \quad \text{for } k > 1.$$

Thus, for all $k \geq 0$ there is an equivalence of categories

$$(Tr_2 X)_k \simeq X_k\,.$$

We can therefore apply Lemma 4.3.2 with $\mathscr{C} = \Delta^{op}$ and conclude that $Tr_2 X$ lifts to a pseudo-functor

$$Tr_2 X \in \mathsf{Ps}[\Delta^{op}, \mathsf{Cat}]$$

and there is a pseudo-natural transformation

$$t_2(X) : Tr_2 X \to X$$

which is a levelwise equivalence of categories. By (10.3), $(Tr_2 X)_0$ is discrete and the Segal maps are isomorphisms. Therefore, by Definition 8.1.2,

$$Tr_2 X \in \mathsf{SegPs}[\Delta^{op}, \mathsf{Cat}]\,.$$

Suppose, inductively, that the theorem holds for $(n-1)$ and let $X \in \mathsf{LTa}^n_{\mathsf{wg}}$. By the definition of $\mathsf{LTa}^n_{\mathsf{wg}}$ (taking $r = 1$ in Definition 9.1.1(b)), for each $\underline{s} \in \Delta^{{n-2}^{op}}$, $j \geq 2$, there is an equivalence of categories

$$X_{j\underline{s}} \simeq X_{1\underline{s}} \times_{X^d_{0\underline{s}}} \overset{j}{\cdots} \times_{X^d_{0\underline{s}}} X_{1\underline{s}}\,. \tag{10.4}$$

Also, by the inductive hypothesis applied to $X_j \in \mathsf{LTa}^{n-1}_{\mathsf{wg}}$, there is an equivalence of categories for all $j \geq 0$ and $\underline{s} \in \Delta^{{n-2}^{op}}$

$$X_{j\underline{s}} \simeq (Tr_{n-1} X_j)_{\underline{s}}\,. \tag{10.5}$$

It follows from (10.4) that for each $j \geq 2$ there is an equivalence of categories

$$X_{1\underline{s}} \times_{X^d_{0\underline{s}}} \overset{j}{\cdots} \times_{X^d_{0\underline{s}}} X_{1\underline{s}} \simeq (Tr_{n-1}X_1)_{\underline{s}} \times_{X^d_{0\underline{s}}} \overset{j}{\cdots} \times_{X^d_{0\underline{s}}} (Tr_{n-1}X_1)_{\underline{s}} \,. \tag{10.6}$$

Thus (10.4), (10.5), (10.6) imply that for each $j \geq 2$, $\underline{s} \in \Delta^{2^{op}}$ there is an equivalence of categories

$$(Tr_{n-1}X_j)_{\underline{s}} \simeq (Tr_{n-1}X_1)_{\underline{s}} \times_{X^d_{0\underline{s}}} \overset{j}{\cdots} \times_{X^d_{0\underline{s}}} (Tr_{n-1}X_1)_{\underline{s}} \,. \tag{10.7}$$

Define $Tr_nX \in [ob(\Delta^{n-1^{op}}), \mathsf{Cat}]$ as follows: for each $\underline{k} = (k_1, \underline{s}) \in \Delta^{n-1^{op}}$ (with $k_1 \in \Delta^{op}$, $\underline{s} \in \Delta^{n-2^{op}}$)

$$(Tr_nX)_{\underline{k}} = \begin{cases} X^d_{\underline{k}} & \text{for } \underline{k} = (0, \underline{s}) \\ (Tr_{n-1}X_1)_{\underline{s}} & \text{for } \underline{k} = (1, \underline{s}) \\ (Tr_{n-1}X_1)_{\underline{s}} \times_{X^d_{0\underline{s}}} \overset{k_1}{\cdots} \times_{X^d_{0\underline{s}}} (Tr_{n-1}X_1)_{\underline{s}} & \text{for } \underline{k} = (k_1, \underline{s}), \\ & k_1 \geq 2. \end{cases} \tag{10.8}$$

We claim that there is an equivalence of categories for all $\underline{k} \in \Delta^{n-1^{op}}$

$$(Tr_nX)_{\underline{k}} \simeq X_{\underline{k}} \,. \tag{10.9}$$

In fact, since $X_{0\underline{s}} \in \mathsf{Cat}_{\mathsf{hd}}$,

$$(Tr_nX)_{0\underline{s}} = X^d_{0\underline{s}} \simeq X_{0\underline{s}} \,.$$

By the inductive hypothesis applied to $Tr_{n-1}X_1$,

$$(Tr_nX)_{1\underline{s}} = (Tr_{n-1}X_1)_{\underline{s}} \simeq X_{1\underline{s}} \,.$$

This implies, when $k_1 \geq 2$,

$$(Tr_nX)_{\underline{k}} = (Tr_{n-1}X_1)_{\underline{s}} \times_{X^d_{0\underline{s}}} \overset{k_1}{\cdots} \times_{X^d_{0\underline{s}}} (Tr_{n-1}X_1)_{\underline{s}} \simeq X_{1\underline{s}} \times_{X^d_{0\underline{s}}} \overset{k_1}{\cdots} \times_{X^d_{0\underline{s}}} X_{1\underline{s}} \,,$$

and together with (10.4) it follows that

$$(Tr_nX)_{\underline{k}} \simeq X_{\underline{k}}$$

when $k_1 \geq 2$. This concludes the proof of (10.9).

We can therefore apply Lemma 4.3.2 with $\mathscr{C} = \Delta^{{n-1}^{op}}$ and conclude that $Tr_n X$ lifts to a pseudo-functor

$$Tr_n X \in \mathsf{Ps}[\Delta^{{n-1}^{op}}, \mathsf{Cat}]$$

with $(Tr_n X)_{\underline{k}}$ as in (10.8).

We now show that $Tr_n X$ is a Segalic pseudo-functor, by checking the conditions in Definition 8.1.2 (see also Remark 8.1.3).

We first check condition (a) that $(Tr_n X)_{\underline{k}(0,i)}$ is discrete for all $1 \leq i \leq n-1$ and all $\underline{k} = (k_1, \ldots, k_n) \in \Delta^{n^{op}}$. By construction (10.8)

$$(Tr_n X)_{\underline{k}(0,1)} = X^d_{\underline{k}(0,1)}$$

is discrete. Also by construction and by the inductive hypothesis, when $k_1 = 1$ and $i > 1$,

$$(Tr_n X)_{\underline{k}(0,i)} = (Tr_{n-1} X_1)_{\underline{s}(0,i-1)}$$

is discrete. Finally, if $k_1 > 1$ and $i > 1$, by construction and by the inductive hypothesis

$$(Tr_n X)_{\underline{k}(0,i)} = (Tr_{n-1} X_1)_{\underline{s}(0,i-1)} \times_{X^d_{0\underline{s}}} \overset{k_1}{\cdots} \times_{X^d_{0\underline{s}}} (Tr_{n-1} X_1)_{\underline{s}(0,i-1)}$$

is discrete. This shows that condition (a) in Definition 8.1.2 is satisfied. We now show condition (b) that for each $\underline{k} \in \Delta^{{n-1}^{op}}$, $1 \leq i \leq n-1$ and $k_i \geq 2$,

$$(Tr_n X)_{\underline{k}} \cong (Tr_n X)_{\underline{k}(1,i)} \times_{(Tr_n X)_{\underline{k}(0,i)}} \overset{k_i}{\cdots} \times_{(Tr_n X)_{\underline{k}(0,i)}} (Tr_n X)_{\underline{k}(1,i)}\,. \tag{10.10}$$

We distinguish various cases:

(i) Let $\underline{k} \in \Delta^{{n-1}^{op}}$ be such that $k_j = 0$ for some $1 \leq j \leq n-1$ and let $k_i \geq 2$ for $1 \leq i \leq n$. Since, by definition of $\mathsf{LTa}^n_{\mathsf{wg}}$ (see also Remark 9.1.2) there is an equivalence of categories

$$X_{\underline{k}} \simeq X_{\underline{k}(1,i)} \times_{X^d_{\underline{k}(0,i)}} \overset{k_i}{\cdots} \times_{X^d_{\underline{k}(0,i)}} X_{\underline{k}(1,i)}$$

and since $X_{\underline{k}} \in \mathsf{Cat}_{\mathsf{hd}}$, $X_{\underline{k}(1,i)} \in \mathsf{Cat}_{\mathsf{hd}}$ (as $k_j = 0$ and $k_i \geq 2$ so $i \neq j$), there is an isomorphism

$$\begin{aligned} X^d_{\underline{k}} &\cong p(X_{\underline{k}(1,i)} \times_{X^d_{\underline{k}(0,i)}} \overset{k_i}{\cdots} \times_{X^d_{\underline{k}(0,i)}} X_{\underline{k}(1,i)}) \\ &\cong X^d_{\underline{k}(1,i)} \times_{X^d_{\underline{k}(0,i)}} \overset{k_i}{\cdots} \times_{X^d_{\underline{k}(0,i)}} X^d_{\underline{k}(1,i)} \end{aligned} \tag{10.11}$$

which, by (10.8), is the same as (10.10) in this case.

(ii) Suppose $k_j \neq 0$ for all $1 \leq j \leq n-1$ and let $i = 1$. Then by (10.8), if $\underline{s} = (k_2, \ldots, k_{n-1})$

$$(Tr_n X)_{\underline{k}(1,1)} = (Tr_{n-1} X_1)_{\underline{s}} \ ,$$
$$(Tr_n X)_{\underline{k}(0,1)} = X^d_{\underline{k}(0,1)} = X^d_{0\underline{s}} \ .$$

Therefore by (10.8), if $k_1 \geq 2$,

$$(Tr_n X)_{\underline{k}} = (Tr_{n-1} X_1)_{\underline{s}} \times_{X^d_{0\underline{s}}} \overset{k_1}{\cdots} \times_{X^d_{0\underline{s}}} (Tr_{n-1} X_1)_{\underline{s}}$$
$$= (Tr_n X)_{\underline{k}(1,1)} \times_{(Tr_n X)_{\underline{k}(0,1)}} \overset{k_1}{\cdots} \times_{(Tr_n X)_{\underline{k}(0,1)}} (Tr_n X)_{\underline{k}(1,1)}$$

which is (10.10) in this case.

(iii) Suppose $k_j \neq 0$ for all $1 \leq j \leq n-1$, $i > 1$ and $k_1 = 1$. Then by (10.8), if $\underline{s} = (k_2, \ldots, k_{n-1})$

$$(Tr_n X)_{\underline{k}} = (Tr_{n-1} X_1)_{\underline{s}}$$

so in particular

$$(Tr_n X)_{\underline{k}(1,i)} = (Tr_{n-1} X_1)_{\underline{s}(1,i-1)} \ ,$$
$$(Tr_n X)_{\underline{k}(0,i)} = (Tr_{n-1} X_1)_{\underline{s}(0,i-1)} \ .$$

By the induction hypothesis applied to X_1, it follows that, since $k_i = s_{i-1} \geq 2$,

$$(Tr_n X)_{\underline{k}} = (Tr_{n-1} X_1)_{\underline{s}}$$
$$= (Tr_{n-1} X_1)_{\underline{s}(1,i-1)} \times_{(Tr_{n-1} X_1)_{\underline{s}(0,i-1)}} \overset{s_{i-1}}{\cdots} \times_{(Tr_{n-1} X_1)_{\underline{s}(0,i-1)}} (Tr_{n-1} X_1)_{\underline{s}(1,i-1)}$$
$$= (Tr_n X)_{\underline{k}(1,i)} \times_{(Tr_n X)_{\underline{k}(0,i)}} \overset{k_i}{\cdots} \times_{(Tr_n X)_{\underline{k}(0,i)}} (Tr_n X)_{\underline{k}(1,i)} \ ,$$

which is (10.10) in this case.

(iv) Suppose $k_j \neq 0$ for all $1 \leq j \leq n-1$, $i > 1$ and $k_1 = 2$. By (10.8), if $\underline{s} = (k_2, \ldots, k_{n-1})$ so that $k_i = s_{i-1}$

$$(Tr_n X)_{\underline{k}} = (Tr_{n-1} X_1)_{\underline{s}} \times_{X^d_{0\underline{s}}} (Tr_{n-1} X_1)_{\underline{s}} \ . \tag{10.12}$$

By the induction hypothesis applied to X_1,

$$(Tr_{n-1} X_1)_{\underline{s}}$$
$$= (Tr_{n-1} X_1)_{\underline{s}(1,i-1)} \times_{(Tr_{n-1} X_1)_{\underline{s}(0,i-1)}} \overset{s_{i-1}}{\cdots} \times_{(Tr_{n-1} X_1)_{\underline{s}(0,i-1)}} (Tr_{n-1} X_1)_{\underline{s}(1,i-1)} \ , \tag{10.13}$$

while by (10.11)

$$X^d_{0\underline{s}} = X^d_{0\underline{s}(1,i-1)} \times_{X^d_{0\underline{s}(0,i-1)}} \overset{s_{i-1}}{\cdots} \times_{X^d_{0\underline{s}(0,i-1)}} X^d_{0\underline{s}(1,i-1)} \,. \tag{10.14}$$

Replacing (10.13) and (10.14) in (10.12), using the commutation of limits and the fact that

$$(Tr_n X)_{\underline{k}(1,i)} = (Tr_{n-1} X_1)_{\underline{s}(1,i-1)} \times_{X^d_{0\underline{s}(1,i-1)}} (Tr_{n-1} X_1)_{\underline{s}(1,i-1)} \,,$$

$$(Tr_n X)_{\underline{k}(0,i)} = (Tr_{n-1} X_1)_{\underline{s}(0,i-1)} \times_{X^d_{0\underline{s}(0,i-1)}} (Tr_{n-1} X_1)_{\underline{s}(0,i-1)} \,,$$

we obtain

$$(Tr_n X)_{\underline{k}} = (Tr_n X)_{\underline{k}(1,i)} \times_{(Tr_n X)_{\underline{k}(0,i)}} \overset{k_i}{\cdots} \times_{(Tr_n X)_{\underline{k}(0,i)}} (Tr_n X)_{\underline{k}(1,i)} \,,$$

which is (10.10) in this case.

(v) Suppose $k_j \neq 0$ for all $1 \leq j \leq n-1$, $i > 1$ and $k_1 > 2$. The proof of (10.10) is completely analogous to the one of case (iv).

This concludes the proof that condition (b) in Definition 8.1.2 is satisfied for Tr_n.

To show that condition (c) in Definition 8.1.2 holds for Tr_n we note that the equivalence of categories (10.9) implies the isomorphism for each $\underline{k} \in \Delta^{n-1^{op}}$,

$$p(Tr_n X)_{\underline{k}} \cong pX_{\underline{k}} = (p^{(n-1)}X)_{\underline{k}} \,.$$

Since $X \in \mathsf{LTa}^{\mathsf{n}}_{\mathsf{wg}}$, $p^{(n-1)}X \in \mathsf{Cat}^{\mathsf{n-1}}_{\mathsf{wg}}$, hence

$$p^{(n-1)} Tr_n X = p^{(n-1)} X \in \mathsf{Cat}^{\mathsf{n-1}}_{\mathsf{wg}} \,,$$

which is condition (c) in Definition 8.1.2. We conclude that

$$Tr_n X \in \mathsf{SegPs}[\Delta^{n-1^{op}}, \mathsf{Cat}\,] \,.$$

By Lemma 4.3.2 there is a morphism in $\mathsf{Ps}[\Delta^{n-1^{op}}, \mathsf{Cat}\,]$

$$t_n(X) : Tr_n X \to X$$

which is levelwise the equivalence of categories

$$(Tr_n X)_{\underline{k}} \simeq X_{\underline{k}} \quad \text{for } \underline{k} \in \Delta^{n-1^{op}} \,.$$

□

Corollary 10.1.2 *Let $X \in \mathsf{LTa}^{\mathsf{n}}_{\mathsf{wg}}$ and $\underline{k} \in \Delta^{n-1^{op}}$ be such that $k_j = 0$ for some $1 \leq j \leq n-1$. Then*

$$(Tr_n X)_{\underline{k}} = X^d_{\underline{k}}\ .$$

Proof By induction on n. When $n = 2$, by definition of $Tr_2 X$,

$$(Tr_2 X)_0 = X^d_0\ .$$

Suppose the statement holds for $(n-1)$ and denote $\underline{s} = (k_2, \ldots, k_{n-1}) \in \Delta^{n-2^{op}}$ so that

$$\underline{k} = (k_1, \underline{s}) \in \Delta^{n-1^{op}}\ .$$

We distinguish three cases:

(i) When $k_1 = 0$, by definition of Tr_n

$$(Tr_n X)_{\underline{k}} = (Tr_n X)_{(0,\underline{s})} = X^d_{0\underline{s}} = X^d_{\underline{k}}\ .$$

(ii) Let $k_1 = 1$ and suppose $s_j = 0$ for some $2 \leq j \leq n-1$. Then by definition of $Tr_n X$ and by the inductive hypothesis applied to $X_1 \in \mathsf{LTa}^{\mathsf{n-1}}_{\mathsf{wg}}$,

$$(Tr_n X)_{\underline{k}} = (Tr_{n-1} X_1)_{\underline{s}} = (X_1)^d_{\underline{s}} = X^d_{\underline{k}}\ .$$

(iii) Let $k_1 > 1$ and suppose $s_j = 0$ for some $2 \leq j \leq n-1$. Then by definition of $Tr_n X$ and by the inductive hypothesis applied to $X_1 \in \mathsf{LTa}^{\mathsf{n-1}}_{\mathsf{wg}}$, we have

$$(Tr_n X)_{\underline{k}} = (Tr_{n-1} X_1)_{\underline{s}} \times_{X^d_{0\underline{s}}} \overset{k_1}{\cdots} \times_{X^d_{0\underline{s}}} (Tr_{n-1} X_1)_{\underline{s}} = X_{1\underline{s}} \times_{X^d_{0\underline{s}}} \overset{k_1}{\cdots} \times_{X^d_{0\underline{s}}} X_{1\underline{s}}\ . \tag{10.15}$$

Since $X \in \mathsf{LTa}^{\mathsf{n}}_{\mathsf{wg}}$ there is an equivalence of categories

$$X_{\underline{k}} = X_{k_1\underline{s}} \simeq X_{1\underline{s}} \times_{X^d_{0\underline{s}}} \overset{k_1}{\cdots} \times_{X^d_{0\underline{s}}} X_{1\underline{s}}$$

and therefore, since $X_{\underline{k}}, X_{1\underline{s}} \in \mathsf{Cat}_{\mathsf{hd}}$,

$$X^d_{\underline{k}} \cong X^d_{1\underline{s}} \times_{X^d_{0\underline{s}}} \overset{k_1}{\cdots} \times_{X^d_{0\underline{s}}} X^d_{1\underline{s}}\ . \tag{10.16}$$

We deduce from (10.15) and (10.16) that

$$(Tr_n X)_{\underline{k}} = X^d_{\underline{k}}\ .$$

□

Example 10.1.3 The functor

$$Tr_3 : \mathsf{LTa}^3_{\mathsf{wg}} \to \mathsf{SegPs}[\Delta^{2^{op}}, \mathsf{Cat}]$$

is given as follows: $X \in \mathsf{LTa}^3_{\mathsf{wg}}$ consists of $X \in [\Delta^{2^{op}}, \mathsf{Cat}]$ such that

(i) $X_k \in \mathsf{Ta}^2_{\mathsf{wg}}$ for all $k \geq 0$ with $X_0 \in \mathsf{Cat}^2_{\mathsf{hd}}$. Thus in particular $X_{k0} \in \mathsf{Cat}_{\mathsf{hd}}$ and the induced Segal maps

$$X_{ks} \to X_{k1} \times_{X^d_{k0}} \overset{s}{\cdots} \times_{X^d_{k0}} X_{k1}$$

are equivalences of categories for all $s \geq 2$.

(ii) $p^{(2)} X \in \mathsf{Cat}^2_{\mathsf{wg}}$.

(iii) For each $s \geq 0$ and $k \geq 2$ the maps

$$X_{ks} \to X_{1s} \times_{X^d_{0s}} \overset{k}{\cdots} \times_{X^d_{0s}} X_{1s}$$

are equivalences of categories.

Below is a picture of the corner of $Tr_3 X$, where the symbol $\cong$ indicates that the squares pseudo-commute, that is, $Tr_3 X$ is not a bisimplicial object in Cat but a pseudo-functor from $\Delta^{2^{op}}$ to Cat.

$$\begin{array}{ccccc}
(X_{11} \times_{X^d_{01}} X_{11}) \times_{(X^d_{10} \times_{X^d_{00}} X^d_{10})} (X_{11} \times_{X^d_{01}} X_{11}) & \Rrightarrow & X_{11} \times_{X^d_{10}} X_{11} & \rightrightarrows & X^d_{01} \times_{X^d_{00}} X^d_{01} \\
\Downarrow & \cong & \Downarrow & \cong & \Downarrow \\
\cdots\ X_{11} \times_{X^d_{01}} X_{11} & \Rrightarrow & X_{11} & \rightrightarrows & X^d_{01} \\
\downdownarrows & \cong & \downdownarrows & \cong & \downdownarrows \\
\cdots\ X^d_{10} \times_{X^d_{00}} X^d_{10} & \Rrightarrow & X^d_{10} & \rightrightarrows & X^d_{00}
\end{array}$$

Note that

$$\begin{aligned}
&(X_{11} \times_{X^d_{01}} X_{11}) \times_{(X^d_{10} \times_{X^d_{00}} X^d_{10})} (X_{11} \times_{X^d_{01}} X_{11}) \\
\cong\ &(X_{11} \times_{X^d_{10}} X_{11}) \times_{(X^d_{10} \times_{X^d_{00}} X^d_{10})} (X_{11} \times_{X^d_{10}} X_{11}) ,
\end{aligned}$$

so that the Segal maps of $Tr_3 X$ in both horizontal and vertical directions are isomorphisms.

The following Lemma will be used in the proof of Proposition 12.2.3. The latter will be crucial in proving the properties of the discretization functor in the proof of Theorem 12.2.5.

Lemma 10.1.4 *Let $X \in \mathsf{Cat}^n_{\mathsf{wg}}$, $Y \in \mathsf{LTa}^n_{\mathsf{wg}}$ be such that $Y_{\underline{k}}$ is discrete for all $\underline{k} \in \Delta^{n-1^{op}}$ such that $k_j = 0$ for some $1 \le j \le n-1$. Let $\underline{k},\ \underline{s} \in \Delta^{n-1^{op}}$ and let $\underline{k} \to \underline{s}$ be a morphism in $\Delta^{n-1^{op}}$. Suppose that the following conditions hold:*

(i) If $k_j, s_j \neq 0$ for all $1 \le j \le n-1$, then $X_{\underline{k}} = Y_{\underline{k}}$, $\quad X_{\underline{s}} = Y_{\underline{s}}$ and the maps

$$X_{\underline{k}} \to X_{\underline{s}}, \quad Y_{\underline{k}} \to Y_{\underline{s}}$$

coincide.

(ii) If $k_j = 0$ for some $1 \le j \le n-1$ and $s_t = 0$ for some $1 \le t \le n-1$, then $X^d_{\underline{k}} = Y_{\underline{k}}$, $\quad X^d_{\underline{s}} = Y_{\underline{s}}$ and the two maps

$$X^d_{\underline{k}} \to X^d_{\underline{s}}, \quad Y_{\underline{k}} \to Y_{\underline{s}}$$

coincide, where $f^d : X^d_{\underline{k}} \to X^d_{\underline{s}}$ is induced by $f : X_{\underline{k}} \to X_{\underline{s}}$ and thus also coincides with the composite

$$X^d_{\underline{k}} \xrightarrow{\gamma'_{X_{\underline{k}}}} X_{\underline{k}} \xrightarrow{f} X_{\underline{s}} \xrightarrow{\gamma_{X_{\underline{s}}}} X^d_{\underline{s}}$$

(where γ is the discretization map and γ' a section), since $f^d = f^d \gamma_{X_k} \gamma'_{X_k} = \gamma_{X_s} f \gamma'_{X_k}$.

(iii) If $k_j \neq 0$ for all $1 \le j \le n-1$ and $s_t = 0$ for some $1 \le t \le n-1$, the following diagram commutes

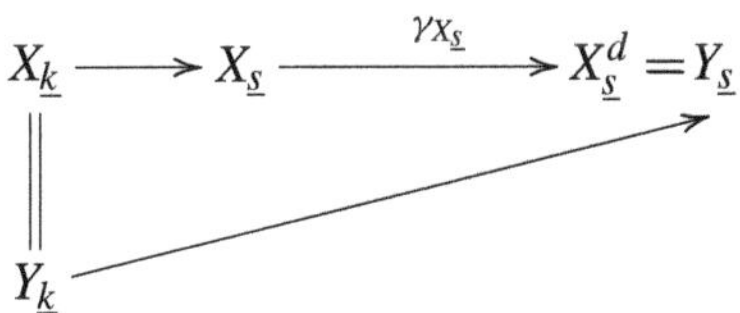

where γ_{X_s} is the discretization map.

(iv) If $k_j = 0$ for some $1 \le j \le n-1$ and $s_t \neq 0$ for all $1 \le t \le n-1$ then the following diagram commutes

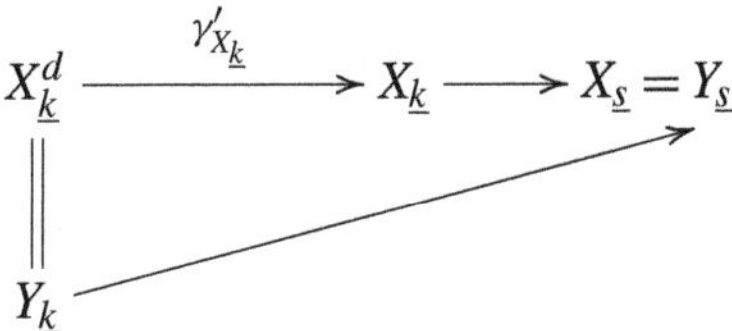

Then

(a) *For all* $\underline{k} \in \Delta^{n-1^{op}}$, $(Tr_n X)_{\underline{k}} = (Tr_n Y)_{\underline{k}}$.
(b) *For all* $\underline{k} \in \Delta^{n-1^{op}}$ *such that* $k_j \neq 0$ *for all* $1 \leq j \leq n-1$*, the maps*

$$(Tr_n X)_{\underline{k}} \rightleftarrows X_{\underline{k}}, \qquad (Tr_n Y)_{\underline{k}} \rightleftarrows Y_{\underline{k}}$$

coincide.
(c) $Tr_n X = Tr_n Y$.

Proof See Appendix A on page 321. □

10.2 Rigidifying Weakly Globular Tamsamani n-Categories

In this section we prove the main result of the chapter, Theorem 10.2.1, establishing the existence of a rigidification functor

$$Q_n : \mathsf{Ta}^{\mathsf{n}}_{\mathsf{wg}} \to \mathsf{Cat}^{\mathsf{n}}_{\mathsf{wg}}$$

replacing $X \in \mathsf{Ta}^{\mathsf{n}}_{\mathsf{wg}}$ with an n-equivalent object $Q_n X \in \mathsf{Cat}^{\mathsf{n}}_{\mathsf{wg}}$.

10.2.1 The Rigidification Functor Q_n: Main Steps

The rigidification functor is half of what is needed for our main comparison result, Theorem 12.2.6, between the categories Ta^{n} and $\mathsf{Cat}^{\mathsf{n}}_{\mathsf{wg}}$. Its construction uses the categories $\mathsf{Ta}^{\mathsf{n}}_{\mathsf{wg}}$, $\mathsf{Cat}^{\mathsf{n}}_{\mathsf{wg}}$, $\mathsf{SegPs}[\Delta^{n-1^{op}}, \mathsf{Cat}]$ whose ideas were discussed in Sects. 6.1.1, 6.3.1, 8.1.2 respectively, as well as the functor Tr_n and the approximation of $\mathsf{Ta}^{\mathsf{n}}_{\mathsf{wg}}$ with $\mathsf{LTa}^{\mathsf{n}}_{\mathsf{wg}}$, discussed informally in Sects. 10.1.1 and 9.2.1 respectively.

The construction of the functor Q_n is inductive and uses three main ingredients:

(a) The approximation up to n-equivalence of an object of $\mathsf{Ta}^{\mathsf{n}}_{\mathsf{wg}}$ by an object of $\mathsf{LTa}^{\mathsf{n}}_{\mathsf{wg}}$ using the pullback construction of Theorem 9.2.4. We showed in Theorem 9.2.4 that if $X \in \mathsf{Ta}^{\mathsf{n}}_{\mathsf{wg}}$ is such that $q^{(n-1)}X$ can be approximated up to $(n-1)$-equivalence by an object of $\mathsf{Cat}^{\mathsf{n-1}}_{\mathsf{wg}}$, then X can be approximated up to an n-equivalence by an object of $\mathsf{LTa}^{\mathsf{n}}_{\mathsf{wg}}$. Thus, given inductively the functor $Q_{(n-1)}$, for each $X \in \mathsf{Ta}^{\mathsf{n}}_{\mathsf{wg}}$ we can approximate $q^{(n-1)}X \in \mathsf{Ta}^{\mathsf{n-1}}_{\mathsf{wg}}$ by $Q_{(n-1)}q^{(n-1)}X \in \mathsf{Cat}^{\mathsf{n-1}}_{\mathsf{wg}}$; thus by the above we can approximate $X \in \mathsf{Ta}^{\mathsf{n}}_{\mathsf{wg}}$ by an object $P_n X \in \mathsf{LTa}^{\mathsf{n}}_{\mathsf{wg}}$ and we obtain a functor

$$P_n : \mathsf{Ta}^{\mathsf{n}}_{\mathsf{wg}} \to \mathsf{LTa}^{\mathsf{n}}_{\mathsf{wg}}.$$

(b) The functor Tr_n from the category $\mathsf{LTa^n_{wg}}$ to the category of Segalic pseudo-functors, which we built in Theorem 10.1.1.
(c) The functor St from Segalic pseudo-functors to weakly globular n-fold categories from Theorem 8.2.3.

We define the rigidification functor Q_2 to be the composite

$$Q_2 : \mathsf{Ta^2_{wg}} \xrightarrow{Tr_2} \mathsf{SegPs}[\Delta^{op}, \mathsf{Cat}] \xrightarrow{St} \mathsf{Cat^2_{wg}} .$$

The rigidification functor Q_n, when $n > 2$, is defined as the composite

$$\mathsf{Ta^n_{wg}} \xrightarrow{P_n} \mathsf{LTa^n_{wg}} \xrightarrow{Tr_n} \mathsf{SegPs}[\Delta^{n-1^{op}}, \mathsf{Cat}] \xrightarrow{St} \mathsf{Cat^n_{wg}} .$$

10.2.2 The Rigidification Functor: The Formal Proof

Theorem 10.2.1 *There is a functor, called rigidification,*

$$Q_n : \mathsf{Ta^n_{wg}} \to \mathsf{Cat^n_{wg}}$$

and for each $X \in \mathsf{Ta^n_{wg}}$ a morphism in $\mathsf{Ta^n_{wg}}$

$$s_n(X) : Q_n X \to X ,$$

natural in X, such that $(s_n(X))_k$ is an $(n-1)$-equivalence for all $k \geq 0$. In particular, $s_n(X)$ is an n-equivalence.

Proof By induction on n. When $n = 2$, let Q_2 be the composite

$$Q_2 : \mathsf{Ta^2_{wg}} \xrightarrow{Tr_2} \mathsf{SegPs}[\Delta^{op}, \mathsf{Cat}] \xrightarrow{St} \mathsf{Cat^2_{wg}} ,$$

where Tr_2 is as in Theorem 10.1.1 and St is as in Theorem 8.2.3. Recall [81] that the strictification functor

$$St : \mathsf{Ps}[\Delta^{op}, \mathsf{Cat}] \to [\Delta^{op}, \mathsf{Cat}]$$

is left adjoint to the inclusion

$$J : [\Delta^{op}, \mathsf{Cat}] \to \mathsf{Ps}[\Delta^{op}, \mathsf{Cat}]$$

and that the components of the unit are equivalences in $\mathsf{Ps}[\Delta^{op}, \mathsf{Cat}]$. By Theorem 10.1.1, for each $X \in \mathsf{Ta^2_{wg}} \subset [\Delta^{op}, \mathsf{Cat}]$ there is a morphism in $\mathsf{Ps}[\Delta^{op}, \mathsf{Cat}]$

$$t_2(X) : Tr_2 X \to JX .$$

By adjunction this morphism corresponds to a morphism in $[\Delta^{op}, \mathsf{Cat}]$

$$Q_2X = St\,Tr_2X \xrightarrow{s_2(X)} X$$

making the following diagram commute

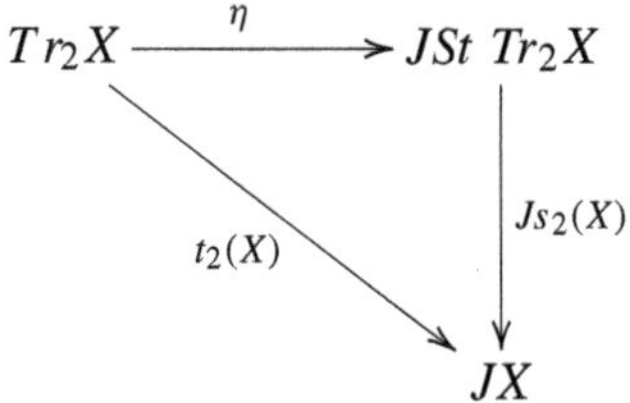

Since η and $t_2(X)$ are levelwise equivalences of categories, so is $Js_2(X)$.

By construction $s_2(X)$ is natural in X. Suppose, inductively, that we have defined Q_{n-1} and s_{n-1}. Define the functor

$$P_n : \mathsf{Ta}^{\mathsf{n}}_{\mathsf{wg}} \to \mathsf{LTa}^{\mathsf{n}}_{\mathsf{wg}}$$

as follows. Given $X \in \mathsf{Ta}^{\mathsf{n}}_{\mathsf{wg}}$, consider the pullback in $[\Delta^{n-1^{op}}, \mathsf{Cat}]$

$$\begin{array}{ccc}
P_nX & \xrightarrow{\quad w_n(X) \quad} & X \\
\Big\downarrow & & \Big\downarrow{\scriptstyle \gamma^{(n-2)}} \\
Q_{n-1}q^{(n-1)}X & \xrightarrow[\quad s_{n-1}(q^{(n-1)}X) \quad]{} & q^{(n-1)}X
\end{array}$$

By Theorem 9.2.4, $P_nX \in \mathsf{LTa}^{\mathsf{n}}_{\mathsf{wg}}$. Define

$$Q_nX = St\,Tr_n P_nX.$$

By Theorem 8.2.3, $Q_nX \in \mathsf{Cat}^{\mathsf{n}}_{\mathsf{wg}}$. Let $s_n(X) : Q_nX \to X$ be the composite

$$s_n(X) : Q_nX \xrightarrow{h_n(P_nX)} P_nX \xrightarrow{w_n(X)} X\,,$$

where the morphism in $[\Delta^{n-1^{op}}, \mathsf{Cat}]$

$$Q_nX = St\,Tr_nP_nX \xrightarrow{h_n(P_nX)} P_nX$$

corresponds by adjunction to the morphism in $\mathsf{Ps}[\Delta^{{n-1}^{op}}, \mathsf{Cat}]$

$$Tr_n P_n X \xrightarrow{t_n(P_n X)} J P_n X$$

(where $t_n(P_n X)$ is as in Theorem 10.1.1) such that the following diagram commutes

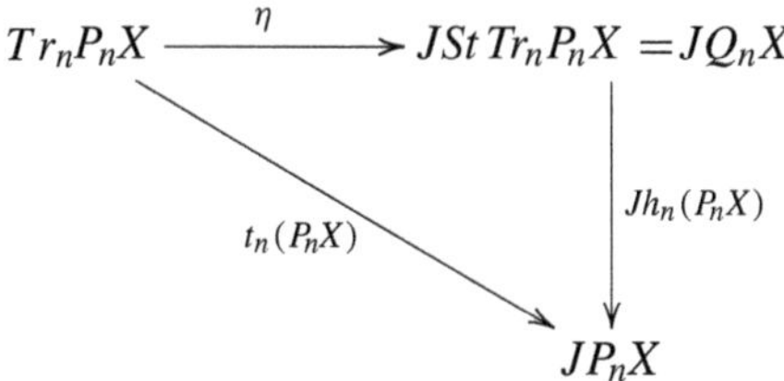

Since by construction and by the inductive hypothesis $w_n(X)$ and $h_n(P_n X)$ are natural in X, so is their composite $s_n(X)$. We need to show that $(s_n(X))_k$ is an $(n-1)$-equivalence. Since η and $t_n(P_n X)$ are levelwise equivalences of categories, so is $h_n(P_n X)$, so in particular $(h_n(P_n X))_k$ is a levelwise equivalence of categories, and thus is an $(n-1)$-equivalence (see Remark 7.1.4).

Since pullbacks in $[\Delta^{{n-1}^{op}}, \mathsf{Cat}]$ are computed pointwise, there is a pullback in $[\Delta^{{n-2}^{op}}, \mathsf{Cat}]$

$$\begin{array}{ccc}
(P_n X)_k & \xrightarrow{(w_n(X))_k} & X_k \\
\big\downarrow & & \big\downarrow \gamma^{(n-1)} \\
(Q_{n-1} q^{(n-1)} X)_k & \xrightarrow[(s_{n-1}(q^{(n-1)} X))_k]{} & q^{(n-2)} X_k
\end{array}$$

where $X_k \in \mathsf{Ta}^{\mathsf{n-1}}_{\mathsf{wg}}$ (since $X \in \mathsf{Ta}^{\mathsf{n}}_{\mathsf{wg}}$) and

$$(Q_{n-1} q^{(n-1)} X)_k \in \mathsf{Cat}^{\mathsf{n-2}}_{\mathsf{wg}}$$

(since $Q_{n-1} q^{(n-1)} X \in \mathsf{Cat}^{\mathsf{n-1}}_{\mathsf{wg}}$) and, by the induction hypothesis, $(s_{n-1}(q^{(n-1)} X))_k$ is an $(n-2)$-equivalence. It follows by Theorem 9.2.4 that $(w_n(X))_k$ is an $(n-1)$-equivalence.

In conclusion, both $(h_n(P_n X))_k$ and $(w_n(X))_k$ are $(n-1)$-equivalences so by Proposition 7.1.2 so is their composite

$$(s_n(X))_k : (Q_n X)_k \xrightarrow{(h_n(P_n X))_k} (P_n X)_k \xrightarrow{(w_n(X))_k} X_k\ ,$$

as required. By Lemma 7.1.3, it follows that $s_n(X)$ is an n-equivalence. □

Remark 10.2.2 It is immediate from Theorem 10.2.1 that Q_n preserves n-equivalences. In fact, given an n-equivalence $f : X \to Y$ in $\mathsf{Ta}^{\mathsf{n}}_{\mathsf{wg}}$, we have the commutative diagram

$$\begin{array}{ccc} Q_nX & \xrightarrow{Q_nf} & Q_nY \\ {\scriptstyle s_n(X)}\downarrow & & \downarrow{\scriptstyle s_n(Y)} \\ X & \xrightarrow{f} & Y \end{array}$$

in which $s_n(X)$ and $s_n(Y)$ are n-equivalences by Theorem 10.2.1. By Proposition 7.1.2(c) and (d) it follows that $Q_n f$ is an n-equivalence.

Corollary 10.2.3 *The functors* $Q_n : \mathsf{Ta}^{\mathsf{n}}_{\mathsf{wg}} \to \mathsf{Cat}^{\mathsf{n}}_{\mathsf{wg}}$ *and the embedding* $i : \mathsf{Cat}^{\mathsf{n}}_{\mathsf{wg}} \hookrightarrow \mathsf{Ta}^{\mathsf{n}}_{\mathsf{wg}}$ *induce an equivalence of categories*

$$\mathsf{Ta}^{\mathsf{n}}_{\mathsf{wg}}/{\sim^n} \simeq \mathsf{Cat}^{\mathsf{n}}_{\mathsf{wg}}/{\sim^n} \tag{10.17}$$

after localization with respect to the n-equivalences.

Proof Given $X \in \mathsf{Ta}^{\mathsf{n}}_{\mathsf{wg}}$, by Theorem 10.2.1 there is an n-equivalence in $\mathsf{Ta}^{\mathsf{n}}_{\mathsf{wg}}$ $iQ_nX \to X$, therefore

$$iQ_nX \cong X$$

in $\mathsf{Ta}^{\mathsf{n}}_{\mathsf{wg}}/{\sim^n}$.

Let $Y \in \mathsf{Cat}^{\mathsf{n}}_{\mathsf{wg}}$, then $iY \in \mathsf{Ta}^{\mathsf{n}}_{\mathsf{wg}}$ so by Theorem 10.2.1 there is an n-equivalence $iQ_niY \to iY$ in $\mathsf{Ta}^{\mathsf{n}}_{\mathsf{wg}}$. Since i is fully faithful, $Q_niY \to Y$ is an n-equivalence in $\mathsf{Cat}^{\mathsf{n}}_{\mathsf{wg}}$. It follows that

$$Q_niY \cong Y$$

in $\mathsf{Cat}^{\mathsf{n}}_{\mathsf{wg}}/{\sim^n}$. In conclusion Q_n and i induce the equivalence of categories (10.17). □

Remark 10.2.4 It follows from Corollary 9.2.5 that given $X \in \mathsf{Cat}^{\mathsf{n}}_{\mathsf{wg}}$, $n > 2$, $P_nX \in \mathsf{Cat}^{\mathsf{n}}_{\mathsf{wg}}$.

Part IV
Weakly Globular n-Fold Categories as a Model of Weak n-Categories

In this Part we construct the discretization functor

$$Disc_n : \mathsf{Cat}^{\mathsf{n}}_{\mathsf{wg}} \to \mathsf{Ta}^{\mathsf{n}}$$

from weakly globular n-fold categories to Tamsamani n-categories, and we prove the final results: the comparison between $\mathsf{Cat}^{\mathsf{n}}_{\mathsf{wg}}$ and Ta^{n}, exhibiting $\mathsf{Cat}^{\mathsf{n}}_{\mathsf{wg}}$ as a model of weak n-categories (Theorem 12.2.6), and the homotopy hypothesis for groupoidal weakly globular n-fold categories (Theorem 12.3.11 and Corollary 12.4.6). A schematic summary of the main results of this Part is contained in Figs. 10.1 and 10.2.

In Chap. 11 we introduce the category $\mathsf{FCat}^{\mathsf{n}}_{\mathsf{wg}}$. This category is a refinement of the category $\mathsf{Cat}^{\mathsf{n}}_{\mathsf{wg}}$ whose objects have homotopically discrete substructures with functorial sections to the discretization maps. The idea of the category $\mathsf{FCat}^{\mathsf{n}}_{\mathsf{wg}}$ is introduced in Sect. 11.3.1, before the formal definitions. The main result of Chap. 11 is Theorem 11.3.6 on the existence of the functor

$$G_n : \mathsf{Cat}^{\mathsf{n}}_{\mathsf{wg}} \to \mathsf{FCat}^{\mathsf{n}}_{\mathsf{wg}}.$$

This functor is constructed inductively using the functor

$$F_n : \mathsf{Cat}^{\mathsf{n}}_{\mathsf{wg}} \to \mathsf{Cat}^{\mathsf{n}}_{\mathsf{wg}}$$

of Proposition 11.2.5. The latter approximates up to n-equivalence a weakly globular n-fold category X with a better behaved one $F_n X$ in which the homotopically discrete object at level 0 admits a functorial section to the discretization map. The construction of the functor F_n is based on a general construction on $X \in \mathsf{Cat}^{\mathsf{n}}_{\mathsf{wg}}$ and $f_0 : Y_0 \to X_0$ given in Proposition 11.1.5, for an appropriate choice of the map $f_0 : Y_0 \to X_0$, given in Proposition 11.2.3.

The ideas of these constructions are explained in Sect. 11.1.1 (for the construction $X(f_0)$), in Sect. 11.2.1 (for the functors V_n and F_n) and in Sect. 11.3.3 (for the functor G_n).

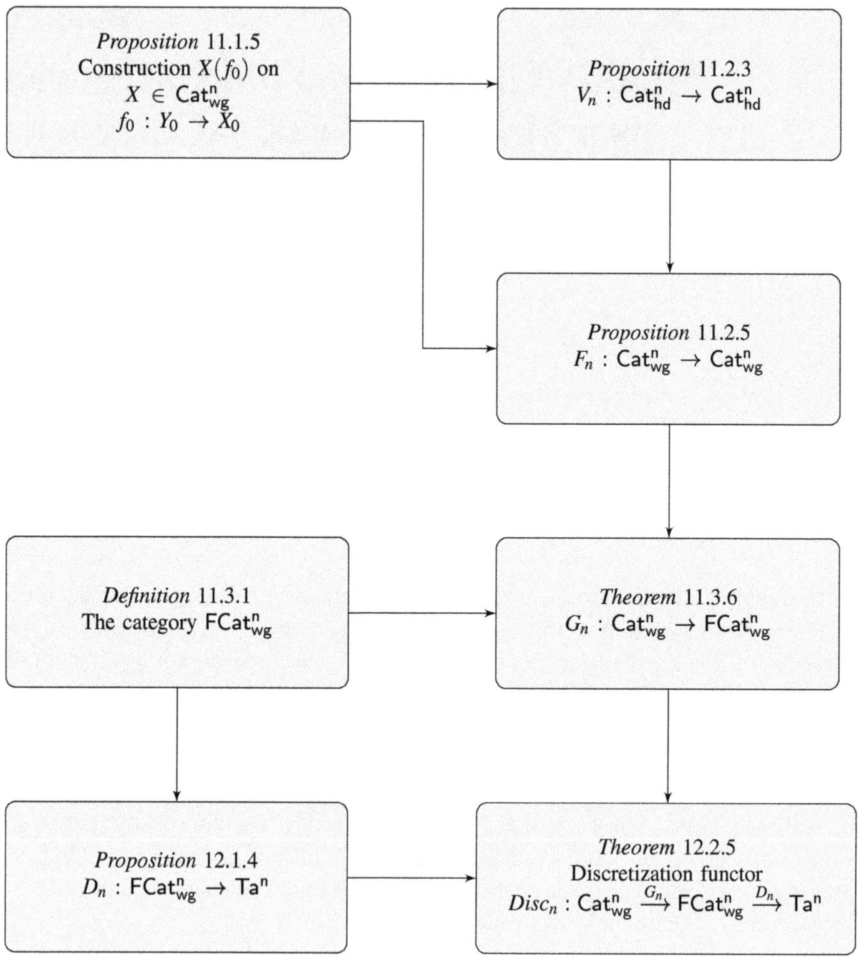

Fig. 10.1 Construction of the discretization functor

In Chap. 12 we define the discretization functor and we obtain the main results of this work. In Proposition 12.1.4 we build a functor

$$D_n : \mathsf{FCat}^n_{\mathsf{wg}} \to \mathsf{Ta}^n$$

which discretizes the homotopically discrete substructures of the objects of $\mathsf{FCat}^n_{\mathsf{wg}}$: because of the properties of the category $\mathsf{FCat}^n_{\mathsf{wg}}$, this can be done in a functorial way. The idea of the functor D_n is explained in Sect. 12.1.1, before the formal definition.

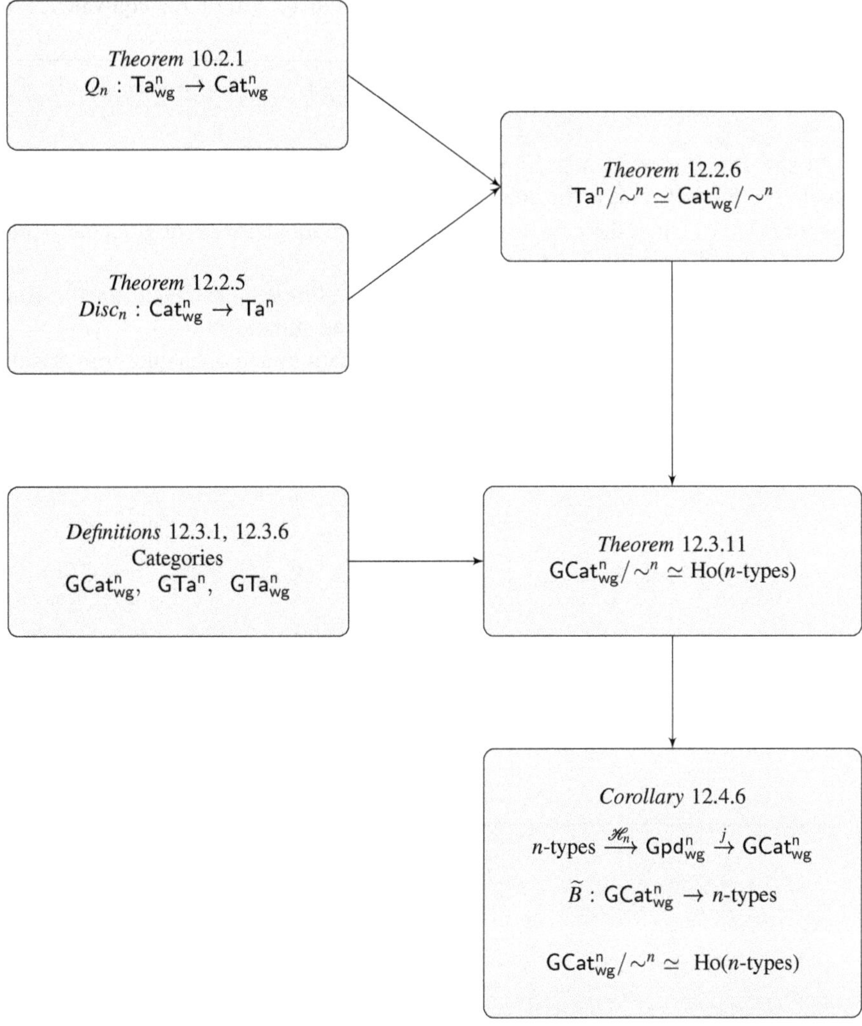

Fig. 10.2 $\mathsf{Cat}^{\mathsf{n}}_{\mathsf{wg}}$ as a model of weak n-categories

In Definition 12.2.1 we define the discretization functor as the composite

$$Disc_n : \mathsf{Cat}^{\mathsf{n}}_{\mathsf{wg}} \xrightarrow{G_n} \mathsf{FCat}^{\mathsf{n}}_{\mathsf{wg}} \xrightarrow{D_n} \mathsf{Ta}^{\mathsf{n}}$$

and we establish its properties in Theorem 12.2.5. The idea of the functor $Disc_n$ is explained in Sect. 12.2, before the formal definition.

The properties of the discretization functor also depends on the properties of the functor rigidification Q_n (see Proposition 12.2.3). Finally, the functors Q_n and

$Disc_n$ lead to the main comparison result (Theorem 12.2.6) on the equivalence of categories

$$\mathsf{Ta}^{\mathsf{n}}/\sim^n \simeq \mathsf{Cat}^{\mathsf{n}}_{\mathsf{wg}}/\sim^n .$$

In the last part of Chap. 12 we define the groupoidal version of the three Segal-type models, that is the categories $\mathsf{GCat}^{\mathsf{n}}_{\mathsf{wg}}$, $\mathsf{GTa}^{\mathsf{n}}_{\mathsf{wg}}$ and $\mathsf{GTa}^{\mathsf{n}}$. We show in Theorem 12.3.11 that the category $\mathsf{GCat}^{\mathsf{n}}_{\mathsf{wg}}$ of groupoidal weakly globular n-fold categories gives a model of n-types.

In Corollary 12.4.6 we exhibit an alternative and more convenient functor from spaces to $\mathsf{GCat}^{\mathsf{n}}_{\mathsf{wg}}$ using the results of Blanc and the author [29].

In Chap. 13 we give an outline of further applications and open questions arising from this work which will be tackled in future projects.

Chapter 11
Functoriality of Homotopically Discrete Objects

Abstract In this chapter we introduce the new category $\mathsf{FCat}^{\mathsf{n}}_{\mathsf{wg}}$. This is a refinement of the category $\mathsf{Cat}^{\mathsf{n}}_{\mathsf{wg}}$ with better behaved homotopically discrete substructures, admitting functorial sections. The main result of this chapter is that there is a functor G_n from $\mathsf{Cat}^{\mathsf{n}}_{\mathsf{wg}}$ to $\mathsf{FCat}^{\mathsf{n}}_{\mathsf{wg}}$ which approximates up to n-equivalence any object of $\mathsf{Cat}^{\mathsf{n}}_{\mathsf{wg}}$ with one of $\mathsf{FCat}^{\mathsf{n}}_{\mathsf{wg}}$. This result is used crucially in Chap. 12 to build the discretization functor from weakly globular n-fold categories to Tamsamani-n categories.

In Chap. 10 we built a rigidification functor from weakly globular Tamsamani n-categories to weakly globular n-fold categories, which in particular affords a functor

$$Q_n : \mathsf{Ta}^{\mathsf{n}} \to \mathsf{Cat}^{\mathsf{n}}_{\mathsf{wg}}$$

producing n-equivalent objects in $\mathsf{Ta}^{\mathsf{n}}_{\mathsf{wg}}$.

To reach the full comparison between Ta^{n} and $\mathsf{Cat}^{\mathsf{n}}_{\mathsf{wg}}$ we need a functor in the other direction, namely a *discretization functor*

$$Disc_n : \mathsf{Cat}^{\mathsf{n}}_{\mathsf{wg}} \to \mathsf{Ta}^{\mathsf{n}} \, .$$

The idea of the functor $Disc_n$ is to replace the homotopically discrete substructures in $X \in \mathsf{Cat}^{\mathsf{n}}_{\mathsf{wg}}$ by their discretizations in order to recover the globularity condition. This affects the Segal maps, which from being isomorphisms in X become $(n-1)$-equivalences in $Disc_n X$.

We illustrate this idea in the case $n = 2$. Given $X \in \mathsf{Cat}^{2}_{\mathsf{wg}}$, by definition $X_0 \in \mathsf{Cat}_{\mathsf{hd}}$, so there is a discretization map $\gamma : X_0 \to X_0^d$ which is an equivalence of categories. Given a choice γ' of pseudo-inverse, we have $\gamma\gamma' = \text{Id}$ since X_0^d is discrete.

We can therefore construct $D_0 X \in [\Delta^{op}, \mathsf{Cat}\,]$ as follows

$$(D_0 X)_k = \begin{cases} X_0^d, \; k = 0 \\ X_k, \; k > 0 \, . \end{cases}$$

S. Paoli, *Simplicial Methods for Higher Categories*, Algebra and Applications 26,
https://doi.org/10.1007/978-3-030-05674-2_11

The face maps

$$(D_0X)_1 \rightrightarrows (D_0X)_0$$

are given by $\gamma\partial_i \;\; i = 0, 1$ (where $\partial_i : X_1 \rightrightarrows X_0$ are face maps of X) while the degeneracy map

$$(D_0X)_0 \to (D_0X)_1$$

is $\sigma_0\gamma'$ (where $\sigma_0 : X_0 \to X_1$ if the degeneracy map of X). All other face and degeneracy maps in D_0X are as in X. Since $\gamma\gamma' = \mathrm{Id}$, all simplicial identities are satisfied for D_0X. By construction, $(D_0X)_0$ is discrete while the Segal maps are given, for each $k \geq 2$, by

$$X_1 \times_{X_0} \overset{k}{\cdots} \times_{X_0} X_1 \to X_1 \times_{X_0^d} \overset{k}{\cdots} \times_{X_0^d} X_1$$

and these are equivalences of categories since $X \in \mathsf{Cat}^2_{\mathsf{wg}}$. Thus, by definition, $D_0X \in \mathsf{Ta}^2$. This construction however does not afford a functor

$$D_0 : \mathsf{Cat}^2_{\mathsf{wg}} \to \mathsf{Ta}^2$$

but only a functor

$$D_0 : \mathsf{Cat}^2_{\mathsf{wg}} \to (\mathsf{Ta}^2)_{\mathsf{ps}} \;,$$

where $(\mathsf{Ta}^2)_{\mathsf{ps}}$ is the full subcategory of $\mathsf{Ps}[\Delta^{op}, \mathsf{Cat}\,]$ whose objects are in Ta^2. This is because, for any morphism $F : X \to Y$ in Ta^2, the diagram in Cat

$$\begin{array}{ccc} X_0^d & \xrightarrow{f^d} & Y_0^d \\ {\scriptstyle \gamma'(X_0)}\downarrow & & \downarrow{\scriptstyle \gamma'(Y_0)} \\ X_0 & \xrightarrow[f]{} & Y_0 \end{array}$$

in general only pseudo-commutes. Hence D_0 cannot be used as a definition of the discretization functor $Disc_2 : \mathsf{Cat}^2_{\mathsf{wg}} \to \mathsf{Ta}^2$.

To overcome this problem we introduce the category $\mathsf{FCat}^n_{\mathsf{wg}}$ whose objects are weakly globular n-fold categories in which there are functorial sections to the discretization maps of the homotopically discrete sub-structures. We then show that we can approximate any object of $\mathsf{Cat}^n_{\mathsf{wg}}$ by an n-equivalent object of $\mathsf{FCat}^n_{\mathsf{wg}}$. Namely we prove in Theorem 11.3.6 that there is a functor

$$G_n : \mathsf{Cat}^n_{\mathsf{wg}} \to \mathsf{FCat}^n_{\mathsf{wg}}$$

and an n-equivalence $G_n X \to X$, natural in X. In the next chapter we build a functor

$$D_n : \mathsf{FCat}^n_{\mathsf{wg}} \to \mathsf{Ta}^n$$

and construct the discretization functor

$$Disc_n : \mathsf{Cat}^n_{\mathsf{wg}} \to \mathsf{Ta}^n$$

as the composite

$$\mathsf{Cat}^n_{\mathsf{wg}} \xrightarrow{G_n} \mathsf{FCat}^n_{\mathsf{wg}} \xrightarrow{D_n} \mathsf{Ta}^n .$$

This chapter is organized as follows. In Sect. 11.1 we develop a general construction on the category $\mathsf{Cat}^n_{\mathsf{wg}}$ that allows us to replace $X \in \mathsf{Cat}^n_{\mathsf{wg}}$ with an n-equivalent $X(f_0) \in \mathsf{Cat}^n_{\mathsf{wg}}$ by modifying $X_0 \in \mathsf{Cat}^{n-1}_{\mathsf{hd}}$ via a map $f_0 : Y_0 \to X_0$ in $\mathsf{Cat}^{n-1}_{\mathsf{hd}}$ satisfying certain properties (see Proposition 11.1.5). In Sect. 11.2 we make an appropriate choice of the map f_0 (see Proposition 11.2.3) to construct in Proposition 11.2.5 a functor

$$F_n : \mathsf{Cat}^n_{\mathsf{wg}} \to \mathsf{Cat}^n_{\mathsf{wg}}$$

such that, for each $X \in \mathsf{Cat}^n_{\mathsf{wg}}$, $(F_n X)_0 \in \mathsf{Cat}^{n-1}_{\mathsf{hd}}$ admits a functorial (that is, natural in X) section to the discretization map $(F_n X)_0 \to (F_n X)^d_0$.

In Sect. 11.3, Theorem 11.3.6, we define the category $\mathsf{FCat}^n_{\mathsf{wg}}$ and we use the functor F_n to build inductively the functor

$$G_n : \mathsf{Cat}^n_{\mathsf{wg}} \to \mathsf{FCat}^n_{\mathsf{wg}}.$$

Namely, we define $G_2 = F_2$ and given G_{n-1},

$$G_n = \overline{G}_{n-1} \circ F_n ,$$

see Definition 11.3.4 and Theorem 11.3.6.

11.1 A Construction on $\mathsf{Cat}^n_{\mathsf{wg}}$

In this section we develop a general construction on the category $\mathsf{Cat}^n_{\mathsf{wg}}$ that allows us to replace $X \in \mathsf{Cat}^n_{\mathsf{wg}}$ with an n-equivalent $X(f_0) \in \mathsf{Cat}^n_{\mathsf{wg}}$ by modifying $X_0 \in \mathsf{Cat}^{n-1}_{\mathsf{hd}}$ in an appropriate way, via a map $f_0 : Y_0 \to X_0$ in $\mathsf{Cat}^{n-1}_{\mathsf{hd}}$ satisfying additional conditions (see Proposition 11.2.5). Proposition 11.2.5 will be used in the next section to functorially approximate up to n-equivalence a weakly globular n-fold category X by a better behaved one in which the homotopically discrete $(n-1)$-fold category X_0 admits a functorial section to the discretization map.

11.1.1 The Idea of the Construction $X(f_0)$

The construction $X(f_0)$, together with the functors $V_n : \mathsf{Cat}^n_{\mathsf{hd}} \to \mathsf{Cat}^n_{\mathsf{hd}}$ and $F_n : \mathsf{Cat}^n_{\mathsf{wg}} \to \mathsf{Cat}^n_{\mathsf{wg}}$ (informally discussed in Sect. 11.2.1) are needed in the construction of the functor $G_n : \mathsf{Cat}^n_{\mathsf{wg}} \to \mathsf{FCat}^n_{\mathsf{wg}}$ (whose idea is explained in Sect. 11.3.3): this will lead to the discretization functor $Disc_n : \mathsf{Cat}^n_{\mathsf{wg}} \to \mathsf{Ta}^n$ (see Sect. 12.2.1), and thus to the main comparison result.

The construction $X(f_0)$ of Proposition 11.2.5 is based on an application of a well-known construction on internal categories (Lemma 11.1.1), and on conditions on the map f_0 to ensure that $X(f_0)$ and X are n-equivalent.

For any internal category $X \in \mathsf{Cat}\,\mathscr{C}$ (where $\mathscr{C}$ has finite limits) and morphism $f_0 : X'_0 \to X_0$, pulling back f_0 along the map $X_1 \xrightarrow{(\partial_0,\partial_1)} X_0 \times X_0$ gives rise to an internal category $X(f_0)$ with $(X(f_0))_k$ given by the pullbacks in $\mathscr{C}$ (11.1) and (11.4). This construction is also well behaved with respect to pullbacks, as spelled out in Lemma 11.1.1.

In Proposition 11.1.5 we apply this general construction to $X \in \mathsf{Cat}^n_{\mathsf{wg}}$, viewed as an internal category in $\mathsf{Cat}^{n-1}_{\mathsf{wg}}$ in direction 1. These pullbacks in $[\Delta^{n-2^{op}}, \mathsf{Cat}]$ are computed levelwise, that is, for each $\underline{k} \in \Delta^{n-2^{op}}$ they give rise to a pullback in Cat.

The additional conditions imposed in the hypotheses of Proposition 11.1.5 are such that the above levelwise pullbacks in Cat are pullbacks along isofibrations which are surjective on objects, and the same is true after application of the functor $p^{(r-1)}$ for each $1 < r < n$.

Two properties of pullbacks in Cat along isofibrations are particularly relevant here:

(a) They are preserved by p (Lemma 4.1.9).
(b) They preserve objects of $\mathsf{Cat}_{\mathsf{hd}}$ (Lemma 11.1.3).

In the proof of Proposition 11.1.5 we show that property (a) implies that the n-fold category $X(f_0)$ satisfies the hypotheses of Proposition 7.2.8 (b), and thus $X(f_0) \in \mathsf{Cat}^n_{\mathsf{wg}}$, while property (b) implies (via Lemma 11.1.4) that $X(f_0) \to X$ is an n-equivalence.

From the fact that isofibrations are stable under pullbacks we also deduce in the proof of Proposition 11.1.5 that the map $V(X) : X(f_0) \to X$ is levelwise an isofibration in Cat which is surjective on objects, and the same holds for $p^{(r-1)}V(X)$ and, under additional conditions on f_0, for $q^{(r-1)}V(X)$. This will be used in Sect. 11.2.2 in the definition of the functors V_n and F_n, where the construction $X(g_0)$ will be used for a map g_0 of the form $V(X)$.

In Corollary 11.1.6 we show that the construction $X(f_0)$ is well behaved with respect to pullbacks. The proof relies on the corresponding property of the construction of Lemma 11.1.1. These properties will be used to study the behaviour with respect to pullbacks of V_n and of F_n in Sect. 11.2.2, in Corollaries 11.2.4 and 11.2.6. In turn, this will play an important role in the construction of the functor G_n in Sect. 11.3.4.

Lemma 11.1.1 *Let $\mathscr{C}$ be a category with finite limits; let $X \in \mathsf{Cat}\,\mathscr{C}$ and $f_0 : X'_0 \to X_0$ be a morphism in $\mathscr{C}$. There is an $X(f_0) \in \mathsf{Cat}\,\mathscr{C}$ with $X(f_0)_0 = X'_0$ and $X(f_0)_1$ given by the pullback in $\mathscr{C}$*

$$\begin{array}{ccc} X(f_0)_1 & \xrightarrow{\ v\ } & X'_0 \times X_0 \\ {\scriptstyle f_1}\downarrow & & \downarrow{\scriptstyle f_0 \times f_0} \\ X_1 & \xrightarrow{(\partial_0,\partial_1)} & X_0 \times X_0 \end{array} \tag{11.1}$$

and a morphism in $\mathsf{Cat}\,\mathscr{C}$

$$V(X) : X(f_0) \to X\ ,$$

with $V(X)_0 = f_0$ and $V(X)_1 = f_1$. Further, given a diagram $X \to Z \leftarrow Y$ in $\mathsf{Cat}\,\mathscr{C}$ and morphisms in $\mathscr{C}$ $f_0 : X'_0 \to X_0$, $g_0 : Y'_0 \to Y_0$, $h_0 : Z'_0 \to Z_0$, $X'_0 \to Z'_0$, $Y'_0 \to Z'_0$ making the following diagram commute

$$\begin{array}{ccccc} X'_0 & \longrightarrow & Z'_0 & \longleftarrow & Y'_0 \\ {\scriptstyle f_0}\downarrow & & \downarrow{\scriptstyle h_0} & & \downarrow{\scriptstyle g_0} \\ X_0 & \longrightarrow & Z_0 & \longleftarrow & Y_0 \end{array}$$

we have

$$(X \times_Z Y)(f_0 \times_{h_0} g_0) = X(f_0) \times_{Z(h_0)} Y(g_0)\ , \tag{11.2}$$

$$V(X \times_Z Y) = V(X) \times_{V(Z)} V(Y)\ . \tag{11.3}$$

Proof Let $\partial'_i = \mathrm{pr}_i\, v : X(f_0)_1 \to X'_0$, $i = 0, 1$ where $\mathrm{pr}_0, \mathrm{pr}_1$ are the two projections, so that $v = (\partial'_0, \partial'_1)$ and $\partial_i f_1 = f_0 \partial'_i$, $i = 0, 1$. We have

$$(\partial_0, \partial_1)c(f_1 \times_{f_0} f_1) = (f_0 \times f_0)((\partial'_0\, \mathrm{pr}_0) \times (\partial'_1\, \mathrm{pr}_1))\ .$$

So there is a map

$$c' : X(f_0)_1 \times_{Y_0} X(f_0)_1 \to X(f_0)_1$$

making the following diagram commute:

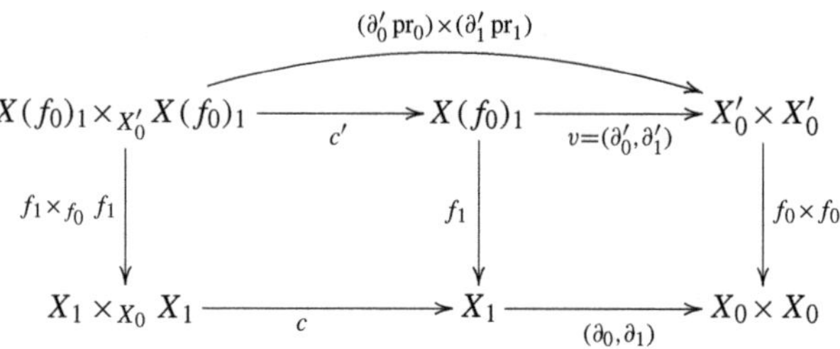

In particular

$$\partial_i' c' = \partial_i' \,\mathrm{pr}_i \qquad i = 0, 1 \, .$$

The other axioms of an internal category for $X(f_0)$ follow immediately from the axioms for X and the universal property of pullbacks. The morphism $V(X) : X(f_0) \to X$ is given by (f_0, f_1).

Given $X \to Z \leftarrow Y$ as in the hypothesis, we have

$$\{(X\times_Z Y)(f_0 \times_{h_0} g_0)\}_0 = X_0' \times_{Z_0'} Y_0' = \{X(f_0)\}_0 \times_{\{Z(h_0)\}_0} \{Y(g_0)\}_0$$

and the pullback in $\mathscr{C}$

$$\begin{array}{ccc} \{(X\times_Z Y)(f_0 \times_{h_0} g_0)\}_1 & \longrightarrow & (X_0' \times_{Z_0'} Y_0') \times (X_0' \times_{Z_0'} Y_0') \\ \big\downarrow & & \big\downarrow \\ X_1 \times_{Z_1} Y_1 & \longrightarrow & (X_0 \times_{Z_0} Y_0) \times (X_0 \times_{Z_0} Y_0) \end{array}$$

Since

$$(X_0 \times_{Z_0} Y_0) \times (X_0 \times_{Z_0} Y_0) = (X_0 \times X_0) \times_{(Z_0 \times Z_0)} (Y_0 \times Y_0)$$

and similarly for X_0', Z_0', Y_0' we conclude that

$$\{(X\times_Z Y)(f_0 \times_{h_0} g_0)\}_1 = \{X(f_0)\}_1 \times_{\{Z(h_0)\}_1} \{Y(g_0)\}_1 \, .$$

Thus (11.2) and (11.3) follow. □

Remark 11.1.2 Note that for each $k \geq 2$, there is a pullback in $\mathscr{C}$

$$\begin{array}{ccc}
X(f_0)_k = X(f_0)_1 \times_{X'_0} \overset{k}{\cdots} \times_{X'_0} X(f_0)_1 & \longrightarrow & X'_0 \times \overset{k+1}{\cdots} \times X'_0 \\
\Big\downarrow {\scriptstyle f_1 \times_{f_0} \overset{k}{\cdots} \times_{f_0} f_1} & & \Big\downarrow {\scriptstyle f_0 \times \overset{k+1}{\cdots} \times f_0} \\
X_k = X_1 \times_{X_0} \overset{k}{\cdots} \times_{X_0} X_1 & \longrightarrow & X_0 \times \overset{k+1}{\cdots} \times X_0
\end{array} \tag{11.4}$$

Lemma 11.1.3 *Let*

$$\begin{array}{ccc}
P & \longrightarrow & C \\
\Big\downarrow & & \Big\downarrow {\scriptstyle f} \\
A & \underset{s}{\longrightarrow} & B
\end{array}$$

be a pullback in Cat *with* f *an isofibration and with* $A, B, C \in \mathsf{Cat}_{\mathsf{hd}}$*. Then* $P \in \mathsf{Cat}_{\mathsf{hd}}$*.*

Proof Since f is an isofibration and $A \simeq A^d$, $B \simeq B^d$, $C \simeq C^d$, by Theorem 4.1.8 we have

$$P \simeq A \overset{ps}{\times}_B C \simeq A^d \times_{B^d} C^d$$

and therefore $P \in \mathsf{Cat}_{\mathsf{hd}}$. □

Lemma 11.1.4 *Let*

$$\begin{array}{ccc}
P & \longrightarrow & C \\
{\scriptstyle h} \Big\downarrow & & \Big\downarrow {\scriptstyle f} \\
A & \underset{g}{\longrightarrow} & B
\end{array}$$

be a pullback in $[\Delta^{n-1^{op}}, \mathsf{Cat}]$ *with* $A, B, C \in \mathsf{Cat}^n_{\mathsf{wg}}$ *and* f *a morphism which is a levelwise isofibration in* Cat *and assume the same holds for* $p^{(r-1)} f$ *for all* $1 < r \leq n$*. Then*

(a) For each $1 < r \leq n$ *there is a pullback in* $[\Delta^{r-2^{op}}, \mathsf{Cat}]$

$$\begin{array}{ccc}
p^{(r-1)} P & \longrightarrow & p^{(r-1)} C \\
\Big\downarrow & & \Big\downarrow \\
p^{(r-1)} A & \longrightarrow & p^{(r-1)} B
\end{array}$$

(b) If f *is an* n*-equivalence, then* h *is an* n*-equivalence.*

(c) *Suppose, further, that for each* $\underline{k} \in \Delta^{n-1^{op}}$ *and* $\underline{s} \in \Delta^{n-2^{op}}$, $B_{\underline{k}}$ *and* $(p^{(n-1)}B)_{\underline{s}}$ *are groupoids and that* $q^{(r-1)}f$ *is a levelwise isofibration in* Cat *for all* $1 < r \leq n$. *Then for each* $1 < r \leq n$ *there is a pullback in* $[\Delta^{r-2^{op}}, \mathsf{Cat}]$

$$\begin{array}{ccc} q^{(r-1)}P & \longrightarrow & q^{(r-1)}C \\ \downarrow & & \downarrow \\ q^{(r-1)}A & \longrightarrow & q^{(r-1)}B \end{array}$$

Proof By induction on n. When $n = 2$, for each $k \in \Delta^{op}$ there is a pullback in Cat

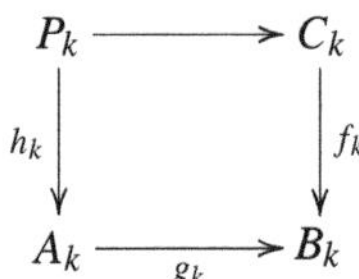

where f_k is an isofibration. Thus, by Lemma 4.1.9 (a), there is a pullback in Set

$$\begin{array}{ccc} (p^{(1)}P)_k = pP_k & \longrightarrow & pC_k = (p^{(1)}C)_k \\ \downarrow & & \downarrow \\ (p^{(1)}A)_k = pA_k & \longrightarrow & pB_k = (p^{(1)}B)_k \end{array}$$

Since this holds for each k, there is a pullback in Cat

$$\begin{array}{ccc} p^{(1)}P & \longrightarrow & p^{(1)}C \\ \downarrow & & \downarrow \\ p^{(1)}A & \longrightarrow & p^{(1)}B \end{array}$$

which proves (a). The proof of (c) is similar using Lemma 4.1.9 (b), whose hypothesis holds since B_k is a groupoid. As for (b), since f is an isofibration, P is equivalent to the pseudo-pullback $A \overset{ps}{\times}_B C$; since f is an equivalence of categories, the latter is equivalent to A. Suppose, inductively, that the lemma holds for $(n-1)$.

(a) For each $\underline{k} \in \Delta^{n-1^{op}}$ there is a pullback in Cat

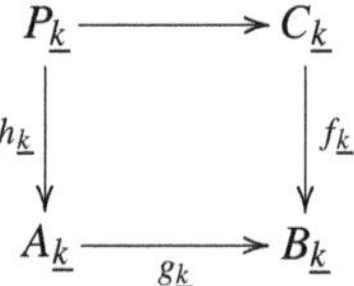

where $f_{\underline{k}}$ is an isofibration. Thus, by Lemma 4.1.9 (a), there is a pullback in Set

$$\begin{array}{ccc} (p^{(n-1)}P)_{\underline{k}} = pP_{\underline{k}} & \longrightarrow & pC_{\underline{k}} = (p^{(n-1)}C)_{\underline{k}} \\ \downarrow & & \downarrow \\ (p^{(n-1)}A)_{\underline{k}} = pA_{\underline{k}} & \longrightarrow & pB_{\underline{k}} = (p^{(n-1)}B)_{\underline{k}} \end{array}$$

Since this holds for each $\underline{k}$, there is a pullback in $[\Delta^{n-2^{op}}, \mathsf{Cat}]$

$$\begin{array}{ccc} p^{(n-1)}P & \longrightarrow & p^{(n-1)}C \\ \downarrow & & \downarrow \\ p^{(n-1)}A & \longrightarrow & p^{(n-1)}B \end{array} \tag{11.5}$$

which is (a) for $r = n$. Since (11.5) satisfies the inductive hypothesis we deduce a pullback in $[\Delta^{r-2^{op}}, \mathsf{Cat}]$ for each $1 < r \leq n-1$

$$\begin{array}{ccc} p^{(r-1)}P & \longrightarrow & p^{(r-1)}C \\ \downarrow & & \downarrow \\ p^{(r-1)}A & \longrightarrow & p^{(r-1)}B \end{array} \tag{11.6}$$

(b) By part (a), there is a pullback in Cat

$$\begin{array}{ccc} p^{(1)}P & \longrightarrow & p^{(1)}C \\ \downarrow & & \downarrow \\ p^{(1)}A & \longrightarrow & p^{(1)}B \end{array}$$

and therefore, at object level, a pullback in Set

$$\begin{array}{ccc} P_0^d & \longrightarrow & C_0^d \\ \downarrow & & \downarrow \\ A_0^d & \longrightarrow & B_0^d \end{array}$$

Let $(a, c), (a', c') \in P_0^d$. Then there is a pullback in $\mathsf{Cat}_{\mathsf{wg}}^{\mathsf{n-1}}$

$$\begin{array}{ccc} P((a,c),(a',c')) & \longrightarrow & C(c,c') \\ {\scriptstyle h((a,c),(a',c'))}\downarrow & & \downarrow \\ A(a,a') & \longrightarrow & B(ga,ga') \end{array}$$

By hypothesis, this satisfies induction hypothesis (b) so $h((a, c), (a', c'))$ is an $(n-1)$-equivalence.

By part (a) we also have the pullback

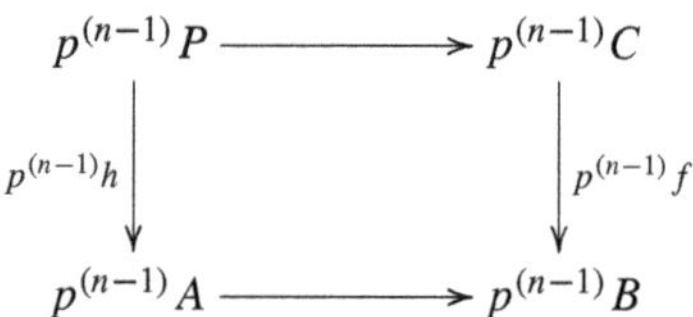

with $p^{(n-1)} f$ an $(n-1)$-equivalence and thus by the inductive hypothesis $p^{(n-1)} h$ is an $(n-1)$-equivalence. We conclude that h is an n-equivalence.

(c) The proof is completely analogous to the one of part (a) using Lemma 4.1.9 (b) (whose hypothesis holds since we assume in part (c) that $B_{\underline{k}}$ and $(p^{(n-1)} B)_{\underline{s}}$ are groupoids for each $\underline{k} \in \Delta^{n-1^{op}}$ and $\underline{s} \in \Delta^{n-2^{op}}$. □

In the following proposition we use the construction of Lemma 11.1.1 for $X \in \mathsf{Cat}_{\mathsf{wg}}^{\mathsf{n}}$ (viewed as an internal category in $\mathsf{Cat}_{\mathsf{wg}}^{\mathsf{n-1}}$ in direction 1) for a particular choice of the map $f_0 : Y_0 \to X_0$ in $\mathsf{Cat}_{\mathsf{hd}}^{\mathsf{n-1}}$ such that the map $V(X) : X(f_0) \to X$ is an n-equivalence and has other desirable properties. This result will be used in the next section in the proofs of Propositions 11.2.3 and 11.2.5.

Proposition 11.1.5 *Let $X \in \mathsf{Cat}_{\mathsf{wg}}^{\mathsf{n}}$ and $f_0 : Y_0 \to X_0$ be a morphism in $\mathsf{Cat}_{\mathsf{hd}}^{\mathsf{n-1}}$ such that, for each $1 < r \leq n$, f_0 and $p^{(r-1)} f_0$ are levelwise isofibrations in* Cat *which are surjective on objects. Then*

(a) $X(f_0) \in \mathsf{Cat}_{\mathsf{wg}}^{\mathsf{n}}$ and

$$p^{(r-1)}(X(f_0)) = (p^{(r-1)} X)(p^{(r-2)} f_0) .$$

(b) The map $V(X) : X(f_0) \to X$ is an n-equivalence.

(c) $V(X)$ is levelwise an isofibration in Cat *surjective on objects, and the same is true for $p^{(r-1)} V(X)$ for all $1 < r \leq n$.*

(d) If $X \in \mathsf{Cat}_{\mathsf{hd}}^{\mathsf{n}}$, $X(f_0) \in \mathsf{Cat}_{\mathsf{hd}}^{\mathsf{n}}$.

(e) For each $1 < r \leq n$,

$$q^{(r-1)}(X(f_0)) = (q^{(r-1)}X)(q^{(r-2)} f_0)$$

and $q^{(r-1)}V(X)$ *is a levelwise isofibration surjective on objects, for all* $1 < r \leq n$.

Proof By induction on n. Let $n = 2$. Since $X(f_0) \in \mathsf{Cat}^2$ and $Y_0 \in \mathsf{Cat}_{\mathsf{hd}}$, to show that $X(f_0) \in \mathsf{Cat}^2_{\mathsf{wg}}$ it is enough to show (by Lemma 7.2.6) that $p^{(1)}X(f_0) \in \mathsf{Cat}$. That is, for each $k \geq 2$,

$$p(X(f_0))_k = p(X(f_0))_1 \times_{p(X(f_0))_0} \overset{k}{\cdots} \times_{p(X(f_0))_0} p(X(f_0))_1 \;.$$

We show this for $k = 2$, the case $k > 2$ being similar. From Remark 11.1.2,

$$X(f_0)_2 = (X_1 \times_{X_0} X_1) \times_{X_0 \times X_0 \times X_0} (Y_0 \times Y_0 \times Y_0) \;.$$

Since f_0 is an isofibration, using Lemma 4.1.9, the fact that $p^{(1)}X \in \mathsf{Cat}$ and the fact that p preserves products, we obtain

$$\begin{aligned} &pX(f_0)_2 = p(X_1 \times_{X_0} X_1) \times_{p(X_0 \times X_0 \times X_0)} p(Y_0 \times Y_0 \times Y_0) \\ &= (pX_1 \times_{pX_0} pX_1) \times_{(pX_0 \times pX_0 \times pX_0)} (pY_0 \times pY_0 \times pY_0) \;. \end{aligned} \tag{11.7}$$

On the other hand,

$$\begin{aligned} & pX(f_0)_1 \times_{pX(f_0)_0} pX(f_0)_1 \\ = \; & (pX_1 \times_{pX_0 \times \, pX_0} (pY_0 \times \, pY_0)) \times_{pX_0 \times_{pX_0} pY_0} (pX_1 \times_{pX_0 \times \, pX_0} (pY_0 \times \, pY_0)) \\ = \; & (pX_1 \times_{pX_0} pX_1) \times_{(pX_0 \times \, pX_0) \times_{pX_0} (pX_0 \times \, pX_0)} ((pY_0 \times \, pY_0) \times_{pY_0} (pY_0 \times \, pY_0)) \\ = \; & p(X_1 \times_{X_0} X_1) \times_{(pX_0 \times pX_0 \times pX_0)} (pY_0 \times pY_0 \times pY_0) \;. \end{aligned}$$

Therefore

$$pX(f_0)_2 = pX(f_0)_1 \times_{pX(f_0)_0} pX(f_0)_1 \;.$$

The case $k > 2$ is similar. This shows $X(f_0) \in \mathsf{Cat}^2_{\mathsf{wg}}$. From (11.7) we see that

$$p^{(1)}(X(f_0)) = (p^{(1)}X)(pf_0).$$

This concludes the proof of (a) when $n = 2$.

We now show that $V(X) : X(f_0) \to X$ is a 2-equivalence. Let $a, b \in Y_0^d$. We have a pullback in Cat

$$\begin{array}{ccc} X(f_0)(a,b) & \longrightarrow & Y_0(a) \times Y_0(b) \\ {\scriptstyle V(X)(a,b)}\downarrow & & \downarrow \\ X_1(f_0 a, f_0 b) & \longrightarrow & X_0(f_0 a) \times X_0(f_0 b) \end{array} \tag{11.8}$$

where $Y_0(a)$ is the fiber at $a \in Y_0^d$ of the discretization map $Y_0 \to Y_0^d$ and $X_0(f_0 a)$ is the fiber at $f_0 a \in X_0^d$ of the discretization map $X_0 \to X_0^d$ (and similarly for $Y_0(b)$ and $X_0(f_0 b)$).

Since $X_0, Y_0 \in \mathsf{Cat_{hd}}$, $Y_0(a) \to X_0(f_0(a))$ is an equivalence of categories hence it is in particular fully faithful. Thus by Lemma 4.1.10, $V(X)(a,b)$ is also fully faithful. Applying Lemma 4.1.9 (a) to (11.8) we obtain a pullback in Set

$$\begin{array}{ccc} pX(f_0)(a,b) & \longrightarrow & pY_0(a) \times pY_0(b) \\ \downarrow & & \downarrow \\ pX_1(f_0 a, f_0 b) & \longrightarrow & pX_0(f_0 a) \times pX_0(f_0 b) \end{array} \tag{11.9}$$

Since, by hypothesis, $Y_0 \to X_0$ is surjective on objects, the right vertical map in (11.9) is surjective, therefore so is the left vertical map in (11.9). Thus

$$X(f_0)(a,b) \to X_1(f_0 a, f_0 b)$$

is essentially surjective on objects and in conclusion it is an equivalence of categories.

To show that $V(X) : X(f_0) \to X$ is a 2-equivalence, it remains to show (by Proposition 7.1.2) that $pV(X)$ is surjective. This follows from the fact that $pV(X)$ is surjective on objects as pf_0 is surjective (since by hypothesis f_0 is surjective on objects). This concludes the proof of (b) in the case $n = 2$.

Using Remark 11.1.2 and the fact that isofibrations are stable under pullbacks, (c) follows.

Finally, if $X \in \mathsf{Cat}^2_{\mathsf{hd}}$, since by (a) $X(f_0) \in \mathsf{Cat}^2_{\mathsf{wg}}$ and $V(X)$ is a 2-equivalence, it follows from Proposition 7.2.1 that $X(f_0) \in \mathsf{Cat}^2_{\mathsf{hd}}$, proving (d) in the case $n = 2$. The proof of part (e) for the case $n = 2$ is completely similar to the one of part (a) using Lemma 4.1.9 (b) (whose hypothesis holds since $X_0(f_0 a)$ is a groupoid as $X_0 \in \mathsf{Cat_{hd}}$).

Suppose, inductively, that the proposition holds for $(n-1)$, let $X \in \mathsf{Cat}^n_{\mathsf{wg}}$ and f_0 be as in the hypothesis.

(a) We show that $X(f_0) \in \mathsf{Cat}^n_{\mathsf{wg}}$ by proving that it satisfies the hypothesis of Proposition 7.2.8 (b). By the general construction of Lemma 11.1.1, $X(f_0) \in \mathsf{Cat}^n$. We first check that $X(f_0)$ satisfies hypothesis (i) of Proposition 7.2.8

(b) that for all $1 \leq r < n$ the $(n-1)$-fold category $X(f_0)^{\{r\}}_0$ is a levelwise equivalence relation. When $r = 1$ we have $X(f_0)^{\{1\}}_0 = X(f_0)_0 = Y_0$ and this is a levelwise equivalence relation since by hypothesis $Y_0 \in \mathsf{Cat}^{\mathsf{n-1}}_{\mathsf{hd}}$. Let $r > 1$. Since pullbacks in $[\Delta^{n-2^{op}}, \mathsf{Cat}]$ are computed pointwise, from the pullbacks in $[\Delta^{n-2^{op}}, \mathsf{Cat}]$ for $k \geq 2$

$$
\begin{array}{ccc}
X(f_0)_1 & \longrightarrow & Y_0 \times Y_0 \\
\downarrow & & \downarrow {\scriptstyle f_0 \times f_0} \\
X_1 & \longrightarrow & X_0 \times X_0
\end{array}
\qquad
\begin{array}{ccc}
X(f_0)_k & \longrightarrow & Y_0 \times \overset{k+1}{\cdots} \times Y_0 \\
\downarrow & & \downarrow {\scriptstyle f_0 \times \overset{k+1}{\cdots} \times f_0} \\
X_k & \longrightarrow & X_0 \times \overset{k+1}{\cdots} \times X_0
\end{array}
\tag{11.10}
$$

we obtain for each $1 < r < n$ pullbacks in $[\Delta^{n-3^{op}}, \mathsf{Cat}]$

$$
\begin{array}{ccc}
(X(f_0)_1)^{\{r\}}_0 & \longrightarrow & (Y_0)^{\{r\}}_0 \times (Y_0)^{\{r\}}_0 \\
\downarrow & & \downarrow \\
(X_1)^{\{r\}}_0 & \longrightarrow & (X_0)^{\{r\}}_0 \times (X_0)^{\{r\}}_0
\end{array}
\qquad
\begin{array}{ccc}
(X(f_0)_k)^{\{r\}}_0 & \longrightarrow & (Y_0)^{\{r\}}_0 \times \overset{k+1}{\cdots} \times (Y_0)^{\{r\}}_0 \\
\downarrow & & \downarrow \\
(X_k)^{\{r\}}_0 & \longrightarrow & (X_0)^{\{r\}}_0 \times \overset{k+1}{\cdots} \times (X_0)^{\{r\}}_0
\end{array}
$$

Since X_1, X_0, Y_0 are in $\mathsf{Cat}^{\mathsf{n-1}}_{\mathsf{wg}}$, by Lemma 7.2.5 $(Y_0)^{\{r\}}_0$, $(X_1)^{\{r\}}_0$, $(X_0)^{\{r\}}_0$ are levelwise equivalence relations; by hypothesis $(f_0)^{\{r\}}_0$ is a levelwise isofibration. Therefore by Lemma 11.1.3 $(X(f_0)_1)^{\{r\}}_0$ and $(X(f_0)_k)^{\{r\}}_0$ are levelwise equivalence relations. In summary, $(X(f_0))^{\{r\}}_0$ is a levelwise equivalence relation, that is, $X(f_0)$ satisfies hypothesis (i) of Proposition 7.2.8 (b).

To show that hypothesis (ii) of Proposition 7.2.8 (b) holds, we need to show that $p^{(n-1)}X(f_0) \in \mathsf{Cat}^{\mathsf{n-1}}_{\mathsf{wg}}$. By hypothesis the pullbacks (11.10) in $[\Delta^{n-2^{op}}, \mathsf{Cat}]$ satisfy the hypothesis of Lemma 11.1.4. It follows that, for each $1 < r \leq n$, there are pullbacks in $[\Delta^{r-2^{op}}, \mathsf{Cat}]$

$$
\begin{array}{ccc}
p^{(r-1)}X(f_0)_1 & \longrightarrow & p^{(r-1)}Y_0 \times p^{(r-1)}Y_0 \\
\downarrow & & \downarrow {\scriptstyle p^{(r-1)}f_0 \times p^{(r-1)}f_0} \\
p^{(r-1)}X_1 & \longrightarrow & p^{(r-1)}X_0 \times p^{(r-1)}X_0
\end{array}
$$

$$
\begin{array}{ccc}
p^{(r-1)}X(f_0)_k & \longrightarrow & p^{(r-1)}Y_0 \times \overset{k+1}{\cdots} \times p^{(r-1)}Y_0 \\
\downarrow & & \downarrow {\scriptstyle p^{(r-1)}f_0 \times \overset{k+1}{\cdots} \times p^{(r-1)}f_0} \\
p^{(r-1)}X_k & \longrightarrow & p^{(r-1)}X_0 \times \overset{k+1}{\cdots} \times p^{(r-1)}X_0
\end{array}
$$

We conclude that

$$p^{(r-1)}(X(f_0)) = (p^{(r-1)}X)(p^{(r-1)}f_0)\ . \tag{11.11}$$

In particular, when $r = n$, $p^{(n-1)}f_0$ satisfies inductive hypothesis (a), therefore $p^{(n-1)}X(f_0) \in \mathsf{Cat}^{n-1}_{\mathsf{wg}}$. Thus $X(f_0)$ satisfies the hypothesis of Proposition 7.2.8 (b) and we conclude that $X(f_0) \in \mathsf{Cat}^{n}_{\mathsf{wg}}$. This completes the proof of (a).

(b) Let $a, b \in Y_0^d$. There is a pullback in $\mathsf{Cat}^{n-1}_{\mathsf{wg}}$

$$\begin{array}{ccc} X(f_0)(a,b) & \longrightarrow & Y_0(a)\times Y_0(b) \\ \downarrow & & \downarrow \\ X(f_0a, f_0b) & \longrightarrow & X_0(f_0a)\times X_0(f_0b) \end{array}$$

Since $X_0(f_0a)$, $Y_0(a) \in \mathsf{Cat}^{n-1}_{\mathsf{hd}}$ and $Y_0(a)^d = \{a\} \cong \{f_0a\} = X_0(f_0a)^d$, the map $Y_0(a) \to X_0(f_0a)$ is an n-equivalence (by Lemma 5.2.6). Also, this map is a levelwise isofibration in Cat, since $f_0 : Y_0 \to X_0$ is, and the same holds after application of $p^{(r-1)}$ (as it is true for $p^{(r-1)}f_0$). It follows from Lemma 11.1.4 (b) that

$$X(f_0)(a,b) \to X(f_0a, f_0b)$$

is an $(n-1)$-equivalence.

Finally, since by hypothesis $p^{(1)}f_0$ is surjective on objects, pf_0 is surjective. By (a), this is the object part of the map $p^{(1)}X(f_0) \to p^{(1)}X$ which implies that $pX(f_0) \to pX$ is surjective. By Proposition 7.1.2, we conclude that $X(f_0) \to X$ is an n-equivalence.

(c) By construction there are pullbacks in $[\Delta^{n-2^{op}}, \mathsf{Cat}\,]$, for each $k \geq 2$

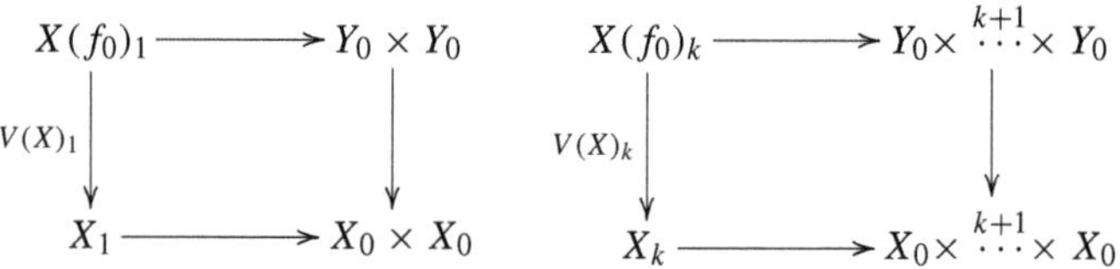

Since pullbacks in $[\Delta^{n-2^{op}}, \mathsf{Cat}\,]$ are computed pointwise, for each $\underline{r} \in \Delta^{n-2^{op}}$ there are pullbacks in Cat

$$\begin{array}{ccc} X(f_0)_{1\underline{r}} & \longrightarrow & Y_{0\underline{r}}\times Y_{0\underline{r}} \\ {\scriptstyle V(X)_{1\underline{r}}}\downarrow & & \downarrow{\scriptstyle f_{0\underline{r}}\times f_{0\underline{r}}} \\ X_{1\underline{r}} & \longrightarrow & Y_{0\underline{r}}\times Y_{0\underline{r}} \end{array} \qquad \begin{array}{ccc} X(f_0)_{k\underline{r}} & \longrightarrow & Y_{0\underline{r}}\times \overset{k+1}{\cdots}\times Y_{0\underline{r}} \\ {\scriptstyle V(X)_{k\underline{r}}}\downarrow & & \downarrow{\scriptstyle f_{0\underline{r}}\times \overset{k+1}{\cdots}\times f_{0\underline{r}}} \\ X_{k\underline{r}} & \longrightarrow & X_{0\underline{r}}\times \overset{k+1}{\cdots}\times X_{0\underline{r}} \end{array}$$

Since, by hypothesis, $f_{0\underline{r}}$ is an isofibration and isofibrations are stable under pullbacks, $V(X)_{1\underline{r}}$ and $V(X)_{k\underline{r}}$ are isofibrations for all $k \geq 2$. Since the object functor $ob : \mathsf{Cat} \to \mathsf{Set}$ preserves pullbacks, we also have pullbacks in Set

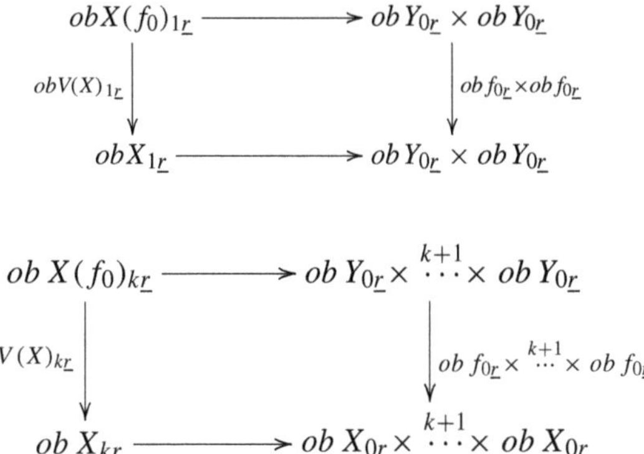

Since, by hypothesis, $ob\, f_{0\underline{r}}$ is surjective, so are $ob\,(V(X)_{1\underline{r}})$ and $ob\,(V(X)_{k\underline{r}})$. In conclusion, $V(X)$ is levelwise an isofibration surjective on objects.

By part (a)

$$p^{(r-1)}(X(f_0)) = (p^{(r-1)}X)(p^{(r-2)} f_0)$$

and by hypothesis $p^{(r-2)} f_0$ is levelwise an isofibration surjective on objects. By the induction hypothesis applied to $p^{(r-1)}X$ we deduce that $p^{(r-1)}V(X)$ is also a levelwise isofibration surjective on objects.

(d) This follows from (a) and (b) using Proposition 7.2.1.

(e) Since f_0 is a morphism in $\mathsf{Cat}^{\mathsf{n-1}}_{\mathsf{hd}}$, $q^{(r-1)} f_0 = p^{(r-1)} f_0$ is a levelwise isofibration in Cat for all $1 < r \leq n$. Further, X_0 and $p^{(n-1)}X_0$ are levelwise groupoids since $X_0 \in \mathsf{Cat}^{\mathsf{n-1}}_{\mathsf{hd}}$. Thus all the hypotheses of Lemma 11.1.4 (c) hold and thus from the pullbacks (11.10) we obtain pullbacks in $[\Delta^{r-2^{op}}, \mathsf{Cat}]$

$$\begin{array}{ccc}
q^{(r-1)}X(f_0)_1 & \longrightarrow & q^{(r-1)}Y_0 \times q^{(r-1)}Y_0 \\
\downarrow & & \downarrow{\scriptstyle q^{(r-1)} f_0 \times q^{(r-1)} f_0} \\
q^{(r-1)}X_1 & \longrightarrow & q^{(r-1)}X_0 \times q^{(r-1)}X_0
\end{array}$$

$$\begin{array}{ccc}
q^{(r-1)}X(f_0)_k & \longrightarrow & q^{(r-1)}Y_0 \times \overset{k+1}{\cdots} \times q^{(r-1)}Y_0 \\
\downarrow & & \downarrow{\scriptstyle q^{(r-1)} f_0 \times \overset{k+1}{\cdots} \times q^{(r-1)} f_0} \\
q^{(r-1)}X_k & \longrightarrow & q^{(r-1)}X_0 \times \overset{k+1}{\cdots} \times q^{(r-1)}X_0
\end{array}$$

This shows that

$$q^{(r-1)}(X(f_0)) = (q^{(r-1)}X)(q^{(r-2)}X_0) \,.$$

By the induction hypothesis applied to $q^{(r-1)}X$ we deduce that $q^{(r-1)}V(X)$ is also a levelwise isofibration surjective on objects. □

We next study the behaviour of the construction of Proposition 11.1.5 with respect to pullbacks.

Corollary 11.1.6

(a) *Suppose that $X \to Z \leftarrow Y$ is a diagram in $\mathsf{Cat}^n_{\mathsf{wg}}$ such that $X\times_Z Y \in \mathsf{Cat}^n_{\mathsf{wg}}$ and suppose we have a commuting diagram in $\mathsf{Cat}^{n-1}_{\mathsf{hd}}$*

$$\begin{array}{ccccc} X'_0 & \longrightarrow & Z'_0 & \longleftarrow & Y'_0 \\ \downarrow{\scriptstyle f_0} & & \downarrow{\scriptstyle h_0} & & \downarrow{\scriptstyle g_0} \\ X_0 & \longrightarrow & Z_0 & \longleftarrow & Y_0 \end{array}$$

such that $X'_0\times_{Z'_0} Y'_0 \in \mathsf{Cat}^{n-1}_{\mathsf{hd}}$. Then

$$(X\times_Z Y)(f_0\times_{h_0} g_0) \cong X(f_0)\times_{Z(h_0)} Y(g_0) \,,$$

$$V(X\times_Z Y) = V(X)\times_{V(Z)} V(Y) \,.$$

Suppose, further, that f_0, g_0, h_0, $p^{(r-1)}f_0$, $p^{(r-1)}g_0$, $p^{(r-1)}h_0$ are levelwise isofibrations in Cat surjective on objects for all $1 < r < n$.

(b) *If, for all $1 \le r \le n$,*

$$p^{(r-1)}(X\times_Z Y) = p^{(r-1)}X\times_{p^{(r-1)}Z} p^{(r-1)}Y \,,$$

then

$$\begin{aligned} &p^{(r-1)}\{(X\times_Z Y)(f_0\times_{h_0} g_0)\} \\ &\cong p^{(r-1)}\{X(f_0)\}\times_{p^{(r-1)}\{Z(h_0)\}} p^{(r-1)}\{Y(g_0)\} \end{aligned} \tag{11.12}$$

(c) *If, for all $1 \le r \le n$,*

$$q^{(r-1)}(X\times_Z Y) = q^{(r-1)}X\times_{q^{(r-1)}Z} q^{(r-1)}Y \,,$$

then

$$q^{(r-1)}\{(X\times_Z Y)(f_0\times_{h_0} g_0)\} \cong q^{(r-1)}\{X(f_0)\}\times_{q^{(r-1)}\{Z(h_0)\}} q^{(r-1)}\{Y(g_0)\} \,.$$

Proof

(a) This follows from Lemma 11.1.1, taking $\mathscr{C} = \mathsf{Cat}^{n-1}_{\mathsf{wg}}$.
(b) By hypothesis and by part (a)

$$\begin{aligned}&(p^{(r-1)}(X\times_Z Y))(p^{(r-2)}f_0\times_{p^{(r-2)}h_0} p^{(r-2)}g_0)\\ \cong\ &(p^{(r-1)}X\times_{p^{(r-1)}Z} p^{(r-1)}Y)(p^{(r-2)}f_0\times_{p^{(r-2)}h_0} p^{(r-2)}g_0)\\ \cong\ &(p^{(r-1)}X)(p^{(r-2)}f_0)\times_{(p^{(r-1)}Z)(p^{(r-2)}h_0)} (p^{(r-1)}Y)(p^{(r-2)}g_0)\,.\end{aligned}$$

By Proposition 11.1.5 (a)

$$(p^{(r-1)}X)(p^{(r-2)}f_0) = p^{(r-1)}(X(f_0))$$

and similarly for Y, Z and $X\times_Z Y$. Therefore from the above we obtain (11.12).
(c) The proof is completely analogous to that of (b), with $q^{(r-1)}$ in place of $p^{(r-1)}$, and using Proposition 11.1.5 (e). □

11.2 Weakly Globular n-Fold Categories and Functoriality of Homotopically Discrete Objects

In this section we show (Proposition 11.2.5) that we can functorially approximate up to n-equivalence a weakly globular n-fold category X by a better behaved one F_nX in which the homotopically discrete $(n-1)$-fold category at level 0 admits a functorial section to the discretization map. The functor F_n will be used in Sect. 11.3 to construct the functor $G_n : \mathsf{Cat}^n_{\mathsf{wg}} \to \mathsf{FCat}^n_{\mathsf{wg}}$, which will then lead in Sect. 12.1 to the discretization functor from $\mathsf{Cat}^n_{\mathsf{wg}}$ to Ta^n.

11.2.1 The Idea of the Functors V_n and F_n

We now discuss the idea of the functors $V_n : \mathsf{Cat}^n_{\mathsf{hd}} \to \mathsf{Cat}^n_{\mathsf{hd}}$ and $F_n : \mathsf{Cat}^n_{\mathsf{wg}} \to \mathsf{Cat}^n_{\mathsf{wg}}$. The functor V_n is used in defining F_n and the latter is needed in the construction of the functor $G_n : \mathsf{Cat}^n_{\mathsf{wg}} \to \mathsf{FCat}^n_{\mathsf{wg}}$ (whose idea is explained in Sect. 11.3.3): this will then lead to the discretization functor $Disc_n : \mathsf{Cat}^n_{\mathsf{wg}} \to \mathsf{Ta}^n$ (see Sect. 12.2.1).

The functor F_n is based on the construction $X(f_0)$ of the previous section, given in Proposition 11.1.5, for an appropriate choice of the map $f_0 : Y_0 \to X_0$.

When $n = 2$ we define

$$F_2X = X(u_{X_0})\ ,$$

where $u_{X_0} : \operatorname{Dec} X_0 \to X_0$ is as in Sect. 2.7. Since u_{X_0} is an isofibration surjective on objects (see Lemma 11.2.1), by Proposition 11.1.5 this replaces X with a 2-equivalent $F_2 X$ in which $(F_2 X)_0 = \operatorname{Dec} X_0$. As observed in Sect. 2.7, $\operatorname{Dec} X_0$ is homotopically discrete and has a functorial section to the discretization map.

When $n > 1$ the construction of $F_n X$ is again based on Proposition 11.1.5 but for a more complex choice of the map $f_0 : Y_0 \to X_0$. We build the appropriate map in Proposition 11.2.3 where we construct a functor

$$V_n : \mathsf{Cat}^{\mathsf{n}}_{\mathsf{hd}} \to \mathsf{Cat}^{\mathsf{n}}_{\mathsf{hd}}$$

together with a map

$$v_n(X) : V_n X \to X$$

for all $X \in \mathsf{Cat}^{\mathsf{n}}_{\mathsf{hd}}$ satisfying the hypotheses needed to apply Proposition 11.1.5 and such that if $h : X \to Y$ is a morphism in $\mathsf{Cat}^{\mathsf{n}}_{\mathsf{hd}}$, the following diagram commutes for appropriate sections to the discretization maps $V_n X \to (V_n X)^d$ and $V_n Y \to (V_n Y)^d$:

$$\begin{array}{ccc} V_n X & \longrightarrow & V_n Y \\ \uparrow & & \uparrow \\ (V_n X)^d & \longrightarrow & (V_n Y)^d \end{array} \tag{11.13}$$

The construction of V_n is inductive on dimension, starting with $V_1 X = \operatorname{Dec} X$ and $v_1(X) = u_X : \operatorname{Dec} X \to X$ as in Sect. 2.7; by Lemma 11.2.1, $v_1(X)$ is an isofibration and is surjective on objects.

Given $X \in \mathsf{Cat}^{\mathsf{n}}_{\mathsf{hd}}$, having defined inductively V_{n-1} and $v_{n-1}(X_0) : V_{n-1}(X_0) \to X_0$, we define

$$F_n X = X(v_{n-1}(X_0))$$

and we define $V_n(X)$ via the pullback in $[\Delta^{{n-1}^{op}}, \mathsf{Cat}]$

$$\begin{array}{ccc} V_n(X) & \xrightarrow{h_n(X)} & F_n X \\ \Big\downarrow{\scriptstyle l} & & \Big\downarrow{\scriptstyle r} \\ \operatorname{Dec} q^{(1)} F_n X & \xrightarrow{u'} & q^{(1)} F_n X \end{array} \tag{11.14}$$

where $r = \gamma^{(1)}_{F_0 X}$ (see Remark 7.2.13) and

$$u' = u_{(q^{(1)} F_n X)}.$$

The map $v_n(X) : V_n X \to X$ is defined by the composite

$$v_n(X) : V_n X \xrightarrow{h_n(X)} F_n X \xrightarrow{f_n(X)} X \ ,$$

where the map $f_n(X)$ is as in Proposition 11.1.5 (we adopt the notation $f_n(X)$ instead of $V(X)$). By the definition of $V_n(X)$ in (11.14), in order to construct the map

$$\gamma'_{V_n X} : (V_n X)^d \to V_n X$$

we need to construct maps

$$t : (V_n X)^d \to \operatorname{Dec} q^{(1)} F_n X, \qquad s : (V_n X)^d \to F_n X$$

such that $u't = rs$, where u', r are as in (11.14). It is not difficult to show that

$$(V_n X)^d = (\operatorname{Dec} q^{(1)} F_n X)^d = (V_{n-1} X_0)^d.$$

This gives rise to natural maps

$$t : (V_n X)^d = (\operatorname{Dec} q^{(1)} F_n X)^d \to \operatorname{Dec} q^{(1)} F_n X$$

and

$$s : (V_n X)^d = (V_{n-1} X_0)^d \xrightarrow{v} (F_n X)_0 \xrightarrow{\alpha_{F_n X}} F_n X \ ,$$

where

$$(F_n X)_0^d = (V_{n-1} X_0)^d \xrightarrow{v} V_{n-1} X_0 = (F_n X)_0$$

is given by the inductive hypothesis and α is the counit of the adjunction $d \dashv ob$. In the proof of Proposition 11.2.3 we show that t and s define $\gamma'_{V_n X} : (V_n X)^d \to V_n X$ with the required properties.

In the course of this proof we thus define a functor

$$F_n : \mathsf{Cat}^n_{\mathsf{hd}} \to \mathsf{Cat}^n_{\mathsf{hd}}$$

by

$$F_n X = X(v_{n-1}(X_0))$$

for $v_{n-1}(X_0) : V_{n-1} X_0 \to X_0$ in $\mathsf{Cat}^{n-1}_{\mathsf{hd}}$. In Proposition 11.2.5 we extend this to a functor

$$F_n : \mathsf{Cat}^n_{\mathsf{wg}} \to \mathsf{Cat}^n_{\mathsf{wg}}$$

again defined by

$$F_n X = X(v_{n-1}(X_0))$$

Since $(F_n X)_0 = V_{n-1} X_0$ the functoriality of the sections to the discretization map of $(F_n X)_0$ follows immediately from (11.13).

The functors V_n and F_n are well behaved with respect to pullbacks, and this plays an important role in the construction of the functor G_n in Sect. 11.3.4 (see also the informal discussion in Sect. 11.3.3 about this point). The proofs of these facts are based on the properties, with respect to pullbacks, of the general constructions of Lemma 11.1.1 and Corollary 11.1.6. In Corollary 11.2.4, we show that given a diagram in $\mathsf{Cat}^n_{\mathsf{hd}}$

$$X \to Z \leftarrow Y$$

such that $X \times_Z Y \in \mathsf{Cat}^n_{\mathsf{hd}}$ and such that this pullback is preserved by $p^{(r-1)}$ and $q^{(r-1)}$ (for each $1 \leq r \leq n$), then this pullback is also preserved by V_n, $p^{(r-1)} V_n$ and $q^{(r-1)} V_n$.

We then deduce in Corollary 11.2.6 a similar property for the functor F_n. Namely, given a diagram in $\mathsf{Cat}^n_{\mathsf{wg}}$

$$X \to Z \leftarrow Y$$

such that $X \times_Z Y \in \mathsf{Cat}^n_{\mathsf{wg}}$ and such that this pullback is preserved by $p^{(r-1)}$ and $q^{(r-1)}$ (for each $1 \leq r \leq n$), then this pullback is also preserved by F_n, $p^{(r-1)} F_n$ and $q^{(r-1)} F_n$.

11.2.2 The Functors V_n and F_n

In this section we construct the functors V_n and F_n and we study their properties.

Lemma 11.2.1 *If $X \in \mathsf{Cat}_{\mathsf{hd}}$, then* Dec $X \in \mathsf{Cat}_{\mathsf{hd}}$ *and the map u_X :* Dec $X \to X$ *as in Sect. 2.7 is an isofibration surjective on objects.*

Proof Since $X \in \mathsf{Cat}_{\mathsf{hd}}$, $X = A[f]$ for a surjective map of sets $f : A \to B$, where $A[f]$ is as in Definition 5.1.8. Thus Dec $X = (A \times_B A)[d_0]$, where $d_0 : A \times_B A \to A$, $d_0(x, y) = x$. The source and target maps

$$\tilde{d}_0, \tilde{d}_1 : (\text{Dec } X)_1 = A \times_B A \times_B A \to (Dec X)_0 = A \times_B A$$

are $\tilde{d}_0(x, y, z) = (x, y)$, $\tilde{d}_1(x, y, z) = (x, z)$.

Given $(x, y) \in (\mathrm{Dec}\, X)_0$ and an isomorphism $(u_X(x, y) = y, z) \in X_1$, we have $(x, y, z) \in (\mathrm{Dec}\, X)_1$ with $\tilde{d}_1(x, y, z) = (x, z)$. Diagramatically:

$$\begin{array}{ccc} (x,y) = \tilde{d}_0(x,y,z) & \xrightarrow{(x,y,z)} & (x,z) = \tilde{d}_1(x,y,z) \\ \downarrow & & \downarrow \\ u_X(x,y) = y & \xrightarrow[(y,z)=u_X(x,y,z)]{} & u_X(x,z) = z \end{array}$$

By definition, this shows that $u_X : \mathrm{Dec}\, X \to X$ is an isofibration. It is also surjective on objects since $(u_X)_1 = pr_2 : A \times_B A \to A$ is surjective. □

Lemma 11.2.2

(a) Let $A \in \mathsf{Cat}^n_{\mathsf{hd}}$, $B, C \in \mathsf{Set}$ and consider the pullback in $[\Delta^{n-1^{op}}, \mathsf{Cat}]$

$$\begin{array}{ccc} Q & \longrightarrow & A \\ \downarrow & & \downarrow \\ C & \longrightarrow & B \end{array}$$

then $Q \in \mathsf{Cat}^n_{\mathsf{hd}}$.

(b) Let $X \in \mathsf{Cat}^n_{\mathsf{hd}}$, $Z \in \mathsf{Cat}_{\mathsf{hd}}$ and consider the pullback in $[\Delta^{n-1^{op}}, \mathsf{Cat}]$

$$\begin{array}{ccc} P & \longrightarrow & X \\ \downarrow & & \downarrow \\ Z & \longrightarrow & q^{(1)}X \end{array}$$

Then $P \in \mathsf{Cat}^n_{\mathsf{hd}}$.

Proof By induction on n. In the case $n = 1$ for (a), since B is discrete, $A \to B$ is an isofibration, therefore

$$Q = A \times_B C \simeq A \overset{ps}{\times}_B C \simeq A^d \times_B C\,.$$

Hence $Q \in \mathsf{Cat}_{\mathsf{hd}}$. The case $n = 2$ for (b) is Lemma 7.1.10. Suppose, inductively, that the lemma holds for $(n-1)$.

(a) For each $k \geq 0$ there is a pullback in $[\Delta^{n-2^{op}}, \mathsf{Cat}]$

$$\begin{array}{ccc} Q_k & \longrightarrow & A_k \\ \downarrow & & \downarrow \\ C_k & \longrightarrow & B_k \end{array}$$

Therefore, by inductive hypothesis (a), $Q_k \in \mathsf{Cat}^{\mathsf{n-1}}_{\mathsf{hd}}$. For each $\underline{r} = (r_1, \dots, r_{n-1}) \in \Delta^{n-1^{op}}$, we have a pullback in Cat

$$\begin{array}{ccc} Q_{\underline{r}} & \longrightarrow & A_{\underline{r}} \\ \downarrow & & \downarrow \\ C & \longrightarrow & B \end{array}$$

Since p commutes with fiber products over discrete objects, we have a pullback in Set

$$\begin{array}{ccc} pQ_{\underline{r}} & \longrightarrow & pA_{\underline{r}} \\ \downarrow & & \downarrow \\ C & \longrightarrow & B \end{array}$$

It follows that there is a pullback in $[\Delta^{n-2^{op}}, \mathsf{Cat}]$

$$\begin{array}{ccc} p^{(n-1)}Q & \longrightarrow & p^{(n-1)}A \\ \downarrow & & \downarrow \\ C & \longrightarrow & B \end{array}$$

By inductive hypothesis (a) we conclude that $p^{(n-1)}Q \in \mathsf{Cat}^{\mathsf{n-1}}_{\mathsf{hd}}$. By Lemma 5.1.6, this means that $Q \in \mathsf{Cat}^{\mathsf{n}}_{\mathsf{hd}}$.

(b) For each $k \geq 0$, there is a pullback in $[\Delta^{n-2^{op}}, \mathsf{Cat}]$

$$\begin{array}{ccc} P_k & \longrightarrow & X_k \\ \downarrow & & \downarrow{\scriptstyle \gamma^{(0)}} \\ Z_k & \longrightarrow & qX_k \end{array}$$

Since $X_k \in \mathsf{Cat}^{\mathsf{n-1}}_{\mathsf{hd}}$ (as $X \in \mathsf{Cat}^{\mathsf{n}}_{\mathsf{hd}}$), by part (a) this implies that $P_k \in \mathsf{Cat}^{\mathsf{n-1}}_{\mathsf{hd}}$. Since $p^{(n-1)}$ commutes with fiber products over discrete objects, we also have a pullback in $[\Delta^{n-2^{op}}, \mathsf{Cat}]$

$$\begin{array}{ccc} p^{(n-1)}P & \longrightarrow & p^{(n-1)}X \\ \downarrow & & \downarrow \\ q^{(1)}Z & \longrightarrow & q^{(1)}X \end{array}$$

where $p^{(n-1)}X \in \mathsf{Cat}^{\mathsf{n-1}}_{\mathsf{hd}}$. By inductive hypothesis (b), $p^{(n-1)}P \in \mathsf{Cat}^{\mathsf{n-1}}_{\mathsf{hd}}$. Hence by Lemma 5.1.6, $P \in \mathsf{Cat}^{\mathsf{n}}_{\mathsf{hd}}$. □

Proposition 11.2.3 *For each $n \geq 1$, there is a functor*

$$V_n : \mathsf{Cat}^{\mathsf{n}}_{\mathsf{hd}} \to \mathsf{Cat}^{\mathsf{n}}_{\mathsf{hd}}$$

with a map

$$v_n(X) : V_n X \to X \ ,$$

natural in $X \in \mathsf{Cat}^{\mathsf{n}}_{\mathsf{hd}}$, such that

(a) $v_n(X)$ is a levelwise isofibration in $\mathsf{Cat}^{\mathsf{n}}_{\mathsf{hd}}$ which is surjective on objects and, for all $1 < r \leq n$, the same holds for $p^{(r-1)} v_n(X)$.
(b) V_n is the identity on discrete objects and preserves pullbacks over discrete objects.
(c) If $h : X \to Y$ is a morphism in $\mathsf{Cat}^{\mathsf{n}}_{\mathsf{hd}}$, the following diagram commutes with the maps $(V_n X)^d \to V_n X$ and $(V_n Y)^d \to V_n Y$ being sections to the discretization maps $V_n X \to (V_n X)^d$ and $V_n Y \to (V_n Y)^d$:

$$\begin{array}{ccc} V_n X & \longrightarrow & V_n Y \\ \uparrow & & \uparrow \\ (V_n X)^d & \longrightarrow & (V_n Y)^d \end{array}$$

Proof By induction on n. For $n = 1$, let $V_1 X = \operatorname{Dec} X$ and $v_1(X) = u_X : \operatorname{Dec} X \to X$ as in Sect. 2.7. By Lemma 11.2.1, $v_1(X)$ is an isofibration and is surjective on objects and it is clearly natural in X. Also Dec preserves pullbacks. Given a morphism $h : X \to Y$ in $\mathsf{Cat}_{\mathsf{hd}}$, we have a commutative diagram

$$\begin{array}{ccc} \operatorname{Dec} X & \xrightarrow{\operatorname{Dec} h} & \operatorname{Dec} Y \\ \uparrow & & \uparrow \\ (\operatorname{Dec} X)^d = X_0 & \xrightarrow{h_0} & (\operatorname{Dec} Y)^d = Y_0 \end{array}$$

where $X_0 \to \operatorname{Dec} X$ and $Y_0 \to \operatorname{Dec} Y$ are the functorial sections to the discretization maps as in Sect. 2.7. This proves the lemma in the case $n = 1$. Suppose, inductively, that it holds for $(n-1)$ and let $X \in \mathsf{Cat}^{\mathsf{n}}_{\mathsf{hd}}$.

(a) Let

$$F_n X = X(v_{n-1}(X_0)) \ ,$$

where $v_{n-1}(X_0) : V_{n-1} X_0 \to X_0$ is given by the inductive hypothesis applied to $X_0 \in \mathsf{Cat}^{\mathsf{n-1}}_{\mathsf{hd}}$ and $X(v_{n-1}(X_0))$ is as in Lemma 11.1.1. By inductive hypothesis (a), $v_{n-1}(X_0)$ satisfies the hypothesis of Proposition 11.1.5 and thus $F_n X \in \mathsf{Cat}^{\mathsf{n}}_{\mathsf{hd}}$.

We define $V_n X$ via the pullback in $[\Delta^{n-1^{op}}, \mathsf{Cat}]$

$$\begin{array}{ccc} V_n X & \xrightarrow{h_n(X)} & F_n X \\ \downarrow l & & \downarrow r \\ \operatorname{Dec} q^{(1)} F_n X & \xrightarrow{u'} & q^{(1)} F_n X \end{array} \tag{11.15}$$

where $r = \gamma^{(1)}_{F_n X}$ (see Remark 7.2.13) and

$$u' = u_{(q^{(1)} F_n X)}$$

as in Sect. 2.7. Since $F_n X \in \mathsf{Cat}^n_{\mathsf{hd}}$, $q^{(1)} F_n X \in \mathsf{Cat}_{\mathsf{hd}}$, hence $\operatorname{Dec} q^{(1)} F_n X \in \mathsf{Cat}_{\mathsf{hd}}$ (see Sect. 2.7). Thus by Lemma 11.2.2, $V_n X \in \mathsf{Cat}^n_{\mathsf{hd}}$.

For each $\underline{k} \in \Delta^{n-1^{op}}$, there is a pullback in Cat

$$\begin{array}{ccc} (V_n X)_{\underline{k}} & \xrightarrow{(h_n(X))_{\underline{k}}} & (F_n X)_{\underline{k}} \\ \downarrow l_{\underline{k}} & & \downarrow r_{\underline{k}} \\ (\operatorname{Dec} q^{(1)} F_n X)_{\underline{k}} & \xrightarrow{u'_{\underline{k}}} & (q^{(1)} F_n X)_{\underline{k}} \, . \end{array} \tag{11.16}$$

The bottom horizontal map is an isofibration since the target is discrete; hence, since isofibrations are stable under pullback, $(h_n(X))_{\underline{k}}$ is also an isofibration. The bottom horizontal map in (11.16) is also surjective on objects since

$$ob(u'_{\underline{k}}) : (\operatorname{Dec} q^{(1)} F_n X)_{k_1} \to (q^{(1)} F_n X)_{k_1}$$

is surjective for all $k_1 \geq 0$. Since $ob((h_n(X))_{\underline{k}})$ is the pullback of $ob(u'_{\underline{k}})$, it follows that $(h_n(X))_{\underline{k}}$ is also surjective on objects.

By inductive hypothesis (a) the map $v_{n-1}(X_0) : V_{n-1} X_0 \to X_0$ satisfies the hypothesis of Proposition 11.1.5, and thus the map (which we now denote by $f_n(X)$ instead of $V(X)$)

$$f_n(X) : F_n X = X(v_{n-1}(X_0)) \to X$$

is a levelwise isofibration surjective on objects. We conclude from the above that the same holds for the composite map

$$v_n(X) : V_n X \xrightarrow{h_n(X)} F_n X \xrightarrow{f_n(X)} X \, .$$

By the induction hypothesis $v_{n-1}(X_0)$ is natural in X, hence so is $f_n(X)$, while $h_n(X)$ is natural in X by construction. Thus $v_n(X)$ is also natural in X. We now show that $p^{(r-1)}v_n(X)$ is a levelwise isofibration surjective on objects for all r. Since, by Proposition 11.1.5 (c), this holds for $p^{(r-1)}f_n(X)$, it is sufficient to show this for $p^{(r-1)}h_n(X)$.

Applying p to (11.16) and using the fact that p commutes with pullbacks over discrete objects, we obtain a pullback in $[\Delta^{r-2^{op}}, \mathsf{Cat}]$

$$\begin{array}{ccc} p^{(r-1)}V_n X & \xrightarrow{p^{(r-1)}h_n(X)} & p^{(r-1)}F_n X \\ \downarrow & & \downarrow \\ \operatorname{Dec} q^{(1)}F_n X & \longrightarrow & q^{(1)}F_n X \end{array}$$

Using an argument similar to the one above we conclude that $p^{(r-1)}h_n(X)$ is a levelwise isofibration surjective on objects.

Finally, by inductive hypothesis (a) the map $v_{n-1}(X_0) : V_{n-1}X_0 \to X_0$ satisfies the hypothesis of Proposition 11.1.5, and thus the map $q^{(r-1)}f_n(X)$ is a levelwise isofibration surjective on objects. To show that the same holds for $q^{(r-1)}v_n(X)$ it is therefore enough to show it for $q^{(r-1)}h_n(X)$.

Applying q to (11.16) and using the fact that q commutes with pullbacks over discrete objects, we obtain a pullback in $[\Delta^{r-2^{op}}, \mathsf{Cat}]$

$$\begin{array}{ccc} q^{(r-1)}V_n X & \xrightarrow{q^{(r-1)}h_n(X)} & q^{(r-1)}F_n X \\ \downarrow & & \downarrow \\ \operatorname{Dec} q^{(1)}F_n X & \longrightarrow & q^{(1)}F_n X. \end{array}$$

Using an argument similar to the one above we conclude that $q^{(r-1)}h_n(X)$ is a levelwise isofibration surjective on objects. This proves (a).

(b) If X is discrete, $F_n X = X = q^{(1)}X$, thus $V_n X = X$. Since, by the inductive hypothesis, V_{n-1} commutes with pullbacks over discrete objets, so does F_n, as is easily seen. Since $q^{(1)}$ commutes with pullbacks over discrete objects and Dec commutes with pullbacks, it follows by construction that V_n commutes with pullbacks over discrete objects.

(c) By the definition of $V_n(X)$ in (11.15), in order to construct the map

$$\gamma'_{V_n X} : (V_n X)^d \to V_n X$$

we need to construct maps

$$t : (V_n X)^d \to \operatorname{Dec} q^{(1)}F_n X, \qquad s : (V_n X)^d \to F_n X$$

such that $u't = rs$, where u', r are as in (11.15). By (11.15) , since $q^{(1)}$ commutes with pullbacks over discrete objects,

$$q^{(1)}V_n X = \operatorname{Dec} q^{(1)} F_n X,$$

so that

$$qV_n X = qq^{(1)}V_n X = q \operatorname{Dec} q^{(1)} F_n X = (q^{(1)} F_n X)_0 = q(F_n X)_0 = qV_{n-1}X_0 \; .$$

It follows that

$$(V_n X)^d = (q^{(1)}V_n X)^d = (\operatorname{Dec} q^{(1)} F_n X)^d \tag{11.17}$$

as well as

$$(V_n X)^d = (F_n X)_0^d = (V_{n-1}X_0)^d \; . \tag{11.18}$$

There is a natural map

$$(\operatorname{Dec} q^{(1)} F_n X)^d \to \operatorname{Dec} q^{(1)} F_n X$$

and therefore, using (11.17), a corresponding map

$$t : (V_n X)^d = (\operatorname{Dec} q^{(1)} F_n X)^d \to \operatorname{Dec} q^{(1)} F_n X \; .$$

Note that t is also natural in X. The composite map

$$(\operatorname{Dec} q^{(1)} F_n X)^d = (q^{(1)} F_n X)_0 \to \operatorname{Dec} q^{(1)} F_n X \to q^{(1)} F_n X$$

is the counit α of the adjunction $d \dashv ob$ (see Remark 2.4.10) at $(q^{(1)} F_n X)_0$, so $u't$ is the corresponding map

$$u't : (V_n X)^d = d(q^{(1)} F_n X)_0 \xrightarrow{\alpha_{(q^{(1)} F_n X)}} (q^{(1)} F_n X) \; . \tag{11.19}$$

Using (11.18) and the inductive hypothesis on X_0, we obtain a natural map

$$v : (F_n X)_0^d = (V_{n-1}X_0)^d \xrightarrow{v} V_{n-1}X_0 = (F_n X)_0 \; . \tag{11.20}$$

We define s to be the composite

$$s : (V_n X)^d = (V_{n-1}X_0)^d \xrightarrow{v} (F_n X)_0 \xrightarrow{\alpha_{F_n X}} F_n X \; .$$

Since α is natural in X (as it is the counit of the adjunction $d \dashv ob$) and v is natural by the induction hypothesis, s is also natural in X.

The map r in (11.15) is natural (since r is levelwise unit of the adjunction $q \dashv d$), therefore we have a commuting diagram

$$\begin{array}{ccc} (F_nX)_0 & \xrightarrow{\alpha_{F_nX}} & F_nX \\ \downarrow{\scriptstyle r'} & & \downarrow{\scriptstyle r} \\ (F_nX)_0^d = (V_nX)^d = (q^{(1)}F_nX)_0 & \xrightarrow{\alpha_{q^{(1)}F_nX}} & q^{(1)}F_nX \end{array}$$

where the equality on the bottom left corner follows from (11.17) and (11.18). Thus $r' = \gamma_{(F_nX)_0}$, where

$$\gamma_{(F_nX)_0} : (F_nX)_0 \to (F_nX)_0^d$$

is the discretization map; the map v as in (11.20) is a section of $r' = \gamma_{(F_nX)_0}$. Hence

$$rs = r\alpha_{F_nX}v = (\alpha_{q^{(1)}F_nX})r'v = \alpha_{q^{(1)}F_nX}\ .$$

Since by (11.19) $u't = \alpha_{(q^{(1)}F_nX)}$, we conclude that $u't = rs$.

Since by (11.17) $(V_nX)^d = (\operatorname{Dec} q^{(1)}F_nX)^d$, by the naturality of the discretization map, there is a commuting diagram

$$\begin{array}{ccc} (V_nX)^d & = & (\operatorname{Dec} q^{(1)}F_nX)^d \\ \uparrow{\scriptstyle \gamma_{V_nX}} & & \uparrow{\scriptstyle \gamma} \\ V_nX & \xrightarrow[\ell]{} & (\operatorname{Dec} q^{(1)}F_nX) \end{array}$$

where γ is the discretization map for $\operatorname{Dec} q^{(1)}F_nX$ and l is as in (11.15). Since by construction of γ'_{V_nX} and t we have $t = \ell\gamma'_{V_nX}$ and $\gamma t = \mathrm{Id}$, we conclude that

$$\gamma_{VnX}\gamma'_{VnX} = \gamma\ell\gamma'_{VnX} = \gamma t = \mathrm{Id}\ .$$

As observed above, by construction both t and s are natural in X, so by elementary properties of pullbacks, so is $\gamma'_{V_n X}$. Thus, given a morphism $h : X \to Y$ in $\mathsf{Cat}^n_{\mathsf{hd}}$, there is a commuting diagram

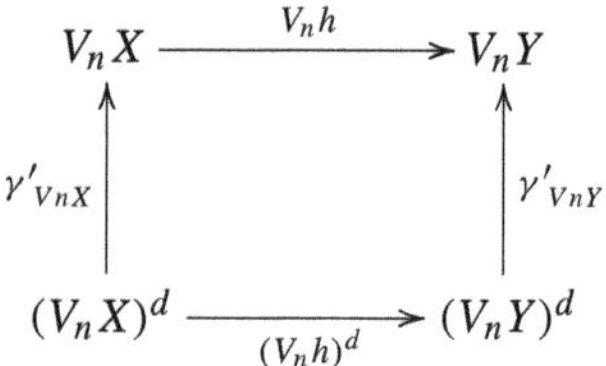

as required. □

In the following corollary we establish the properties of V_n with respect to certain pullbacks.

Corollary 11.2.4 *Let $X \to Z \leftarrow Y$ be a diagram in $\mathsf{Cat}^n_{\mathsf{hd}}$ such that $X \times_Z Y \in \mathsf{Cat}^n_{\mathsf{hd}}$ and such that, for all $1 \leq r \leq n$,*

$$p^{(r-1)}(X \times_Z Y) = p^{(r-1)} X \times_{p^{(r-1)} Z} p^{(r-1)} Y \,, \tag{11.21}$$

$$q^{(r-1)}(X \times_Z Y) = q^{(r-1)} X \times_{q^{(r-1)} Z} q^{(r-1)} Y \,. \tag{11.22}$$

Then

(a) For all $n \geq 1$,

$$V_n(X \times_Z Y) \cong V_n X \times_{V_n Z} V_n Y \,,$$
$$\upsilon_n(X \times_Z Y) \cong \upsilon_n(X) \times_{\upsilon_n(Z)} \upsilon_n(Y) \,.$$

(b) For all $1 \leq r \leq n$,

$$p^{(r-1)}\{V_n(X \times_Z Y)\} = p^{(r-1)}(V_n X) \times_{p^{(r-1)}(V_n Z)} p^{(r-1)}(V_n Y) \,,$$
$$q^{(r-1)}\{V_n(X \times_Z Y)\} = q^{(r-1)}(V_n X) \times_{q^{(r-1)}(V_n Z)} q^{(r-1)}(V_n Y) \,.$$

Proof

(a) Since $V_1 = \mathrm{Dec}$ commutes with pullbacks, (a) holds for $n = 1$. Suppose, inductively, that it holds for $(n-1)$. We claim that

$$\begin{aligned} F_n(X \times_Z Y) &= F_n X \times_{F_n Z} F_n Y \,, \\ f_n(X \times_Z Y) &= f_n(X) \times_{f_n(Z)} f_n(Y) \,, \end{aligned} \tag{11.23}$$

where F_n and f_n are as in the proof of Proposition 11.2.3 (a). In fact, by construction

$$F_n(X\times_Z Y) = (X\times_Z Y)(v_{n-1}(X_0\times_{Z_0} Y_0))\ ,$$

where

$$v_{n-1}(X_0\times_{Z_0} Y_0) : V_{n-1}(X_0\times_{Z_0} Y_0) \to X_0\times_{Z_0} Y_0\ .$$

Observe that the diagram in $\mathsf{Cat}^{n-1}_{\mathsf{hd}}$ $X_0 \to Z_0 \leftarrow Y_0$ satisfies the inductive hypothesis. In fact, since $X\times_Z Y \in \mathsf{Cat}^{n}_{\mathsf{hd}}$,

$$(X\times_Z Y)_0 = X_0\times_{Z_0} Y_0 \in \mathsf{Cat}^{n-1}_{\mathsf{hd}}\ ,$$

while hypotheses (11.21), (11.22) imply

$$\begin{aligned} &p^{(r-2)}(X_0\times_{Z_0} Y_0) = \{p^{(r-1)}(X\times_Z Y)\}_0 \\ &= \{p^{(r-1)}X\times_{p^{(r-1)}Z}\, p^{(r-1)}Y)\}_0 = \{p^{(r-2)}X_0\times_{p^{(r-2)}Z_0}\, p^{(r-2)}Y_0)\}\ , \end{aligned}$$

and similarly for $q^{(r-2)}$. By the induction hypothesis we deduce that

$$\begin{aligned} V_{n-1}(X_0\times_{Z_0} Y_0) &= V_{n-1}X_0\times_{V_{n-1}Z_0} V_{n-1}Y_0\ , \\ v_{n-1}(X_0\times_{Z_0} Y_0) &= v_{n-1}(X_0)\times_{v_{n-1}(Z_0)} v_{n-1}(Y_0)\ . \end{aligned}$$

From above, the commutative diagram in $\mathsf{Cat}^{n-1}_{\mathsf{hd}}$

$$\begin{array}{ccccc} V_{n-1}X_0 & \longrightarrow & V_{n-1}Z_0 & \longleftarrow & V_{n-1}Y_0 \\ {\scriptstyle v_{n-1}(X_0)}\downarrow & & {\scriptstyle v_{n-1}(Y_0)}\downarrow & & \downarrow{\scriptstyle v_{n-1}(Z_0)} \\ X_0 & \longrightarrow & Z_0 & \longleftarrow & Y_0 \end{array}$$

is such that $V_{n-1}X_0\times_{V_{n-1}Z_0} V_{n-1}Y_0 \in \mathsf{Cat}^{n-1}_{\mathsf{hd}}$ and $X_0\times_{Z_0} Y_0 \in \mathsf{Cat}^{n-1}_{\mathsf{hd}}$. Also, by Proposition 11.2.3 (a), for all $1 < r < n$ the maps

$$\begin{array}{lll} v_{n-1}(X_0), & v_{n-1}(Z_0), & v_{n-1}(Y_0), \\ p^{(r-1)}v_{n-1}(X_0), & p^{(r-1)}v_{n-1}(Z_0), & p^{(r-1)}v_{n-1}(Y_0) \end{array}$$

are levelwise isofibrations in Cat, surjective on objects. Thus, together with hypotheses (11.21) and (11.22), we see that all the hypotheses of Corollary 11.1.6 are satisfied and we conclude that

$$\begin{aligned} &F_n(X\times_Z Y) = (X\times_Z Y)(v_{n-1}(X_0\times_{Z_0} Y_0)) \\ &= X(v_{n-1}(X_0))\times_{Z(v_{n-1}(Z_0))} Y(v_{n-1}(Y_0)) = F_nX\times_{F_nZ} F_nY \end{aligned}$$

and

$$f_n(X \times_Z Y) = f_n(X) \times_{f_n(Z)} f_n(Y) ,$$

which is (11.23). Further, by Corollary 11.1.6,

$$\begin{aligned} & q^{(1)} F_n(X \times_Z Y) = q^{(1)}\{(X \times_Z Y)(f_{X_0} \times_{f_{Z_0}} f_{Y_0})\} \\ \cong\, & q^{(1)}\{X(f_{X_0})\} \times_{q^{(1)}\{Z(f_{Z_0})\}} q^{(1)}\{Y(f_{Y_0})\} \\ \cong\, & q^{(1)} F_n X \times_{q^{(1)} F_n Z} q^{(1)} F_n Y . \end{aligned} \tag{11.24}$$

Since Dec commutes with pullbacks, it also follows that

$$\begin{aligned} & \operatorname{Dec} q^{(1)} F_n(X \times_Z Y) = \operatorname{Dec}(q^{(1)} F_n X \times_{q^{(1)} F_n Z} q^{(1)} F_n Y) \\ = & \operatorname{Dec} q^{(1)} F_n X \times_{\operatorname{Dec} q^{(1)} F_n Z} \operatorname{Dec} q^{(1)} F_n Y . \end{aligned} \tag{11.25}$$

Since, by construction,

$$V_n(X \times_Z Y) = \operatorname{Dec} q^{(1)} F_n(X \times_Z Y) \times_{q^{(1)} F_n(X \times_Z Y)} F_n(X \times_Z Y)$$

using (11.23), (11.24), (11.25) and the commutation of pullbacks we conclude that

$$V_n(X \times_Z Y) = V_n X \times_{V_n Z} V_n Y ,$$

as well as

$$h_n(X \times_Z Y) = h_n(X) \times_{h_n(Z)} h_n(Y) ,$$

where $h_n(X) : V_n X \to F_n X$ is as in the proof of Proposition 11.2.3 (a).

Since, by definition, $v_n(X) = f_n(X) h_n(X)$, by (11.23) we deduce that

$$v_n(X \times_Z Y) = v_n(X) \times_{v_n(Z)} v_n(Y) .$$

This concludes the proof of (a).

(b) By (a) and Corollary 11.1.6 (whose hypotheses hold because of Proposition 11.2.3) we have

$$\begin{aligned} & p^{(r-1)}\{F_n(X \times_Z Y)\} = p^{(r-1)}\{(X \times_Z Y)(\nu_{n-1}(X_0 \times_{Z_0} Y_0))\} \\ = & p^{(r-1)}\{X(\nu_{n-1}(X_0))\} \times_{p^{(r-1)}\{Z(\nu_{n-1}(Z_0))\}} p^{(r-1)}\{Y(\nu_{n-1}(Y_0))\} \\ = & p^{(r-1)} F_n X \times_{p^{(r-1)} F_n Z} p^{(r-1)} F_n Y . \end{aligned} \tag{11.26}$$

Similarly, using Corollary 11.1.6 (c) one shows

$$q^{(r-1)}\{F_n(X \times_Z Y)\} = q^{(r-1)} F_n X \times_{q^{(r-1)} F_n Z} q^{(r-1)} F_n Y . \tag{11.27}$$

In particular this holds for $r = 2$, from which we deduce

$$\operatorname{Dec} q^{(1)}\{F_n(X \times_Z Y)\} = \operatorname{Dec} q^{(1)} F_n X \times_{\operatorname{Dec} q^{(1)} F_n Z} \operatorname{Dec} q^{(1)} F_n Y \,. \tag{11.28}$$

As in the proof of Proposition 11.2.3 (a), from the definition of $V_n X$, we obtain a pullback in $[\Delta^{r-2^{op}}, \mathsf{Cat}]$

$$\begin{array}{ccc} p^{(r-1)} V_n X & \longrightarrow & p^{(r-1)} F_n X \\ \downarrow & & \downarrow \\ \operatorname{Dec} q^{(1)} F_n X & \longrightarrow & q^{(1)} F_n X \end{array}$$

So in particular,

$$p^{(r-1)} V_n(X \times_Z Y) = \operatorname{Dec} q^{(1)}\{F_n(X \times_Z Y)\} \times_{q^{(1)} F_n(X \times_Z Y)} p^{(r-1)}\{F_n(X \times_Z Y)\} \,.$$

Using (11.26), (11.27), (11.28) and the commutation of pullbacks we deduce

$$p^{(r-1)} V_n(X \times_Z Y) = p^{(r-1)} V_n X \times_{p^{(r-1)} V_n Z} p^{(r-1)} V_n Y \,.$$

Similarly, as in the proof of Proposition 11.2.3 (a), we obtain a pullback in $[\Delta^{r-2^{op}}, \mathsf{Cat}]$

$$\begin{array}{ccc} q^{(r-1)} V_n X & \longrightarrow & q^{(r-1)} F_n X \\ \downarrow & & \downarrow \\ \operatorname{Dec} q^{(1)} F_n X & \longrightarrow & q^{(1)} F_n X \end{array}$$

Using (11.27), (11.28) and the commutation of pullbacks we deduce

$$q^{(r-1)} V_n(X \times_Z Y) = q^{(r-1)} V_n X \times_{q^{(r-1)} V_n Z} q^{(r-1)} V_n Y \,.$$

□

In the proof of Proposition 11.2.3 we defined a functor

$$F_n : \mathsf{Cat}^{\mathsf{n}}_{\mathsf{hd}} \to \mathsf{Cat}^{\mathsf{n}}_{\mathsf{hd}}$$

by

$$F_n X = X(v_{n-1}(X_0)) \,,$$

where the map $v_{n-1}(X_0) : V_{n-1}X_0 \to X_0$ in $\mathsf{Cat}^{n-1}_{\mathsf{hd}}$ is as in Proposition 11.2.3. We now extend this to a functor

$$F_n : \mathsf{Cat}^{n}_{\mathsf{wg}} \to \mathsf{Cat}^{n}_{\mathsf{wg}}$$

defined by

$$F_n X = X(v_{n-1}(X_0)) \ ,$$

where $v_{n-1}(X_0) : V_{n-1}X_0 \to X_0$ is as in Proposition 11.2.3.

Proposition 11.2.5 *For each $n \geq 2$, there is a functor*

$$F_n : \mathsf{Cat}^{n}_{\mathsf{wg}} \to \mathsf{Cat}^{n}_{\mathsf{wg}}$$

and a map

$$f_n(X) : F_n X \to X \ ,$$

natural in $X \in \mathsf{Cat}^{n}_{\mathsf{wg}}$, such that

(i) $f_n(X)$ is an n-equivalence.
(ii) F_n is the identity on discrete objects and preserves pullbacks over discrete objects.
(iii) If $f : X \to Y$ is a morphism in $\mathsf{Cat}^{n}_{\mathsf{wg}}$ the following diagram commutes where the maps $(F_n X)^d_0 \to (F_n X)_0$ and $(F_n Y)^d_0 \to (F_n Y)_0$ are sections to the discretization maps $(F_n X)_0 \to (F_n X)^d_0$ and $(F_n Y)_0 \to (F_n Y)^d_0$

$$\begin{array}{ccc} (F_n X)_0 & \xrightarrow{f_0} & (F_n Y)_0 \\ \uparrow & & \uparrow \\ (F_n X)^d_0 & \xrightarrow{f^d_0} & (F_n Y)^d_0 \end{array}$$

(iv) If $X \in \mathsf{Cat}^{n}_{\mathsf{hd}}$, then $F_n X \in \mathsf{Cat}^{n}_{\mathsf{hd}}$.

Proof Given $X \in \mathsf{Cat}^{n}_{\mathsf{wg}}$, since $X_0 \in \mathsf{Cat}^{n-1}_{\mathsf{hd}}$, by Proposition 11.2.3 there is a map in $\mathsf{Cat}^{n-1}_{\mathsf{hd}}$

$$v_{n-1}(X_0) : V_{n-1}X_0 \to X_0 \ ,$$

natural in X, such that $v_{n-1}(X_0)$ and $p^{(r-1)}v_{n-1}(X_0)$ are levelwise isofibrations in Cat surjective on objects for all $1 < r \leq (n-1)$. Let

$$F_n X = X(v_{n-1}(X_0)).$$

By Proposition 11.1.5, $F_nX \in \mathsf{Cat}^n_{\mathsf{wg}}$ and there is an n-equivalence which we now denote by $f_n(X)$ (instead of $V(X)$ as in Proposition 11.1.5),

$$f_n(X) : F_nX \to X \ ,$$

proving (i). Since $v_{n-1}(X_0)$ is natural in X, so is $f_n(X)$.

If X is discrete, so is X_0, thus by Proposition 11.2.3 $v_{n-1}(X_0) = \mathrm{Id}$ and therefore $F_nX = X$.

Let $X \to Z \leftarrow Y$ be a pullback in $\mathsf{Cat}^n_{\mathsf{wg}}$ with Z discrete. By Proposition 11.2.3,

$$V_{n-1}(X_0 \times_Z Y_0) = V_{n-1}X_0 \times_Z V_{n-1}Y_0 \ ,$$

and therefore, as easily checked,

$$(F_n(X \times_Z Y))_1 = (F_nX)_1 \times_Z (F_nY)_1 \ .$$

It follows that

$$F_n(X \times_Z Y) = F_nX \times_Z F_nY \ ,$$

which is (ii).

Since $(F_nX)_0 = V_{n-1}X_0$, (iii) follows from Proposition 11.2.3. Since, by (i), $f_n(X)$ is an n-equivalence, (iv) follows from Proposition 7.2.1. □

We next study the behaviour of F_n with respect to certain pullbacks.

Corollary 11.2.6 *Let $X \to Z \leftarrow Y$ be a diagram in $\mathsf{Cat}^n_{\mathsf{wg}}$ such that $X \times_Z Y \in \mathsf{Cat}^n_{\mathsf{wg}}$ and such that, for each $1 \le r \le n$,*

$$p^{(r-1)}(X \times_Z Y) = p^{(r-1)}X \times_{p^{(r-1)}Z} p^{(r-1)}Y \ , \tag{11.29}$$

$$q^{(r-1)}(X \times_Z Y) = q^{(r-1)}X \times_{q^{(r-1)}Z} q^{(r-1)}Y \ . \tag{11.30}$$

Then

(a) For all n,

$$F_n(X \times_Z Y) \cong F_nX \times_{F_nZ} F_nY \ ,$$

$$f_n(X \times_Z Y) \cong f_n(X) \times_{f_n(Z)} f_n(Y).$$

(b) For all $1 \le r \le n$,

$$p^{(r-1)}\{F_n(X \times_Z Y)\} \cong p^{(r-1)}F_nX \times_{p^{(r-1)}F_nZ} p^{(r-1)}F_nY \ ,$$

$$q^{(r-1)}\{F_n(X \times_Z Y)\} \cong q^{(r-1)}F_nX \times_{q^{(r-1)}F_nZ} q^{(r-1)}F_nY \ .$$

Proof

(a) By construction $F_n X = X(f_{X_0})$, where $v_{n-1}(X_0) : V_{n-1}X_0 \to X_0$. Since $X\times_Z Y \in \mathsf{Cat}^n_{\mathsf{wg}}$, $X_0\times_{Z_0} Y_0 \in \mathsf{Cat}^{n-1}_{\mathsf{hd}}$ and it satisfies the hypotheses of Corollary 11.2.4. Therefore

$$V_{n-1}(X_0\times_{Z_0} Y_0) \cong V_{n-1}X_0\times_{V_{n-1}Z_0} V_{n-1}Y_0\ ,$$

so that

$$v_{n-1}(X_0\times_{Z_0} Y_0) = v_{n-1}(X_0)\times_{v_{n-1}(Z_0)} v_{n-1}(Y_0)\ .$$

By Corollary 11.1.6 (a) it follows that

$$\begin{aligned} F_n(X\times_Z Y) &= (X\times_Z Y)(v_{n-1}(X_0\times_{Z_0} Y_0)) \\ &= X(v_{n-1}(X_0))\times_{Z(v_{n-1}(Z_0))} Y(v_{n-1}(Y_0)) = F_nX\times_{F_nZ} F_nY \end{aligned}$$

and

$$f_n(X\times_Z Y) \cong f_n(X)\times_{f_n(Z)} f_n(Y).$$

(b) By Proposition 11.2.3 (a) the maps

$$v_{n-1}(X_0),\quad v_{n-1}(Y_0),\quad v_{n-1}(Z_0),\quad v_{n-1}(X_0\times_{Z_0} Y_0)$$

satisfy the hypotheses of Corollary 11.1.6 (b). Further, by hypothesis (11.29)

$$p^{(r-2)}(X_0\times_{Z_0} Y_0) = p^{(r-2)}X_0\times_{p^{(r-2)}Z_0} p^{(r-2)}Y_0\ .$$

Thus all the hypotheses of Corollary 11.1.6 (b) are satisfied and we conclude that

$$\begin{aligned} & p^{(r-1)}\{F_n(X\times_Z Y)\} = p^{(r-1)}\{(X\times_Z Y)(v_{n-1}(X_0)\times_{v_{n-1}(Z_0)} v_{n-1}(Y_0))\} \\ &= p^{(r-1)}\{X(v_{n-1}(X_0))\}\times_{p^{(r-1)}\{Z(v_{n-1}(Z_0))\}} p^{(r-1)}\{Y(v_{n-1}(Y_0))\} \\ &= p^{(r-1)}F_nX\times_{p^{(r-1)}F_nZ} p^{(r-1)}F_nY\ . \end{aligned}$$

The proof for $q^{(r-1)}$ is similar (using Corollary 11.1.6 (c)). □

11.3 The Category $\mathsf{FCat}^{\mathsf{n}}_{\mathsf{wg}}$

In this section we introduce the category $\mathsf{FCat}^{\mathsf{n}}_{\mathsf{wg}}$. We then show in Theorem 11.3.6 that there is a functor

$$G_n : \mathsf{Cat}^{\mathsf{n}}_{\mathsf{wg}} \to \mathsf{FCat}^{\mathsf{n}}_{\mathsf{wg}}$$

and an n-equivalence, $G_n X \to X$, for each $X \in \mathsf{Cat}^{\mathsf{n}}_{\mathsf{wg}}$. The construction of G_n is inductive and uses the functor F_n built in Sect. 11.1. Namely, we define $G_2 = F_2$ and given G_{n-1},

$$G_n = \overline{G}_{n-1} \circ F_n \ ,$$

see Definition 11.3.4 and Theorem 11.3.6. In the next chapter, G_n will be used to define the discretization functor $Disc_n : \mathsf{Cat}^{\mathsf{n}}_{\mathsf{wg}} \to \mathsf{Ta}^{\mathsf{n}}$ via a composite $\mathsf{Cat}^{\mathsf{n}}_{\mathsf{wg}} \xrightarrow{G_n} \mathsf{FCat}^{\mathsf{n}}_{\mathsf{wg}} \xrightarrow{D_n} \mathsf{Ta}^{\mathsf{n}}$.

11.3.1 The Idea of the Category $\mathsf{FCat}^{\mathsf{n}}_{\mathsf{wg}}$

The idea of the construction of the discretization functor from $\mathsf{Cat}^{\mathsf{n}}_{\mathsf{wg}}$ to Ta^{n} is to replace the homotopically discrete substructures in $X \in \mathsf{Cat}^{\mathsf{n}}_{\mathsf{wg}}$ by their discretization. As outlined in the introduction to this chapter, this cannot be done in a functorial way unless there are functorial sections to the discretization maps of the homotopically discrete substructures. For this reason, we introduce in this section a new category $\mathsf{FCat}^{\mathsf{n}}_{\mathsf{wg}}$.

The category $\mathsf{FCat}^{\mathsf{n}}_{\mathsf{wg}}$ is a refinement of the category $\mathsf{Cat}^{\mathsf{n}}_{\mathsf{wg}}$: the idea of the latter goes back to Sect. 6.3.1. This category offers the advantage over $\mathsf{Cat}^{\mathsf{n}}_{\mathsf{wg}}$ of having objects with better behaved homotopically discrete substructures, admitting functorial sections to their discretization maps as well as other functoriality properties with respect to maps. In Sect. 11.3.4 we build a functor $G_n : \mathsf{Cat}^{\mathsf{n}}_{\mathsf{wg}} \to \mathsf{FCat}^{\mathsf{n}}_{\mathsf{wg}}$ using the functor F_n, whose idea was discussed in Sect. 11.2.1.

The defining properties of $\mathsf{FCat}^{\mathsf{n}}_{\mathsf{wg}}$ are obtained by abstracting the properties that are seen to occur in the essential image of the functor $G_n : \mathsf{Cat}^{\mathsf{n}}_{\mathsf{wg}} \to [\Delta^{op}, \mathsf{Cat}^{\mathsf{n-1}}_{\mathsf{wg}}]$ defined inductively by $G_2 = F_2$ and $G_n = \overline{G}_{n-1} \circ F_n$ (see Definition 11.3.4). The category $\mathsf{FCat}^{\mathsf{n}}_{\mathsf{wg}}$ plays an essential role in constructing the functors $D_n : \mathsf{FCat}^{\mathsf{n}}_{\mathsf{wg}} \to \mathsf{Ta}^{\mathsf{n}}$ and $Disc_n : \mathsf{Cat}^{\mathsf{n}}_{\mathsf{wg}} \to \mathsf{Ta}^{\mathsf{n}}$, whose idea is discussed in Sects. 12.1.1 and 12.2.1 respectively.

The idea of the category $\mathsf{FCat}^{\mathsf{n}}_{\mathsf{wg}}$ is to modify objects and morphisms in the category $\mathsf{Cat}^{\mathsf{n}}_{\mathsf{wg}}$ by imposing extra structure giving functorial sections to the discretization maps of the homotopically discrete substructures of objects of $\mathsf{Cat}^{\mathsf{n}}_{\mathsf{wg}}$.

So a map $f : A \to B$ of homotopically discrete k-fold categories (for the appropriate dimension k) gives rise to a corresponding commuting diagram

$$\begin{array}{ccc} A & \xrightarrow{f} & B \\ {\scriptstyle \gamma'(A)}\uparrow & & \uparrow{\scriptstyle \gamma'(B)} \\ A^d & \xrightarrow[f^d]{} & B^d \end{array}$$

where $\gamma(A)\gamma'(A) = \mathrm{Id}$, $\gamma(B)\gamma'(B) = \mathrm{Id}$, $\gamma(A) : A \to A^d$ and $\gamma(B) : B \to B^d$ being the discretization maps.

The maps $f : A \to B$ with respect to which we require this functorial behavior are as follows. Given $X \in \mathsf{Cat}^{\mathsf{n}}_{\mathsf{wg}}$, $\underline{k}, \underline{r} \in \Delta^{s^{op}}$, and a morphism $\underline{k} \to \underline{r}$ in $\Delta^{s^{op}}$ we have a corresponding morphism in $\mathsf{Cat}^{\mathsf{n-s-1}}_{\mathsf{hd}}$

$$f : X_{\underline{k}0} \to X_{\underline{r}0} \,. \tag{11.31}$$

In the definition of the category $\mathsf{FCat}^{\mathsf{n}}_{\mathsf{wg}}$ (see Definition 11.3.1), we impose the commutativity of diagram (11.35): that is, there are sections to the discretization maps of $X_{\underline{k}0}$ and $X_{\underline{r}0}$ that behave functorially with respect to the map (11.31). We next consider the maps

$$\gamma'(X_0) : X_0^d \to X_0, \qquad \gamma'(X_{\underline{k}0}) : X_{\underline{k}0}^d \to X_{\underline{k}0}$$

and the corresponding maps of homotopically discrete objects

$$(\gamma'(X_0))_0 : X_0^d \to X_{00}, \qquad (\gamma'(X_{\underline{k}0}))_{\underline{h}0} : X_{\underline{k}0}^d \to X_{\underline{k}0\underline{h}0} \tag{11.32}$$

for each $\underline{h} \in \Delta^{t^{op}}$, $1 \le t \le n-3$. We then impose the functoriality of the sections to the discretization maps with respect to the maps (11.32). That is, we require the commutativity of diagrams (11.36) and (11.37) in Definition 11.3.1. This defines objects of $\mathsf{FCat}^{\mathsf{n}}_{\mathsf{wg}}$.

As for morphisms of $\mathsf{FCat}^{\mathsf{n}}_{\mathsf{wg}}$, given a morphism $F : X \to Y$ in $\mathsf{Cat}^{\mathsf{n}}_{\mathsf{wg}}$ and $\underline{k} \in \Delta^{s^{op}}$ we obtain maps of homotopically discrete structures

$$F_0 : X_0 \to Y_0, \qquad F_{\underline{k}0} : X_{\underline{k}0} \to Y_{\underline{k}0} \tag{11.33}$$

for all $\underline{k} \in \Delta^{s^{op}}$. We impose functoriality of the sections to the discretization maps with respect to the maps (11.33). This translates into the commutativity of diagrams (11.38) and (11.39) in Definition 11.3.1.

In Theorem 11.3.6 we show that all these conditions are satisfied by the essential image of the functor G_n (Definition 11.3.4) so that $G_n : \mathsf{Cat}^{\mathsf{n}}_{\mathsf{wg}} \to \mathsf{FCat}^{\mathsf{n}}_{\mathsf{wg}}$. As we will see in the next chapter, the definition of $\mathsf{FCat}^{\mathsf{n}}_{\mathsf{wg}}$ is exactly what is needed to functorially discretize the homotopically discrete substructures and thus build a functor $D_n : \mathsf{FCat}^{\mathsf{n}}_{\mathsf{wg}} \to \mathsf{Ta}^{\mathsf{n}}$. In the next chapter this will be used to define the discretization functor as the composite $\mathsf{Cat}^{\mathsf{n}}_{\mathsf{wg}} \xrightarrow{G_n} \mathsf{FCat}^{\mathsf{n}}_{\mathsf{wg}} \xrightarrow{D_n} \mathsf{Ta}^{\mathsf{n}}$, where G_n is as in Theorem 11.3.6.

11.3.2 The Formal Definition of the Category $\mathsf{FCat}^{\mathsf{n}}_{\mathsf{wg}}$

We now give the formal definition of the category $\mathsf{FCat}^{\mathsf{n}}_{\mathsf{wg}}$ and we establish its properties.

Definition 11.3.1 Define the category $\mathsf{FCat}^{\mathsf{n}}_{\mathsf{wg}}$ as follows. Let $\mathsf{FCat}^{1}_{\mathsf{wg}} = \mathsf{Cat}$. Let $\mathsf{FCat}^{2}_{\mathsf{wg}}$ have the following objects and morphisms:

(i) Objects of $\mathsf{FCat}^{2}_{\mathsf{wg}}$ are objects of $X \in \mathsf{Cat}^{2}_{\mathsf{wg}}$ such that the discretization map $\gamma : X_0 \to X_0^d$ has a section $\gamma\,'$ which is natural in X.
(ii) A morphism $F : X \to Y$ in $\mathsf{FCat}^{2}_{\mathsf{wg}}$ is a morphism in $\mathsf{Cat}^{2}_{\mathsf{wg}}$ such that the following diagram commutes

$$\begin{array}{ccc} X_0 & \xrightarrow{F_0} & Y_0 \\ {\scriptstyle \gamma'(X_0)}\uparrow & & \uparrow{\scriptstyle \gamma'(Y_0)} \\ X_0^d & \xrightarrow[F_0^d]{} & Y_0^d \end{array} \tag{11.34}$$

where $\gamma\,'(X_0)$ and $\gamma\,'(Y_0)$ are sections to the discretization maps.

For each $n > 2$, let $\mathsf{FCat}^{\mathsf{n}}_{\mathsf{wg}}$ have the following objects and morphisms: an object X of $\mathsf{FCat}^{\mathsf{n}}_{\mathsf{wg}}$ consists of $X \in \mathsf{Cat}^{\mathsf{n}}_{\mathsf{wg}}$ such that for all $\underline{k} = (k_1, \dots, k_s) \in \Delta^{s^{op}}$, $\underline{r} = (r_1, \dots, r_s) \in \Delta^{s^{op}}$, $(1 \le s \le n-2)$ and morphisms $\underline{k} \to \underline{r}$ in $\Delta^{s^{op}}$, the corresponding morphisms

$$f : X_{\underline{k}0} \to X_{\underline{r}0}$$

in $\mathsf{Cat}^{\mathsf{n-s-1}}_{\mathsf{hd}}$ are such that there are sections to the discretization maps

$$\gamma(X_{\underline{k}0}) : X_{\underline{k}0} \to X^d_{\underline{k}0}\,,$$

$$\gamma(X_{\underline{r}0}) : X_{\underline{r}0} \to X^d_{\underline{r}0}$$

making the following diagrams commute

(i)

$$\begin{array}{ccc} X_{\underline{k}0} & \xrightarrow{\ f\ } & X_{\underline{r}0} \\ {\scriptstyle \gamma'(X_{\underline{k}0})}\uparrow & & \uparrow{\scriptstyle \gamma'(X_{\underline{r}0})} \\ X^d_{\underline{k}0} & \xrightarrow[\ f^d\ ]{} & X^d_{\underline{r}0} \end{array} \tag{11.35}$$

(ii) For all $\underline{h} = (h_1, \dots, h_t) \in \Delta^{t^{op}}$, $1 \le t \le n-3$,

$$\begin{array}{ccc} X^d_0 & \xrightarrow{(\gamma'(X_0))_0} & X_{00} \\ \| & & \uparrow{\scriptstyle \gamma'(X_{00})} \\ X^d_0 & \xrightarrow[(\gamma'(X_0))^d_0]{} & X^d_{00} \end{array} \qquad \begin{array}{ccc} X^d_0 & \xrightarrow{(\gamma'(X_0))_{\underline{h}0}} & X_{0\underline{h}0} \\ \| & & \uparrow{\scriptstyle \gamma'(X_{0\underline{h}0})} \\ X^d_0 & \xrightarrow[(\gamma'(X_0))^d_{\underline{h}0}]{} & X^d_{0\underline{h}0} \end{array} \tag{11.36}$$

$$\begin{array}{ccc} X^d_{\underline{k}0} & \xrightarrow{(\gamma'(X_{\underline{k}0}))_0} & X_{\underline{k}00} \\ \| & & \uparrow{\scriptstyle \gamma'(X_{\underline{k}00})} \\ X^d_{\underline{k}0} & \xrightarrow[(\gamma'(X_{\underline{k}0}))^d_0]{} & X^d_{\underline{k}00} \end{array} \qquad \begin{array}{ccc} X^d_{\underline{k}0} & \xrightarrow{(\gamma'(X_{\underline{k}0}))_{\underline{h}0}} & X_{\underline{k}0\underline{h}0} \\ \| & & \uparrow{\scriptstyle \gamma'(X_{\underline{k}0\underline{h}0})} \\ X^d_{\underline{k}0} & \xrightarrow[(\gamma'(X_{\underline{k}0}))^d_{\underline{h}0}]{} & X^d_{\underline{k}0\underline{h}0} \end{array} \tag{11.37}$$

A morphism $F : X \to Y$ in $\mathsf{FCat}^{\mathsf{n}}_{\mathsf{wg}}$ is a morphism in $\mathsf{Cat}^{\mathsf{n}}_{\mathsf{wg}}$ such that the following diagram commutes

$$\begin{array}{ccc} X_0 & \xrightarrow{\ F_0\ } & Y_0 \\ {\scriptstyle \gamma'(X_0)}\uparrow & & \uparrow{\scriptstyle \gamma'(Y_0)} \\ X^d_0 & \xrightarrow[\ F^d_0\ ]{} & Y^d_0 \end{array} \tag{11.38}$$

and such that for all $\underline{k} = (k_1, \dots, k_s) \in \Delta^{s^{op}}$, $1 \le s \le n-2$, the following diagram commutes

$$\begin{array}{ccc} X_{\underline{k}0} & \xrightarrow{\ F_{\underline{k}0}\ } & Y_{\underline{k}0} \\ {\scriptstyle \gamma'(X_{\underline{k}0})}\uparrow & & \uparrow{\scriptstyle \gamma'(Y_{\underline{k}0})} \\ X^d_{\underline{k}0} & \xrightarrow[\ F^d_{\underline{k}0}\ ]{} & Y^d_{\underline{k}0} \end{array} \tag{11.39}$$

where $\gamma'(X_0)$, $\gamma'(Y_0)$, $\gamma'(X_{\underline{k}0})$, $\gamma'(Y_{\underline{k}0})$ are sections to the corresponding discretization maps.

Remark 11.3.2 It is immediate from Definition 11.3.1 that

$$\mathsf{FCat^n_{wg}} \subset [\Delta^{^{op}}, \mathsf{FCat^{n-1}_{wg}}] \,.$$

Also, from Definition 11.3.1 we see that $\gamma'(X_0) : X_0^d \to X_0$ is a map in $\mathsf{FCat^{n-1}_{wg}}$ and for all $\underline{k} \in \Delta^{s^{op}}$, $1 \leq s \leq n-2$ $\gamma'(X_{\underline{k}0}) : X_{\underline{k}0}^d \to X_{\underline{k}0}$ is a map in $\mathsf{FCat^{n-s-1}_{wg}}$.

Further, we observe that if $X \in \mathsf{FCat^n_{wg}}$ and $X \in \mathsf{Cat^n_{hd}}$, the discretization map $\gamma : X \to X^d$ is a map in $\mathsf{FCat^n_{wg}}$. In fact, by naturality of the discretization map, for each $\underline{k} \in \Delta^{s^{op}}$ $(1 \leq s \leq n-2)$ the following diagrams commute

$$\begin{array}{ccc} X_0 & \xrightarrow{\gamma_0} & (X^d)_0 \\ {\scriptstyle \delta_0}\downarrow & & \| \\ X_0^d & \xrightarrow[\gamma_0^d]{} & (X^d)_0^d \end{array} \qquad \begin{array}{ccc} X_{\underline{k}0} & \xrightarrow{\gamma_{\underline{k}0}} & (X^d)_{\underline{k}0} \\ {\scriptstyle \delta_{\underline{k}0}}\downarrow & & \| \\ X_{\underline{k}0}^d & \xrightarrow[\gamma_{\underline{k}0}^d]{} & (X^d)_{\underline{k}0}^d \end{array}$$

where δ_0 (resp. $\delta_{\underline{k}0}$) is the discretization map for X_0 (resp. $X_{\underline{k}0}$). Thus we have the following commuting diagrams

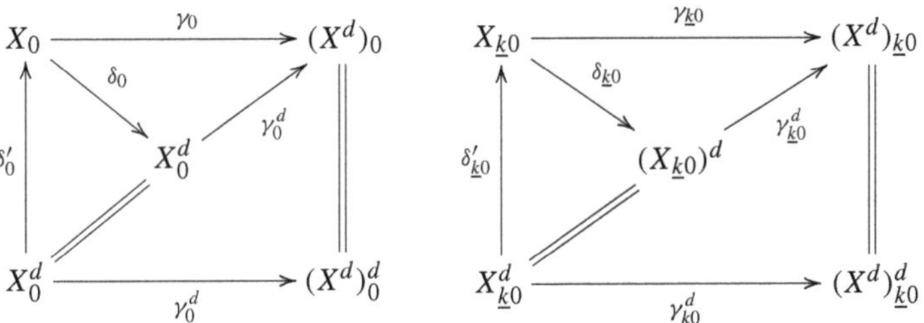

By Definition 11.3.1, the commutativity of the outer part of these diagrams mean that γ is a morphism in $\mathsf{FCat^n_{wg}}$.

Lemma 11.3.3 *The functors* $p^{(n-1)}, q^{(n-1)} : \mathsf{Cat^n_{wg}} \to \mathsf{Cat^{n-1}_{wg}}$ *induce functors*

$$p^{(n-1)}, q^{(n-1)} : \mathsf{FCat^n_{wg}} \to \mathsf{FCat^{n-1}_{wg}} \,.$$

Proof Let $X \in \mathsf{FCat^n_{wg}}$ and $\underline{k} \to \underline{r}$ be a morphism in $\Delta^{s^{op}}$. By applying the functor $p^{(n-s-2)}$ to the commuting diagram (11.35) and using the fact that

$$p^{(n-s-2)} X_{\underline{k}0} = (p^{(n-1)} X)_{\underline{k}0}, \qquad p^{(n-s-2)} X_{\underline{k}0}^d = X_{\underline{k}0}^d = (p^{(n-1)} X)_{\underline{k}0}^d,$$

we obtain the commuting diagram

$$\begin{CD}
(p^{(n-1)}X)_{\underline{k}0} @>{p^{(n-s-2)}f}>> (p^{(n-1)}X)_{\underline{r}0}\\
@AAA @AAA\\
(p^{(n-1)}X)^d_{\underline{k}0} @>>{p^{(n-s-2)}f^d}> (p^{(n-1)}X)^d_{\underline{r}0}
\end{CD} \tag{11.40}$$

By applying $p^{(n-3)}$ to the diagram on the left of (11.36) and $p^{(n-t-3)}$ to the diagram on the right of (11.36) and using the fact that

$$p^{(n-3)}X_{00} = (p^{(n-1)}X)_{00}, \qquad p^{(n-t-3)}X_{0\underline{h}0} = (p^{(n-1)}X)_{0\underline{h}0},$$

we obtain commuting diagrams

$$\begin{CD}
(p^{(n-1)}X)^d_0 @>>> (p^{(n-1)}X)_{00}\\
@| @AAA\\
(p^{(n-1)}X)^d_0 @>>> (p^{(n-1)}X)^d_{00}
\end{CD}
\qquad
\begin{CD}
(p^{(n-1)}X)^d_0 @>>> (p^{(n-1)}X)_{0\underline{h}0}\\
@| @AAA\\
(p^{(n-1)}X)^d_0 @>>> (p^{(n-1)}X)^d_{0\underline{h}0}
\end{CD} \tag{11.41}$$

Similarly, applying $p^{(n-s-3)}$ to the diagram on the left of (11.37) and $p^{(n-s-t-3)}$ to the diagram on the right of (11.37) and using the fact that

$$p^{(n-s-3)}X_{\underline{k}00} = (p^{(n-1)}X)_{\underline{k}00}, \qquad p^{(n-s-t-3)}X_{\underline{k}0\underline{h}0} = (p^{(n-1)}X)_{\underline{k}0\underline{h}0},$$

we obtain commuting diagrams

$$\begin{CD}
(p^{(n-1)}X)^d_{\underline{k}0} @>>> (p^{(n-1)}X)_{\underline{k}00}\\
@| @AAA\\
(p^{(n-1)}X)^d_{\underline{k}0} @>>> (p^{(n-1)}X)^d_{\underline{k}00}
\end{CD}
\qquad
\begin{CD}
(p^{(n-1)}X)^d_{\underline{k}0} @>>> (p^{(n-1)}X)_{\underline{k}0\underline{h}0}\\
@| @AAA\\
(p^{(n-1)}X)^d_{\underline{k}0} @>>> (p^{(n-1)}X)^d_{\underline{k}0\underline{h}0}
\end{CD} \tag{11.42}$$

Together with (11.40), (11.41) and (11.42) mean by definition that $p^{(n-1)}X \in \mathsf{FCat}^{n-1}_{\mathsf{wg}}$.

Given $F : X \to Y$ in $\mathsf{FCat}^{\mathsf{n}}_{\mathsf{wg}}$, by applying $p^{(n-2)}$ to the commuting diagram (11.38) we obtain the commuting diagram

$$\begin{array}{ccc} (p^{(n-1)}X)_0 & \xrightarrow{(p^{(n-1)}F)_0} & (p^{(n-1)}Y)_0 \\ \uparrow & & \uparrow \\ (p^{(n-1)}X)^d_0 & \xrightarrow[(p^{(n-1)}F)^d_0]{} & (p^{(n-1)}Y)^d_0 \end{array}$$

By applying $p^{(n-s-2)}$ to the commuting diagram (11.39) we obtain the commuting diagram

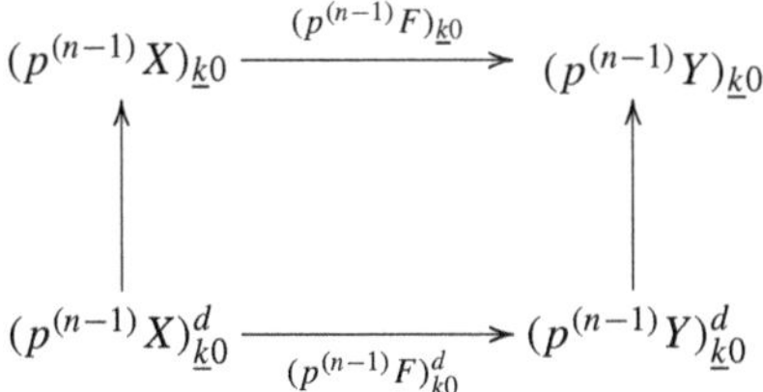

By definition this means that $p^{(n-1)}F \in \mathsf{FCat}^{\mathsf{n-1}}_{\mathsf{wg}}$.

The proof for $q^{(n-1)}$ is analogous. □

11.3.3 The Idea of the Functor G_n

The functor G_n uses the category $\mathsf{FCat}^{\mathsf{n}}_{\mathsf{wg}}$, whose idea was discussed in Sect. 11.3.1, and is the key to building the functors $D_n : \mathsf{FCat}^{\mathsf{n}}_{\mathsf{wg}} \to \mathsf{Ta}^{\mathsf{n}}$ and $Disc_n : \mathsf{Cat}^{\mathsf{n}}_{\mathsf{wg}} \to \mathsf{Ta}^{\mathsf{n}}$, whose idea is discussed in Sects. 12.1.1 and 12.2.1, respectively.

As shown in Proposition 11.2.5, the functor $F_n : \mathsf{Cat}^{\mathsf{n}}_{\mathsf{wg}} \to \mathsf{Cat}^{\mathsf{n}}_{\mathsf{wg}}$ replaces $X \in \mathsf{Cat}^{\mathsf{n}}_{\mathsf{wg}}$ with an n-equivalent $F_n X$ in which $(F_n X)_0$ admits a functorial section to the discretization map. By definition of $\mathsf{FCat}^{2}_{\mathsf{wg}}$, the functor $F_2 : \mathsf{Cat}^{2}_{\mathsf{wg}} \to \mathsf{Cat}^{2}_{\mathsf{wg}}$ is in fact a functor $F_2 : \mathsf{Cat}^{2}_{\mathsf{wg}} \to \mathsf{FCat}^{2}_{\mathsf{wg}}$ and we define $G_2 = F_2$.

When $n > 2$, recall that by definition objects of $\mathsf{FCat}^{\mathsf{n}}_{\mathsf{wg}}$ are such that the homotopically discrete substructures have functorial sections to the discretization maps, as well as other functoriality properties. The idea of the functor G_n is to inductively apply F_s to every sub-simplicial dimension s. That is, we define inductively

$$G_n : \mathsf{Cat}^{\mathsf{n}}_{\mathsf{wg}} \to [\Delta^{^{op}}, \mathsf{Cat}^{\mathsf{n-1}}_{\mathsf{wg}}]$$

by $G_2 = F_2$; when $n > 2$, given G_{n-1} we let

$$G_n = \overline{G}_{n-1} \circ F_n \ ,$$

where the notation $\overline{G}_{n-1}$ is as in Definition 2.1.1.

Showing that G_n lands in fact in $\mathsf{FCat}^n_{\mathsf{wg}}$ involves several steps, as shown in the proof of Theorem 11.3.6. First we show that $G_n X \in \mathsf{Cat}^n_{\mathsf{wg}}$ by proving that it satisfies the hypotheses of Lemma 7.2.10.

For this we first check that $G_n X \in \mathsf{Ta}^n_{\mathsf{wg}}$. The main ingredients here are the inductive hypotheses on G_{n-1} that it preserves $(n-1)$-equivalences and pullbacks over discrete objects and it is the identity on discrete objects. This easily implies that there is an $(n-1)$-equivalence

$$(G_n X)_0 = G_{n-1}(F_n X)_0 \to G_{n-1}(F_n X)_0^d = (F_n X)_0^d \ ,$$

so that $(G_n X)_0$ is homotopically discrete. It also implies that the induced Segal maps of $G_n X$ are $(n-1)$-equivalences, thus in conclusion $X \in \mathsf{Ta}^n_{\mathsf{wg}}$.

Hypothesis (a) of Lemma 7.2.10 is immediate by induction since $(G_n X)_k = G_{n-1}(F_n X)_k \in \mathsf{FCat}^{n-1}_{\mathsf{wg}}$ so in particular $(G_n X)_k \in \mathsf{Cat}^{n-1}_{\mathsf{wg}}$.

As for hypotheses (b) and (c) in Lemma 7.2.10, these are shown to hold for $G_n X$ by assuming, inductively, that G_{n-1} satisfies the same properties of F_n with respect to pullbacks established in Corollary 11.2.6. Recall that this states that given a diagram in $\mathsf{Cat}^n_{\mathsf{wg}}$

$$X \to Z \leftarrow Y$$

such that $X \times_Z Y \in \mathsf{Cat}^n_{\mathsf{wg}}$ and such that this pullback is preserved by $p^{(r-1)}$ and $q^{(r-1)}$ (for each $1 \leq r \leq n$), then this pullback is also preserved by F_n, $p^{(r-1)} F_n$ and $q^{(r-1)} F_n$.

We require the same property to hold, inductively, for G_{n-1}. We then apply G_{n-1} to the diagram in $\mathsf{Cat}^{n-1}_{\mathsf{wg}}$

$$(F_n X)_1 \xrightarrow{\partial_0} (F_n X)_0 \xleftarrow{\partial_0} (F_n X)_1 \ . \tag{11.43}$$

Since $F_n X \in \mathsf{Cat}^n_{\mathsf{wg}}$, this diagram is such that

$$(F_n X)_2 \cong (F_n X)_1 \times_{(F_n X)_0} (F_n X)_1 \in \mathsf{Cat}^{n-1}_{\mathsf{wg}}$$

and it commutes with $p^{(r-1)}$ and $q^{(r-1)}$. Thus from the inductive hypothesis,

$$\begin{aligned}(G_n X)_2 &= G_{n-1}(F_n X)_2 = G_{n-1}((F_n X)_1 \times_{(F_n X)_0} (F_n X)_1) \\ &\cong G_{n-1}(F_n X)_1 \times_{G_{n-1}(F_n X)_0} G_{n-1}(F_n X)_1 \cong (G_n X)_1 \times_{(G_n X)_0} (G_n X)_1 \ .\end{aligned}$$

Similarly one shows that, for each $s \geq 2$,

$$(G_n X)_s \cong (G_n X)_1 \times_{(G_n X)_0} \overset{s}{\cdots} \times_{(G_n X)_0} (G_n X)_1 \ .$$

This proves that $G_n X$ satisfies hypothesis (b) of Lemma 7.2.10. Hypothesis (c) of Lemma 7.2.10 is checked similarly and we conclude that $G_n X \in \mathsf{Cat}^{\mathsf{n}}_{\mathsf{wg}}$.

For the inductive step we then need to show that G_n satisfies the commutation properties with respect to pullbacks at step n. This is done by using Corollary 11.2.6, as detailed in the proof of Theorem 11.3.6 (e).

In the proof of Theorem 11.3.6 we also check the remaining functoriality conditions and show that $G_n : \mathsf{Cat}^{\mathsf{n}}_{\mathsf{wg}} \to \mathsf{FCat}^{\mathsf{n}}_{\mathsf{wg}}$.

11.3.4 The Functor G_n: The Formal Proof

Definition 11.3.4 Define inductively

$$G_n : \mathsf{Cat}^{\mathsf{n}}_{\mathsf{wg}} \to [\Delta^{op}, \mathsf{Cat}^{\mathsf{n-1}}_{\mathsf{wg}}]$$

by $G_2 = F_2 : \mathsf{Cat}^{2}_{\mathsf{wg}} \to \mathsf{Cat}^{2}_{\mathsf{wg}}$; given $G_{n-1} : \mathsf{Cat}^{\mathsf{n-1}}_{\mathsf{wg}} \to \mathsf{Cat}^{\mathsf{n-1}}_{\mathsf{wg}}$, let G_n be the composite

$$G_n = \overline{G}_{n-1} \circ F_n : \mathsf{Cat}^{\mathsf{n}}_{\mathsf{wg}} \xrightarrow{F_n} \mathsf{Cat}^{\mathsf{n}}_{\mathsf{wg}} \hookrightarrow [\Delta^{op}, \mathsf{Cat}^{\mathsf{n-1}}_{\mathsf{wg}}] \xrightarrow{\overline{G}_{n-1}} [\Delta^{op}, \mathsf{Cat}^{\mathsf{n-1}}_{\mathsf{wg}}] \ ,$$

where F_n is as in Proposition 11.2.5 and the embedding $\mathsf{Cat}^{\mathsf{n}}_{\mathsf{wg}} \hookrightarrow [\Delta^{op}, \mathsf{Cat}^{\mathsf{n-1}}_{\mathsf{wg}}]$ is the nerve functor in direction 1. That is, given $X \in \mathsf{Cat}^{\mathsf{n}}_{\mathsf{wg}}$ and $k \in \Delta^{op}$, $(G_n X)_k = G_{n-1}(F_n X)_k$.

Lemma 11.3.5 *For* $(k_1, \ldots, k_s) \in \Delta^{s^{op}}$, $1 \leq s \leq n-2$, $Y \in \mathsf{Cat}^{\mathsf{n}}_{\mathsf{wg}}$ *define* $Z_{k_1 k_2 \ldots k_s} Y \in \mathsf{Cat}^{\mathsf{n-s}}_{\mathsf{wg}}$ *inductively by*

$$Z_{k_1} Y = (F_n Y)_{k_1}$$

and for $1 < i \leq s-1$, *given* $Z_{k_1 k_2 \ldots k_{i-1}} Y$, *let*

$$Z_{k_1 k_2 \ldots k_i} Y = (F_{n-i+1} Z_{k_1 k_2 \ldots k_{i-1}} Y)_{k_i} \ . \tag{11.44}$$

Then for each $2 \leq s \leq n-2$

$$(G_n Y)_{k_1 \ldots k_s} = G_{n-s}(F_{n-s+1} Z_{k_1 \ldots k_{s-1}} Y)_{k_s} \ ,$$

where G_n *is as in Definition 11.3.4.*

Proof We show this by induction on s. By Definition of G_n,

$$(G_nY)_{k_1} = G_{n-1}(F_nY)_{k_1}\ ,$$
$$(G_nY)_{k_1k_2} = G_{n-2}(F_{n-1}(F_nY)_{k_1})_{k_2} = G_{n-2}(F_{n-1}Z_{k_1}Y)_{k_2}\ .$$

Suppose, inductively, that the lemma holds for $s-1$. Then, by the inductive hypothesis, the definition of G_n and by (11.44),

$$(G_nY)_{k_1\dots k_s} = \{(G_nY)_{k_1\dots k_{s-1}}\}_{k_s} = \{G_{n-s+1}(F_{n-s+2}Z_{k_1\dots k_{s-2}}Y)_{k_{s-1}}\}_{k_s}$$
$$= (G_{n-s+1}Z_{k_1\dots k_{s-1}}Y)_{k_s} = G_{n-s}(F_{n-s+1}Z_{k_1\dots k_{s-1}}Y)_{k_s}$$

proving the lemma. □

Theorem 11.3.6 *Let G_n be as in Definition 11.3.4. Then*

(a) $G_n : \mathsf{Cat}^n_{\mathsf{wg}} \to \mathsf{FCat}^n_{\mathsf{wg}}$.
(b) There is an n-equivalence

$$g_n(X) : G_nX \to X\ ,$$

natural in $X \in \mathsf{Cat}^n_{\mathsf{wg}}$, such that $(g_n(X))^d_0 : (G_nX)^d_0 \to X^d_0$ is surjective.
(c) G_n preserves n-equivalences.
(d) G_n is the identity on discrete objects and preserves pullbacks over discrete objects.
(e) Let $X \to Z \leftarrow Y$ be a diagram in $\mathsf{Cat}^n_{\mathsf{wg}}$ such that $X\times_Z Y \in \mathsf{Cat}^n_{\mathsf{wg}}$ and such that, for each $1 \le r \le n$,

$$p^{(r-1)}(X\times_Z Y) = p^{(r-1)}X\times_{p^{(r-1)}Z} p^{(r-1)}Y\ ,$$

$$q^{(r-1)}(X\times_Z Y) = q^{(r-1)}X\times_{q^{(r-1)}Z} q^{(r-1)}Y\ .$$

Then

$$G_n(X\times_Z Y) = G_nX\times_{G_nZ} G_nY\ ,$$

$$g_n(X\times_Z Y) = g_n(X)\times_{g_n(Z)} g_n(Y)\ ,$$

$$p^{(r-1)}G_n(X\times_Z Y) \cong p^{(r-1)}G_nX\times_{p^{(r-1)}G_nZ} p^{(r-1)}G_nY\ ,$$

$$q^{(r-1)}G_n(X\times_Z Y) \cong q^{(r-1)}G_nX\times_{q^{(r-1)}G_nZ} q^{(r-1)}G_nY\ .$$

(f) If $Y \in \mathsf{Cat}^n_{\mathsf{hd}}$ and $\gamma' : Y^d \to Y$ is a section to the discretization map, then $G_n\gamma'$ is a morphism in $\mathsf{FCat}^n_{\mathsf{wg}}$.

Proof By induction on n. For $n = 2$, by Proposition 11.2.5 and Corollary 11.2.6 the functor $G_2 = F_2 : \mathsf{Cat}^2_{\mathsf{wg}} \to \mathsf{Cat}^2_{\mathsf{wg}}$ is in fact a functor $\mathsf{Cat}^2_{\mathsf{wg}} \to \mathsf{FCat}^2_{\mathsf{wg}}$

satisfying (a)–(f), where to show (b) we use the fact that $(G_2X)^d_0 = X_{00} \to X^d_0$ is surjective.
Suppose we have defined G_{n-1} satisfying the above properties and let $X \in \mathsf{Cat}^{\mathsf{n}}_{\mathsf{wg}}$.

(a) First we show that $G_nX \in \mathsf{Cat}^{\mathsf{n}}_{\mathsf{wg}}$ using the criterion given in Lemma 7.2.10.

We first check that $G_nX \in \mathsf{Ta}^{\mathsf{n}}_{\mathsf{wg}}$. For each $k \geq 0$, by the inductive hypothesis we have

$$(G_nX)_k = G_{n-1}(F_nX)_k \in \mathsf{Cat}^{\mathsf{n-1}}_{\mathsf{wg}} .$$

By construction we have

$$(G_nX)_0 = G_{n-1}(F_nX)_0 .$$

Since $(F_nX)_0 \in \mathsf{Cat}^{\mathsf{n-1}}_{\mathsf{hd}}$ (as $F_nX \in \mathsf{Cat}^{\mathsf{n}}_{\mathsf{wg}}$), there is an $(n-1)$-equivalence $(F_nX) \to (F_nX)^d_0$. Thus by inductive hypothesis (c) and (d) this induces an $(n-1)$-equivalence

$$(G_nX)_0 = G_{n-1}(F_nX)_0 \to G_{n-1}(F_nX)^d_0 = (F_nX)^d_0 .$$

Therefore by Proposition 7.2.1 $(G_nX)_0 \in \mathsf{Cat}^{\mathsf{n-1}}_{\mathsf{hd}}$ and

$$(G_nX)^d_0 = (F_nX)^d_0 .$$

To show that $G_nX \in \mathsf{Ta}^{\mathsf{n}}_{\mathsf{wg}}$ it remains to prove that the induced Segal maps are $(n-1)$-equivalences. Since $F_nX \in \mathsf{Cat}^{\mathsf{n}}_{\mathsf{wg}}$ there are $(n-1)$-equivalences

$$(F_nX)_2 \to (F_nX)_1 \times_{(F_nX)^d_0} (F_nX)_1 .$$

Using the induction hypotheses (c) and (d) this induces an $(n-1)$-equivalence

$$\begin{aligned}(G_nX)_2 = G_{n-1}(F_nX)_2 &\to G_{n-1}\{(F_nX)_1 \times_{(F_nX)^d_0} (F_nX)_1\}\\ &\cong (G_nX)_1 \times_{(G_nX)^d_0} (G_nX)_1 .\end{aligned}$$

Similarly one shows that all other induced Segal maps for G_nX are $(n-1)$-equivalences. We conclude that $G_nX \in \mathsf{Ta}^{\mathsf{n}}_{\mathsf{wg}}$.

We now check the rest of the hypotheses in Lemma 7.2.10. Hypothesis (a) holds since, as seen above, for each $k \geq 0$, $(G_nX)_k \in \mathsf{Cat}^{\mathsf{n-1}}_{\mathsf{wg}}$.

Note that the diagram in $\mathsf{Cat}^{\mathsf{n-1}}_{\mathsf{wg}}$

$$(F_nX)_1 \xrightarrow{\partial_0} (F_nX)_0 \xleftarrow{\partial_0} (F_nX)_1 \tag{11.45}$$

satisfies inductive hypothesis (e). In fact, since $F_n X \in \mathsf{Cat}^n_{\mathsf{wg}}$,

$$(F_n X)_2 \cong (F_n X)_1 \times_{(F_n X)_0} (F_n X)_1 \in \mathsf{Cat}^{n-1}_{\mathsf{wg}} \ .$$

The rest of inductive hypothesis (e) for the diagram (11.45) follows from the fact that $F_n X \in \mathsf{Cat}^n_{\mathsf{wg}}$ using Lemma 6.3.8 and Remark 7.2.13. Thus by inductive hypothesis (e) we obtain

$$\begin{aligned}(G_n X)_2 &= G_{n-1}(F_n X)_2 = G_{n-1}((F_n X)_1 \times_{(F_n X)_0} (F_n X)_1) \\ &\cong G_{n-1}(F_n X)_1 \times_{G_{n-1}(F_n X)_0} G_{n-1}(F_n X)_1 \cong (G_n X)_1 \times_{(G_n X)_0} (G_n X)_1 \ .\end{aligned}$$

Similarly one shows that, for each $s \geq 2$,

$$(G_n X)_s \cong (G_n X)_1 \times_{(G_n X)_0} \overset{s}{\cdots} \times_{(G_n X)_0} (G_n X)_1 \ .$$

This proves that $G_n X$ satisfies hypothesis (b) of Lemma 7.2.10. By inductive hypothesis (e) applied to the diagram (11.45) we obtain, for each $1 \leq r < n$,

$$\begin{aligned}& p^{(r-1)}(G_n X)_2 = p^{(r-1)} G_{n-1}(F_n X)_2 \\ &\cong p^{(r-1)} G_{n-1}\{(F_n X)_1 \times_{(F_n X)_0} (F_n X)_1\} \\ &\cong p^{(r-1)} G_{n-1}(F_n X)_1 \times_{p^{(r-1)} G_{n-1}(F_n X)_0} p^{(r-1)} G_{n-1}(F_n X)_1 \\ &= p^{(r-1)}(G_n X)_1 \times_{p^{(r-1)}(G_n X)_0} p^{(r-1)}(G_n X)_1 \ .\end{aligned}$$

Similarly one shows that, for each $s \geq 2$,

$$p^{(r-1)}(G_n X)_s \cong p^{(r-1)}(G_n X)_1 \times_{p^{(r-1)}(G_n X)_0} \overset{s}{\cdots} \times_{p^{(r-1)}(G_n X)_0} p^{(r-1)}(G_n X)_1 \ ,$$

which is hypothesis (c) of Lemma 7.2.10 for $G_n X$. Thus all hypotheses of Lemma 7.2.10 hold and we conclude that $G_n X \in \mathsf{Cat}^n_{\mathsf{wg}}$.

To show that $G_n X \in \mathsf{FCat}^n_{\mathsf{wg}}$ we need to prove the commutativity of the diagrams in (i) and (ii) of Definition 11.3.1. Let $\underline{k} = (k_1, \dots, k_s)$, $\underline{r} = (r_1, \dots, r_s)$ in $\Delta^{s^{op}}$, denote $\underline{k}' = (k_2, \dots, k_s)$, $\underline{r}' = (r_2, \dots, r_s)$ and suppose we have a morphism $\underline{k} \to \underline{r}$ in $\Delta^{s^{op}}$. By factoring this as

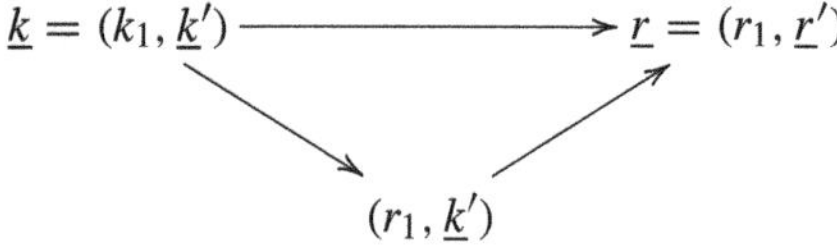

we obtain a factorization

$$\begin{array}{ccc} (G_n X)_{\underline{k}0} = \{G_{n-1}(F_n X)_{k_1}\}_{\underline{k}'0} & \longrightarrow & \{G_{n-1}(F_n X)_{r_1}\}_{\underline{r}'0} = (G_n X)_{\underline{r}0} \\ & \searrow \quad \nearrow & \\ & \{G_{n-1}(F_n X)_{r_1}\}_{\underline{k}'0} & \end{array} \tag{11.46}$$

Consider the morphism $(F_n X)_{k_1} \to (F_n X)_{r_1}$ in $\mathsf{Cat}^{\mathsf{n-1}}_{\mathsf{wg}}$. Since, by induction hypothesis (a), $G_{n-1} : \mathsf{Cat}^{\mathsf{n-1}}_{\mathsf{wg}} \to \mathsf{FCat}^{\mathsf{n-1}}_{\mathsf{wg}}$, there is a commuting diagram

$$\begin{array}{ccc} (G_n X)_{\underline{k}0} = \{G_{n-1}(F_n X)_{k_1}\}_{\underline{k}'0} & \longrightarrow & \{G_{n-1}(F_n X)_{r_1}\}_{\underline{k}'0} \\ \uparrow & & \uparrow \\ (G_n X)^d_{\underline{k}0} = \{G_{n-1}(F_n X)_{k_1}\}^d_{\underline{k}'0} & \longrightarrow & \{G_{n-1}(F_n X)_{r_1}\}^d_{\underline{k}'0} \end{array} \tag{11.47}$$

Since, by the induction hypothesis, $G_{n-1}(F_n X)_{r_1} \in \mathsf{FCat}^{\mathsf{n-1}}_{\mathsf{wg}}$ we also have a commuting diagram

$$\begin{array}{ccc} \{G_{n-1}(F_n X)_{r_1}\}_{\underline{k}'0} & \longrightarrow & \{G_{n-1}(F_n X)_{r_1}\}_{\underline{r}'0} = (G_n X)_{\underline{r}0} \\ \uparrow & & \uparrow \\ \{G_{n-1}(F_n X)_{r_1}\}^d_{\underline{k}'0} & \longrightarrow & \{G_{n-1}(F_n X)_{r_1}\}^d_{\underline{r}'0} = (G_n X)^d_{\underline{r}0} \end{array} \tag{11.48}$$

Combining (11.46), (11.47), (11.48) we obtain a commuting diagram

$$\begin{array}{ccc} (G_n X)_{\underline{k}0} & \longrightarrow & (G_n X)_{\underline{r}0} \\ \uparrow & & \uparrow \\ (G_n X)^d_{\underline{k}0} & \longrightarrow & (G_n X)^d_{\underline{r}0} \end{array} \tag{11.49}$$

Thus $G_n X$ satisfies condition (11.35) in the definition of $\mathsf{FCat}^{\mathsf{n}}_{\mathsf{wg}}$.

Applying inductive hypothesis (f) to the map in $\mathsf{Cat}^{\mathsf{n-1}}_{\mathsf{hd}}$ $(F_n X)^d_0 \to (F_n X)_0$, we see that the map

$$(G_n X)^d_0 = (F_n X)^d_0 = G_{n-1}(F_n X)^d_0 \to G_{n-1}(F_n X)_0 = (G_n X)_0$$

is a morphism in $\mathsf{FCat}^{\mathsf{n-1}}_{\mathsf{wg}}$. This means that the two diagrams (11.36) in (ii) of Definition 11.3.1 commute for $G_n X$.

Applying inductive hypothesis (f) to the map in $\mathsf{Cat}^{\mathsf{n-s}}_{\mathsf{hd}}$

$$(F_{n-s+1} Z_{\underline{k}}(X))^d_0 \to (F_{n-s+1} Z_{\underline{k}}(X))_0$$

and using the fact that, by Lemma 11.3.5, we have

$$(G_n X)_{\underline{k}0} = G_{n-s}(F_{n-s+1} Z_{\underline{k}}(X))_0 \ ,$$

we deduce that the map

$$(G_n X)^d_{\underline{k}0} \to (G_n X)_{\underline{k}0}$$

is a morphism in $\mathsf{FCat}^{\mathsf{n-s-1}}_{\mathsf{wg}}$. This means that the two diagrams (11.37) in (ii) of Definition 11.3.1 commute for $G_n X$. Together with (11.49) we conclude that $G_n X \in \mathsf{FCat}^{\mathsf{n}}_{\mathsf{wg}}$.

Finally, let $F : X \to Y$ be a morphism in $\mathsf{Cat}^{\mathsf{n}}_{\mathsf{wg}}$. Then

$$(F_n F)_0 : (F_n X)_0 \to (F_n Y)_0$$

is a morphism in $\mathsf{Cat}^{\mathsf{n-1}}_{\mathsf{wg}}$. Thus by the induction hypothesis it induces a morphism in $\mathsf{FCat}^{\mathsf{n-1}}_{\mathsf{wg}}$

$$G_{n-1}(F_n X)_0 \to G_{n-1}(F_n Y)_0$$

such that the following diagram commutes:

$$\begin{array}{ccc} (G_n X)_0 = G_{n-1}(F_n X)_0 & \longrightarrow & G_{n-1}(F_n Y)_0 = (G_n Y)_0 \\ \uparrow & & \uparrow \\ (G_n X)^d_0 & \longrightarrow & (G_n Y)^d_0 \end{array}$$

It also induces a morphism

$$G_{n-1}(F_n X)_{k_1} \to G_{n-1}(F_n Y)_{k_1}$$

such that the following diagram commutes:

$$\begin{array}{ccc} (G_n X)_{\underline{k}0} = \{G_{n-1}(F_n X)_{k_1}\}_{\underline{k}'0} & \longrightarrow & \{G_{n-1}(F_n Y)_{k_1}\}_{\underline{k}'0} = (G_n Y)_{\underline{k}0} \\ \uparrow & & \uparrow \\ (G_n X)^d_{\underline{k}0} & \longrightarrow & (G_n Y)^d_{\underline{k}0} \end{array}$$

By Definition 11.3.1 this shows that $G_n F$ is a morphism in $\mathsf{FCat}^{\mathsf{n}}_{\mathsf{wg}}$. In conclusion

$$G_n : \mathsf{Cat}^{\mathsf{n}}_{\mathsf{wg}} \to \mathsf{FCat}^{\mathsf{n}}_{\mathsf{wg}} \ .$$

(b) The morphism $g_n(X) : G_n X \to X$ is given levelwise by

$$(g_n(X))_k = g_{n-1}(X_k) : (G_n X)_k = G_{n-1}X_k \to X_k \; .$$

Since, by the inductive hypothesis, g_{n-1} is natural in X, so is g_n.

By the inductive hypothesis this is an $(n-1)$-equivalence for each k, hence $g_n(X)$ is an n-equivalence by Lemma 7.1.3.

We now show that $(g_n(X))_0^d$ is surjective. As in the proof of (a), $(G_n X)_0^d = (F_n X)_0^d$ and by Proposition 11.2.5, $(F_n X)_0^d = (V_{n-1}X_0)^d$. So we need to show that

$$(V_{n-1}X_0)^d \to X_0^d$$

is surjective. By Proposition 11.2.3 applied to $X_0 \in \mathsf{Cat}^{\mathsf{n-1}}_{\mathsf{hd}}$, the functor

$$p^{(1)}V_{n-1}X_0 \to p^{(1)}X_0$$

is surjective on objects. Hence

$$(V_{n-1}X_0)^d = pp^{(1)}V_{n-1}X_0 \to pp^{(1)}X_0 = X_0^d$$

is surjective, as required.

(c) Let $F : X \to Y$ be an n-equivalence in $\mathsf{Cat}^{\mathsf{n}}_{\mathsf{wg}}$. By naturality of g_n, there is a commuting diagram

$$\begin{array}{ccc} G_n X & \xrightarrow{G_n F} & G_n Y \\ {\scriptstyle g_n(X)}\downarrow & & \downarrow{\scriptstyle g_n(Y)} \\ X & \xrightarrow[F]{} & Y \end{array}$$

in which, by (b), the vertical maps and the bottom horizontal map are n-equivalences. By Proposition 7.1.2 (c) it follows that $G_n F$ is also an n-equivalence.

(d) This follows immediately by the analogous properties of F_n and by the inductive hypothesis.

(e) By hypothesis the diagram in $\mathsf{Cat}^{\mathsf{n}}_{\mathsf{wg}}$ $X \to Z \leftarrow Y$ satisfies the hypotheses of Corollary 11.2.6 so that

$$F_n(X\times_Z Y) \cong F_n X \times_{F_n Z} F_n Y \; , \tag{11.50}$$

$$p^{(r-1)}F_n(X\times_Z Y) \cong p^{(r-1)}F_n X \times_{p^{(r-1)}F_n Z} p^{(r-1)}F_n Y \; , \tag{11.51}$$

$$q^{(r-1)}F_n(X\times_Z Y) \cong q^{(r-1)}F_n X \times_{q^{(r-1)}F_n Z} q^{(r-1)}F_n Y \; . \tag{11.52}$$

We claim that the diagram in $\mathsf{Cat}^{n-1}_{\mathsf{wg}}$

$$(F_n X)_k \to (F_n Z)_k \leftarrow (F_n Y)_k \tag{11.53}$$

satisfies inductive hypothesis (e). In fact,

$$(F_n X)_k \times_{(F_n Z)_k} (F_n Y)_k = (F_n (X \times_Z Y))_k \in \mathsf{Cat}^{n-1}_{\mathsf{wg}} \;,$$

since $F_n(X \times_Z Y) \in \mathsf{Cat}^{n}_{\mathsf{wg}}$. Also, taking the kth component in (11.51) we obtain

$$\begin{aligned} &p^{(r-2)}\{(F_n X)_k \times_{(F_n Z)_k} (F_n Y)_k\} = p^{(r-2)}(F_n(X \times_Z Y))_k \\ &\cong p^{(r-2)}(F_n X)_k \times_{p^{(r-2)}(F_n Z)_k} p^{(r-2)}(F_n Y)_k \end{aligned}$$

and similarly for $q^{(r-2)}$. Thus by inductive hypothesis (e) applied to (11.53) we obtain

$$\begin{aligned} &(G_n(X \times_Z Y))_k = G_{n-1}(F_n(X \times_Z Y))_k = G_{n-1}((F_n X)_k \times_{(F_n Z)_k} (F_n Y)_k) \\ &= G_{n-1}(F_n X)_k \times_{G_{n-1}(F_n Z)_k} G_{n-1}(F_n Y)_k = (G_n X)_k \times_{(G_n Z)_k} (G_n Y)_k \end{aligned}$$

and

$$\begin{aligned} &(g_n(X \times_Z Y))_k = g_{n-1}(F_n(X \times_Z Y))_k = g_{n-1}((F_n X)_k \times_{(F_n Z)_k} (F_n Y)_k) \\ &= g_{n-1}((F_n X)_k) \times_{g_{n-1}((F_n Z)_k)} g_{n-1}((F_n Y)_k) = (g_n(X))_k \times_{(g_n(Z))_k} (g_n(Y))_k \;. \end{aligned}$$

Since this holds for each $k \geq 0$ it follows that

$$G_n(X \times_Z Y) = G_n X \times_{G_n Z} G_n Y$$

and

$$g_n(X \times_Z Y) = g_n(X) \times_{g_n(Z)} g_n(Y) \;.$$

By inductive hypothesis (e) applied to (11.53) we also obtain

$$\begin{aligned} &\{p^{(r-1)} G_n(X \times_Z Y)\}_k = p^{(r-2)} G_{n-1}(F_n(X \times_Z Y))_k \\ &= p^{(r-2)} G_{n-1}((F_n X)_k \times_{(F_n Z)_k} (F_n Y)_k) \\ &= p^{(r-2)} G_{n-1}(F_n X)_k \times_{p^{(r-2)} G_{n-1}(F_n Z)_k} p^{(r-2)} G_{n-1}(F_n Y)_k \\ &= (p^{(r-1)} G_n X)_k \times_{(p^{(r-1)} G_n Z)_k} (p^{(r-1)} G_n Y)_k \;. \end{aligned}$$

Since this holds for each $k \geq 0$ it follows that

$$p^{(r-1)}G_n(X \times_Z Y) = p^{(r-1)}G_n X \times_{p^{(r-1)}G_n Z} p^{(r-1)}G_n Y \ .$$

Similarly, taking the kth component in (11.52) and using inductive hypothesis (e) on (11.53) we obtain

$$q^{(r-1)}G_n(X \times_Z Y) = q^{(r-1)}G_n X \times_{q^{(r-1)}G_n Z} q^{(r-1)}G_n Y \ .$$

This concludes the proof of (e) at step n.

(f) By Proposition 11.2.5 the morphism $\gamma' : Y^d \to Y$ induces a commuting diagram

$$\begin{array}{ccc} (F_n Y^d)_0 = Y^d & \xrightarrow{\gamma'_0} & (F_n Y)_0 \\ \| & & \uparrow \\ (F_n Y^d)^d_0 = Y^d & \xrightarrow{\gamma'^d_0} & (F_n Y)^d_0 \end{array}$$

Thus applying G_{n-1} to this diagram and using the fact (from above) that

$$(G_n Y)^d_0 = (F_n Y)^d_0 = G_{n-1}(F_n Y)^d_0$$

we obtain

$$\begin{array}{ccc} (G_n Y^d)_0 = G_{n-1}(F_n Y^d)_0 & \xrightarrow{(G_n \gamma')_0} & G_{n-1}(F_n Y)_0 = (G_n Y)_0 \\ \| & & \uparrow \\ (G_n Y^d)^d_0 = G_{n-1}(F_n Y^d)_0 & \xrightarrow{(G_n \gamma')^d_0} & G_{n-1}(F_n Y)^d_0 = (G_n Y)^d_0 \end{array} \tag{11.54}$$

For each $\underline{k} = (k_1, \ldots, k_{s-1}) \in \Delta^{s-1^{op}}$, by Lemma 11.3.5,

$$(G_n Y)_{\underline{k}0} = G_{n-s}(F_{n-s+1} Z_{\underline{k}} Y)_0 \ ,$$

where we abbreviated $Z_{k_1 \ldots k_{s-1}} Y$ by $Z_{\underline{k}} Y$. The map $Y^d \to Y$ induces the map

$$Z_{\underline{k}} Y^d \to Z_{\underline{k}} Y$$

and by Proposition 11.2.5 this induces the commuting diagram

$$\begin{array}{ccc} (F_{n-s+1}Z_{\underline{k}}Y^d)_0 & \longrightarrow & (F_{n-s+1}Z_{\underline{k}}Y)_0 \\ \| & & \uparrow \\ (F_{n-s+1}Z_{\underline{k}}Y^d)_0 & \longrightarrow & (F_{n-s+1}Z_{\underline{k}}Y)_0^d \end{array}$$

Applying G_{n-s} to this diagram and recalling that by Lemma 11.3.5 and the above

$$\begin{aligned} (G_nY)_{\underline{k}0} &= G_{n-s}(F_{n-s+1}Z_{\underline{k}}Y)_0 \ , \\ (G_nY)^d_{\underline{k}0} &= (F_{n-s+1}Z_{\underline{k}}Y)^d_0 \ , \end{aligned}$$

we obtain the commuting diagram

$$\begin{array}{ccc} (G_nY^d)_{\underline{k}0} & \xrightarrow{(G_n\gamma')_{\underline{k}0}} & (G_nY)_{\underline{k}0} \\ \| & & \uparrow \\ (G_nY^d)^d_{\underline{k}0} & \xrightarrow{(G_n\gamma')^d_{\underline{k}0}} & (G_nY)^d_{\underline{k}0} \end{array} \qquad (11.55)$$

By definition, (11.54) and (11.55) mean that $G_n\gamma'$ is a morphism in $\mathsf{FCat}^n_{\mathsf{wg}}$. □

Chapter 12
Weakly Globular n-Fold Categories as a Model of Weak n-Categories

Abstract In this chapter we prove the main results of this work: the equivalence after localization of the categories $\mathsf{Cat}^{\mathsf{n}}_{\mathsf{wg}}$ and Ta^{n} and the proof of the homotopy hypothesis for $\mathsf{Cat}^{\mathsf{n}}_{\mathsf{wg}}$. This exhibits weakly globular n-fold categories as a model of weak n-categories. The two comparison functors are the rigidification functor $Q_n : \mathsf{Ta}^{\mathsf{n}} \to \mathsf{Cat}^{\mathsf{n}}_{\mathsf{wg}}$ built in Chap. 10 and the discretization functor $Disc_n : \mathsf{Cat}^{\mathsf{n}}_{\mathsf{wg}} \to \mathsf{Ta}^{\mathsf{n}}$ of this chapter. The construction of the latter uses the category $\mathsf{FCat}^{\mathsf{n}}_{\mathsf{wg}}$ of Chap. 11. The proof of the homotopy hypothesis requires the introduction of a groupoidal version of the three Segal-type models.

In this chapter we prove that the category $\mathsf{Cat}^{\mathsf{n}}_{\mathsf{wg}}$ of weakly globular n-fold categories constitutes a model of weak n-categories. We show this by proving that there is an equivalence of categories between the localizations of Ta^{n} and of $\mathsf{Cat}^{\mathsf{n}}_{\mathsf{wg}}$ with respect to the n-equivalences. This shows a type of equivalence (up to higher categorical equivalence) between $\mathsf{Cat}^{\mathsf{n}}_{\mathsf{wg}}$ and Ta^{n}.

We also show that the category $\mathsf{Cat}^{\mathsf{n}}_{\mathsf{wg}}$ satisfies the homotopy hypothesis. As explained in Part I, the latter is one of the main desiderata for a model of weak n-categories, while the comparison with the Tamsamani model is a contribution to the still largely open problem of comparing between different models of higher categories.

The homotopy hypothesis is shown by introducing the full subcategory

$$\mathsf{GCat}^{\mathsf{n}}_{\mathsf{wg}} \subset \mathsf{Cat}^{\mathsf{n}}_{\mathsf{wg}}$$

of groupoidal weakly globular n-fold categories and showing (Theorem 12.3.11) that there is an equivalence of categories

$$\mathsf{GCat}^{\mathsf{n}}_{\mathsf{wg}}/\sim^n \;\simeq\; \mathrm{Ho}\,(n\text{-types}). \tag{12.1}$$

In Corollary 12.4.6 the equivalence of categories (12.1) is realized by a different pair of functors, that uses the functor $\mathsf{Top} \to \mathsf{GCat}^{\mathsf{n}}_{\mathsf{wg}}$ of Blanc and Paoli [29]:

S. Paoli, *Simplicial Methods for Higher Categories*, Algebra and Applications 26,
https://doi.org/10.1007/978-3-030-05674-2_12

this provides a more explicit form for the fundamental groupoidal weakly globular n-fold groupoid of a space, which is independent of [126].

Our main result, Theorem 12.2.6, is that there are comparison functors

$$Q_n : \mathsf{Ta}^{\mathsf{n}} \leftrightarrows \mathsf{Cat}^{\mathsf{n}}_{\mathsf{wg}} : Disc_n$$

inducing equivalences of categories

$$\mathsf{Ta}^{\mathsf{n}}/{\sim^n} \simeq \mathsf{Cat}^{\mathsf{n}}_{\mathsf{wg}}/{\sim^n}$$

after localization with respect to the n-equivalences. The rigidification functor Q_n was defined in Theorem 10.2.1 while the functor $Disc_n$, called the *discretization functor*, is built in this chapter.

The idea of the functor $Disc_n$ is to replace the homotopically discrete substructures in $X \in \mathsf{Cat}^{\mathsf{n}}_{\mathsf{wg}}$ by their discretizations in order to recover the globularity condition. This affects the Segal maps, which from being isomorphisms in X become $(n-1)$-equivalences in $Disc_n X$.

However, as outlined in the introduction to Chap. 11, for this method to work the discretization maps of the homotopically discrete substructures in a weakly globular n-fold category need to have functorial sections.

For this reason, we use the category $\mathsf{FCat}^{\mathsf{n}}_{\mathsf{wg}}$ introduced in Definition 11.3.1. We build in Proposition 12.1.4 a functor

$$D_n : \mathsf{FCat}^{\mathsf{n}}_{\mathsf{wg}} \to \mathsf{Ta}^{\mathsf{n}}$$

which discretizes the homotopically discrete substructures in the objects of $\mathsf{FCat}^{\mathsf{n}}_{\mathsf{wg}}$. We showed in Theorem 11.3.6 that there is a functor

$$G_n : \mathsf{Cat}^{\mathsf{n}}_{\mathsf{wg}} \to \mathsf{FCat}^{\mathsf{n}}_{\mathsf{wg}}$$

and an n-equivalence $G_n X \to X$ for each $X \in \mathsf{Cat}^{\mathsf{n}}_{\mathsf{wg}}$. The discretization functor

$$Disc_n : \mathsf{Cat}^{\mathsf{n}}_{\mathsf{wg}} \to \mathsf{Ta}^{\mathsf{n}}$$

is defined to be the composite

$$\mathsf{Cat}^{\mathsf{n}}_{\mathsf{wg}} \xrightarrow{G_n} \mathsf{FCat}^{\mathsf{n}}_{\mathsf{wg}} \xrightarrow{D_n} \mathsf{Ta}^{\mathsf{n}}\,.$$

This chapter is organized as follows. In Sect. 12.1, Proposition 12.1.4, we define the functor $D_n : \mathsf{FCat}^{\mathsf{n}}_{\mathsf{wg}} \to \mathsf{Ta}^{\mathsf{n}}$ and establish its properties. As a consequence, and using the previous results of Proposition 7.1.2, we show in Corollary 12.1.6 that n-equivalences in $\mathsf{Ta}^{\mathsf{n}}_{\mathsf{wg}}$ have the 2-out-of-3 property.

In Sect. 12.2 the functor D_n is used in Definition 12.2.1 to build the discretization functor from $\mathsf{Cat}^{\mathsf{n}}_{\mathsf{wg}}$ to Ta^{n}. Together with the results of Chaps. 10 and 11 this leads to the main comparison result Theorem 12.2.6.

In Sect. 12.3 we define groupoidal weakly globular n-fold categories and, using the results of the previous sections we show in Theorem 12.3.11 that they are an algebraic model of n-types. In Sect. 12.4 we realize the equivalence of categories of Theorem 12.3.11 through a different pair of functors, using the results of Blanc and Paoli [29]. This provides a more convenient model for the fundamental groupoidal weakly globular n-fold groupoid of a space, which is very explicit and is independent of [126]. We illustrate this with some examples in low dimensions.

12.1 From $\mathsf{FCat}^{\mathsf{n}}_{\mathsf{wg}}$ to Tamsamani n-Categories

In this section we define a functor

$$D_n : \mathsf{FCat}^{\mathsf{n}}_{\mathsf{wg}} \to \mathsf{Ta}^{\mathsf{n}}$$

and we study its properties. As a corollary, using our previous results, we also establish that n-equivalences in $\mathsf{Ta}^{\mathsf{n}}_{\mathsf{wg}}$ have the 2-out-of-3 property.

12.1.1 *The Idea of the Functor D_n*

The functor D_n is a discretization functor built from the category $\mathsf{FCat}^{\mathsf{n}}_{\mathsf{wg}}$ whose idea was discussed in Sect. 11.3.1. When pre-composed with the functor G_n, it will give rise to the discretization functor $Disc_n : \mathsf{Cat}^{\mathsf{n}}_{\mathsf{wg}} \to \mathsf{Ta}^{\mathsf{n}}$.

The idea of the functor D_n is to replace the homotopically discrete substructures in $X \in \mathsf{FCat}^{\mathsf{n}}_{\mathsf{wg}}$ by their discretizations, thus recovering the globularity condition. From the definition of $\mathsf{FCat}^{\mathsf{n}}_{\mathsf{wg}}$, this can be done in a functorial way. This discretization process is at the expense of the Segal maps, which from being isomorphisms in $\mathsf{FCat}^{\mathsf{n}}_{\mathsf{wg}}$ become higher categorical equivalences, so we obtain objects of Ta^{n}.

The construction of D_n is inductive, and we first discretize the structure at level 0 via a functor

$$R_0 : \mathsf{FCat}^{\mathsf{n}}_{\mathsf{wg}} \to [\Delta^{op}, \mathsf{FCat}^{\mathsf{n-1}}_{\mathsf{wg}}]$$

such that $(R_0 X)_0$ is discrete for all $X \in \mathsf{FCat}^{\mathsf{n}}_{\mathsf{wg}}$.

The definition of R_0 is based on the following general construction. Given a category $\mathscr{C}$ with finite limits, let $Y \in [\Delta^{^{op}}, \mathscr{C}]$, $Y_0^d \in \mathscr{C}$ and suppose there are maps in $\mathscr{C}$

$$\gamma(Y_0) : Y_0 \to Y_0^d \qquad \gamma'(Y_0) : Y_0^d \to Y_0 \ ,$$

natural in Y, such that $\gamma(Y_0)\gamma'(Y_0) = \mathrm{Id}$ and such that any morphism $F : Y \to Z$ in $[\Delta^{^{op}}, \mathscr{C}]$ induces commuting diagrams

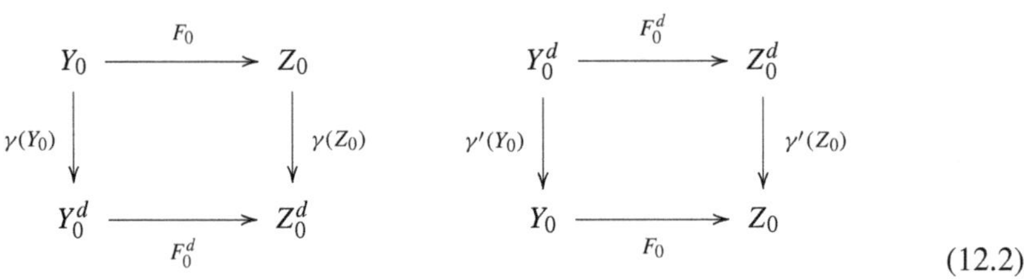

(12.2)

Define $R_0 Y$ as follows:

$$(R_0 Y)_k = \begin{cases} Y_0^d, & k = 0 \\ Y_k, & k > 0 \ . \end{cases}$$

The face operators $\partial_0', \partial_1' : Y_1 \rightrightarrows Y_0^d$ are given by $\partial_i' = \gamma(Y_0)\partial_i \;\; i = 0, 1$ and the degeneracy operator $\sigma' : Y_0^d \to Y_1$ by $\sigma' = \sigma\gamma'(Y_0)$, where $\partial_0, \partial_1 : Y_1 \rightrightarrows Y_0$ and $\sigma : Y_0 \to Y_1$, are the face and degeneracy operators of Y. All other face and degeneracy operators of $R_0 Y$ are as in Y. Since $\gamma(Y_0)\gamma'(Y_0) = \mathrm{Id}$, all simplicial identities for $R_0 Y$ hold so that $R_0 Y \in [\Delta^{^{op}}, \mathscr{C}]$.

Let $F : Y \to Z$ be a map in $[\Delta^{^{op}}, \mathscr{C}]$. From the commutativity of (12.2) this induces a map in $[\Delta^{^{op}}, \mathscr{C}]$, $R_0 F : R_0 Y \to R_0 Z$, so R_0 is a functor

$$R_0 : [\Delta^{^{op}}, \mathscr{C}] \to [\Delta^{^{op}}, \mathscr{C}] \ .$$

We apply this construction to the case where $Y = N^{(1)} X$ with $X \in \mathsf{FCat}^{\mathsf{n}}_{\mathsf{wg}}$, $\gamma : X_0 \to X_0^d$ is the discretization map and $\gamma' = X_0^d \to X_0$ a functorial section. As observed in Remark 11.3.2, $N^{(1)} X \in [\Delta^{^{op}}, \mathsf{FCat}^{\mathsf{n-1}}_{\mathsf{wg}}]$ and γ, γ' are maps in $\mathsf{FCat}^{\mathsf{n-1}}_{\mathsf{wg}}$; thus by definition of morphism in $\mathsf{FCat}^{\mathsf{n}}_{\mathsf{wg}}$ (see Definition 11.3.1) a morphism $F : X \to Y$ in $\mathsf{FCat}^{\mathsf{n}}_{\mathsf{wg}}$ induces commuting diagrams as in (12.2).

So all the conditions to apply the previous construction are met and we define the functor

$$R_0 : \mathsf{FCat}^{\mathsf{n}}_{\mathsf{wg}} \to [\Delta^{^{op}}, \mathsf{FCat}^{\mathsf{n-1}}_{\mathsf{wg}}]$$

(see Definition 12.1.1). The effect of R_0 is to discretize the object X_0 to X_0^d. Except when $n = 2$, this construction, however, does not yet produce an object of Ta^{n} since, for $k > 0$, $(R_0 X)_k = X_k$ is in $\mathsf{FCat}^{\mathsf{n-1}}_{\mathsf{wg}}$, not in $\mathsf{Ta}^{\mathsf{n-1}}$.

We perform the rest of the discretization of X inductively. Namely, we define inductively

$$D_n : \mathsf{FCat}^{\mathsf{n}}_{\mathsf{wg}} \to \mathsf{Ta}^{\mathsf{n}}$$

by

$$D_2 = R_0, \qquad D_n = \overline{D}_{n-1} R_0 \ .$$

The effect of D_n is to discretize all the homotopically discrete substructures of $X \in \mathsf{FCat}^{\mathsf{n}}_{\mathsf{wg}}$, thus recovering the globularity condition. The proof that $D_n X \in \mathsf{Ta}^{\mathsf{n}}$ requires us to check the Segal maps condition, and this is done inductively in the proof of Proposition 12.1.4.

12.1.2 The Functor D_n: Definition and Properties

Definition 12.1.1 Let $R_0 : \mathsf{FCat}^{\mathsf{n}}_{\mathsf{wg}} \to [\Delta^{^{op}}, \mathsf{FCat}^{\mathsf{n-1}}_{\mathsf{wg}}]$ be defined by

$$(R_0 X)_k = \begin{cases} X_0^d, \ k = 0 \\ X_k, \ k > 0 \ . \end{cases}$$

The face operators $\partial'_0, \partial'_1 : X_1 \rightrightarrows X_0^d$ are given by $\partial'_i = \gamma \partial_i$, $i = 0, 1$ and the degeneracy $\sigma' : X_0^d \to X_1$ by $\sigma' = \sigma \gamma'$, where $\partial_0, \partial_1 : X_1 \rightrightarrows X_0$ and $\sigma : X_0 \to X_1$ are the corresponding face and degeneracy operators of X, $\gamma : X_0 \to X_0^d$ is the discretization map and $\gamma' : X_0^d \to X_0$ is a functorial section. All other face and degeneracy maps of $R_0 X$ are as in X.

Note that by definition of $\mathsf{FCat}^{\mathsf{n}}_{\mathsf{wg}}$ and by Remark 11.3.2 the maps γ and γ' are morphisms in $\mathsf{FCat}^{\mathsf{n-1}}_{\mathsf{wg}}$, therefore so are ∂'_i and σ'. Since $\gamma\gamma' = id$, all simplicial identities are satisfied, thus $R_0 X \in [\Delta^{^{op}}, \mathsf{FCat}^{\mathsf{n-1}}_{\mathsf{wg}}]$.

Remark 12.1.2 By definition of $\mathsf{FCat}^{\mathsf{n}}_{\mathsf{wg}}$, given $f : X \to Y$ in $\mathsf{FCat}^{\mathsf{n}}_{\mathsf{wg}}$ there is a commuting diagram

$$\begin{array}{ccc} X_0 & \xrightarrow{f_0} & Y_0 \\ {\scriptstyle \gamma'(X_0)} \uparrow & & \uparrow {\scriptstyle \gamma'(Y_0)} \\ X_0^d & \longrightarrow & Y_0^d \end{array} \qquad (12.3)$$

and this induces a morphism in $[\Delta^{op}, \mathsf{FCat}^{n-1}_{wg}]$

$$R_0 f : R_0 X \to R_0 Y.$$

Thus R_0 is a functor. Note that while $R_0 X$ could be defined for any $X \in \mathsf{Ta}^n_{wg}$, given a morphism f in Ta^n_{wg} since in general (12.3) does not commute, one cannot define $R_0 f$ as above.

In what follows, the notation $\overline{p}, \overline{q}$ is as in Definition 2.1.1.

Lemma 12.1.3 *Let R_0 be as in Definition 12.1.1. Then:*

a) R_0 is the identity on objects and commutes with pullbacks over discrete objects.
b) $p^{(1)} = \overline{p} R_0$, $q^{(1)} = \overline{q} R_0$ while for $n > 2$
$\overline{p^{(n-2)}} R_0 = R_0 p^{(n-1)}$, $\overline{q^{(n-2)}} R_0 = R_0 q^{(n-1)}$.
c) For each $X \in \mathsf{FCat}^n_{wg}$ the Segal maps of $R_0 X$

$$(R_0 X)_k \to (R_0 X)_1 \times_{(R_0 X)_0} \overset{k}{\cdots} \times_{(R_0 X)_0} (R_0 X)_1$$

are $(n-1)$-equivalences for all $k \geq 2$.

Proof

a) This is immediate by the definition of R_0 since, if $X \to Z \leftarrow Y$ is a pullback in FCat^n_{wg} with Z discrete, $(X \times_Z Y)^d_0 = X^d_0 \times_Z Y^d_0$ by Corollary 5.2.2.
b) If $X \in \mathsf{FCat}^2_{wg}$,

$$(\overline{p} R_0 X)_0 = p X^d_0 = X^d_0 = (p^{(1)} X)_0 \ ,$$

while for $k > 1$,

$$(\overline{p} R_0 X)_k = p X_k = (p^{(1)} X)_k \ ,$$

so that $\overline{p} R_0 X = p^{(1)} X$. Similarly one shows that $\overline{q} R_0 X = q^{(1)} X$.
If $X \in \mathsf{FCat}^n_{wg}$ for $n > 2$, we have

$$(\overline{p^{(n-2)}} R_0 X)_0 = X^d_0 = (p^{(n-1)} X)^d_0 = (R_0 p^{(n-1)} X)_0 \ ,$$

while for $k > 0$,

$$(\overline{p^{(n-2)}} R_0 X)_k = p^{(n-2)} (R_0 X)_k = p^{(n-2)} X_k = R_0 (p^{(n-1)} X)_k \ .$$

In conclusion $\overline{p^{(n-2)}} R_0 X = R_0 p^{(n-1)} X$. Similarly one shows that

$$\overline{q^{(n-2)}} R_0 X = R_0 q^{(n-1)} X.$$

c) For each $k \geq 2$ the Segal maps for $R_0 X$ are

$$(R_0X)_k = X_k \to X_1 \times_{X_0^d} \overset{k}{\cdots} \times_{X_0^d} X_1 = (R_0X)_1 \times_{(R_0X)_0} \overset{k}{\cdots} \times_{(R_0X)_0} (R_0X)_1$$

and these are $(n-1)$-equivalences since $X \in \mathsf{FCat}^{\mathsf{n}}_{\mathsf{wg}}$. □

Proposition 12.1.4 *There is a functor*

$$D_n : \mathsf{FCat}^{\mathsf{n}}_{\mathsf{wg}} \to \mathsf{Ta}^{\mathsf{n}}$$

defined inductively by

$$D_2 = R_0, \qquad D_n = \overline{D}_{n-1} \circ R_0 \qquad \textit{for } n > 2$$

where R_0 is as in Definition 12.1.1, such that

a) D_n is the identity on discrete objects and commutes with pullbacks over discrete objects.

b) $p^{(1)} D_2 = p^{(1)}$, $q^{(1)} D_2 = q^{(1)}$, while for $n > 2$,

$$p^{(n-1)} D_n = D_{n-1} p^{(n-1)}, \qquad q^{(n-1)} D_n = D_{n-1} q^{(n-1)} .$$

c) For each $X \in \mathsf{FCat}^{\mathsf{n}}_{\mathsf{wg}}$ and $a, b \in X_0^d$,

$$(D_n X)(a, b) = D_{n-1}(X(a, b)) .$$

d) D_n preserves and reflects n-equivalences.

Proof By induction on n. When $n = 2$, $D_2 = R_0$ is such that for each $X \in \mathsf{FCat}^{2}_{\mathsf{wg}}$ $(D_2 X)_0 = X_0^d$ is discrete and, by Lemma 12.1.3 c), the Segal maps are equivalences of categories. Thus $D_2 X \in \mathsf{Ta}^2$. Note also that by Lemma 12.1.3 b)

$$p^{(1)} D_2 X = \overline{p} R_0 X = p^{(1)} X .$$

Similarly, $q^{(1)} D_2 = q^{(1)}$, so b) holds. By Lemma 12.1.3 a), D_2 satisfies a).

Let $f : X \to Y$ be a 2-equivalence. For all $a, b \in (D_2 X)_0^d = X_0^d$, $(D_2 f)(a, b) = f(a, b)$ is an equivalence of categories. Also, $p^{(1)} D_2 f = p^{(1)} f$ is an equivalence of categories. Thus by definition $D_2 f$ is a 2-equivalence.

Suppose that $f : X \to Y$ is such that $D_2 f$ is a 2-equivalence. Then for all $a, b \in (D_2 X)_0^d = X_0^d$, $(D_2 f)(a, b) = f(a, b)$ is an equivalence of categories. Also, $p^{(1)} D_2 f = p^{(1)} f$ is an equivalence of categories. Thus by definition f is a 2-equivalence. This completes the proof of c) and d) when $n = 2$.

Suppose, inductively, that the proposition holds for $(n-1)$ and let $X \in \mathsf{FCat}^{\mathsf{n}}_{\mathsf{wg}}$. We first show that $D_n X \in \mathsf{Ta}^{\mathsf{n}}$. By induction hypothesis a)

$$(D_n X)_k = \begin{cases} D_{n-1}X_0^d = X_0^d & k=0 \\ D_{n-1}X_k, & k>0\,. \end{cases}$$

Thus by the induction hypothesis $(D_n X)_k \in \mathsf{Ta}^{\mathsf{n-1}}$ for all $k \geq 0$ with $(D_n X)_0$ discrete.

By Lemma 6.2.3, to show that $D_n X \in \mathsf{Ta}^{\mathsf{n}}$ it remains to show that the Segal maps are $(n-1)$-equivalences. Since $X \in \mathsf{FCat}^{\mathsf{n}}_{\mathsf{wg}}$, for each $k \geq 2$ the map

$$\mu_k : X_k \to X_1 \times_{X_0^d} \overset{k}{\cdots} \times_{X_0^d} X_1$$

is an $(n-1)$-equivalence. By inductive hypotheses a) and d) this induces a $(n-1)$-equivalence

$$D_{n-1}\mu_k : D_{n-1}X_k = (D_n X)_k \to D_{n-1}(X_1 \times_{X_0^d} \overset{k}{\cdots} \times_{X_0^d} X_1)$$
$$\cong (D_n X)_1 \times_{(D_n X)_0} \overset{k}{\cdots} \times_{(D_n X)_0} (D_n X)_1\,.$$

This shows that the Segal maps of $D_n X$ are $(n-1)$-equivalences. We conclude that $D_n X \in \mathsf{Ta}^{\mathsf{n}}$. We now prove the inductive step for points a)–d).

a) This follows from Lemma 12.1.3 and the inductive hypothesis.
b) Recalling that $p^{(n-1)} = \overline{p^{(n-2)}} : \mathsf{Ta}^{\mathsf{n}} \to \mathsf{Ta}^{\mathsf{n-1}}$, using the inductive hypothesis, the definition of D_n, Proposition 12.1.4 b) and Lemma 12.1.3 b) we obtain

$$p^{(n-1)} D_n = \overline{p^{(n-2)}}\,\overline{D_{n-1}} R_0 = \overline{D_{n-2}}\,\overline{p^{(n-2)}} R_0 = \overline{D_{n-2}} R_0 p^{(n-1)} = D_{n-1} p^{(n-1)}\,.$$

The proof for $q^{(n-1)} D_n$ is similar.

c) By definition of $X(a,b)$, we have a pullback in $[\Delta^{n-2^{op}}, \mathsf{Cat}]$

$$\begin{array}{ccc} X(a,b) & \longrightarrow & \{a\} \times \{b\} \\ \downarrow & & \downarrow \\ X_1 & \longrightarrow & X_0^d \times X_0^d \end{array}$$

Therefore, by a), we also have a pullback

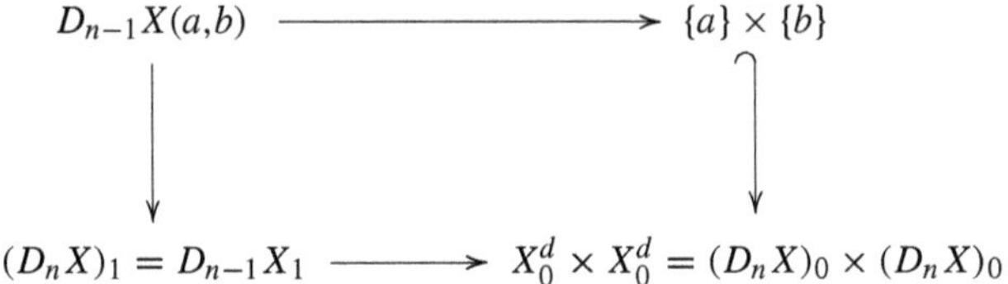

so that

$$(D_n X)(a,b) = D_{n-1}X(a,b)\ .$$

d) Let $f : X \to Y$ be an n-equivalence in $\mathsf{FCat}^{\mathsf{n}}_{\mathsf{wg}}$. By c), for each $a, b \in (D_n X)_0 = X_0^d$ we have

$$(D_n f)(a,b) = D_{n-1} f(a,b)$$

and this is an $(n-1)$-equivalence by the inductive hypothesis applied to the $(n-1)$-equivalence $f(a,b)$. Further, by b),

$$p^{(n-1)} D_n f = D_{n-1} p^{(n-1)} f$$

is also an $(n-1)$-equivalence by the inductive hypothesis applied to the $(n-1)$-equivalence $p^{(n-1)} f$. This shows that $D_n f$ is an n-equivalence.

Let $f : X \to Y$ be such that $D_n f$ is an n-equivalence. By c), for each $a, b \in (D_n X)_0 = X_0^d$ we have

$$(D_n f)(a,b) = D_{n-1} f(a,b)$$

and this is an $(n-1)$-equivalence. By the inductive hypothesis we conclude that $f(a,b)$ is an $(n-1)$-equivalence. Further, by b),

$$p^{(n-1)} D_n f = D_{n-1} p^{(n-1)} f$$

is also an $(n-1)$-equivalence. So by the inductive hypothesis $p^{(n-1)} f$ is an $(n-1)$-equivalence. In conclusion f is an n-equivalence. □

Example 12.1.5 Let $X \in \mathsf{FCat}^{\mathsf{3}}_{\mathsf{wg}}$, so that $X \in [\Delta^{2^{op}}, \mathsf{Cat}]$. A picture of the corner of X is found in Fig. 12.1 on page 284 where the structures in red are homotopically discrete and they are equipped with functorial sections to the discretization maps. In Fig. 12.2 on page 284 we depict the corner of $D_3 X \in [\Delta^{2^{op}}, \mathsf{Cat}]$.

We see that the homotopically discrete substructures in X have been replaced by discrete ones (also in red).

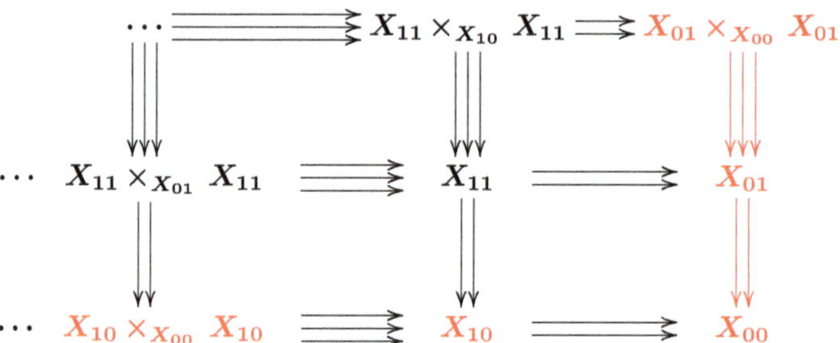

Fig. 12.1 Corner of $X \in \mathsf{FCat}^3_{\mathsf{wg}} \subset [\Delta^{2^{op}}, \mathsf{Cat}]$

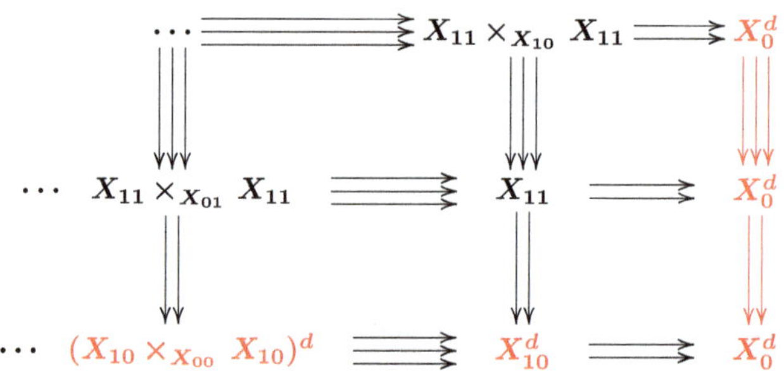

Fig. 12.2 Corner of $D_3 X \in [\Delta^{2^{op}}, \mathsf{Cat}]$ for $X \in \mathsf{FCat}^3_{\mathsf{wg}}$

We explore one of the consequences of the previous results, which is the 2-out-of-3 property of n-equivalences in $\mathsf{Ta}^n_{\mathsf{wg}}$. Several cases of this property have already been proved in Proposition 7.1.2, the remaining case is shown in the next corollary.

Corollary 12.1.6 *n-Equivalences in $\mathsf{Ta}^n_{\mathsf{wg}}$ have the 2-out-of-3 property.*

Proof By Proposition 7.1.2 the only case that remains to be checked is when we have morphisms

$$X \xrightarrow{g} Z \xrightarrow{h} Y$$

in $\mathsf{Ta}^n_{\mathsf{wg}}$ such that hg and g are n-equivalences, but (unlike in Proposition 7.1.2 c)), no further assumptions are required on g. We need to show that h is an n-equivalence.

Recall that, by Remark 10.2.2, Theorem 11.3.6 and Proposition 12.1.4, the functors Q_n, G_n, D_n preserve n-equivalences. We therefore have morphisms in Ta^n

$$D_n G_n Q_n X \xrightarrow{D_n G_n Q_n g} D_n G_n Q_n Z \xrightarrow{D_n G_n Q_n h} D_n G_n Q_n Y$$

in which $D_nG_nQ_nh$ and the composite $(D_nG_nQ_nh)(D_nG_nQ_ng)$ are n-equivalences

Since by Simpson [119] n-equivalences in $\mathsf{Ta^n}$ have the 2-out-of-3 property, this implies that $D_nG_nQ_nh$ is an n-equivalence. Since, by Proposition 12.1.4 d), D_n reflects n-equivalences, we conclude that G_nQ_nh is an n-equivalence. On the other hand, we have a commuting diagram (with g_n as in Theorem 11.3.6 and s_n as in Theorem 10.2.1).

$$\begin{array}{ccc} G_nQ_nZ & \xrightarrow{G_nQ_nh} & G_nQ_nY \\ {\scriptstyle g_n(Q_nZ)}\downarrow & & \downarrow{\scriptstyle g_n(Q_nY)} \\ Q_nZ & & Q_nY \\ {\scriptstyle s_n(Z)}\downarrow & & \downarrow{\scriptstyle s_n(Y)} \\ Z & \xrightarrow[h]{} & Y \end{array}$$

in which the vertical maps are n-equivalences (by Proposition 7.1.2 d)) and the top horizontal map is an n-equivalence. Thus by Proposition 7.1.2 d) we conclude that the composite

$$G_nQ_nZ \xrightarrow{s_n(Z)g_n(Q_nZ)} Z \xrightarrow{h} Y \tag{12.4}$$

is an n-equivalence. By Theorem 11.3.6 b), the map $(g_n(Q_nZ))_0^d : (G_nQ_nZ)_0^d \to (Q_nZ)_0^d$ is surjective. By Theorem 10.2.1 the map $s_n(Z)_0 : (Q_nZ)_0 \to Z_0$ is an $(n-1)$-equivalence in $\mathsf{Cat}^{\mathsf{n-1}}_{\mathsf{hd}}$, thus by Lemma 5.2.6 $(s_n(Z))_0^d : (Q_nZ)_0^d \to Z_0^d$ is an isomorphism. Hence the composite map $(g_n(Q_nZ) \circ s_n(Z))_0^d$ is surjective. Thus the morphisms (12.4) satisfy the hypotheses of Proposition 7.1.2 e) and we conclude that h is an n-equivalence. □

12.2 The Discretization Functor and the Comparison Result

In this section we define the discretization functor

$$Disc_n : \mathsf{Cat}^{\mathsf{n}}_{\mathsf{wg}} \to \mathsf{Ta^n}$$

and we establish the main result of this work, Theorem 12.2.6, asserting that the functors $Disc_n$ and $Q_n : \mathsf{Ta^n} \to \mathsf{Cat}^{\mathsf{n}}_{\mathsf{wg}}$ induce an equivalence of categories

$$\mathsf{Ta^n}/{\sim^n} \simeq \mathsf{Cat}^{\mathsf{n}}_{\mathsf{wg}}/{\sim^n} \ .$$

12.2.1 The Idea of the Functor $Disc_n$

The discretization functor is the second half of what we need to establish our comparison result (Theorem 12.2.6) between weakly globular n-fold categories and Tamsamani n-categories, the other half being the rigidification functor Q_n, whose idea was discussed in Sect. 10.2.1.

The idea of the discretization functor

$$Disc_n : \mathsf{Cat}^n_{\mathsf{wg}} \to \mathsf{Ta}^n$$

is to discretize the homotopically discrete sub-structures of objects of $\mathsf{Cat}^n_{\mathsf{wg}}$ to recover the globularity condition. As discussed in Chap. 11, this needs functorial sections to the discretization maps of the homotopically discrete sub-structures of weakly globular n-fold categories. For this reason we introduced the category $\mathsf{FCat}^n_{\mathsf{wg}}$ and the discretization process from this category is the functor

$$D_n : \mathsf{FCat}^n_{\mathsf{wg}} \to \mathsf{Ta}^n$$

of Sect. 12.1.1.

We define the discretization functor to be the composite

$$Disc_n : \mathsf{Cat}^n_{\mathsf{wg}} \xrightarrow{G_n} \mathsf{FCat}^n_{\mathsf{wg}} \xrightarrow{D_n} \mathsf{Ta}^n\ ,$$

where G_n is as in Theorem 11.3.6 and D_n as in Proposition 12.1.4.

This realizes the idea of discretizing the homotopically discrete sub-structures in each $X \in \mathsf{Cat}^n_{\mathsf{wg}}$, but after replacing X with the n-equivalent $G_n X \in \mathsf{FCat}^n_{\mathsf{wg}}$.

The main property of $Disc_n$ is that, for each $X \in \mathsf{Cat}^n_{\mathsf{wg}}$, $Disc_n X$ and X are suitably equivalent in $\mathsf{Ta}^n_{\mathsf{wg}}$. We show this fact in Theorem 12.2.5, where we prove that there is a zig-zag of n-equivalences in $\mathsf{Ta}^n_{\mathsf{wg}}$, of the form

$$Disc_n X \leftarrow Q_n Disc_n X \to X. \tag{12.5}$$

This relies on Proposition 12.2.3, establishing that, for each $X \in \mathsf{FCat}^n_{\mathsf{wg}}$, $Q_n D_n X = Q_n X$. When applied to $G_n X$ (for $X \in \mathsf{Cat}^n_{\mathsf{wg}}$) this fact implies

$$Q_n Disc_n X = Q_n D_n G_n X = Q_n G_n X.$$

The zig-zag (12.5) is then obtained using the maps $s_n(Disc_n X)$, $s_n(G_n X)$, $g_n(X)$ as follows

$$Disc_n X \xleftarrow{s_n(Disc_n X)} Q_n Disc_n X = Q_n G_n X \xrightarrow{s_n(G_n X)} G_n X \xrightarrow{g_n(X)} X\ .$$

The proof of Proposition 12.2.3 relies on the definition of Q_n as well as on the property of the functor Tr_n established in Lemma 10.1.4. The latter gives conditions on $X \in \mathsf{Cat}^{\mathsf{n}}_{\mathsf{wg}}$ and $Y \in \mathsf{LTa}^{\mathsf{n}}_{\mathsf{wg}}$ under which $Tr_n X = Tr_n Y$. Using the definition of $P_n : \mathsf{Ta}^{\mathsf{n}}_{\mathsf{wg}} \to \mathsf{LTa}^{\mathsf{n}}_{\mathsf{wg}}$ in Theorem 10.2.1(see also Remark 10.2.4), we show that, given $X \in \mathsf{Cat}^{\mathsf{n}}_{\mathsf{wg}}$, $P_n D_n X \in \mathsf{LTa}^{\mathsf{n}}_{\mathsf{wg}}$ and $P_n X \in \mathsf{Cat}^{\mathsf{n}}_{\mathsf{wg}}$ satisfy these conditions, and therefore

$$Tr_n P_n D_n X = Tr_n P_n X.$$

In turn, this implies

$$Q_n D_n X = St\, Tr_n P_n D_n X = St\, Tr_n P_n X = Q_n X \ ,$$

which is Proposition 12.2.3.

12.2.2 The Comparison Result

In this section we prove our main comparison result between Tamsamani n-categories and weakly globular n-fold categories. We first need to establish a number of properties about the functor D_n of the previous section, see Proposition 12.2.3 and Lemma 12.2.4 below.

Definition 12.2.1 Define the discretization functor

$$Disc_n : \mathsf{Cat}^{\mathsf{n}}_{\mathsf{wg}} \to \mathsf{Ta}^{\mathsf{n}}$$

to be the composite

$$\mathsf{Cat}^{\mathsf{n}}_{\mathsf{wg}} \xrightarrow{G_n} \mathsf{FCat}^{\mathsf{n}}_{\mathsf{wg}} \xrightarrow{D_n} \mathsf{Ta}^{\mathsf{n}} \ ,$$

where G_n is as in Theorem 11.3.6 and D_n as in Proposition 12.1.4.

The following lemmas and proposition establish some facts about the functor D_n of Proposition 12.1.4, which will be needed to study the properties of the discretization functor $Disc_n$.

Lemma 12.2.2 *Let* $X \in \mathsf{FCat}^{\mathsf{n}}_{\mathsf{wg}}$.

a) If $\underline{k} \in \Delta^{n-1^{op}}$ *is such that* $k_j \neq 0$ *for all* $1 \leq j \leq n-1$. *Then* $(D_n X)_{\underline{k}} = X_{\underline{k}}$.
b) If $\underline{k} \to \underline{s}$ *is a morphism in* $\Delta^{n-1^{op}}$ *with* $k_j = 0$ *for some* $1 \leq j \leq n-1$ *and* $s_i \neq 0$ *for all* $1 \leq i \leq n-1$, *then the map* $(D_n X)_{\underline{k}} \to (D_n X)_{\underline{s}} = X_{\underline{s}}$ *factors as*

$$(D_n X)_{\underline{k}} \to X_{\underline{k}} \to X_{\underline{s}}.$$

Proof By induction on n. When $n = 2$, $D_2X = R_0X$ has $(D_2X)_k = X_k$ for all $k \neq 0$, proving a). By construction of R_0X, if $s > 0$, the map

$$(D_2X)_0 = X_0^d \to (D_2X)_s = X_s$$

factors as

$$X_0^d \to X_0 \to X_s \ ,$$

proving b).

Suppose, inductively, that the lemma holds for $(n-1)$.

a) Let $\underline{r} = (k_2, \ldots, k_{n-1})$. Then by the inductive hypothesis applied to X_{k_1} we have

$$(D_nX)_{\underline{k}} = (D_{n-1}X_{k_1})_{\underline{r}} = (X_{k_1})_{\underline{r}} = X_{\underline{k}} \ .$$

b) By the construction of R_0X, for each $s_1 > 0$ the map

$$(R_0X)_0 = X_0^d \to (R_0X)_{s_1} = X_{s_1}$$

factors as

$$X_0^d \to X_0 \to X_{s_1}.$$

Thus applying D_{n-1} we see that the map

$$(D_nX)_0 = X_0^d \to (D_nX)_{s_1} = D_{n-1}X_{s_1}$$

factors as

$$X_0^d \to D_{n-1}X_0 \to D_{n-1}X_{s_1} \ . \tag{12.6}$$

Let $\underline{k} \to \underline{s}$ be a morphism in $\Delta^{{n-1}^{op}}$ satisfying hypotheses b), and let $\underline{k} = (k_1, \underline{r})$, $\underline{s} = (s_1, \underline{v})$, so we have a corresponding morphism $\underline{r} \to \underline{v}$ in $\Delta^{{n-2}^{op}}$. We distinguish two cases:

i) Consider the case $k_1 = 0$. By naturality, (12.6) gives a commuting diagram in Cat

$$\begin{array}{ccccc} (D_nX)_{\underline{k}} = X_0^d & \longrightarrow & (D_{n-1}X_0)_{\underline{v}} & \longrightarrow & (D_{n-1}X_{s_1})_{\underline{v}} = (D_nX)_{\underline{s}} \\ & \searrow & \uparrow & & \uparrow \\ & & (D_{n-1}X_0)_{\underline{r}} & \longrightarrow & (D_{n-1}X_{s_1})_{\underline{r}} \end{array} \tag{12.7}$$

where we used the fact that, since X_0^d is discrete, $(X_0^d)_{\underline{v}} = (X_0^d)_{\underline{s}} = X_0^d$, and since $k_1 = 0$, $(D_n X)_{\underline{k}} = X_0^d$.

Suppose that $r_i \neq 0$ for all $1 \leq i \leq (n-2)$. Then by part a) $(D_{n-1} X_0)_{\underline{r}} = X_{0\underline{r}} = X_{\underline{k}}$. So by diagram (12.7) we see that the map $(D_n X)_{\underline{k}} \to (D_n X)_{\underline{s}}$ factors through $X_{\underline{k}}$.

Suppose $r_i = 0$ for some $1 \leq i \leq (n-2)$. Then, since by hypothesis $v_j \neq 0$ for all $1 \leq j \leq (n-2)$, we can apply the inductive hypothesis to X_0 and deduce that the map

$$(D_{n-1} X_0)_{\underline{r}} \to (D_{n-1} X_0)_{\underline{v}}$$

factors through $X_{0\underline{r}} = X_{\underline{k}}$. From the commuting diagram (12.7) we conclude that the map $(D_n X)_{\underline{k}} \to (D_n X)_{\underline{s}}$ factors through $X_{\underline{k}}$.

ii) Consider the case $k_1 > 0$. By hypothesis $r_j = 0$ for some $1 \leq j \leq (n-2)$ and $(D_n X)_{\underline{k}} = D_{n-1} X_{k_1}$. The morphism $(D_n X)_{\underline{k}} \to (D_n X)_{\underline{s}}$ factors as

$$(D_n X)_{\underline{k}} = (D_n X)_{(k_1, \underline{r})} \to (D_n X)_{(k_1, \underline{v})} \to (D_n X)_{(s_1, \underline{v})} = (D_n X)_{\underline{s}}. \tag{12.8}$$

But $(D_n X)_{(k_1, \underline{r})} = (D_{n-1} X_{k_1})_{\underline{r}}$ and $(D_n X)_{(k_1, \underline{v})} = (D_{n-1} X_{k_1})_{\underline{v}}$. By the induction hypothesis applied to X_{k_1} the map

$$(D_{n-1} X_{k_1})_{\underline{r}} \to (D_{n-1} X_{k_1})_{\underline{v}}$$

factors through $X_{k_1 \underline{r}} = X_{\underline{k}}$. Thus by (12.8) we see that the map $(D_n X)_{\underline{k}} \to (D_n X)_{\underline{s}}$ factors through $X_{\underline{k}}$. □

Proposition 12.2.3 *Let $X \in \mathsf{FCat}^n_{\mathsf{wg}}$, then $Q_n D_n X = Q_n X$.*

Proof By induction on n. Let $X \in \mathsf{FCat}^2_{\mathsf{wg}}$. It is immediate that $R_0 X$ and X satisfy the hypotheses of Lemma 10.1.4, so that

$$Tr_2 R_0 X = Tr_2 X .$$

Hence

$$Q_2 D_2 X = St\, Tr_2 R_0 X = St\, Tr_2 X = Q_2 X .$$

Suppose, inductively, the statement holds for $(n-1)$ and let $X \in \mathsf{FCat}^n_{\mathsf{wg}}$.

The strategy of the proof is to show that $P_n D_n X$ and $P_n X$ satisfy the hypotheses of Lemma 10.1.4, where $P_n : \mathsf{Ta}^n_{\mathsf{wg}} \to \mathsf{LTa}^n_{\mathsf{wg}}$ is as in Theorem 10.2.1.

We first need a number of preliminary observations. Note that since $X \in \mathsf{FCat}^n_{\mathsf{wg}}$, in particular $X \in \mathsf{Cat}^n_{\mathsf{wg}}$, so by Remark 10.2.4 $P_n X \in \mathsf{Cat}^n_{\mathsf{wg}}$. We now show that $(P_n D_n X)_{\underline{k}}$ is discrete for all $\underline{k} \in \Delta^{n-1^{op}}$ such that $k_j = 0$ for some $1 \leq j \leq (n-1)$.

By definition of P_n there is a pullback in $[\Delta^{n-1^{op}}, \mathsf{Cat}]$

$$\begin{array}{ccc} P_n D_n X & \longrightarrow & D_n X \\ {\scriptstyle z}\downarrow & & \downarrow{\scriptstyle \gamma^{(n-1)}} \\ Q_{n-1} q^{(n-1)} D_n X & \longrightarrow & q^{(n-1)} D_n X \end{array} \tag{12.9}$$

On the other hand, by Proposition 12.1.4 and the inductive hypothesis

$$Q_{n-1} q^{(n-1)} D_n X = Q_{n-1} D_{n-1} q^{(n-1)} X = Q_{n-1} q^{(n-1)} X$$

so that (12.9) coincides with

$$\begin{array}{ccc} P_n D_n X & \longrightarrow & D_n X \\ {\scriptstyle z}\downarrow & & \downarrow{\scriptstyle \gamma^{(n-1)}} \\ Q_{n-1} q^{(n-1)} X & \longrightarrow & q^{(n-1)} D_n X \end{array} \tag{12.10}$$

Since pullbacks in $[\Delta^{n-1^{op}}, \mathsf{Cat}]$ are computed pointwise, for each $\underline{k} \in \Delta^{n-1^{op}}$ the diagram (12.10) gives rise to a pullback in Cat

$$\begin{array}{ccc} (P_n D_n X)_{\underline{k}} & \longrightarrow & (D_n X)_{\underline{k}} \\ {\scriptstyle z_{\underline{k}}}\downarrow & & \downarrow{\scriptstyle \gamma^{(0)}} \\ (Q_{n-1} q^{(n-1)} X)_{\underline{k}} & \longrightarrow & q(D_n X)_{\underline{k}} \end{array} \tag{12.11}$$

If $k_j = 0$ for some $1 \leq j \leq (n-1)$, then $(D_n X)_{\underline{k}}$ is discrete (since $D_n X \in \mathsf{Ta^n}$) hence the right vertical map in (12.11) is the identity, and thus so is the left vertical map in (12.11). That is

$$(P_n D_n X)_{\underline{k}} = (Q_{n-1} q^{(n-1)} X)_{\underline{k}}$$

and

$$p(P_n D_n X)_{\underline{k}} = (Q_{n-1} q^{(n-1)} X)_{\underline{k}}\ . \tag{12.12}$$

We also have a pullback in $[\Delta^{n-1^{op}}, \mathsf{Cat}]$

$$\begin{array}{ccc} P_n X & \longrightarrow & X \\ {\scriptstyle t}\downarrow & & \downarrow{\scriptstyle \gamma^{(n-1)}} \\ Q_{n-1}q^{(n-1)}X & \longrightarrow & q^{(n-1)}X \end{array} \tag{12.13}$$

and for each $\underline{k} \in \Delta^{n-1^{op}}$ a pullback in Cat

$$\begin{array}{ccc} (P_n X)_{\underline{k}} & \longrightarrow & X_{\underline{k}} \\ {\scriptstyle t_{\underline{k}}}\downarrow & & \downarrow{\scriptstyle \gamma^{(0)}} \\ (Q_{n-1}q^{(n-1)}X)_{\underline{k}} & \longrightarrow & qX_{\underline{k}} \end{array} \tag{12.14}$$

We now check hypotheses i) through iv) of Lemma 10.1.4 for $P_n X \in \mathsf{Cat}^{\mathsf{n}}_{\mathsf{wg}}$ and $P_n D_n X \in \mathsf{LTa}^{\mathsf{n}}_{\mathsf{wg}}$.

i) Let $\underline{k} \in \Delta^{n-1^{op}}$ and $\underline{s} \in \Delta^{n-1^{op}}$ be such that $k_j \neq 0$ and $s_j \neq 0$ for all $1 \leq j \leq (n-1)$. Then by Lemma 12.2.2 $(D_n X)_{\underline{k}} = X_{\underline{k}}$. Hence the right vertical maps in (12.11) and (12.14) coincide. It follows that $z_{\underline{k}} = t_{\underline{k}}$ and

$$(P_n D_n X)_{\underline{k}} = (P_n X)_{\underline{k}}\ .$$

Similarly $(P_n D_n X)_{\underline{s}} = (P_n X)_{\underline{s}}$. Given a morphism $\underline{k} \to \underline{s}$ in $\Delta^{n-1^{op}}$, clearly the maps

$$(P_n X)_{\underline{k}} \to (P_n X)_{\underline{s}}, \qquad (P_n D_n X)_{\underline{k}} \to (P_n D_n X)_{\underline{s}}$$

coincide.

ii) Let $\underline{k} \in \Delta^{n-1^{op}}$ and $\underline{s} \in \Delta^{n-1^{op}}$ be such that $k_j = 0$ for some $1 \leq j \leq (n-1)$ and $s_i = 0$ for some $1 \leq i \leq (n-1)$. Then $X_{\underline{k}} \in \mathsf{Cat}_{\mathsf{hd}}$ so $qX_{\underline{k}} = pX_{\underline{k}}$. Thus from (12.14), using the fact that p commutes with pullbacks over discrete objects, we obtain

$$p(P_n X)_{\underline{k}} = (Q_{n-1}q^{(n-1)}X)_{\underline{k}}\ . \tag{12.15}$$

It follows from (12.12) and (12.15) that

$$(P_n D_n X)^d_{\underline{k}} = p(P_n D_n X)_{\underline{k}} = p(P_n X)_{\underline{k}} = (P_n X)^d_{\underline{k}}\ .$$

Similarly $(P_n D_n X)^d_{\underline{s}} = (P_n X)^d_{\underline{s}}$. Given a morphism $\underline{k} \to \underline{s}$ in $\Delta^{n-1^{op}}$, the maps

$$(P_n X)^d_{\underline{k}} \to (P_n X)^d_{\underline{s}}, \qquad (P_n D_n X)_{\underline{k}} \to (P_n D_n X)_{\underline{s}}$$

coincide, and they are equal to the maps

$$(Q_{n-1}q^{(n-1)}X)_{\underline{k}} \to (Q_{n-1}q^{(n-1)}X)_{\underline{s}}.$$

iii) Let $\underline{k} \to \underline{s}$ be a morphism in $\Delta^{n-1^{op}}$ and suppose that $k_j \neq 0$ for all $1 \leq j \leq n-1$ and $s_i = 0$ for some $1 \leq i \leq n-1$. From above, $z_{\underline{s}} = \mathrm{Id}$ while $t_{\underline{s}} : (P_n X)_{\underline{s}} \to (P_n X)^d_{\underline{s}}$ is the discretization map. By construction we have commuting diagrams

$$\begin{array}{ccc} (P_nD_nX)_{\underline{k}} & \longrightarrow & (P_nD_nX)_{\underline{s}} \\ {\scriptstyle z_{\underline{k}}}\downarrow & & \downarrow{\scriptstyle z_{\underline{s}}=\mathrm{Id}} \\ (Q_{n-1}q^{(n-1)}X)_{\underline{k}} & \to & (Q_{n-1}q^{(n-1)}X)_{\underline{s}} \end{array} \qquad \begin{array}{ccc} (P_nX)_{\underline{k}} & \longrightarrow & (P_nX)_{\underline{s}} \\ {\scriptstyle t_{\underline{k}}}\downarrow & & \downarrow{\scriptstyle t_{\underline{s}}} \\ (Q_{n-1}q^{(n-1)}X)_{\underline{k}} & \to & (Q_{n-1}q^{(n-1)}X)_{\underline{s}} \end{array} \tag{12.16}$$

and, from above, $(P_n D_n X)_{\underline{k}} = (P_n X)_{\underline{k}}$, $z_{\underline{k}} = t_{\underline{k}}$ while

$$(P_n D_n X)_{\underline{s}} = (Q_{n-1}q^{(n-1)}X)_{\underline{s}} = (P_n X)^d_{\underline{s}}.$$

We therefore see from (12.16) that the map $(P_n D_n X)_{\underline{k}} \to (P_n D_n X)_{\underline{s}}$ factors as

$$(P_n D_n X)_{\underline{k}} = (P_n X)_{\underline{k}} \xrightarrow{z_{\underline{k}}=t_{\underline{k}}} (Q_{n-1}q^{(n-1)}X)_{\underline{k}} \to (Q_{n-1}q^{(n-1)}X)_{\underline{s}}\,,$$

which is the same as

$$(P_n D_n X)_{\underline{k}} = (P_n X)_{\underline{k}} \to (P_n X)_{\underline{s}} \xrightarrow{t_{\underline{s}}} (P_n X)^d_{\underline{s}} = (Q_{n-1}q^{(n-1)}X)_{\underline{s}}\,.$$

This proves hypothesis iii) in Lemma 10.1.4 for $P_n X$ and $P_n D_n X$.

iv) Let $\underline{k} \to \underline{s}$ be a morphism in $\Delta^{n-1^{op}}$ and suppose that $k_j = 0$ for some $1 \leq j \leq n-1$ and $s_i \neq 0$ for all $1 \leq i \leq n-1$. Since, by Lemma 12.2.2 b), the map $(D_n X)_{\underline{k}} \to (D_n X)_{\underline{s}} = X_{\underline{s}}$ factors as

$$(D_n X)_{\underline{k}} \to X_{\underline{k}} \to X_{\underline{s}}\,,$$

by the definitions it follows that the map of pullbacks

$$
\begin{array}{c}
(P_n X)^d_{\underline{k}} = (P_n D_n X)_{\underline{k}} = (Q_{n-1} q^{(n-1)} X)_{\underline{k}} \times_{q(D_n X)_{\underline{k}}} (D_n X)_{\underline{k}} \\
\downarrow \\
(P_n D_n X)_{\underline{s}} = (P_n X)_{\underline{s}} = (Q_{n-1} q^{(n-1)} X)_{\underline{s}} \times_{q X_{\underline{s}}} X_{\underline{s}}
\end{array}
$$

factors through

$$(P_n X)_{\underline{k}} = (Q_{n-1} q^{(n-1)} X)_{\underline{k}} \times_{q X_{\underline{k}}} X_{\underline{k}}.$$

Thus all the hypotheses of Lemma 10.1.4 are satisfied for $P_n X \in \mathsf{Cat}^{\mathsf{n}}_{\mathsf{wg}}$ and $P_n D_n X \in \mathsf{LTa}^{\mathsf{n}}_{\mathsf{wg}}$ and we conclude that

$$Tr_n P_n D_n X = Tr_n P_n X \ ,$$

which implies

$$Q_n D_n X = St\, Tr_n P_n D_n X = St\, Tr_n P_n X = Q_n X \ .$$

□

Lemma 12.2.4 *Let $f : Z \to X$ be a map in $\mathsf{Ta}^{\mathsf{n}}_{\mathsf{wg}}$ with $Z \in \mathsf{FCat}^{\mathsf{n}}_{\mathsf{wg}}$ and $X \in \mathsf{Ta}^{\mathsf{n}}$. Then*

a) There is a map in Ta^{n} $g : D_n Z \to X$, natural in $Z \to X$.
b) If f is an n-equivalence, then so is $g : D_n Z \to X$.

Proof Denote by ∂_i, σ_i the face and degeneracy operators of Z, and ∂'_i, σ'_i those of X. Let $\gamma' : Z^d_0 \to Z_0$ be the functorial section to the discretization map $\gamma : Z_0 \to Z^d_0$. Let $f^d_0 : Z^d_0 \to X^d_0 = X_0$. Then

$$
\begin{aligned}
f^d_0 \gamma &= \mathrm{Id}\, f_0 = f_0 \ , \\
f^d_0 &= f^d_0 \gamma\gamma' = f_0 \gamma' \ .
\end{aligned}
\tag{12.17}
$$

This implies

$$
\begin{aligned}
f^d_0 (\gamma \partial_i) &= f_0 \partial_i = \partial'_i f_1 \ , \\
f_1 (\sigma_0 \gamma') &= \sigma'_0 f_0 \gamma' = \sigma'_0 f^d_0 \ .
\end{aligned}
\tag{12.18}
$$

We prove the lemma by induction on n. When $n = 2$, $D_2 = R_0$ and we define

$$g_k : (R_0 Z)_k \to X_k$$

to be f^d_0 when $k = 0$ and f_k when $k > 0$. From (12.18), this is a simplicial map $g : R_0 Z \to X$.

Suppose $f : Z \to X$ is a 2-equivalence. Then, for each $a, b \in Z_0^d$, there are equivalences of categories

$$(R_0Z)(a, b) = Z(a, b) \simeq X(fa, fb) ,$$
$$p^{(1)}R_0Z = p^{(1)}Z \simeq p^{(1)}X ,$$

so that g is also a 2-equivalence. Suppose, inductively, that the lemma holds for $n - 1$.

a) By (12.18) there is a map

$$h : R_0Z \to X$$

given by $h_0 = f_0^d$, $h_k = f_k$ when $k > 0$. By the induction hypothesis, we have maps for each $k > 0$

$$(D_nZ)_k = (\overline{D_{n-1}}R_0Z)_k = D_{n-1}Z_k \xrightarrow{v_k} X_k ,$$
$$(D_nZ)_0 = Z_0^d = (\overline{D_{n-1}}R_0Z)_0 \xrightarrow{f_0^d} X_0 .$$

Therefore, since $h : R_0Z \to X$ is a simplicial map and v_k is natural in $Z_k \to X_k$, we obtain a map

$$g : D_nZ \to X$$

given by $g_k = v_k$ for $k > 0$, $g_0 = f_0^d$.

b) If f is an n-equivalence, for all $a, b \in Z_0^d$, $Z(a, b) \to X(fa, fb)$ is an $(n-1)$-equivalence, thus by the inductive hypothesis and by Proposition 12.1.4 c), so is

$$(D_nZ)(a, b) = D_{n-1}Z(a, b) \to X(fa, fb) .$$

Since f is an n-equivalence, $p^{(n-1)}Z \to p^{(n-1)}X$ is an $(n-1)$-equivalence, so using Proposition 12.1.4 b) and the inductive hypothesis we obtain an $(n-1)$-equivalence

$$p^{(n-1)}D_nZ = D_{n-1}p^{(n-1)}Z \to p^{(n-1)}X .$$

By definition we conclude that $D_nZ \to X$ is an n-equivalence. □

We now establish the main properties of the discretization functor $Disc_n$: $\mathsf{Cat}^n_{\mathsf{wg}} \to \mathsf{Ta}^n$ of Definition 12.2.1. The proof of this result relies on the properties of D_n established in this chapter, as well as on the properties of the functor G_n studied in Chap. 11 and of the functor Q_n studied in Chap. 10.

Theorem 12.2.5 *Let $Disc_n : \mathsf{Cat}^{\mathsf{n}}_{\mathsf{wg}} \to \mathsf{Ta}^{\mathsf{n}}$ be as in Definition 12.2.1. Then*

a) $Disc_n$ is the identity on discrete objects and commutes with pullbacks over discrete objects.
b) For each $X \in \mathsf{Cat}^{\mathsf{n}}_{\mathsf{wg}}$ there is a zig-zag of n-equivalences in $\mathsf{Ta}^{\mathsf{n}}_{\mathsf{wg}}$ between X and $Disc_n X$.
c) $Disc_n$ preserves n-equivalences.

Proof

a) This follows from the fact that the same is true for G_n and D_n (see Theorem 11.3.6 and Proposition 12.1.4).
b) Let $X \in \mathsf{Cat}^{\mathsf{n}}_{\mathsf{wg}}$, then by Proposition 12.2.3

$$Q_n Disc_n X = Q_n D_n G_n X = Q_n G_n X \ .$$

Hence by Theorem 10.2.1 there are n-equivalences in $\mathsf{Ta}^{\mathsf{n}}_{\mathsf{wg}}$

$$Disc_n X \xleftarrow{s_n(Disc_n X)} Q_n Disc_n X = Q_n G_n X \xrightarrow{s_n(G_n X)} G_n X.$$

On the other hand by Theorem 11.3.6 there is an n-equivalence

$$G_n X \xrightarrow{g_n(X)} X.$$

So by composition we obtain a zig-zag of n-equivalences

$$Disc_n X \xleftarrow{s_n(Disc_n X)} Q_n Disc_n X \xrightarrow{g_n(X) s_n(G_n X)} X$$

as required.
c) This follows from the fact that, by Theorem 11.3.6 and Proposition 12.1.4, the same is true for G_n and D_n. □

We now prove our main comparison result between weakly globular n-fold categories and Tamsamani n-categories.

Theorem 12.2.6 *The functors*

$$Q_n : \mathsf{Ta}^{\mathsf{n}} \leftrightarrows \mathsf{Cat}^{\mathsf{n}}_{\mathsf{wg}} : Disc_n$$

induce an equivalence of categories after localization with respect to the n-equivalences

$$\mathsf{Ta}^{\mathsf{n}}/\!\sim^n \ \simeq \ \mathsf{Cat}^{\mathsf{n}}_{\mathsf{wg}}/\!\sim^n \ .$$

Proof Let $X \in \mathsf{Cat}^n_{wg}$. By Proposition 12.2.3, Theorem 10.2.1 and Theorem 11.3.6 there are n-equivalences

$$Q_n Disc_n X = Q_n D_n G_n X = Q_n G_n X \to G_n X \to X\ .$$

So there is an n-equivalence in Cat^n_{wg}

$$\beta_X : Q_n Disc_n X \to X\ .$$

It follows that $Q_n Disc_n X \cong X$ in $\mathsf{Cat}^n_{wg}/{\sim^n}$.

Let $Y \in \mathsf{Ta}^n$. By Theorem 10.2.1 and the above there are n-equivalences in Ta^n_{wg}

$$Disc_n Q_n Y \xleftarrow{s_n(Disc_n Q_n Y)} Q_n Disc_n Q_n Y \xrightarrow{\beta_{Q_n Y}} Q_n Y \xrightarrow{s_n(Y)} Y\ .$$

Composing this with the n-equivalence

$$Z = G_n Q_n Disc_n Q_n Y \xrightarrow{g_n(Q_n Disc_n Q_n Y)} Q_n Disc_n Q_n Y$$

we obtain n-equivalences in Ta^n_{wg}

$$Disc_n Q_n Y \overset{a}{\leftarrow} Z \overset{b}{\rightarrow} Y\ , \tag{12.19}$$

where

$$a = s_n(Disc_n Q_n Y) g_n(Q_n Disc_n Q_n Y), \quad b = s_n(Y)\beta_{Q_n Y} g_n(Q_n Disc_n Q_n Y).$$

Since $Z \in \mathsf{FCat}^n_{wg}$, $Disc_n Q_n Y \in \mathsf{Ta}^n$ and $Y \in \mathsf{Ta}^n$, applying Lemma 12.2.4 to a and b in (12.19) we obtain a zig-zag of n-equivalences in Ta^n

$$Disc_n Q_n Y \leftarrow D_n Z \to Y\ .$$

It follows that $Disc_n Q_n Y \cong Y$ in $\mathsf{Ta}^n/{\sim^n}$. □

Remark 12.2.7 The proof of Theorem 12.2.6 also implies that the equivalence of categories $\mathsf{Ta}^n/\sim^n \simeq \mathsf{Cat}^n_{wg}/\sim^n$ extends to an equivalence of $(\infty, 1)$-categories between the Dwyer–Kan localizations on both sides.

To see this, we recall the model of $(\infty, 1)$-categories given by Barwick–Kan consisting of relative categories [9]. A relative category [9, §3.1] is a category $\mathscr{C}$ with a subcategory weC whose maps are called weak equivalences, such that weC contains all the objects of $\mathscr{C}$. Thus Ta^n and Cat^n_{wg} with the n-equivalences are relative categories and the functors

$$Q_n : (\mathsf{Ta}^n, \sim^n) \leftrightarrows (Disc_n, \sim^n)$$

are relative functors in the sense of [9, §3.1] since Q_n and $Disc_n$ preserve n-equivalences. From the proof of Theorem 12.2.6 we can show that Q_n and $Disc_n$ are homotopy equivalences of relative categories, in the following sense.

Recall from [9, §3.3] that two maps $f, g : \mathscr{C} \to \mathscr{D}$ of relative categories are strictly homotopic if there is a relative functor

$$h : \mathscr{C} \times [1]_{max} \to \mathscr{D}$$

(where $[1]_{max}$ is the arrow category $0 \to 1$ with all maps weak equivalences) such that for all $X \in \mathscr{C}$

$$h(X, 0) = fX \qquad h(X, 1) = gX\ .$$

More generally, two maps $f, g : \mathscr{C} \to \mathscr{D}$ of relative categories are called homotopic if they can be connected by a finite zig-zag of strict homotopies. A map $f : \mathscr{C} \to \mathscr{D}$ of relative categories is called a homotopy equivalence if there exists a map $f' : \mathscr{D} \to \mathscr{C}$ such that $f'f$ and ff' are homotopic to the identity maps of $\mathscr{C}$ and $\mathscr{D}$ respectively.

Let $Y \in \mathsf{Ta^n}$. From the proof of Theorem 12.2.6 there is a zig-zag of natural n-equivalences in $\mathsf{Ta^n}$

$$Disc_n Q_n Y \to D_n G_n Q_n Disc_n Q_n Y \leftarrow Y\ .$$

This means that there is a zig-zag of strict homotopies between the maps $\mathsf{Ta^n} \to \mathsf{Ta^n}$:

$$Disc_n Q_n, \qquad D_n G_n Q_n Disc_n Q_n, \qquad \mathrm{Id}_{\mathsf{Ta^n}}\ .$$

That is, $Disc_n Q_n$ and $\mathrm{Id}_{\mathsf{Ta^n}}$ are homotopic.

Let $X \in \mathsf{Cat^n_{wg}}$; by the proof of Theorem 12.2.6, there is a natural n-equivalence $Q_n Disc_n X \to X$, which means that $Q_n Disc_n$ and $\mathrm{Id}_{\mathsf{Cat^n_{wg}}}$ are (strictly) homotopic.

By definition, we conclude that Q_n and $Disc_n$ are homotopy equivalences of relative categories. From [9, Theorem 6.1 and Proposition 7.5] it follows that Q_n and $Disc_n$ are weak equivalences in the Quillen model structure on relative categories, which is Quillen equivalent to complete Segal spaces ([9] and [26, Corollary 8.3.15]). Further, in [8] it is shown that the weak equivalences in this model category of relative categories are exactly the maps between relative categories which induce a weak equivalence between their simplicial localizations.

So we conclude that the Dwyer–Kan simplicial localizations of $\mathsf{Ta^n}$ and $\mathsf{Cat^n_{wg}}$ are equivalent as $(\infty, 1)$-categories.

Remark 12.2.8 From Corollary 10.2.3 and Theorem 12.2.6 we have equivalences of categories

$$\mathsf{Ta^n_{wg}}/\sim^n \simeq \mathsf{Cat^n_{wg}}/\sim^n \simeq \mathsf{Ta^n}/\sim^n\ . \tag{12.20}$$

This means all three Segal-type models are equivalent after localization by the n-equivalences. Since both Ta^{n} and $\mathsf{Cat}^{\mathsf{n}}_{\mathsf{wg}}$ are embedded in $\mathsf{Ta}^{\mathsf{n}}_{\mathsf{wg}}$ this can be interpreted as a kind of partial strictification result for the larger model $\mathsf{Ta}^{\mathsf{n}}_{\mathsf{wg}}$. Namely, in $\mathsf{Ta}^{\mathsf{n}}_{\mathsf{wg}}$ the weakening occurs in two ways: with the weakening of the Segal maps and with the weak globularity condition. The equivalences of categories (12.20) shows that only one of these two weakenings is necessary to obtain a model of weak n-categories: the weak globularity condition only, giving rise to the model $\mathsf{Cat}^{\mathsf{n}}_{\mathsf{wg}}$ or the Segal maps condition only, giving rise to the model Ta^{n}.

12.3 Groupoidal Weakly Globular n-Fold Categories

In this section we introduce the subcategory $\mathsf{GCat}^{\mathsf{n}}_{\mathsf{wg}} \subset \mathsf{Cat}^{\mathsf{n}}_{\mathsf{wg}}$ of groupoidal weakly globular n-fold categories and we show that it is an algebraic model of n-types. This means that weakly globular n-fold categories satisfy the homotopy hypothesis.

Definition 12.3.1 The full subcategory $\mathsf{GTa}^{\mathsf{n}}_{\mathsf{wg}} \subset \mathsf{Ta}^{\mathsf{n}}_{\mathsf{wg}}$ of groupoidal weakly globular Tamsamani n-categories is defined inductively as follows.

For $n = 1$, $\mathsf{GTa}^{1}_{\mathsf{wg}} = \mathsf{Gpd}$. Note that $\mathsf{Cat}_{\mathsf{hd}} \subset \mathsf{GTa}^{1}_{\mathsf{wg}}$. Suppose inductively we have defined $\mathsf{GTa}^{\mathsf{n-1}}_{\mathsf{wg}} \subset \mathsf{Ta}^{\mathsf{n-1}}_{\mathsf{wg}}$. We define $X \in \mathsf{GTa}^{\mathsf{n}}_{\mathsf{wg}} \subset \mathsf{Ta}^{\mathsf{n}}_{\mathsf{wg}}$ such that

i) $X_k \in \mathsf{GTa}^{\mathsf{n-1}}_{\mathsf{wg}}$ for all $k \geq 0$.
ii) $p^{(n-1)}X \in \mathsf{GTa}^{\mathsf{n-1}}_{\mathsf{wg}}$.

Remark 12.3.2 From the definitions, it is immediate that

$$\mathsf{Cat}^{\mathsf{n}}_{\mathsf{hd}} \subset \mathsf{GTa}^{\mathsf{n}}_{\mathsf{wg}} .$$

In fact, this holds for $n = 1$ since $\mathsf{Cat}^{1}_{\mathsf{hd}} \subset \mathsf{Gpd} = \mathsf{GTa}^{1}_{\mathsf{wg}}$. If, inductively, it holds for $n - 1$ and $X \in \mathsf{Cat}^{\mathsf{n}}_{\mathsf{hd}}$ then $X_k \in \mathsf{Cat}^{\mathsf{n-1}}_{\mathsf{hd}}$ for each $k \geq 0$ and $p^{(n-1)}X \in \mathsf{Cat}^{\mathsf{n-1}}_{\mathsf{hd}}$. Thus, by induction, $X_k \in \mathsf{GTa}^{\mathsf{n-1}}_{\mathsf{wg}}$ and $p^{(n-1)}X \in \mathsf{GTa}^{\mathsf{n-1}}_{\mathsf{wg}}$, so by definition $X \in \mathsf{GTa}^{\mathsf{n}}_{\mathsf{wg}}$.

Remark 12.3.3 If $X \in \mathsf{GTa}^{\mathsf{n}}_{\mathsf{wg}}$, then by definition $X_1 \in \mathsf{GTa}^{\mathsf{n-1}}_{\mathsf{wg}}$, thus for each $a, b \in X^d_0$, $X(a, b) \in \mathsf{GTa}^{\mathsf{n-1}}_{\mathsf{wg}}$. It is easily checked that $\mathsf{GTa}^{\mathsf{n}}_{\mathsf{wg}}$ has the same closure properties of $\mathsf{Ta}^{\mathsf{n}}_{\mathsf{wg}}$ (see Definition 6.1.8).

Lemma 12.3.4 *Let $f : X \to Y$ be an n-equivalence in* $\mathsf{Ta}^{\mathsf{n}}_{\mathsf{wg}}$

i) If $Y \in \mathsf{GTa}^{\mathsf{n}}_{\mathsf{wg}}$ then $X \in \mathsf{GTa}^{\mathsf{n}}_{\mathsf{wg}}$.
ii) If $X \in \mathsf{GTa}^{\mathsf{n}}_{\mathsf{wg}}$ then $Y \in \mathsf{GTa}^{\mathsf{n}}_{\mathsf{wg}}$.

Proof By induction on n. The case $n = 1$ holds since a category equivalent to a groupoid is itself a groupoid. Suppose, inductively, that the lemma holds for $n-1$ and let $f : X \to Y$ be an n-equivalence.

i) For each $a, b \in X_0^d$ the map

$$f(a,b) : X(a,b) \to Y(fa, fb)$$

is an $(n-1)$-equivalence in $\mathsf{Ta}_{\mathsf{wg}}^{\mathsf{n-1}}$ with $Y(fa, fb) \in \mathsf{GTa}_{\mathsf{wg}}^{\mathsf{n-1}}$ (see Remark 12.3.3). So by the induction hypothesis $X(a,b) \in \mathsf{GTa}_{\mathsf{wg}}^{\mathsf{n-1}}$. Since

$$X_1 = \coprod_{a,b \in X_0^d} X(a,b)$$

it follows that $X_1 \in \mathsf{GTa}_{\mathsf{wg}}^{\mathsf{n-1}}$. By the closure properties of $\mathsf{GTa}_{\mathsf{wg}}^{\mathsf{n-1}}$ (see Remark 12.3.3), since $X_1 \in \mathsf{GTa}_{\mathsf{wg}}^{\mathsf{n-1}}$ and X_0^d is discrete,

$$X_1 \times_{X_0^d} \overset{k}{\cdots} \times_{X_0^d} X_1 \in \mathsf{GTa}_{\mathsf{wg}}^{\mathsf{n-1}}$$

for all $k \geq 2$. On the other hand, there is an $(n-1)$-equivalence in $\mathsf{Ta}_{\mathsf{wg}}^{\mathsf{n-1}}$

$$X_k \to X_1 \times_{X_0^d} \overset{k}{\cdots} \times_{X_0^d} X_1 .$$

Thus by the inductive hypothesis we conclude that $X_k \in \mathsf{GTa}_{\mathsf{wg}}^{\mathsf{n-1}}$ for all $k \geq 0$.

By definition there is an $(n-1)$-equivalence

$$p^{(n-1)} f : p^{(n-1)} X \to p^{(n-1)} Y$$

with $p^{(n-1)} Y \in \mathsf{GTa}_{\mathsf{wg}}^{\mathsf{n-1}}$ since by hypothesis $Y \in \mathsf{GTa}_{\mathsf{wg}}^{\mathsf{n}}$. Hence by the inductive hypothesis $p^{(n-1)} X \in \mathsf{GTa}_{\mathsf{wg}}^{\mathsf{n-1}}$. We conclude that $X \in \mathsf{GTa}_{\mathsf{wg}}^{\mathsf{n-1}}$.

ii) The proof is completely similar to that of i). □

Remark 12.3.5 It follows immediately from the definition of $\mathsf{GTa}_{\mathsf{wg}}^{\mathsf{n}}$ that the embedding

$$\mathsf{Ta}_{\mathsf{wg}}^{\mathsf{n}} \hookrightarrow [\Delta^{n-1^{op}}, \mathsf{Cat}]$$

restricts to the embedding

$$\mathsf{GTa}_{\mathsf{wg}}^{\mathsf{n}} \hookrightarrow [\Delta^{n-1^{op}}, \mathsf{Gpd}] .$$

Since $p = q : \mathsf{Gpd} \to \mathsf{Set}$ it follows that for each $X \in \mathsf{GTa}^{\mathsf{n}}_{\mathsf{wg}}$ there is a morphism, natural in X,

$$X \to p^{(n-1)}X\ .$$

Definition 12.3.6 The category $\mathsf{GCat}^{\mathsf{n}}_{\mathsf{wg}} \subset \mathsf{Cat}^{\mathsf{n}}_{\mathsf{wg}}$ of groupoidal weakly globular n-fold categories is the full subcategory of $\mathsf{Cat}^{\mathsf{n}}_{\mathsf{wg}}$ whose objects X are in $\mathsf{GTa}^{\mathsf{n}}_{\mathsf{wg}}$.

The category $\mathsf{GTa}^{\mathsf{n}} \subset \mathsf{Ta}^{\mathsf{n}}$ of groupoidal Tamsamani n-categories is the full subcategory of Ta^{n} whose objects X are in $\mathsf{GTa}^{\mathsf{n}}_{\mathsf{wg}}$.

Remark 12.3.7 Since $\mathsf{Cat}^{\mathsf{n}}_{\mathsf{hd}} \subset \mathsf{Cat}^{\mathsf{n}}_{\mathsf{wg}}$ and by Remark 12.3.2 $\mathsf{Cat}^{\mathsf{n}}_{\mathsf{hd}} \subset \mathsf{GTa}^{\mathsf{n}}_{\mathsf{wg}}$, it follows from Definition 12.3.6 that $\mathsf{Cat}^{\mathsf{n}}_{\mathsf{hd}} \subset \mathsf{GCat}^{\mathsf{n}}_{\mathsf{wg}}$.

Remark 12.3.8 The following facts are immediate from the definitions:

a) $X \in \mathsf{GCat}^{\mathsf{n}}_{\mathsf{wg}}$ (resp. $X \in \mathsf{GTa}^{\mathsf{n}}$) if and only if for each $k \geq 0$, $X_k \in \mathsf{GCat}^{\mathsf{n-1}}_{\mathsf{wg}}$ (resp. $X_k \in \mathsf{GTa}^{\mathsf{n-1}}$) and $p^{(n-1)}X \in \mathsf{GCat}^{\mathsf{n-1}}_{\mathsf{wg}}$ (resp. $p^{(n-1)}X \in \mathsf{GTa}^{\mathsf{n-1}}$).
b) Let $f : X \to Y$ be an n-equivalence in $\mathsf{Ta}^{\mathsf{n}}_{\mathsf{wg}}$ and suppose that $Y \in \mathsf{GTa}^{\mathsf{n}}_{\mathsf{wg}}$. Then if $X \in \mathsf{Cat}^{\mathsf{n}}_{\mathsf{wg}}$ we have $X \in \mathsf{GCat}^{\mathsf{n}}_{\mathsf{wg}}$ and if $X \in \mathsf{Ta}^{\mathsf{n}}$ then $X \in \mathsf{GTa}^{\mathsf{n}}$. Similarly if f is an n-equivalence in $\mathsf{Ta}^{\mathsf{n}}_{\mathsf{wg}}$ and $X \in \mathsf{GTa}^{\mathsf{n}}_{\mathsf{wg}}$.

Corollary 12.3.9 *The following facts hold:*

a) The functor

$$Q_n : \mathsf{Ta}^{\mathsf{n}}_{\mathsf{wg}} \to \mathsf{Cat}^{\mathsf{n}}_{\mathsf{wg}}$$

restricts to a functor

$$Q_n : \mathsf{GTa}^{\mathsf{n}}_{\mathsf{wg}} \to \mathsf{GCat}^{\mathsf{n}}_{\mathsf{wg}}$$

such that for each $X \in \mathsf{GTa}^{\mathsf{n}}_{\mathsf{wg}}$ there is an n-equivalence in $\mathsf{GTa}^{\mathsf{n}}_{\mathsf{wg}}$ $s_n(X) : Q_nX \to X$.

b) The functor

$$Disc_n : \mathsf{Cat}^{\mathsf{n}}_{\mathsf{wg}} \to \mathsf{Ta}^{\mathsf{n}}$$

restricts to a functor

$$Disc_n : \mathsf{GCat}^{\mathsf{n}}_{\mathsf{wg}} \to \mathsf{GTa}^{\mathsf{n}}$$

such that for each $X \in \mathsf{GCat}^{\mathsf{n}}_{\mathsf{wg}}$ there is a zig-zag of n-equivalences in $\mathsf{GTa}^{\mathsf{n}}_{\mathsf{wg}}$ between X and $Disc_nX$.

Proof

a) Let $X \in \mathsf{GTa}^{\mathsf{n}}_{\mathsf{wg}}$. By Theorem 10.2.1 there is an n-equivalence in $\mathsf{Ta}^{\mathsf{n}}_{\mathsf{wg}}$

$$s_n(X) : Q_n X \to X.$$

Since $X \in \mathsf{GTa}^{\mathsf{n}}_{\mathsf{wg}}$ and $Q_n X \in \mathsf{Cat}^{\mathsf{n}}_{\mathsf{wg}}$, by Lemma 12.3.4 and Remark 12.3.8 $Q_n X \in \mathsf{GCat}^{\mathsf{n}}_{\mathsf{wg}}$.

b) Let $X \in \mathsf{Cat}^{\mathsf{n}}_{\mathsf{wg}}$. By Theorem 12.2.5 there is a zig-zag of n-equivalences between X and $Disc_n X$. Since $X \in \mathsf{GTa}^{\mathsf{n}}_{\mathsf{wg}}$, by Lemma 12.3.4 this is a zig-zag of n-equivalences in $\mathsf{GTa}^{\mathsf{n}}_{\mathsf{wg}}$, and since $X \in \mathsf{Ta}^{\mathsf{n}}$, by Remark 12.3.8 $Disc_n X \in \mathsf{GTa}^{\mathsf{n}}$. □

In the next proposition we specialize the comparison result of Theorem 12.2.6 to the higher groupoidal setting.

Proposition 12.3.10 *The functors*

$$Q_n : \mathsf{GTa}^{\mathsf{n}} \leftrightarrows \mathsf{GCat}^{\mathsf{n}}_{\mathsf{wg}} : Disc_n$$

induce an equivalence of categories after localization with respect to the n-equivalences

$$\mathsf{GTa}^{\mathsf{n}}/\!\sim^n \;\simeq\; \mathsf{GCat}^{\mathsf{n}}_{\mathsf{wg}}/\!\sim^n \ .$$

Proof Let $X \in \mathsf{GCat}^{\mathsf{n}}_{\mathsf{wg}}$. As in the proof of Theorem 12.2.6 there is an n-equivalence in $\mathsf{Cat}^{\mathsf{n}}_{\mathsf{wg}}$

$$Q_n Disc_n X \to X \ .$$

Since $X \in \mathsf{GCat}^{\mathsf{n}}_{\mathsf{wg}}$, by Remark 12.3.8, $Q_n Disc_n X \in \mathsf{GCat}^{\mathsf{n}}_{\mathsf{wg}}$, so this is an n-equivalence in $\mathsf{GCat}^{\mathsf{n}}_{\mathsf{wg}}$. It follows that there is an isomorphism in $\mathsf{GCat}^{\mathsf{n}}_{\mathsf{wg}}/\!\sim^n$

$$Q_n Disc_n X \cong X.$$

Let $Y \in \mathsf{GTa}^{\mathsf{n}}$. By the proof of Theorem 12.2.6 there is a zig-zag of n-equivalences in Ta^{n}

$$Disc_n Q_n Y \leftarrow D_n Z \to Y \ .$$

Since $Y \in \mathsf{GTa}^{\mathsf{n}}$, by Remark 12.3.8 this is a zig-zag of n-equivalences in $\mathsf{GTa}^{\mathsf{n}}$. It follows that there is an isomorphism in $\mathsf{GTa}^{\mathsf{n}}/\!\sim^n$

$$Disc_n Q_n Y \cong Y \ .$$

□

As a consequence of the previous proposition and of the results of Tamsamani [126], we obtain that groupoidal weakly globular n-fold categories are an algebraic model of n-types. This shows that our model $\mathsf{Cat}^{\mathsf{n}}_{\mathsf{wg}}$ of weak n-categories satisfies the homotopy hypothesis. In what follows let

$$\mathsf{GTa}^{\mathsf{n}} \underset{B}{\overset{\mathscr{T}_n}{\rightleftarrows}} n\text{-types}$$

be the fundamental Tamsamani n-groupoid functor $\mathscr{T}_n$ and the classifying space functor B be as in [119].

Theorem 12.3.11 *The functors*

$$\mathsf{GCat}^{\mathsf{n}}_{\mathsf{wg}} \underset{B \circ Disc_n}{\overset{Q_n \circ \mathscr{T}_n}{\rightleftarrows}} n\text{-}types$$

induce an equivalence of categories

$$\mathsf{GCat}^{\mathsf{n}}_{\mathsf{wg}}/{\sim^n} \simeq \mathrm{Ho}(n\text{-}types)\ .$$

Proof By Simpson [119] the functors $\mathscr{T}_n$ and B induce an equivalence of categories

$$\mathsf{GTa}^{\mathsf{n}}/{\sim^n} \simeq \mathrm{Ho}(n\text{-types}) \tag{12.21}$$

while, by Proposition 12.3.10, the functors Q_n and $Disc_n$ induce an equivalence of categories

$$\mathsf{GCat}^{\mathsf{n}}_{\mathsf{wg}}/{\sim^n} \simeq \mathsf{GTa}^{\mathsf{n}}/{\sim^n}\ . \tag{12.22}$$

By (12.21) and (12.22) the result follows. □

Remark 12.3.12 We call a map f in $\mathsf{GCat}^{\mathsf{n}}_{\mathsf{wg}}$ a geometric weak equivalence if $(B \circ Disc_n)(f)$ is a weak homotopy equivalence of spaces. We note that a map f in $\mathsf{GCat}^{\mathsf{n}}_{\mathsf{wg}}$ is an n-equivalence if and only if it is a geometric weak equivalence. In fact, if f is an n-equivalence, it is an isomorphism in $\mathsf{GCat}^{\mathsf{n}}_{\mathsf{wg}}/ \sim^n$ so by Theorem 12.3.11, $(B \circ Disc_n)(f)$ is an isomorphism in Ho(n-types), thus it is a weak homotopy equivalence in n-types.

Conversely, if f is a geometric weak equivalence, $(B \circ Disc_n)(f)$ is an isomorphism in Ho(n-types), so by Theorem 12.3.11 (since equivalence of categories reflect isomorphisms), f is an isomorphism in $\mathsf{GCat}^{\mathsf{n}}_{\mathsf{wg}}/ \sim^n$, so f is an n-equivalence in $\mathsf{GCat}^{\mathsf{n}}_{\mathsf{wg}}$.

We finally observe that, as an immediate consequence of our results, all three Segal-type models of this work are models of weak n-categories satisfying the homotopy hypothesis. In what follows $\mathsf{GSeg_n}$ denotes any of the three groupoidal Segal-type models $\mathsf{GCat^n_{wg}}$, $\mathsf{GTa^n}$, $\mathsf{GTa^n_{wg}}$.

Corollary 12.3.13 *Each of the three Segal-type models* $\mathsf{Seg_n}$ *is a model of weak* n*-categories satisfying the homotopy hypothesis, that is, there is an equivalence of categories*

$$\mathsf{GSeg_n}/\!\sim^n \;\simeq\; \mathrm{Ho}(n\text{-}types)\ .$$

Proof In the case $\mathsf{Seg_n} = \mathsf{Ta^n}$ this is the result of [126]. When $\mathsf{Seg_n} = \mathsf{Cat^n_{wg}}$ this is the content of Theorems 12.2.6 and 12.3.11. In the case $\mathsf{Seg_n} = \mathsf{Ta^n_{wg}}$, by Corollary 10.2.3 there is an equivalence of categories

$$\mathsf{Ta^n_{wg}}/\!\sim^n \;\simeq\; \mathsf{Cat^n_{wg}}/\!\sim^n\ . \tag{12.23}$$

Thus by Remark 12.3.8 this restricts to an equivalence of categories

$$\mathsf{GTa^n_{wg}}/\!\sim^n \;\simeq\; \mathsf{GCat^n_{wg}}/\!\sim^n\ ,$$

so by Theorem 12.3.11 we conclude that there is an equivalence of categories

$$\mathsf{GTa^n_{wg}}/\!\sim^n \;\simeq\; \mathrm{Ho}(n\text{-types})\ .$$

□

12.4 An Alternative Fundamental Functor

Theorem 12.3.11 exhibits the fundamental groupoidal weakly globular n-fold category functor

$$\mathcal{G}_n : n\text{-types} \to \mathsf{GCat^n_{wg}} \tag{12.24}$$

as the composite $Q_n \circ \mathcal{T}_n$, where $\mathcal{T}_n$ is the Tamsamani n-groupoid functor from [119] and Q_n is the rigidification functor. Using the results of Blanc and Paoli [29] we exhibit an alternative functor

$$j\mathcal{H}_n : n\text{-types} \to \mathsf{GCat^n_{wg}}\ ,$$

which is simpler than $\mathcal{G}_n$ and whose definition is independent of [119].

Using our previous results, we show in Corollary 12.4.6 that $j\mathcal{H}_n$ and the classifying space functor $\widetilde{B} : \mathsf{GCat^n_{wg}} \to n$-types induce an equivalence of categories

$$\mathsf{GCat^n_{wg}}/\!\sim^n \;\simeq\; \mathrm{Ho}(n\text{-types})\ .$$

Thus $j\mathcal{H}_n$ can be used as a fundamental functor instead of $\mathcal{G}_n$.

12.4.1 The Functor $\mathcal{H}_n$

In [29, Definition 3.19] Blanc and Paoli introduced the category $\mathsf{Gpd}^n_{\mathsf{wg}}$ of weakly globular n-fold groupoids, which is a full subcategory of the category Gpd^n of n-fold groupoids of Sect. 2.5. It is immediate from the definitions in [29] that $\mathsf{Gpd}^n_{\mathsf{wg}}$ is a full subcategory of $\mathsf{GCat}^n_{\mathsf{wg}}$ and that a map in $\mathsf{Gpd}^n_{\mathsf{wg}}$ is an n-equivalence if and only if it is so in $\mathsf{GCat}^n_{\mathsf{wg}}$.

As in the case of $\mathsf{GCat}^n_{\mathsf{wg}}$ (see Remark 12.3.12), it was shown in [29] that n-equivalences in $\mathsf{Gpd}^n_{\mathsf{wg}}$ are the same as geometric weak equivalences. Let

$$\mathcal{S} : \mathsf{Top} \to [\Delta^{^{op}}, \mathsf{Set}]$$

be the singular functor, which produces fibrant simplicial sets. Let

$$\mathrm{Or}_{(n)} : [\Delta^{^{op}}, \mathsf{Set}] \to [\Delta^{n^{op}}, \mathsf{Set}]$$

be the functor induced by the ordinal sum $or_n : \Delta^n \to \Delta$. Thus

$$(\mathrm{Or}_{(n)}\, X)_{p_1 \dots p_n} = X_{n-1+p_1+\dots+p_n} \,.$$

Remark 12.4.1 The functor $\mathrm{Or}_{(n)}$ produces an n-fold simplicial resolution of a simplicial set, since it can be shown [29, Lemma 2.13] that for any simplicial set X, there is a natural weak equivalence

$$\varepsilon_{(n)} : Diag_n\, \mathrm{Or}_{(n)}\, X \to X$$

where $Diag_n : [\Delta^{n^{op}}, \mathsf{Set}] \to [\Delta^{^{op}}, \mathsf{Set}]$ is the multi-diagonal functor, given for each $m \in \Delta^{op}$ by

$$(Diag_n X)_m = X_{m,m,\dots,m} \,.$$

We also showed in [29, Section 2.9] that

$$\mathrm{Or}_{(n)}\, Y = \overline{\mathrm{Or}}^{(2)}_{(n-1)}\, \mathrm{Or}_{(2)}\, Y \tag{12.25}$$

and we proved in [29, Lemma 2.28] that if Y is a Kan complex, for each $n \geq 2$,

$$\overline{\mathrm{Or}}^{(2)}_{(n-1)} N^{(2)} \hat{\pi}^{(2)}_1\, \mathrm{Or}_{(2)}\, X \cong N^{(n)} \hat{\pi}^{(n)}_1\, \mathrm{Or}_{(n)}\, X \,, \tag{12.26}$$

where $\hat{\pi}_1^{(i)}$ denotes the fundamental groupoid functor in the ith direction.

Let

$$\mathscr{P}_n : [\Delta^{n^{op}}, \mathsf{Set}] \to \mathsf{Gpd}^n$$

be the left adjoint to the n-fold nerve $N_{(n)} : \mathsf{Gpd}^n \to [\Delta^{n^{op}}, \mathsf{Set}]$.

Definition 12.4.2 ([29, Definition 2.30]) The fundamental weakly globular n-fold groupoid functor is given by the composite

$$\mathscr{H}_n : n\text{-types} \xrightarrow{\mathscr{S}} [\Delta^{op}, \mathsf{Set}] \xrightarrow{\mathrm{Or}_{(n)}} [\Delta^{n^{op}}, \mathsf{Set}] \xrightarrow{\mathscr{P}_n} \mathsf{Gpd}^n . \tag{12.27}$$

For a general n-fold simplicial set Y, $\mathscr{P}_n Y$ does not have a simple and explicitly computable expression. However, we showed in [29] that, given a space X, the fibrancy of $\mathscr{S}X$ induces a property of $\mathrm{Or}_{(n)}\, \mathscr{S}X$ which we called in [29] $(n, 2)$-fibrancy (see [29, Definition 2.3.1 and Proposition 2.3.9]). We then showed that to apply $\mathscr{P}_n$ to an $(n, 2)$-fibrant n-fold simplicial set we need only apply the usual fundamental groupoid in each of the $(n-1)$-simplicial directions. Thus we have

Theorem 12.4.3 ([29, Theorem 2.40]) *Let $\mathscr{H}_n$ be as in Definition 12.4.2 and X be a space. Then*

$$\mathscr{H}_n X = \hat{\pi}^{(1)} \hat{\pi}^{(2)} \cdots \hat{\pi}^{(n)}\, \mathrm{Or}_{(n)}\, \mathscr{S}X . \tag{12.28}$$

Using this explicit description of $\mathscr{H}_n$ we showed in [29] that the functor $\mathscr{H}_n$ in fact lands in $\mathsf{Gpd}^{\mathsf{n}}_{\mathsf{wg}}$. Further, we proved

Theorem 12.4.4 ([29, Theorem 4.32]) *The functors*

$$\mathscr{H}_n : n\text{-types} \to \mathsf{Gpd}^{\mathsf{n}}_{\mathsf{wg}}, \qquad B : \mathsf{Gpd}^{\mathsf{n}}_{\mathsf{wg}} \to n\text{-types}$$

induce functors

$$\mathrm{Ho}(n\text{-types}) \leftrightarrows \mathsf{Gpd}^{\mathsf{n}}_{\mathsf{wg}}/{\sim^n}$$

with $B\mathscr{H}_n \cong \mathrm{Id}$.

Remark 12.4.5 Let

$$\widetilde{B} = B \circ Disc_n : \mathsf{GCat}^{\mathsf{n}}_{\mathsf{wg}} \to n\text{-types}.$$

From [29], the following diagram commutes

$$\begin{array}{ccc} \mathsf{GCat}^{\mathsf{n}}_{\mathsf{wg}}/{\sim^n} & \xrightarrow{\widetilde{B}} & \mathrm{Ho}(n\text{-types}) \\ \Big\uparrow j & \nearrow_{B} & \\ \mathsf{Gpd}^{\mathsf{n}}_{\mathsf{wg}}/{\sim^n} & & \end{array}$$

Using our previous results we deduce the following corollary, which shows that $j\mathscr{H}_n$ can be used as an alternative fundamental functor from n-types to $\mathsf{GCat}^{\mathsf{n}}_{\mathsf{wg}}$.

Corollary 12.4.6 *Let $j\mathscr{H}_n$ be the composite*

$$j\mathscr{H}_n : n\text{-types} \xrightarrow{\mathscr{S}} [\Delta^{op}, \mathsf{Set}] \xrightarrow{\mathrm{Or}_{(n)}} [\Delta^{n^{op}}, \mathsf{Set}] \xrightarrow{\mathscr{P}_n} \mathsf{Gpd}^{\mathsf{n}}_{\mathsf{wg}} \xhookrightarrow{j} \mathsf{GCat}^{\mathsf{n}}_{\mathsf{wg}}$$

and let $\widetilde{B} : \mathsf{GCat}^{\mathsf{n}}_{\mathsf{wg}} \to n$-types be as in Remak 12.4.5. Then $j\mathscr{H}_n$ and $\widetilde{B}$ induce an equivalence of categories

$$\mathsf{GCat}^{\mathsf{n}}_{\mathsf{wg}}/{\sim^n} \simeq \mathrm{Ho}(n\text{-}types)\ . \tag{12.29}$$

Proof Let $X \in \mathsf{GCat}^{\mathsf{n}}_{\mathsf{wg}}$; by Theorem 12.3.11, $\widetilde{B}X$ is an n-type and by Theorem 12.4.4 and Remark 12.4.5

$$\widetilde{B}j\mathscr{H}_n\widetilde{B}X \cong B\mathscr{H}_n\widetilde{B}X \cong \widetilde{B}X \tag{12.30}$$

in Ho(n-types). Since both X and $j\mathscr{H}_n\widetilde{B}X$ are in $\mathsf{GCat}^{\mathsf{n}}_{\mathsf{wg}}$, the equivalence of categories

$$\mathsf{GCat}^{\mathsf{n}}_{\mathsf{wg}}/{\sim^n} \cong \mathrm{Ho}(n\text{-types})$$

of Theorem 12.3.11 induced by $Q_n\mathscr{T}_n$ and $\widetilde{B}$, together with (12.30), imply that

$$j\mathscr{H}_n\widetilde{B}X \cong X \tag{12.31}$$

in $\mathsf{GCat}^{\mathsf{n}}_{\mathsf{wg}}/{\sim^n}$.

Let $Y \in$ Ho(n-types). By Theorem 12.4.4 and Remark 12.4.5

$$\widetilde{B}j\mathscr{H}_nY \cong B\mathscr{H}_nY \cong Y \tag{12.32}$$

in Ho(n-types). By (12.31) and (12.32) we conclude that $j\mathscr{H}_n$ and $\widetilde{B}$ induce the equivalence of categories (12.29). □

The following corollary shows that the functor

$$p^{(n-1)} : \mathsf{GCat}^{\mathsf{n}}_{\mathsf{wg}} \to \mathsf{GCat}^{\mathsf{n-1}}_{\mathsf{wg}}$$

is the algebraic version of the Postnikov truncation functor

$$n\text{-types} \to (n-1)\text{-types} .$$

Corollary 12.4.7 *Let* $X \in \mathsf{GCat}^{\mathsf{n}}_{\mathsf{wg}}$*. The map*

$$\gamma^{(n-1)} : X \to q^{(n-1)}X = p^{(n-1)}X$$

induces a map of spaces

$$\widetilde{B}\gamma^{(n-1)} : \widetilde{B}X \to \widetilde{B}p^{(n-1)}X$$

such that, for each $0 \le i \le n-1$, $x \in \widetilde{B}X$,

$$\pi_i(\widetilde{B}X, x) \cong \pi_i(\widetilde{B}p^{(n-1)}X, x) .$$

Proof It is shown in [29] that the functor

$$p^{(n-1)} : \mathsf{Gpd}^{\mathsf{n}}_{\mathsf{wg}} \to \mathsf{Gpd}^{\mathsf{n-1}}_{\mathsf{wg}}$$

(denoted $\Pi_0^{(n)}$ in [29]) is such that, for each $Y \in \mathsf{Gpd}^{\mathsf{n}}_{\mathsf{wg}}$, the map $\gamma^{(n-1)} : Y \to p^{(n-1)}Y$ induces a map of spaces

$$\widetilde{B}\gamma^{(n-1)} : BY \to Bp^{(n-1)}Y$$

such that, for each $0 \le i \le n-1$, $y \in BY$,

$$\pi_i(BY, y) \cong \pi_i(Bp^{(n-1)}Y, y) . \tag{12.33}$$

Let $X \in \mathsf{GCat}^{\mathsf{n}}_{\mathsf{wg}}$. By Corollary 12.4.7 there is a zig-zag of n-equivalences between X and $j\mathcal{H}_n\widetilde{B}X$, and thus also a zig-zag of $(n-1)$-equivalences between $p^{(n-1)}X$ and $p^{(n-1)}j\mathcal{H}_n\widetilde{B}X = jp^{(n-1)}\mathcal{H}_n\widetilde{B}X$. This implies that there are zig-zags of weak homotopy equivalences between $\widetilde{B}X$ and $\widetilde{B}j\mathcal{H}_n\widetilde{B}X = B\mathcal{H}_n\widetilde{B}X$ as well as between $\widetilde{B}p^{(n-1)}X$ and $\widetilde{B}jp^{(n-1)}\mathcal{H}_n\widetilde{B}X = Bp^{(n-1)}\mathcal{H}_n\widetilde{B}X$. Therefore, for all $x \in \widetilde{B}X$,

$$\begin{aligned} \pi_i(\widetilde{B}X, x) &\cong \pi_i(B\mathcal{H}_n\widetilde{B}X, x') , \\ \pi_i(\widetilde{B}p^{(n-1)}X, x) &\cong \pi_i(Bp^{(n-1)}\mathcal{H}_n\widetilde{B}X, x') . \end{aligned} \tag{12.34}$$

By (12.33), taking $Y = \mathscr{H}_n \widetilde{B} X \in \mathsf{Gpd}^{\mathsf{n}}_{\mathsf{wg}}$, we obtain for each $0 \leq i \leq n-1$

$$\pi_i(B\mathscr{H}_n\widetilde{B}X, x') = \pi_i(Bp^{(n-1)}\mathscr{H}_n\widetilde{B}X, x') \, .$$

By (12.34) this implies, for all $0 \leq i \leq n-1$,

$$\pi_i(\widetilde{B}X, x) \cong \pi_i(\widetilde{B}p^{(n-1)}X, x) \, .$$

□

12.4.2 Some Examples

We now illustrate the fundamental groupoidal weakly globular n-fold category of a space in some low-dimensional cases. This shows how explicit and convenient the use of the functor $j\mathscr{H}_n$ is.

For each $n > 1$ denote by $\mathscr{R}_n$ the composite

$$[\Delta^{^{op}}, \mathsf{Set}] \xrightarrow{\mathrm{Or}_{(n)}} [\Delta^{n^{op}}, \mathsf{Set}] \xrightarrow{\mathscr{P}_n} \mathsf{Gpd}^{\mathsf{n}}_{\mathsf{wg}} \tag{12.35}$$

and let $\mathscr{R}_1 = \hat{\pi}_1 : [\Delta^{^{op}}, \mathsf{Set}] \to \mathsf{Gpd}$ be the fundamental groupoid, so that, using our previous notation

$$j\mathscr{H}_n = j\mathscr{R}_n\mathscr{S} : n\text{-types} \to \mathsf{GCat}^{\mathsf{n}}_{\mathsf{wg}} \, .$$

In [29, Lemma 4.14] we give an iterative description of $\mathscr{R}_n Y$ for a Kan complex Y which is more transparent than the formula (12.28). More precisely, let u_Y : $\mathrm{Dec}\, Y \to Y$ be as in Sect. 2.7 and consider the corresponding internal equivalence relation

$$(\mathrm{Dec}\, Y)[u] \in \mathsf{Gpd}([\Delta^{^{op}}, \mathsf{Set}])$$

as in Definition 5.1.8. Denote by

$$\mathscr{L}_\bullet Y \in [\Delta^{^{op}}, [\Delta^{^{op}}, \mathsf{Set}]]$$

the nerve of $(\mathrm{Dec}\, Y)[u]$, so that

$$\mathscr{L}_k Y = \begin{cases} \mathrm{Dec}\, Y, & \text{if } k = 0\,, \\ \mathrm{Dec}\, Y \times_Y \overset{k+1}{\cdots} \times_Y \mathrm{Dec}\, Y, & \text{if } k \geq 1\,. \end{cases} \tag{12.36}$$

A picture of the corner of $\mathscr{L}_\bullet Y$ is given in Fig 12.3 on page 309.

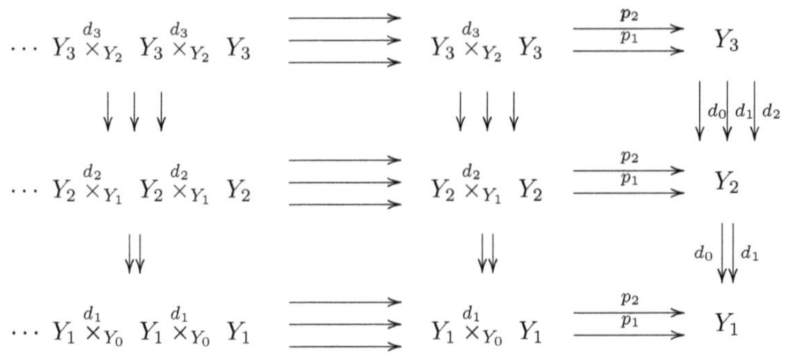

$$
\begin{array}{ccccc}
\cdots\; Y_3 \overset{d_3}{\times}_{Y_2} Y_3 \overset{d_3}{\times}_{Y_2} Y_3 & \substack{\longrightarrow\\ \longrightarrow\\ \longrightarrow} & Y_3 \overset{d_3}{\times}_{Y_2} Y_3 & \substack{\xrightarrow{p_2}\\ \xrightarrow{p_1}} & Y_3 \\
\downarrow\downarrow\downarrow & & \downarrow\downarrow\downarrow & & \downarrow d_0 \downarrow d_1 \downarrow d_2 \\
\cdots\; Y_2 \overset{d_2}{\times}_{Y_1} Y_2 \overset{d_2}{\times}_{Y_1} Y_2 & \substack{\longrightarrow\\ \longrightarrow\\ \longrightarrow} & Y_2 \overset{d_2}{\times}_{Y_1} Y_2 & \substack{\xrightarrow{p_2}\\ \xrightarrow{p_1}} & Y_2 \\
\downarrow\downarrow & & \downarrow\downarrow & & d_0 \downarrow\downarrow d_1 \\
\cdots\; Y_1 \overset{d_1}{\times}_{Y_0} Y_1 \overset{d_1}{\times}_{Y_0} Y_1 & \substack{\longrightarrow\\ \longrightarrow\\ \longrightarrow} & Y_1 \overset{d_1}{\times}_{Y_0} Y_1 & \substack{\xrightarrow{p_2}\\ \xrightarrow{p_1}} & Y_1
\end{array}
$$

Fig. 12.3 Corner of $\mathscr{L}_\bullet Y$

In what follows, $N^{(n)}$ denotes the nerve functor in the nth direction (see Definition 2.4.8).

Proposition 12.4.8 ([29, Lemma 4.14]) *Let Y be a Kan complex.*

a) For each $k \geq 0$

$$(N^{(n)}\mathscr{R}_n Y)_k \cong \mathscr{R}_{n-1}\mathscr{L}_n Y \; . \tag{12.37}$$

Thus, for each $k \geq 1$,

$$\mathscr{R}_n\mathscr{L}_n Y \cong \mathscr{R}_{n-1}\mathscr{L}_1 Y \times_{\mathscr{R}_{n-1} \operatorname{Dec} Y} \overset{k}{\cdots} \times_{\mathscr{R}_{n-1} \operatorname{Dec} Y} \mathscr{R}_{n-1}\mathscr{L}_1 Y \; . \tag{12.38}$$

b) If Y is homotopically trivial for $k \geq 1$, then

$$\mathscr{R}_n\mathscr{L}_n Y \cong \mathscr{R}_n \operatorname{Dec} Y \times_{\mathscr{R}_n Y} \overset{k+1}{\cdots} \times_{\mathscr{R}_n Y} \mathscr{R}_n \operatorname{Dec} Y \; .$$

We also proved in [29, Proposition 4.28] that for every Kan complex Y,

$$p^{(n-1)}\mathscr{R}_n Y \cong \mathscr{R}_{n-1} Y \tag{12.39}$$

and thus, if X is an n-type and $Y = \mathscr{S}X$,

$$p^{(n-1)} j \mathscr{H}_n X = p^{(n-1)} j \mathscr{R}_n \mathscr{S} X = j \mathscr{H}_{n-1} X \; . \tag{12.40}$$

Example 12.4.9 The fundamental groupoidal weakly globular double category of a space.

Let X be a space and $Y = \mathscr{S}X$ be its singular simplicial set. The bisimplicial set $\operatorname{Or}_{(2)} Y$ can be described as follows (Fig. 12.4).

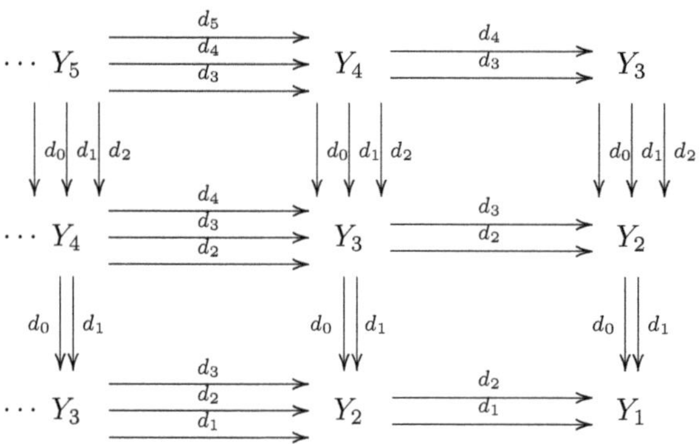

Fig. 12.4 Corner of $\mathrm{Or}_{(2)}\, Y$

Let Dec and Dec′ be the two décalage comonads as in Sect. 2.7. The comonad Dec yields a simplicial resolution $Z \in [\Delta^{op}, [\Delta^{op}, \mathsf{Set}]]$ for any $Y \in [\Delta^{op}, \mathsf{Set}]$ with

$$Z_{k-1} = \mathrm{Dec}^k\, Y = \underbrace{\mathrm{Dec}(\mathrm{Dec}\ldots\ \mathrm{Dec}\ldots)}_{k} \in [\Delta^{op}, \mathsf{Set}]\ .$$

It can be shown (see for instance [73]) that

$$\mathrm{Or}_{(2)}\, Y = Z\ .$$

The bisimplicial set $\mathrm{Or}_{(2)}\, Y$ is depicted in Fig. 12.4 on page 310, viewed as a horizontal simplicial object in $[\Delta^{op}, \mathsf{Set}]$ (the degeneracy maps are not shown). The corresponding resolution using Dec′ is also depicted in Fig. 12.4, viewed as a vertical simplicial object in $[\Delta^{op}, \mathsf{Set}]$.

From Theorem 12.4.3 we have

$$j\mathscr{H}_n X = \mathscr{P}_2\, \mathrm{Or}_{(2)}\, \mathscr{S} X = \hat{\pi}_1^{(1)} \hat{\pi}_1^{(2)}\, \mathrm{Or}_{(2)}\, Y\ , \tag{12.41}$$

where $\hat{\pi}_1^{(1)}$ and $\hat{\pi}_1^{(2)}$ are the fundamental groupoids in the two simplicial directions. Since $Y = \mathscr{S} X$ is a Kan complex, such are $\mathrm{Dec}\, Y$ and $\mathrm{Dec}'\, Y$, so $(\mathrm{Or}_{(2)}\, Y)_k$ and $(\mathrm{Or}_{(2)}\, Y)_k^{\{2\}}$ are Kan complexes for all k, and taking their fundamental groupoids amounts to dividing out the 1-simplices by the relations given by the 2-simplices. Using the formula (12.37) we obtain

$$(N^{(2)} j\mathscr{H}_2 X)_0 = \hat{\pi}_1\, \mathrm{Dec}\, Y\ ,$$

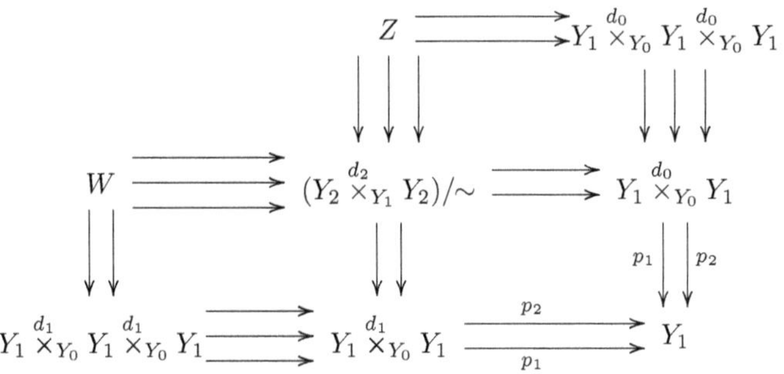

Fig. 12.5 Corner of the double nerve of $\mathscr{H}_2 X$ for $Y = \mathscr{S}X$ where:

$$Z = (Y_2 \overset{d_2}{\times}_{Y_1} Y_2)/\sim \times_{(Y_1 \overset{d_1}{\times}_{Y_0} Y_1)} (Y_2 \overset{d_2}{\times}_{Y_1} Y_2)/\sim$$

$$W = (Y_2 \overset{d_2}{\times}_{Y_1} Y_2)/\sim \times_{(Y_1 \overset{d_0}{\times}_{Y_0} Y_1)} (Y_2 \overset{d_2}{\times}_{Y_1} Y_2)/\sim$$

which is the homotopically discrete groupoid corresponding to the surjective map of sets $d_1 : X_1 \to X_0$. By (12.37) we also have

$$(N^{(2)} j\mathscr{H}_2 X)_k = \hat{\pi}_1 \mathscr{L}_k Y ,$$

where $\mathscr{L}_k Y$ is as in Fig. 12.3.

In Fig. 12.5 on page 311 we display the corner of the double nerve of $j\mathscr{H}_2 X$, where $(Y_2 \times_{Y_1} Y_2)/\sim$ denotes the result dividing out by the relations of the 2-simplices.

Note that $(j\mathscr{H}_2 X)_0$ is homotopically discrete while $\bar{p} j\mathscr{H}_2 X$ is the nerve of the groupoid $\hat{\pi}_1 Y$. Therefore, by Lemma 7.2.6, $j\mathscr{H}_2 X \in \mathsf{Cat}^2_{\mathsf{wg}}$ and, given the groupoidal structures, $j\mathscr{H}_2 X \in \mathsf{GCat}^2_{\mathsf{wg}}$. In fact this is also a double groupoid, so that $\mathscr{H}_2 X \in \mathsf{Gpd}^2_{\mathsf{wg}}$.

Example 12.4.10 The fundamental groupoidal weakly globular 3-fold category of a space.

Let X be a space and $Y = \mathscr{S}X$ its singular simplicial set. By (12.25)

$$\mathrm{Or}_{(3)}\, Y = \overline{\mathrm{Or}}^{(2)}_{(2)}\, \mathrm{Or}_{(2)}\, Y .$$

See Fig. 12.6 on page 312 for a picture of the corner of $\mathrm{Or}_{(3)}\, Y$. In Fig. 12.7 on page 312 we have a picture of $j\mathscr{H}_3 X$, where $\mathscr{S}X = Y$.

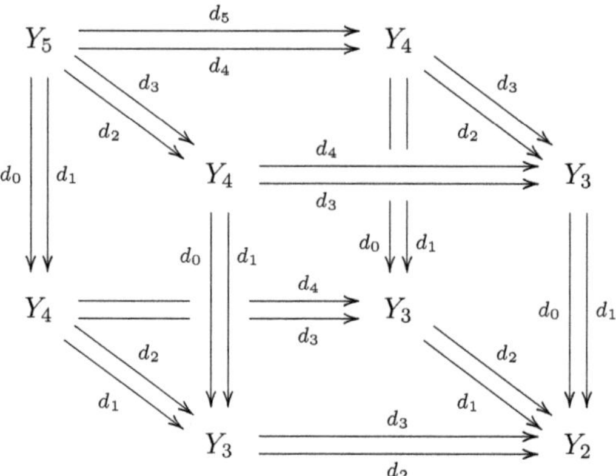

Fig. 12.6 Corner of $\mathrm{Or}_{(3)}\, Y$

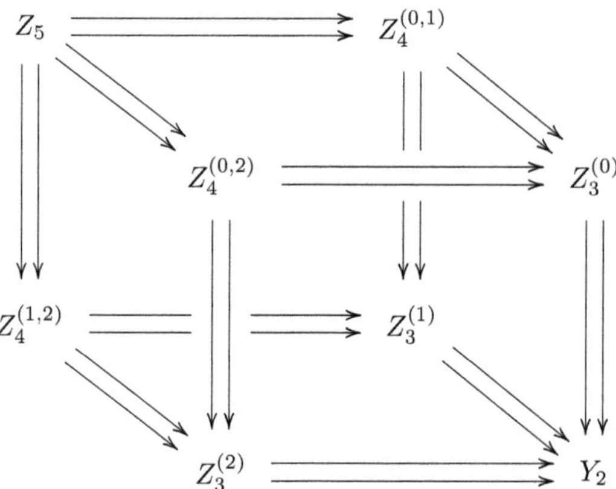

Fig. 12.7 Corner of the 3-fold nerve of $\mathscr{H}_3 X$, with $Y = \mathscr{S} X$

$$Z_3^{(0)} = Y_2 \overset{d_0}{\times}_{Y_1} Y_2, \quad Z_3^{(1)} = Y_2 \overset{d_1}{\times}_{Y_1} Y_2, \quad Z_3^{(2)} = Y_2 \overset{d_2}{\times}_{Y_1} Y_2 \,,$$

$$Z_4^{(0,1)} = (Y_2 \overset{d_1}{\times}_{Y_1} Y_2) \times_{\underset{(Y_1 \times_{Y_0} Y_1)}{d_0}} (Y_2 \overset{d_1}{\times}_{Y_1} Y_2) \cong (Y_2 \overset{d_0}{\times}_{Y_1} Y_2) \times_{\underset{(Y_1 \times_{Y_0} Y_1)}{d_0}} (Y_2 \overset{d_0}{\times}_{Y_1} Y_2) \,,$$

$$Z_4^{(0,2)} = (Y_2 \overset{d_2}{\times}_{Y_1} Y_2) \times_{\underset{(Y_1 \times_{Y_0} Y_1)}{d_0}} (Y_2 \overset{d_2}{\times}_{Y_1} Y_2) \cong (Y_2 \overset{d_0}{\times}_{Y_1} Y_2) \times_{\underset{(Y_1 \times_{Y_0} Y_1)}{d_1}} (Y_2 \overset{d_0}{\times}_{Y_1} Y_2) \,,$$

$$Z_4^{(1,2)} = (Y_2 \overset{d_2}{\times}_{Y_1} Y_2) \times_{\underset{(Y_1 \times_{Y_0} Y_1)}{d_1}} (Y_2 \overset{d_2}{\times}_{Y_1} Y_2) \cong (Y_2 \overset{d_1}{\times}_{Y_1} Y_2) \times_{\underset{(Y_1 \times_{Y_0} Y_1)}{d_1}} (Y_2 \overset{d_1}{\times}_{Y_1} Y_2) \,,$$

$$Z_5 = Z_4^{(0,1)}/{\sim} \cong Z_4^{(0,2)}/{\sim} \cong Z_4^{(1,2)}/{\sim} \;.$$

The isomorphisms describing $Z_4^{(0,1)}$, $Z_4^{(0,2)}$, $Z_4^{(1,2)}$ in Fig. 12.7 are derived from the simplicial identities. Namely, the simplicial identity $d_0 d_0 = d_0 d_1$ implies that the limit of the following diagrams are isomorphic:

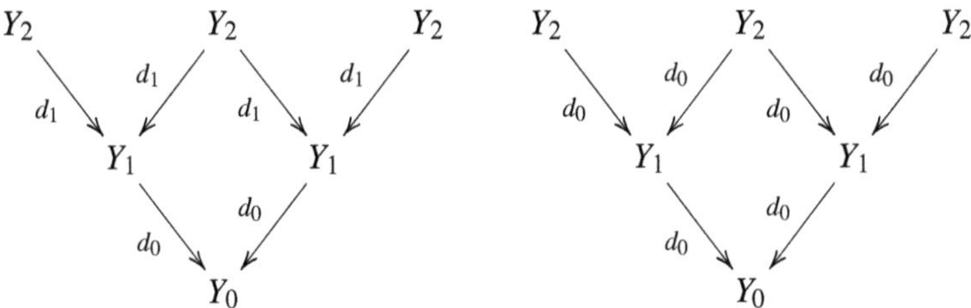

That is,

$$(Y_2 \overset{d_1}{\times}_{Y_1} Y_2) \times_{\overset{d_0}{(Y_1 \times_{Y_0} Y_1)}} (Y_2 \overset{d_1}{\times}_{Y_1} Y_2) \cong (Y_2 \overset{d_0}{\times}_{Y_1} Y_2) \times_{\overset{d_0}{(Y_1 \times_{Y_0} Y_1)}} (Y_2 \overset{d_0}{\times}_{Y_1} Y_2)\,.$$

Similarly, the simplicial identity $d_0 d_2 = d_1 d_0$ implies the isomorphism

$$(Y_2 \overset{d_2}{\times}_{Y_1} Y_2) \times_{\overset{d_0}{(Y_1 \times_{Y_0} Y_1)}} (Y_2 \overset{d_2}{\times}_{Y_1} Y_2) \cong (Y_2 \overset{d_0}{\times}_{Y_1} Y_2) \times_{\overset{d_1}{(Y_1 \times_{Y_0} Y_1)}} (Y_2 \overset{d_0}{\times}_{Y_1} Y_2)$$

and the simplicial identity $d_1 d_2 \cong d_1 d_1$ implies the isomorphism

$$(Y_2 \overset{d_2}{\times}_{Y_1} Y_2) \times_{\overset{d_1}{(Y_1 \times_{Y_0} Y_1)}} (Y_2 \overset{d_2}{\times}_{Y_1} Y_2) \cong (Y_2 \overset{d_1}{\times}_{Y_1} Y_2) \times_{\overset{d_1}{(Y_1 \times_{Y_0} Y_1)}} (Y_2 \overset{d_1}{\times}_{Y_1} Y_2)\,.$$

The face operators indicated in the picture are the respective projections, while we omitted drawing the degeneracies.

Chapter 13
Conclusions and Further Directions

Abstract In this chapter we give a sketch of possible further developments and open questions arising from this work. One area of application is homotopy theory. We plan to use our new model of n-types consisting of groupoidal weakly globular n-fold categories to gain an algebraic understanding of the k-invariants of spaces and of simplicial categories; an interesting open problem is to build a Quillen model structure on weakly globular n-fold categories. We also envisage the use of our new model to tackle some long term open problems in higher category theory. We end with a discussion of a possible extension of the weakly globular approach to the (∞, n)-case.

In this book we have laid the foundations of a new approach to weak n-categories based on a new paradigm to weaken higher categorical structures, which is the notion of weak globularity.

We now give a sketch of several lines of further developments and applications of this new approach that we envisage, and point out some open questions. Each of these topics will be explored in future projects and goes beyond the scope of this work, whose aim is to establish the foundations of this theory.

One of the main novelties of our approach is the use of an entirely rigid structure, namely a subcategory of n-fold categories, to model weak n-categories.

The terminology 'rigid structure' refers to the fact that n-fold categories, being iterated internal categories, have compositions in n different simplicial directions and these compositions are associative and unital. In this sense, n-fold categories are a strict higher categorical structure, though not the same as strict n-categories, since they do not have sets of higher cells in dimensions 0 to up n. Instead, the higher cells in dimension k have themselves an $(n-1+k)$-fold categorical structure.

By imposing the weak globularity condition that these sub-structures are homotopically discrete we recover the notion of 'sets of higher cells' in an n-fold category: the latter correspond to the underlying sets of the discretizations of the homotopically discrete sub-structures. The induced Segal map condition regulates the behaviour of the compositions, which are weakly associative and unital. This leads to the category $\mathsf{Cat}^n_{\mathsf{wg}}$ of weakly globular n-fold categories.

S. Paoli, *Simplicial Methods for Higher Categories*, Algebra and Applications 26,
https://doi.org/10.1007/978-3-030-05674-2_13

In this book we have shown that this is a model of weak n-categories by proving that it is suitably equivalent (after localization) to the Tamsamani–Simpson model and by showing that it satisfies the homotopy hypothesis.

The n-fold nature of our model is a new feature in the literature on modelling of weak n-categories which opens the possibility of new developments and applications, as sketched below.

13.1 Algebraic Description of Postnikov Systems

An application of our new model of weak n-categories relates to an open problem in homotopy theory, which is the algebraic-categorical modelling of Postnikov systems. Postnikov systems are a classical tool for studying topological spaces [63] by breaking them into smaller pieces, the Postnikov sections (which are n-types) and their k-invariants, defined in terms of cohomology classes of spaces.

More recently, homotopy theories themselves have become fundamental objects of study in algebraic topology, besides spaces. There are several models for the 'homotopy theory of homotopy theories' [23]; the classical one is categories enriched in simplicial sets, also called simplicial categories. The Postnikov tower for the simplicial mapping spaces in a simplicial category gives rise to a Postnikov tower for simplicial categories. Its Postnikov sections are categories enriched in simplicial n-types. The corresponding k-invariants are described in terms of a cohomology theory defined by Dwyer, Kan and Smith [51].

The Postnikov systems, both of spaces and of simplicial categories, are defined topologically. One of the main questions at the origin of the development of higher categorical structures is how to describe Postnikov systems by purely algebraic and categorical means.

The algebraic-categorical modelling of Postnikov systems thus involves two steps: modelling the Postnikov sections and modelling the k-invariants. The former is quite well understood, and connects to the homotopy hypothesis for different models of weak n-category. However, an algebraic description of the k-invariants of spaces and of simplicial categories is unknown. In particular, the cohomology theories used to describe these k-invariants are poorly understood from an algebraic viewpoint, with the exception of some special low-dimensional cases.

In the case of spaces, an algebraic description of the cohomology of connected 2-types with coefficients in a π_1-module was developed by Paoli in [100, 101]. The non-path connected case is open even in dimension $n = 2$, and the general case $n > 2$ is also unknown (both in the path-connected and non-path connected cases).

For the case of simplicial categories, the k-invariants are defined topologically in terms of Dwyer–Kan–Smith cohomology. An algebraic description for the cohomology giving the first k-invariant under certain assumptions was given by Baues and Wirsching [20] and, one dimension up, by Blanc and Paoli [28]. The case of general dimension n is open.

The key to the algebraic description of these cohomology theories is to find an algebraic resolution of the algebraic models of the Postnikov sections, to be fed into an algebraic André–Quillen cohomology (as in [100]) or a nerve approach (as in [20] and [28]). The n-fold nature of our model of general n-types given by the category $\mathsf{GCat}^n_{\mathsf{wg}}$ of groupoidal weakly globular n-fold groupoids will make it possible to build such resolutions, as partially seen in some restricted contexts (see [58, 100, 101]). In low dimension (for categories enriched in groupoids) some progress in the general case was recently made by Blanc and Paoli [30].

13.2 Model Comparisons

Another area of application of weakly globular n-fold categories we envisage is within higher category theory itself, leading to further applications in different areas. The comparison between different models of weak n-categories is still largely open, and it involves connecting sometimes completely different types of combinatorics. Two models of weak n-categories in the literature are particularly developed both in terms of foundations and applications: the Tamsamani–Simpson model of weak n-categories, [119, 126], denoted Ta^n in this work, and the higher operadic model of Batanin [12], also studied by Berger [22], Leinster [84], Batanin and Weber [16] and others. It was conjectured from the early days of higher category theory that the comparison between these two models should be reminiscent of the classical result in homotopy theory on the comparison between the Segalic and operadic approach to iterated loop spaces established by May and Thomason [97].

Establishing a suitable comparison between the Segalic and the higher operadic models of weak n-categories is also important in term of applications, as it will allow the transfer of techniques in areas where both these models are used: for instance E_n-structures, loop space theory, and Hochschild cohomology for the Batanin model [13–15, 125], and non-abelian cohomology and higher stacks for the Tamsamani–Simpson model [119].

We propose to use weakly globular n-fold categories as a way to bridge between the Tamsamani–Simpson model and the higher operadic model of Batanin, eventually leading to a comparison between the two. A comparison between $\mathsf{Cat}^n_{\mathsf{wg}}$ and Ta^n has been established in this work. The n-fold nature of the category $\mathsf{Cat}^n_{\mathsf{wg}}$ is again crucial to the comparison with Batanin's model.

The category $\mathsf{Cat}^n_{\mathsf{wg}}$ is not cocomplete, so cannot admit a direct description as algebras over a monad as in Batanin's model [12]. However, we seek a monadic description of a larger subcategory of n-fold categories satisfying the weak globularity condition but not the induced Segal maps condition: this will build a first bridge with the higher operadic models.

This approach to the comparison problem will also bring the inductive nature of the Segal type models into the higher operadic models, therefore potentially linking to the approach to higher operads of Batanin, Cisinski and Weber [17]. The latter is not a fully weak model of higher categories, since it has strict units and weak composites, and no equivalence to the fully weak models has yet been proved. Our comparison result could lead to the proof of such a coherence theorem.

13.3 Intermediate Levels of Weakness

From the early days of higher category theory it was conjectured that there should exist a class of weak n-categories with a minimal amount of weakness beyond strict n-categories, called 'semistrict n-categories'. The semistrictification hypothesis is that there should be a coherence theorem proving that the semistrict structures are suitably equivalent to the fully weak ones. The power of such a coherence theorem is that it would reduce the complexity of weak higher categories, therefore making them easier to study and use in applications. Until now, this hypothesis has been proved only in low dimensions: for $n = 2$, where bicategories are equivalent to strict 2-categories [21], and for $n = 3$, where it was proved by Gordon, Power and Street [64] that tricategories (which are fully weak 3-categories) are equivalent to Gray categories (which are semistrict 3-categories).

In fact, it is conceivable that there should be models of higher categories whose level of weakness is intermediate between the fully weak and completely strict categories, with the 'semistrict' categories being the smallest beyond strict n-categories, and a corresponding coherence theorem saying that the fully weak categories should be suitably equivalent to these intermediate ones. Once again, a result of this type would reduce the complexity of the structure in their use in applications.

In [118], Simpson formulated a version of this hypothesis called the 'weak units conjecture'. This states that it should be possible to build a model of weak n-categories in which compositions are associative but units are weak, and this should be suitably equivalent to the fully weak models. Kock in [80] formulated a precise model that encodes the notions of weak units and strictly associative compositions, called 'fair n-categories', and formulated Simpson's weak units conjecture as stating that fair n-categories should be equivalent to Tamsamani n-categories. To date, this conjecture is open.

The Kock model is built inductively in a way that is similar to the Tamsamani model, but the simplicial category Δ is replaced by the category 'fat delta' $\underline{\Delta}$ of colored ordinals. Similarly to multi-simplicial objects in a category $\mathscr{C}$, functors from $\underline{\Delta}^{n^{op}}$ into $\mathscr{C}$ admit a notion of Segal maps, and these are used in [80] to encode in the model the associative compositions and weak units axioms.

One of the novelties of this work is the use of pseudo-functors to model higher structures. More precisely, a class of pseudo-functors from $\Delta^{n^{op}}$ to Cat called Segalic pseudo-functors was shown in Chap. 7 to connect the Tamsamani model to weakly globular n-fold categories. We envisage using pseudo-functors in the Kock model as a new tool to tackle the weak units conjecture.

13.4 Model Structures

An important open problem arising from this work is the model category theoretic treatment of weakly globular n-fold categories. Simpson in [119] gave a model category theoretic treatment of Tamsamani n-categories by putting a model structure on a larger (co)complete category and then recovering objects of Ta^{n} by a process of fibrant cofibrant replacement.

It would be desirable to provide a similar model category theoretic approach to the category $\mathsf{Cat}^{\mathsf{n}}_{\mathsf{wg}}$ and obtain a Quillen equivalence with Ta^{n}. Simpson's work relies on a very non-trivial technique of 'generators and relations' to build the model structure, and it would be interesting to extend this to our setting.

13.5 A Weakly Globular Approach to (∞, n)-Categories

Another interesting direction of future investigation is the extension of the current weakly globular approach to weak n-categories to the (∞, n)-case, and a subsequent comparison with the existing models of (∞, n)-categories. This extension is highly non-trivial.

The current Segal-type models of weak n-categories are developed inside the category Cat of small categories, that is, $\mathsf{Seg_n} \subset [\Delta^{n-1^{op}}, \mathsf{Cat}]$. A possible approach to extending to the (∞, n)-case is to replace Cat with the category $[\Delta^{^{op}}, \mathsf{Set}]$ of simplicial sets. A simplicial set X would be called homotopically discrete if the natural map $X \to qX$ (as in Lemma 2.2.7) is a weak equivalence. We can then build a category $\mathsf{Cat}^{\mathsf{n-1}}_{\mathsf{hd}}[\Delta^{^{op}}, \mathsf{Set}]$ of homotopically discrete $(n-1)$-fold categories in simplicial sets by iteratively using internal equivalence relations as in Sect. 5.1.3. This category could be used to formulate the weak globularity condition in the ∞-context. Next, we could build categories $\mathsf{Cat}^{\mathsf{n-1}}_{\mathsf{wg}}[\Delta^{^{op}}, \mathsf{Set}]$ and $\mathsf{Ta}^{\mathsf{n-1}}_{\mathsf{wg}}[\Delta^{^{op}}, \mathsf{Set}]$ in a way formally analogous to $\mathsf{Cat}^{\mathsf{n}}_{\mathsf{wg}}$ and $\mathsf{Ta}^{\mathsf{n}}_{\mathsf{wg}}$.

The notion in this book which is particularly difficult to extend to the (∞, n)-case is the rigidification functor. In this book, the rigidification functor from $\mathsf{Ta}^{\mathsf{n}}_{\mathsf{wg}}$ to $\mathsf{Cat}^{\mathsf{n}}_{\mathsf{wg}}$ factors through the category $\mathsf{SegPs}[\Delta^{n-1^{op}}, \mathsf{Cat}]$ of Segalic pseudo-functors. The latter does not extend in a straightforward way by replacing Cat with simplicial sets. Thus the (∞, n)-case calls for a different set of techniques to rigidify weak higher categorical structures into n-fold weakly globular ones.

Appendix A
Proof of Lemma 10.1.4

Lemma 10.1.4 *Let* $X \in \mathsf{Cat}^n_{\mathsf{wg}}$, $Y \in \mathsf{LTa}^n_{\mathsf{wg}}$ *be such that* $Y_{\underline{k}}$ *is discrete for all* $\underline{k} \in \Delta^{n-1^{op}}$ *such that* $k_j = 0$ *for some* $1 \leq j \leq n-1$. *Let* $\underline{k}$, $\underline{s} \in \Delta^{n-1^{op}}$ *and let* $\underline{k} \to \underline{s}$ *be a morphism in* $\Delta^{n-1^{op}}$. *Suppose that the following conditions hold:*

i) If $k_j, s_j \neq 0$ *for all* $1 \leq j \leq n-1$, *then* $X_{\underline{k}} = Y_{\underline{k}}$, $\quad X_{\underline{s}} = Y_{\underline{s}}$ *and the maps*

$$X_{\underline{k}} \to X_{\underline{s}}, \quad Y_{\underline{k}} \to Y_{\underline{s}}$$

coincide.

ii) If $k_j = 0$ *for some* $1 \leq j \leq n-1$ *and* $s_t = 0$ *for some* $1 \leq t \leq n-1$, *then* $X^d_{\underline{k}} = Y_{\underline{k}}$, $\quad X^d_{\underline{s}} = Y_{\underline{s}}$ *and the two maps*

$$X^d_{\underline{k}} \to X^d_{\underline{s}}, \quad Y_{\underline{k}} \to Y_{\underline{s}}$$

coincide, where $f^d : X^d_{\underline{k}} \to X^d_{\underline{s}}$ *is induced by* $f : X_{\underline{k}} \to X_{\underline{s}}$ *and thus also coincides with the composite*

$$X^d_{\underline{k}} \xrightarrow{\gamma'_{X_k}} X_{\underline{k}} \xrightarrow{f} X_{\underline{s}} \xrightarrow{\gamma_{X_s}} X^d_{\underline{s}}$$

(where γ *is the discretization map and* γ' *a section), since* $f^d = f^d \gamma_{X_k} \gamma'_{X_k} = \gamma_{X_s} f \gamma'_{X_k}$.

S. Paoli, *Simplicial Methods for Higher Categories*, Algebra and Applications 26,
https://doi.org/10.1007/978-3-030-05674-2

iii) If $k_j \neq 0$ for all $1 \leq j \leq n-1$ and $s_t = 0$ for some $1 \leq t \leq n-1$, the following diagram commutes

$$\begin{array}{ccccc} X_{\underline{k}} & \longrightarrow & X_{\underline{s}} & \xrightarrow{\gamma_{X_{\underline{s}}}} & X^d_{\underline{s}} = Y_{\underline{s}} \\ \Vert & & & \nearrow & \\ Y_{\underline{k}} & & & & \end{array}$$

where γ_{X_s} is the discretization map.

iv) If $k_j = 0$ for some $1 \leq j \leq n-1$ and $s_t \neq 0$ for all $1 \leq t \leq n-1$ then the following diagram commutes

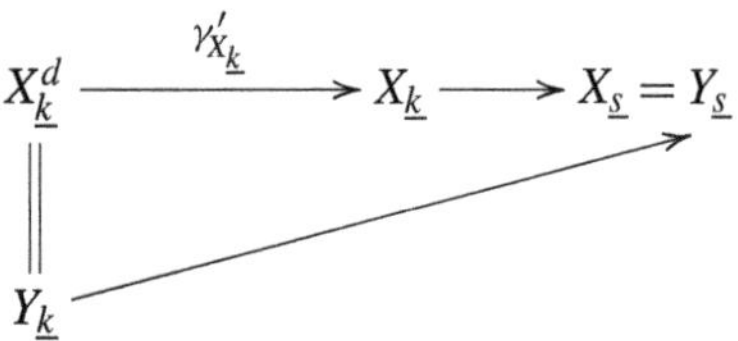

Then

a) For all $\underline{k} \in \Delta^{n-1^{op}}$, $(Tr_n X)_{\underline{k}} = (Tr_n Y)_{\underline{k}}$.

b) For all $\underline{k} \in \Delta^{n-1^{op}}$ such that $k_j \neq 0$ for all $1 \leq j \leq n-1$, the maps

$$(Tr_n X)_{\underline{k}} \rightleftarrows X_{\underline{k}}, \qquad (Tr_n Y)_{\underline{k}} \rightleftarrows Y_{\underline{k}}$$

coincide.

c) $Tr_n X = Tr_n Y$.

Proof By induction on n. Let $n = 2$. By definition of Tr_2 and by conditions i) and ii) in the hypothesis,

$$(Tr_2 X)_0 = X^d_0 = Y_0 = (Tr_2 Y)_0 \ ,$$

$$(Tr_2 X)_1 = X_1 = Y_1 = (Tr_2 Y)_1 \ .$$

Further, by hypothesis iii) the maps $\partial'_0, \partial'_1 : Y_1 \to Y_0$ are the composites

$$Y_1 = X_1 \underset{\partial_0}{\overset{\partial_0}{\rightrightarrows}} X_0 \xrightarrow{\gamma_{X_0}} X^d_0 = Y_0$$

that is, $\partial'_i = \gamma_{X_0} \partial_i$, $i = 0, 1$. This implies that for each $k \geq 2$

$$(Tr_2 X)_k = X_1 \times_{X^d_0} \overset{k}{\cdots} \times_{X^d_0} X_1 = Y_1 \times_{Y_0} \overset{k}{\cdots} \times_{Y_0} Y_1 = (Tr_2 Y)_k \ .$$

This proves a) when $n = 2$.

We now prove b) when $n = 2$. The map

$$(Tr_2 X)_1 = X_1 \to X_1$$

is the identity and by hypothesis i) coincides with

$$(Tr_2 Y)_1 = Y_1 \to Y_1 = X_1 \ .$$

When $k > 1$, the maps

$$(Tr_2 X)_k = X_1 \times_{X_0^d} \overset{k}{\cdots} \times_{X_0^d} X_1 \rightleftarrows X_1 \times_{X_0} \overset{k}{\cdots} \times_{X_0} X_1 \tag{A.1}$$

are the induced Segal maps for X and their pseudo-inverses. The induced Segal maps of X arise from the commuting diagram (see also Definition 2.1.3).

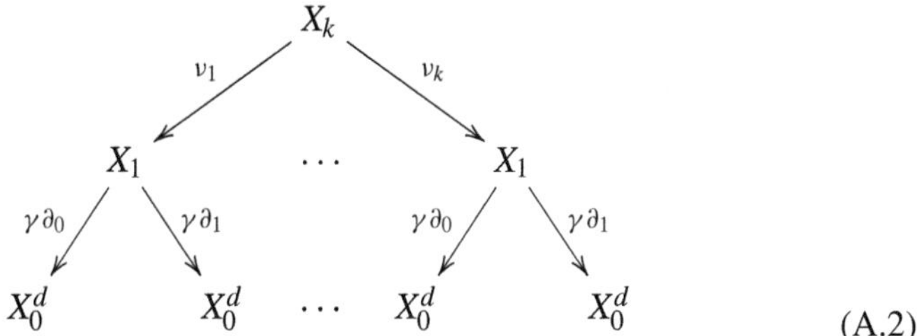

(A.2)

By hypothesis i) the maps

$$\nu_i : X_k \to X_1 \qquad \nu_i : Y_k \to Y_1$$

coincide; by hypothesis iii) the maps

$$X_1 \xrightarrow{\gamma\partial_i} X_0^d \qquad Y_1 \xrightarrow{\partial_i'} Y_0$$

coincide. Thus (A.2) coincides with

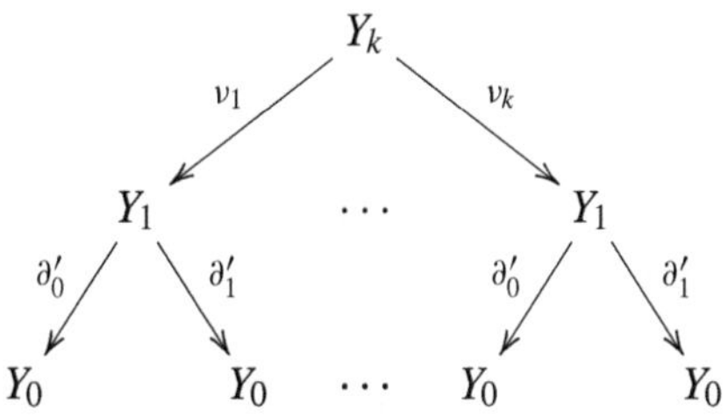

so the induced Segal maps of X and Y coincide. So by (A.1) the maps

$$(Tr_2X)_k \rightleftarrows X_k, \qquad (Tr_2Y)_k \rightleftarrows Y_k$$

coincide, proving b) when $n = 2$.

To show c) when $n = 2$ we first show that, for each morphism $k \to s$ in Δ^{op}, the maps

$$(Tr_2X)_k \to (Tr_2X)_s, \qquad (Tr_2Y)_k \to (Tr_2Y)_s \tag{A.3}$$

coincide. By the proof of Lemma 4.3.2 these maps are the composites

$$(Tr_2X)_k \to X_k \to X_s \to (Tr_2X)_s \ , \tag{A.4}$$

$$(Tr_2Y)_k \to Y_k \to Y_s \to (Tr_2Y)_s \ . \tag{A.5}$$

Let $k > 0$ and $s > 0$. Then by b) and by hypothesis i), the maps (A.4) and (A.5) coincide.

Let $k = 0$ and $s > 0$. Then (A.4) and (A.5) are given by

$$(Tr_2X)_0 = X_0^d \to X_0 \to X_s = (Tr_2X)_s \ ,$$

$$(Tr_2Y)_0 = Y_0 \to Y_s \ ,$$

and these coincide by hypothesis iv).

Suppose $k > 0$ and $s = 0$. Then (A.4) and (A.5) are given by

$$(Tr_2X)_k \to X_k \to X_0 \to X_0^d = (Tr_2X)_0 \ , \tag{A.6}$$

$$(Tr_2Y)_k \to Y_k \to Y_0 = (Tr_2Y)_0 \ . \tag{A.7}$$

By b), the maps

$$(Tr_2X)_k \to X_k, \qquad (Tr_2Y)_k \to Y_k$$

coincide while by hypothesis iii) the maps

$$X_k \to X_0 \to X_0^d, \qquad Y_k \to Y_0$$

coincide. Therefore (A.6) and (A.7) coincide.

If $k = s = 0$, the composite

$$(Tr_2X)_0 = X_0^d \to X_0 \xrightarrow{f_{X_0}} X_0 \to X_0^d$$

is equal to $f^d_{X_0}$, which coincides with $f^d_{Y_0}$ by hypothesis ii).

We conclude that (A.4) and (A.5) always coincide.

By the definition of pseudo-functor (see Definition 4.2.3) in order to prove that $Tr_2X = Tr_2Y$ it remains to show that, given morphisms $k \to s \to r$ in Δ^{op}, the 2-dimensional pasting diagrams

$$\begin{array}{c}(Tr_2X)_k \longrightarrow (Tr_2X)_s \longrightarrow (Tr_2X)_r \\ \Downarrow \end{array}$$

$$\begin{array}{c}(Tr_2Y)_k \longrightarrow (Tr_2Y)_s \longrightarrow (Tr_2Y)_r \\ \Downarrow \end{array}$$

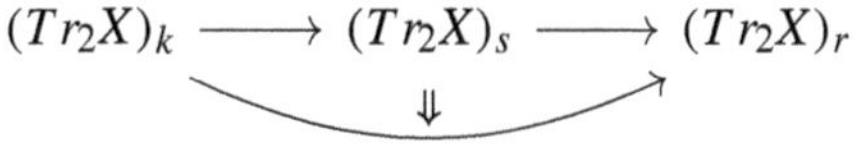

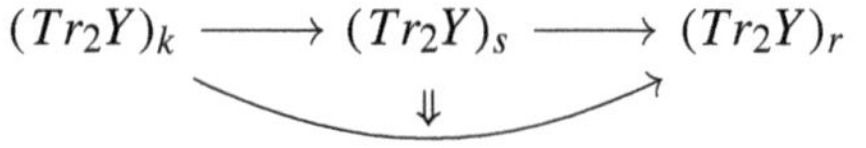

coincide and, given $id_k : k \to k$ in Δ^{op}, the 2-dimensional pasting diagrams

$$\begin{array}{c}(Tr_2X)_k \xrightarrow{(Tr_2X)(\mathrm{Id}_k)} (Tr_2X)_k \\ \Downarrow \\ \mathrm{Id}\end{array}$$

$$\begin{array}{c}(Tr_2Y)_k \xrightarrow{(Tr_2Y)(\mathrm{Id}_k)} (Tr_2Y)_k \\ \Downarrow \\ \mathrm{Id}\end{array}$$

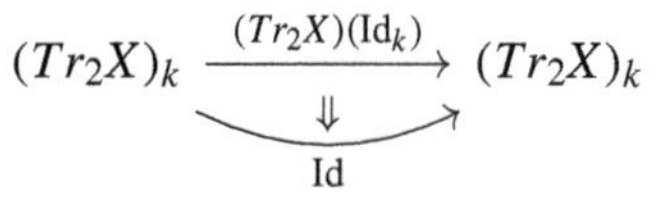

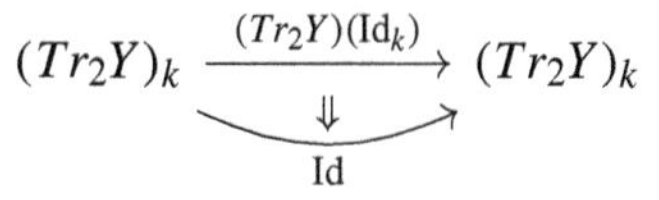

coincide.

The proof of this is as in the inductive step on pages 329 to 330. In fact, the proof of these parts of the inductive step only uses a) and b) and the equality of the maps (A.3), all of which have been proved for the case $n = 2$, but it does not use the equality of the 2-dimensional pasting diagrams at step $(n-1)$. We therefore refer the reader to the later part of this proof for this step. This concludes the proof of the lemma in the case $n = 2$.

Suppose, inductively, that the lemma holds for $(n-1)$ and let X, Y be as in the hypothesis.

a) By definition of Tr_n and by hypothesis ii), for all $\underline{s} \in \Delta^{n-2^{op}}$,

$$(Tr_nX)_{(0,\underline{s})} = X^d_{(0,\underline{s})} = Y^d_{(0,\underline{s})} = (Tr_nY)_{(0,\underline{s})} \ .$$

Clearly $X_1 \in \mathsf{Cat}^{n-1}_{\mathsf{wg}}$ and $Y_1 \in \mathsf{LTa}^{n-1}_{\mathsf{wg}}$ satisfy the inductive hypothesis. Thus, using the definition of Tr_n and inductive hypothesis a) on X_1, Y_1 we obtain

$$(Tr_nX)_{(1,\underline{s})} = (Tr_{n-1}X)_{\underline{s}} = (Tr_{n-1}Y)_{\underline{s}} = (Tr_nY)_{(1,\underline{s})} \ .$$

We claim that, for all $\underline{s} \in \Delta^{n-2^{op}}$, the maps

$$(Tr_{n-1}X_1)_{\underline{s}} \to X_{1\underline{s}} \to X_{0\underline{s}} \to X^d_{0\underline{s}} \tag{A.8}$$

$$(Tr_{n-1}Y_1)_{\underline{s}} \to Y_{1\underline{s}} \longrightarrow Y_{0\underline{s}} \tag{A.9}$$

coincide. In fact, suppose $s_j \neq 0$ for all $1 \leq j \leq n-2$. Then by inductive hypothesis b) applied to X_1, Y_1 the maps

$$(Tr_{n-1}X_1)_{\underline{s}} \to X_{1\underline{s}}, \qquad (Tr_{n-1}Y_1)_{\underline{s}} \to Y_{1\underline{s}}$$

coincide, while by hypothesis iii) the maps

$$X_{1\underline{s}} \to X_{0\underline{s}} \to X^d_{0\underline{s}}, \qquad Y_{1\underline{s}} \to Y_{0\underline{s}}$$

coincide. Thus the composites (A.8) and (A.9) coincide.

Suppose $s_j = 0$ for some $1 \leq j \leq n-2$. Then by Corollary 10.1.2

$$(Tr_{n-1}X_1)_{\underline{s}} = X^d_{1\underline{s}}, \qquad (Tr_{n-1}Y_1)_{\underline{s}} = Y^d_{1\underline{s}}\,.$$

Thus the maps (A.8) and (A.9) are given by

$$X^d_{1\underline{s}} \to X_{1\underline{s}} \to X_{0\underline{s}} \to X^d_{0\underline{s}}, \qquad Y_{1\underline{s}} \to Y_{0\underline{s}}$$

and these coincide by hypothesis ii). This proves the claim. From this claim, inductive hypothesis a) on X_1, Y_1 and the definition of Tr_n it follows that, for each $k_1 > 1$, $\underline{k} = (k_1, \underline{s})$

$$\begin{aligned}(Tr_nX)_{\underline{k}} &= (Tr_{n-1}X_1)_{\underline{s}} \times_{X^d_{0\underline{s}}} \overset{k_1}{\cdots} \times_{X^d_{0\underline{s}}} (Tr_{n-1}X_1)_{\underline{s}} \\ &= (Tr_{n-1}Y_1)_{\underline{s}} \times_{Y_0} \overset{k_1}{\cdots} \times_{Y_0} (Tr_{n-1}Y_1)_{\underline{s}} = (Tr_nY)_{\underline{k}}\,.\end{aligned}$$

This proves a).

b) Let $\underline{k} = (k_1, \underline{s})$ with $k_1 \neq 0$ and $s_j \neq 0$ for all $1 \leq j \leq n-2$. By induction hypothesis b) applied to X_1 and Y_1, the maps

$$\begin{aligned}(Tr_nX)_{(1,\underline{s})} &= (Tr_{n-1}X_1)_{\underline{s}} \rightleftarrows X_{1\underline{s}}\,, \\ (Tr_nY)_{(1,\underline{s})} &= (Tr_{n-1}Y_1)_{\underline{s}} \rightleftarrows Y_{1\underline{s}}\end{aligned}$$

coincide. This implies that, for each $k_1 > 1$, the maps

$$(Tr_n X)_{\underline{k}} = (Tr_{n-1}X_1)_{\underline{s}} \times_{X^d_{0\underline{s}}} \overset{k_1}{\cdots} \times_{X^d_{0\underline{s}}} (Tr_{n-1}X_1)_{\underline{s}} \to X_{1\underline{s}} \times_{X^d_{0\underline{s}}} \overset{k_1}{\cdots} \times_{X^d_{0\underline{s}}} X_{1\underline{s}}\,, \tag{A.10}$$

$$(Tr_n Y)_{\underline{k}} = (Tr_{n-1}Y_1)_{\underline{s}} \times_{Y_{0\underline{s}}} \overset{k_1}{\cdots} \times_{Y_{0\underline{s}}} (Tr_{n-1}Y_1)_{\underline{s}} \to Y_{1\underline{s}} \times_{Y_{0\underline{s}}} \overset{k_1}{\cdots} \times_{Y_{0\underline{s}}} Y_{1\underline{s}} \tag{A.11}$$

coincide. On the other hand, reasoning as in the case $n = 2$, we see that the maps

$$X_{1\underline{s}} \times_{X^d_{0\underline{s}}} \overset{k_1}{\cdots} \times_{X^d_{0\underline{s}}} X_{1\underline{s}} \rightleftarrows X_{1\underline{s}} \times_{X_{0\underline{s}}} \overset{k_1}{\cdots} \times_{X_{0\underline{s}}} X_{1\underline{s}} = X_{k_1\underline{s}}\,, \tag{A.12}$$

$$Y_{1\underline{s}} \times_{Y_{0\underline{s}}} \overset{k_1}{\cdots} \times_{Y_{0\underline{s}}} Y_{1\underline{s}} \rightleftarrows Y_{k_1\underline{s}} \tag{A.13}$$

coincide. Composing (A.10) with (A.12) and (A.11) with (A.13) we therefore conclude that the maps

$$(Tr_n X)_{\underline{k}} \rightleftarrows X_{\underline{k}}, \qquad (Tr_n Y)_{\underline{k}} \rightleftarrows Y_{\underline{k}}$$

coincide for each $\underline{k}$ such that $k_j \neq 0$ for all $1 \le j \le n-1$. This proves b).

c) By a) and by the definition of pseudo-functor (see Definition 4.2.3), in order to prove that $Tr_n X = Tr_n Y$ it remains to show that:

   I) For each morphism $\underline{k} \to \underline{s}$ in $\Delta^{n-1^{op}}$, the maps

   $$(Tr_n X)_{\underline{k}} \to (Tr_n X)_{\underline{s}}, \qquad (Tr_n Y)_{\underline{k}} \to (Tr_n Y)_{\underline{s}}$$

   coincide.

   II) Given morphisms $\underline{k} \to \underline{s} \to \underline{r}$ in $\Delta^{n-1^{op}}$ the 2-dimensional pasting diagrams

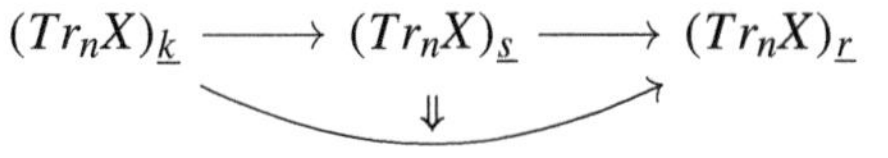

   $$\begin{array}{ccccc} (Tr_n X)_{\underline{k}} & \longrightarrow & (Tr_n X)_{\underline{s}} & \longrightarrow & (Tr_n X)_{\underline{r}} \\ & \searrow & \Downarrow & \nearrow & \end{array} \tag{A.14}$$

   $$\begin{array}{ccccc} (Tr_n Y)_{\underline{k}} & \longrightarrow & (Tr_n Y)_{\underline{s}} & \longrightarrow & (Tr_n Y)_{\underline{r}} \\ & \searrow & \Downarrow & \nearrow & \end{array} \tag{A.15}$$

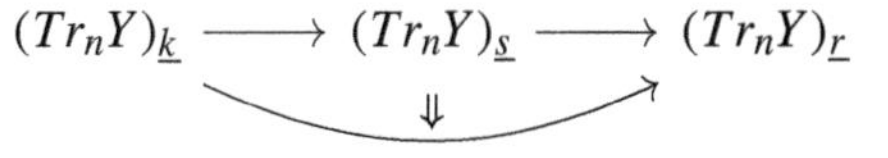

   coincide.

III) Given $id_{\underline{k}} : \underline{k} \to \underline{k}$ in $\Delta^{n-1^{op}}$, the 2-dimensional pasting diagrams

$$(Tr_nX)_{\underline{k}} \underset{\text{Id}}{\overset{(Tr_nX)(\text{Id}_{\underline{k}})}{\underset{\Downarrow}{\rightrightarrows}}} (Tr_nX)_{\underline{k}} \tag{A.16}$$

$$(Tr_nY)_{\underline{k}} \underset{\text{Id}}{\overset{(Tr_nY)(\text{Id}_{\underline{k}})}{\underset{\Downarrow}{\rightrightarrows}}} (Tr_nY)_{\underline{k}} \tag{A.17}$$

coincide.

I) By the proof of Lemma 4.3.2, these maps are given as composites

$$(Tr_nX)_{\underline{k}} \to X_{\underline{k}} \to X_{\underline{s}} \to (Tr_nX)_{\underline{s}}\ , \tag{A.18}$$

$$(Tr_nY)_{\underline{k}} \to Y_{\underline{k}} \to Y_{\underline{s}} \to (Tr_nY)_{\underline{s}}\ . \tag{A.19}$$

Suppose that $k_j \neq 0$, $s_j \neq 0$ for all $1 \leq j \leq n-1$. Then by b) and by hypothesis i), (A.18) and (A.19) coincide.

Suppose that $k_j = 0$ for some $1 \leq j \leq n-1$ and $s_t = 0$ for some $1 \leq t \leq n-1$. Then by Corollary 10.1.2, (A.18) and (A.19) are given by the composites

$$\begin{aligned} &(Tr_nX)_{\underline{k}} = X^d_{\underline{k}} \to X_{\underline{k}} \to X_{\underline{s}} \to (Tr_nX)_{\underline{s}} = X^d_{\underline{s}}\ , \\ &(Tr_nY)_{\underline{k}} = Y_{\underline{k}} \longrightarrow Y_{\underline{s}} \end{aligned}$$

and these coincide by hypothesis ii). Suppose that $k_j = 0$ for some $1 \leq j \leq n-1$ and $s_i \neq 0$ for all $1 \leq i \leq n-1$. Then (A.18) and (A.19) are given by the composites

$$(Tr_nX)_{\underline{k}} = X^d_{\underline{k}} \to X_{\underline{k}} \to X_{\underline{s}} \to (Tr_nX)_{\underline{s}}\ , \tag{A.20}$$

$$(Tr_nY)_{\underline{k}} = Y_{\underline{k}} \longrightarrow Y_{\underline{s}} \to (Tr_nY)_{\underline{s}}\ . \tag{A.21}$$

By a) the maps

$$X_{\underline{s}} \to (Tr_nX)_{\underline{s}}, \qquad Y_{\underline{s}} \to (Tr_nY)_{\underline{s}}$$

coincide; and by hypothesis iv) the maps

$$X^d_{\underline{k}} \to X_{\underline{k}} \to X_{\underline{s}}, \qquad Y_{\underline{k}} \to Y_{\underline{s}}$$

coincide. Hence by composing, we deduce that (A.20) and (A.21) coincide. In conclusion the maps (A.18) and (A.19) always coincide, proving i).

II) We distinguish the following eight cases. For each we refer to diagrams on pages 331 and 332. Using a), b) and the hypotheses we see that in each case the left and right pasting diagrams coincide.

Case 1: $k_i \neq 0$ for all $1 \leq i \leq (n-1)$; $s_j \neq 0$ for all $1 \leq j \leq (n-1)$; $r_t \neq 0$ for all $1 \leq t \leq (n-1)$.

Case 2: $k_i = 0$ for some $1 \leq i \leq (n-1)$; $s_j \neq 0$ for all $1 \leq j \leq (n-1)$; $r_t \neq 0$ for all $1 \leq t \leq (n-1)$.
Note that by hypothesis iv) the map $Y_{\underline{k}} \to Y_{\underline{s}}$ factors as

$$Y_{\underline{k}} = X^d_{\underline{k}} \to X_{\underline{k}} \to X_{\underline{s}} = Y_{\underline{s}}.$$

Case 3: $k_i \neq 0$ for all $1 \leq i \leq (n-1)$; $s_j = 0$ for some $1 \leq j \leq (n-1)$; $r_t \neq 0$ for all $1 \leq t \leq (n-1)$.
Note that by hypotheses iii) and iv) the maps

$$Y_{\underline{k}} \to Y_{\underline{s}}, \qquad Y_{\underline{s}} \to Y_{\underline{r}}$$

factor as

$$Y_{\underline{k}} = X_{\underline{k}} \to X_{\underline{s}} \to X^d_{\underline{s}} = Y_{\underline{s}}, \quad Y_{\underline{s}} = X^d_{\underline{s}} \to X_{\underline{s}} \to X_{\underline{r}} = Y_{\underline{r}}.$$

Case 4: $k_i \neq 0$ for all $1 \leq i \leq (n-1)$; $s_j \neq 0$ for all $1 \leq j \leq (n-1)$; $r_t = 0$ for some $1 \leq t \leq (n-1)$.
Note that by hypothesis iv) the map $Y_{\underline{s}} \to Y_{\underline{r}}$ factors as

$$Y_{\underline{s}} = X_{\underline{s}} \to X_{\underline{r}} \to X^d_{\underline{r}} = Y_{\underline{r}}.$$

Case 5: $k_i \neq 0$ for all $1 \leq i \leq (n-1)$; $s_j = 0$ for some $1 \leq j \leq (n-1)$; $r_t = 0$ for some $1 \leq t \leq (n-1)$.
Note that by hypotheses ii) and iv) the maps

$$Y_{\underline{k}} \to Y_{\underline{s}}, \qquad Y_{\underline{s}} \to Y_{\underline{r}}$$

factor as

$$Y_{\underline{k}} = X_{\underline{k}} \to X_{\underline{s}} \to X^d_{\underline{s}} = Y_{\underline{s}}, \quad Y_{\underline{s}} = X^d_{\underline{s}} \to X_{\underline{s}} \to X_{\underline{r}} \to X^d_{\underline{r}} = Y_{\underline{r}}.$$

Case 6: $k_i = 0$ for some $1 \le i \le (n-1)$; $s_j \neq 0$ for all $1 \le j \le (n-1)$; $r_t = 0$ for some $1 \le t \le (n-1)$.

Note that by hypotheses iii) and iv) the maps

$$Y_{\underline{k}} \to Y_{\underline{s}}, \qquad Y_{\underline{s}} \to Y_{\underline{r}}$$

factor as

$$Y_{\underline{k}} = X^d_{\underline{k}} \to X_{\underline{k}} \to X_{\underline{s}} = Y_{\underline{s}}, \qquad Y_{\underline{s}} = X_{\underline{s}} \to X_{\underline{r}} \to X^d_{\underline{r}} = Y_{\underline{r}}.$$

Case 7: $k_i = 0$ for some $1 \le i \le (n-1)$; $s_j = 0$ for some $1 \le j \le (n-1)$; $r_t \neq 0$ for all $1 \le t \le (n-1)$.

Note that by hypotheses ii) and iv) the maps

$$Y_{\underline{k}} \to Y_{\underline{s}}, \qquad Y_{\underline{s}} \to Y_{\underline{r}}$$

factor as

$$Y_{\underline{k}} = X^d_{\underline{k}} \to X_{\underline{k}} \to X_{\underline{s}} \to X^d_{\underline{s}} = Y_{\underline{s}}, \qquad Y_{\underline{s}} = X^d_{\underline{s}} \to X_{\underline{s}} \to X_{\underline{r}} = Y_{\underline{r}}.$$

Case 8: $k_i = 0$ for some $1 \le i \le (n-1)$; $s_j = 0$ for some $1 \le j \le (n-1)$; $r_t = 0$ for some $1 \le t \le (n-1)$.

Note that by hypothesis ii) the maps

$$Y_{\underline{k}} \to Y_{\underline{s}}, \qquad Y_{\underline{s}} \to Y_{\underline{r}}$$

factor as

$$Y_{\underline{k}} = X^d_{\underline{k}} \to X_{\underline{k}} \to X_{\underline{s}} \to X^d_{\underline{s}} = Y_{\underline{s}}, \quad Y_{\underline{s}} = X^d_{\underline{s}} \to X_{\underline{s}} \to X_{\underline{r}} \to X^d_{\underline{r}} = Y_{\underline{r}}.$$

III) Suppose that $k_i \neq 0$ for all $1 \le i \le (n-1)$. Then A.16 and A.17 are given by

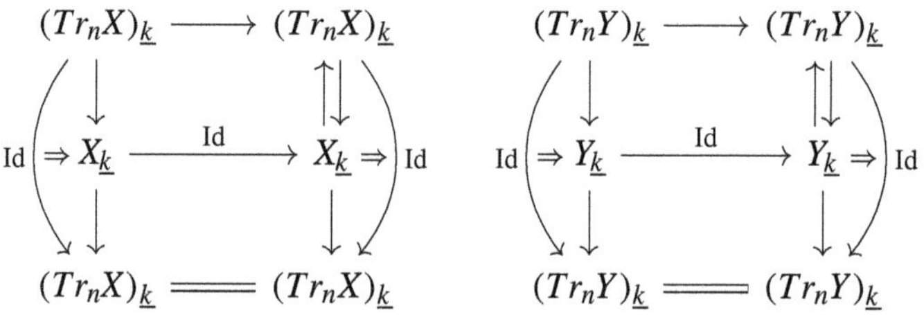

and these coincide by a), b) and hypothesis i).

Suppose that $k_i = 0$ for some $1 \leq i \leq (n-1)$. Then A.16 and A.17 are given by

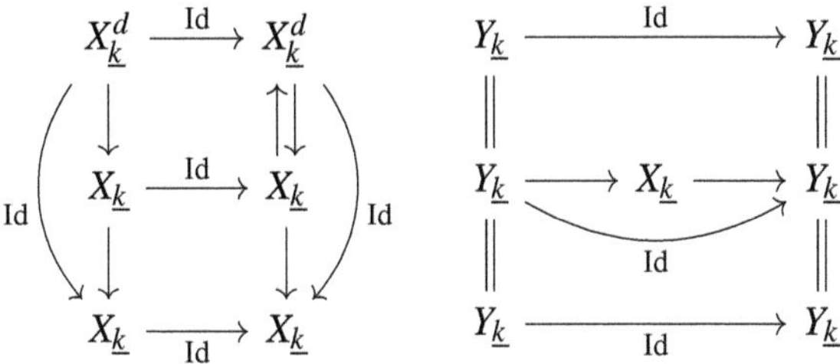

and they coincide as $Y_{\underline{k}} = X^d_{\underline{k}}$.

Case 1

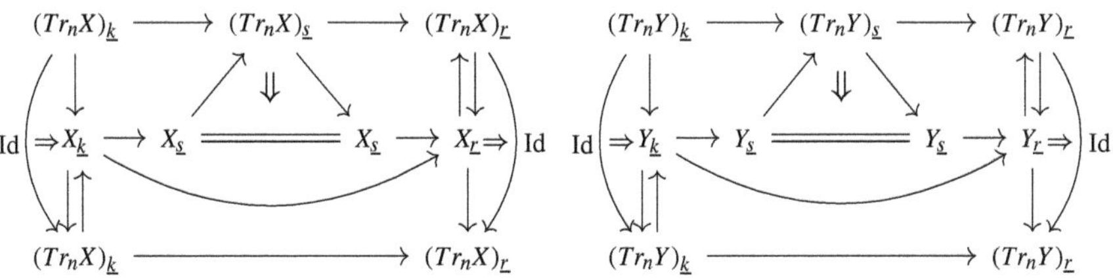

Case 2

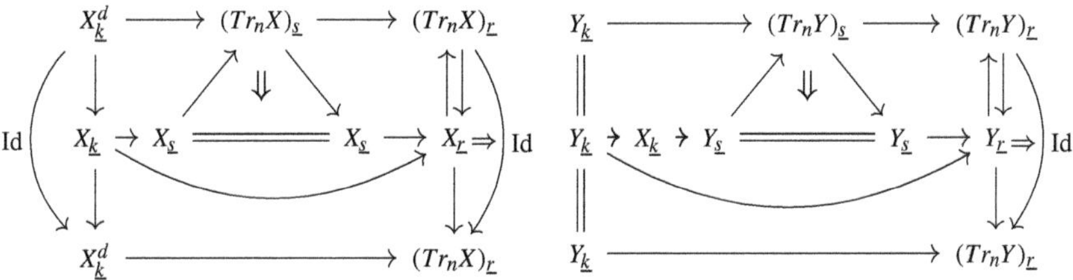

Case 3

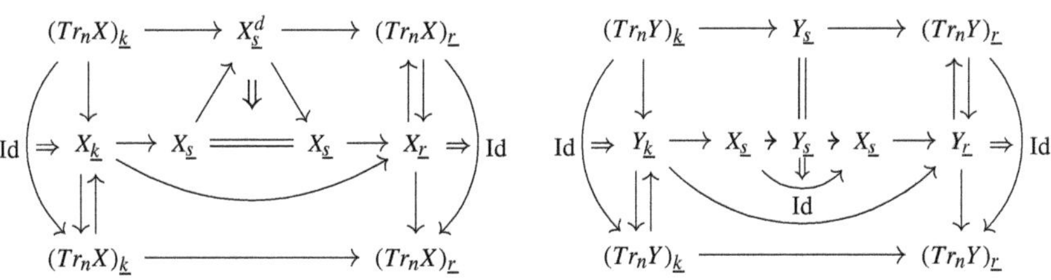

Case 4

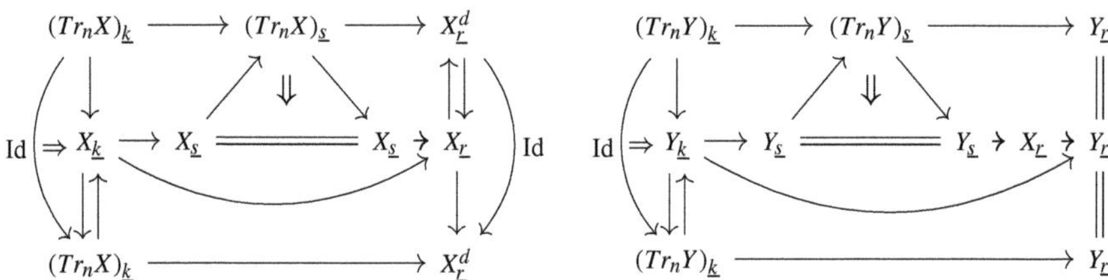

Case 5

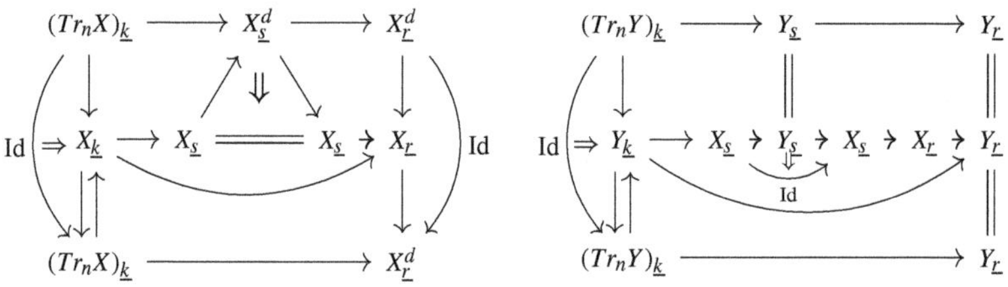

Case 6

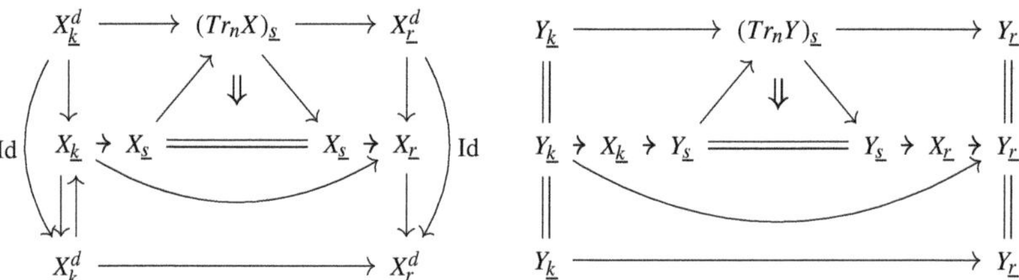

Case 7

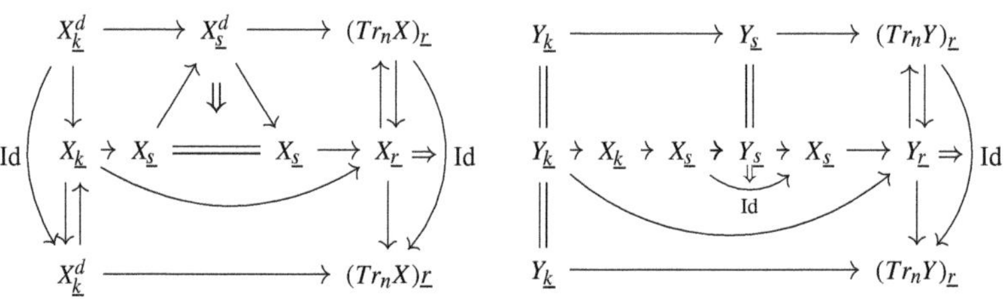

Case 8

$X^d_{\underline{k}}$
$X^d_{\underline{s}}$
$X^d_{\underline{r}}$
Id
$X_{\underline{k}}$
$X_{\underline{s}}$
$X_{\underline{s}}$
$X_{\underline{r}}$
Id
$X^d_{\underline{k}}$
$X^d_{\underline{r}}$
$Y_{\underline{k}}$
$Y_{\underline{s}}$
$Y_{\underline{r}}$
$Y_{\underline{k}}$
$X_{\underline{k}}$
$X_{\underline{s}}$
$Y_{\underline{s}}$
$X_{\underline{s}}$
$X_{\underline{r}}$
$Y_{\underline{r}}$
Id
$Y_{\underline{k}}$
$Y_{\underline{r}}$

□

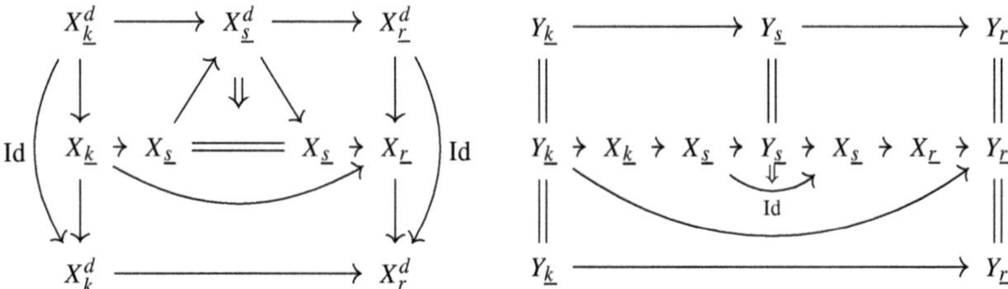

References

1. Ara, D.: Higher quasi-categories vs higher Rezk spaces. J. K-Theory **14**(3), 701–749 (2014)
2. Awodey, S., Warren, M.A.: Homotopy theoretic models of identity types. Math. Proc. Cambridge Philos. Soc. **146**(1), 45–55 (2009)
3. Baez, J.C., Dolan, J.: Higher dimensional algebra and topological quantum field theory. J. Math. Phys **36**, 6073–6105 (1995)
4. Baez, J.C., Dolan, J.: Higher dimensional algebra III: n-categories and the algebra of opetopes. Adv. Math. **135**, 145–206 (1998)
5. Bar, K., Vicary, J.: Data structures for quasistrict higher categories. In: Proceedings on 32th Annual ACM/IEEE Symposium on Logic and Computer Science (2017)
6. Bar, K., Kissinger, A., Vicary, J.: Globular: an online proof assistant for higher-dimensional rewriting. Logical Methods Comput. Sci. **14**(1), paper no. 8 (2018)
7. Barwick, C.: From operator categories to higher operads. Geom. Topol. **22**(4), 1893–1959 (2018)
8. Barwick, C., Kan, D.M.: A characterization of simplicial localization functors and a discussion of DK equivalences. Indag. Math. **23**, 69–79 (2012)
9. Barwick, C., Kan, D.M.: Relative categories: another model for the homotopy theory of homotopy theories. Indag. Math. (N.S.) **23**, 42–68 (2012)
10. Barwick, C., Kan, D.M.: n-relative categories: a model for the homotopy theory of n-fold homotopy theories. Homology Homotopy Applic. **15**(2), 281–300 (2013)
11. Barwick, C., Schommer-Pries, C.: On the unicity of the homotopy theory of higher categories. ArXiv:1112.0040v4
12. Batanin, M.A.: Monoidal globular categories as a natural environment for the theory of weak n-categories. Adv. Math. **136**(1), 39–103 (1998)
13. Batanin, M.A.: Symmetrization of n-operads and compactification of real configuration spaces. Adv. Math. **211**(2), 684–725 (2007)
14. Batanin, M.A.: The Eckmann-Hilton argument and higher operads. Adv. Math. **217**(1), 334–385 (2008)
15. Batanin, M., Markl, M.: Operadic categories and duoidal Deligne's conjecture. Adv. Math. **285**, 1630–1687 (2015)
16. Batanin, M., Weber, M.: Algebras of higher operads as enriched categories. Appl. Categ. Structures **19**(1), 93–135 (2011)
17. Batanin, M., Cisinski, D.C., Weber, M.: Multitensor lifting and strictly unital higher category theory. Theory Appl. Categ. **28**(25), 804–856 (2013)
18. Baues, H.J.: Combinatorial homotopy and 4-dimensional complexes. In: De Gruyter Expositions in Mathematics, vol. 2. Walter de Gruyter, Berlin (1991)

S. Paoli, *Simplicial Methods for Higher Categories*, Algebra and Applications 26,
https://doi.org/10.1007/978-3-030-05674-2

19. Baues, H.J.: The algebra of secondary cohomology operations. In: Progress in Mathematics, vol. 247. Birkhäuser-Springer, Basel (2006)
20. Baues, H., Wirsching, G.: The cohomology of small categories. J. Pure Appl. Alg. **38**, 187–211 (1985)
21. Bénabou, J.: Introduction to bicategories. In: Reports of the Midwest Category Seminar. Springer, Berlin (1967)
22. Berger, C.: Iterated wreath product of the simplex category and iterated loop spaces. Adv. Math. **213**(1), 230–270 (2007)
23. Bergner, J.: Three models for the homotopy theory of homotopy theories. Topology **46**, 397–436 (2007)
24. Bergner, J.E.: A survey of $(\infty, 1)$-categories. towards higher categories. IMA Math. Appl. **152**, 69–83 (2010)
25. Bergner, J.: Models for (∞, n)-categories and the cobordism hypothesis. In: Proceedings of Symposia in Pure Mathematics, vol. 83. American Mathematical Society, Providence (2011)
26. Bergner, J.E.: The homotopy theory of $(\infty, 1)$-categories. In: London Mathematical Society Student Texts, vol. 90. Cambridge University Press, Cambridge (2018)
27. Bergner, J., Rezk, C.: Comparison of models for (∞, n)-categories 1. Geom. Topol. **17**(4), 2163–2202 (2013)
28. Blanc, D., Paoli, S.: Two-track categories. J. K-theory **8**(1), 59–106 (2011)
29. Blanc, D., Paoli, S.: Segal-type algebraic models of n-types. Algebr. Geom. Topol. **14**, 3419–3491 (2014)
30. Blanc, D., Paoli, S.: Comonad cohomology of track categories. J. Homotopy Relat. Struct. (to appear)
31. Boardman, J.M., Vogt, R.M.: Homotopy Invariant Algebraic Structures on Topological Spaces. Lecture Notes in Mathematics, vol. 347. Springer, Berlin (1973)
32. Borceux, F.: Handbook of Categorical Algebra. Cambridge University Press, Cambridge (1994)
33. Bourn, D.: The shift functor and the comprehensive factorization for internal groupoids. Cah. Top. Géom. Différ. Catég. **28**(3), 197–226 (1987)
34. Brown, R., Loday, J.L.: Van Kampen theorems for diagrams of spaces. Topology **26**(3), 311–335 (1987)
35. Brown, R., Spencer, C.B.: Double groupoids and crossed modules. Cah. Top. Géom. Différ. Catég. **17**(4), 343–362 (1976)
36. Brown, R., Higgins, P.J., Sivera, R.: Nonabelian algebraic topology : filtered spaces, crossed complexes, cubical homotopy groupoids. With contributions by Christopher D. Wensley and Sergei V. Soloviev. Eur. Math. Soc. Tracks Math. **15** (2011)
37. Calanque, D., Scheimbauer, C.: A note on the (∞, n)-category of bordisms (2018). Preprint arXiv:1509.08906V2
38. Carrasco, P.C., Cegarra, A.M.: Group theoretic algebraic models for homotopy types. J. Pure Appl. Alg. **75**, 195–235 (1991)
39. Cheng, E.: Weak n-categories: opetopic and multitopic foundations. J. Pure Appl. Alg. **186**(2), 109–137 (2004)
40. Cheng, E.: Comparing operadic theories of n-category. Homology Homotopy Applic. **13**(2), 217–249 (2011)
41. Chu, H., Haugseng, R., Heuts, G.: Two models for the homotopy theory of ∞-operads. J. Topol. **11**(4), 856–872 (2018)
42. Cisinski, D.C., Moerdijk, I.: Dendroidal sets as models for homotopy operads. J. Topol. **4**(2), 257–299 (2011)
43. Cisinski, D.C., Moerdijk, I.: Dendroidal sets and simplicial operads. J. Topol. **6**(3), 705–756 (2013)
44. Coecke, B., Kissinger, A.: Picturing Quantum Processes: A First Course in Quantum Theory and Diagrammatic Reasoning. Cambridge University Press, Cambridge (2017)
45. Conduché, D.: Modules croisés généralisés de longueur 2. J. Pure Appl. Alg. **34**(2-3), 155–178 (1984)

46. Dawson, R., Pare, R., Pronk, D.: The span construction. Theory Appl. Categ. **24**(13), 302–377 (2010)
47. Duskin, J.W.: Simplicial matrices and the nerves of weak n-categories, 1: nerves of bicategories. Theory Appl. Categ. **9**, 198–308 (2000)
48. Dwyer, W., Kan, D.: Calculating simplicial localizations. J. Pure Appl. Alg. **18**, 17–35 (1980)
49. Dwyer, W., Kan, D.: Function complexes in homotopical algebra. J. Pure Appl. Alg. **19**, 427–440 (1980)
50. Dwyer, W., Kan, D.: Simplicial localizations of categories. J. Pure Appl. Alg. **17**(3), 267–284 (1980)
51. Dwyer, W., Kan, D., Smith, J.: An obstruction theory for simplicial categories. Proc. Kon. Ned. Akad. Wet. Ind. Math. **89**(2), 153–161 (1986)
52. Dwyer, W., Kan, D., Smith, J.: Homotopy commutative diagrams and their realizations. J. Pure Appl. Alg. **57**, 5–24 (1989)
53. Ehresmann, C.: Catégories doubles et catégories structurées. C. R. Acad. Sci. Paris **256**, 1198–1201 (1963)
54. Ehresmann, A., Ehresmann, C.: Multiple functors II. The monoidal closed category of multiple categories. Cah. Top. Géom. Differ. Catég. **19**(3), 295–334 (1978)
55. Ehresmann, A., Ehresmann, C.: Multiple functors III. The cartesian close category cat^n. Cah. Top. Géom. Differ. Catég. **19**(4), 387–443 (1978)
56. Ehresmann, A., Ehresmann, C.: Multiple functors IV. Monoidal closed structures on cat^n. Cah. Top. Géom. Differ. Catég. **20**(1), 59–104 (1979)
57. Ellis, G.J., Steiner, R.J.: Higher-dimensional crossed modules and the homotopy groups of $(n + 1)$-ads. J. Pure Appl. Alg. **46**, 117–136 (1987)
58. Everaert, T., Gran, M.: Homology of n-fold groupoids. Theory Appl. Categ. **23**(2), 22–41 (2010)
59. Fiore, T.M., Paoli, S.: A Thomason model structure on the category of small n-fold categories. Algebr. Geom. Topol. **10**, 1933–2008 (2010)
60. Fiore, T.M., Paoli, S., Pronk, D.A.: Model structures on the category of small double categories. Algebr. Geom. Topol. **8**, 1855–1959 (2008)
61. Garner, R.: Understanding the small object argument. Appl. Categ. Structures **20**(2), 103–141 (2012)
62. Garner, R., Gurski, N.: The low-dimensional structures formed by tricategories. Math. Proc. Camb. Philos. Soc. **146**(3), 551–589 (2009)
63. Goerss, P.G., Jardine, J.F.: Simplicial Homotopy Theory. Birkhauser, Basel (2009)
64. Gordon, R., Power, A.J., Street, R.: Coherence for tricategories. Mem. AMS **117**(558), 558 (1995)
65. Grandis, M., Paré, R.: Limits in double categories. Cah. Top. Géom. Differ. Catég. **40**(3), 162–220 (1999)
66. Grandis, M., Paré, R.: Adjoints for double categories. Cah. Top. Géom. Differ. Catég. **45**, 193–240 (2004)
67. Grothendieck, A.: Pursuing stacks. Available at http://www.math.jussieu.fr/maltsin/ps.html
68. Gurski, N.: Coherence in Three-Dimensional Category Theory. Cambridge Tracts in Mathematics. Cambridge University Press, Cambridge (2013)
69. Hackney, P., Robertson, M., Yau, D.: Infinity Properads and Infinity Wheeled Properads. Lecture Notes in Mathematics, vol. 2147. Springer, Cham (2015)
70. Heuts, G., Hinich, V., Moerdijk, I.: On the equivalence between Lurie's model and the dendroidal model for infinity-operads. Adv. Math. **302**, 869–1043 (2016)
71. Hirschowitz, A., Simpson, C.: Descente pour les n-champs. Preprint arXiv:math/9807049
72. Horel, G.: A model structure on internal categories in simplicial sets. Theory Appl. Categ. **30**(20), 704–750 (2015)
73. Illusie, L.: Complexe cotangent et déformations. II. Lecture Notes in Mathematics, vol. 283. Springer, Berlin (1972)
74. Joyal, A.: Quasi-categories and Kan complexes. J. Pure Appl. Alg. **175**(1–3), 207–222 (2002)

75. Joyal, A., Street, R.: Pullbacks equivalent to pseudo pullbacks. Cah. Top. Géom. Différ. Catég. **34**, 153–156 (1993)
76. Kapulkin, C., Lumsdaine, P.L.F.: The simplicial model of univalent foundations (after Voevodsky). Preprint arXiv:1211.2851
77. Kelly, G.: Basic Concepts of Enriched Category Theory. LMS Lecture Notes, vol. 64. London Mathematical Society, London (1982)
78. Kelly, G.M., Lack, S.: Monoidal functors generated by adjunctions, with applications to transport of structure. In: Galois Theory, Hopf Algebras, and Semiabelian Categories, pp. 319–340. Fields Institute Communications, R. American Mathematical Society, Providence (2004)
79. Kelly, G.M., Street, R.: Review of the elements of 2-categories. In: Category Seminar, Lec. Notes Math., vol. 420. Springer, Berlin (1974)
80. Kock, J.: Weak identity arrows in higher categories. IMRP Int. Math. Res. Pap., pp. 1–54 (2006)
81. Lack, S.: Codescent objects and coherence. J. Pure Appl. Alg. **175**, 223–241 (2002)
82. Lack, S., Paoli, S.: 2-nerves for bicategories. K-Theory **38**(2), 153–175 (2008)
83. Leinster, T.: A survey of definitions of n-category. Theory Appl. Categ. **10**(1), 1–70 (2002)
84. Leinster, T.: Higher Operads, Higher Categories. London Mathematical Society Lecture Note Series, vol. 298. Cambridge University Press, Cambridge (2004)
85. Loday, J.L.: Spaces with finitely many nontrivial homotopy groups. J. Pure Appl. Alg. **24**, 179–202 (1982)
86. Loday, J.L.: La renaissance des opérades. Astérisque **1994/95**, 47–74 (1996)
87. Lumsdaine, P.L.: Weak ω-categories from intensional type theory. Log. Methods Comput. Sci. **6**(3), 3–24 (2010)
88. Lurie, J.: On the classification of topological field theories. Curr. Dev. Math. **2008**(2009), 129–280 (2008)
89. Lurie, J.: Higher Topos Theory. Annals of Mathematics Studies, vol. 170. Princeton University Press, Princeton (2009)
90. Lurie, J.: Higher algebra. Available at http://www.math.harvard.edu/lurie/papers/HA.pdf
91. MacLane, S.: Categories for the working mathematician. In: Graduate Texts in Mathematics, 2nd edn. Springer, Berlin (1998)
92. MacLane, S., Whitehead, J.H.C.: On the 3-type of a complex. Proc. Natl. Acad. Sci. USA **36**, 41–48 (1950)
93. Markl, M., Shnider, S., Stasheff, J.: Operads in algebra, topology and physics. In: Mathematical Surveys and Monographs, vol. 96. American Mathematical Society, Providence (2002)
94. May, J.P.: Simplicial Objects in Algebraic Topology. Chicago Lectures in Mathematics Series. University of Chicago Press, Chicago (1967)
95. May, J.P.: The Geometry of Iterated Loop Spaces. Lecture Notes in Mathematics, vol. 271. Springer, Berlin (1972)
96. May, J.P.: H_∞ Ring Spaces and H_∞ Ring Spectra. Lecture Notes in Mathematics, vol. 577. Springer, Berlin (1977)
97. May, J.P., Thomason, R.: The uniqueness of infinite loop space machines. Topology **17**(3), 205–224 (1978)
98. Moerdijk, I., Töen, B.: Simplicial Methods for Operads and Algebraic Geometry. Advanced Courses in Mathematics. Birkhäuser/Springer Basel AG (2010)
99. Moerdijk, I., Weiss, I.: On inner Kan complexes in the category of dendroidal sets. Adv. Math. **221**(2), 343–389 (2009)
100. Paoli, S.: (Co)homology of crossed modules with coefficients in a π_1-module. Homology Homotopy Appl. **5**(1), 261–296 (2003)
101. Paoli, S.: Developments in the (co)-homology of crossed modules. Ph.D. thesis, University of Warwick (2003)
102. Paoli, S.: Weakly globular $\mathrm{cat}^{\mathrm{n}}$-groups and Tamsamani's model. Adv. Math. **222**, 621–727 (2009)

103. Paoli, S., Pronk, D.: A double categorical model of weak 2-categories. Theory Appl. Categ. **27**, 933–980 (2013)
104. Porter, T.: N-types of simplicial groups and crossed n-cubes. Topology **32**(1), 5–24 (1993)
105. Postnikov, M.M.: Determination of the homology groups of a space by means of the homotopy invariants. Doklady Akad. Nauk SSSR (N.S.) **76**, 359–362 (1951)
106. Power, A.J.: A general coherence result. J. Pure Appl. Alg. **57**(2), 165–173 (1989)
107. Pridham, J.P.: Presenting higher stacks as simplicial schemes. Adv. Math. **238**, 184–245 (2013)
108. Quillen, D.G.: Homotopical Algebra. Lecture Notes in Mathematics, vol. 43. Springer, Berlin (1967)
109. Rezk, C.: A model for the homotopy theory of homotopy theory. Trans. Amer. Math. Soc. **353**(3), 973–1007 (2001)
110. Rezk, C.: A cartesian presentation of weak n-categories. Geom. Topol. **14**(1), 521–571 (2010)
111. Riehl, E.: Categorical homotopy theory. In: New Mathematical Monographs, vol. 24. University Press, Cambridge (2014)
112. Riehl, E., Verity, D.: The 2-category theory of quasi-categories. Adv. Math. **280**, 549–642 (2015)
113. Riehl, E., Verity, D.: Homotopy coherent adjunctions and the formal theory of monads. Adv. Math. **286**, 802–888 (2016)
114. Schommer-Pries, C.: The classification of two-dimensional extended topological field theories. Ph.D. thesis, University of California, Berkeley (2009)
115. Schommer-Pries, C., Christopher, J.: Dualizability in low - dimensional higher category theory. In: Contemp. Math. Topology and field theories, pp. 111–176. American Mathematical Society, Providence (2004)
116. Segal, G.: Classifying spaces and spectral sequences. Inst. Hautes Études Sci. Publ. Math. **34**(1), 105–112 (1968)
117. Segal, G.: Categories and cohomology theories. Topology **13**, 293–312 (1974)
118. Simpson, C.: Homotopy types of strict 3-groupoids (1988). Preprint, arXiv:math/9810059V1
119. Simpson, C.: Homotopy theory of higher categories. In: New Math. Monographs, vol. 19. Cambridge University Press, Cambridge (2012)
120. Spanier, E.H.: Algebraic Topology. Springer, Berlin (1966)
121. Stasheff, J.D.: Homotopy associativity of h-spaces. I, II. Trans. Amer. Math. Soc. **108**, 275–292, ibid. 293–312 (1963)
122. Street, R.: Two constructions of lax functors. Cah. Top. Géom. Différ. Catég. **13**(1972), 217–264 (1972)
123. Street, R.: The algebra of oriented simplexes. J. Pure Appl. Alg. **59**(3), 283–335 (1987)
124. Sugawara, M.: On the homotopy-commutativity of groups and loop spaces. Mem. Coll. Sci. Univ. Kyoto Ser. A Math. **33**, 257–269 (1960/1961)
125. Tamarkin, D.: What do dg-categories form? Compos. Math. **143**(5), 1335–1358 (2007)
126. Tamsamani, Z.: Sur des notions de n-catégorie et n-groupoide non-strictes via des ensembles multi-simpliciaux. K-theory **16**, 51–99 (1999)
127. Thomason, R.W.: Homotopy colimits in the category of small categories. Math. Proc. Camb. Philos. Soc. **85**(1), 91–109 (1979)
128. Thomason, R.W.: Cat as a closed model category. Cah. Top. Géom. Différ. Catég. **21**, 305–324 (1980)
129. Toën, B.: Higher and derived stacks: a global overview. In: Algebraic geometry—Seattle 2005. Proceedings of Symposia in Pure Mathematics, Part 1, vol. 80, pp. 435–487. R. American Mathematical Society, Providence (2009)
130. Toën, B.: Derived algebraic geometry. EMS Surv. Math. Sci. **1**(2), 153–240 (2014)
131. Toën, B., Vezzosi, G.: Homotopical algebraic geometry. I. Topos theory. Adv. Math. **193**(2), 257–372 (2005)
132. Toën, B., Vezzosi, G.: Homotopical algebraic geometry. II. Geometric stacks and applications. Mem. Amer. Math. Soc. **193**(902) (2008)

133. Univalent Foundations Program, T.: Homotopy Type Theory: Univalent Foundations of Mathematics. https//homotopyrtpetheory.org/book (2013)
134. van den Berg, B., Garner, R.: Types are weak ω-groupoids. Proc. Lond. Math. Soc. 2 **102**, 370–394 (2011)
135. Verity, D.: Weak complicial sets II: nerves of complicial Gray categories. Comtemp. Math **431** (2007)
136. Verity, D.: Complicial sets characterising the simplicial nerves of strict ω-categories. Mem. Amer. Math. Soc. **193**, 184 (2008)
137. Verity, D.: Weak complicial sets I: Basic homotopy theory. Adv. Math **219**, 1081–1149 (2008)
138. Vicary, J.: Higher quantum theory. Preprint arXiv:1207.4563
139. Voevodsky, V.: An experimental library of formalized mathematics based on the univalent foundations. Math. Struct. Comput. Sci. **25**(5), 1278–1294 (2015)
140. Weizhe, Z.: Gluing pseudo functors via n-fold categories. J. Homotopy Relat. Struct. **12**, 189–271 (2017)

Index

A
Adams spectral sequence, 7
Algebraic model of n-types, 12, 298

B
Baez–Dolan conjectures, 5
Bicategory, 8
Bisimplicial object, 22

C
Cat^n-groups , 14
Cells, 4
Classifying space, 105, 302, 303
Closure properties, 27
- of $\mathsf{Cat}^{\mathsf{n}}_{\mathsf{hd}}$, 100
- of $\mathsf{Cat}^{\mathsf{n}}_{\mathsf{wg}}$, 120
- of $\mathsf{Seg}_{\mathsf{n}}$, 58
- of Ta^{n}, 118
- of $\mathsf{Ta}^{\mathsf{n}}_{\mathsf{wg}}$, 109

Cobordism hypothesis, 7
Coherence axioms, 8
Complete Segal spaces, 6
Complicial sets, 7
Connected
- components functor, 72
- n-types, 13

Crossed
- modules, 14
- n-complexes, 55
- N-cubes, 13

D
Décalage functor, 46
Deligne conjecture, 7
Dendroidal sets, 5
Diagonal, 105
Discrete
- internal category, 33
- internal n-fold category, 30
- simplicial object, 19

Discretization
- functor, 287
- map, 94

Double
- category, 37
- nerve, 37

E
Enriched category, 39, 40
Equivalence relation, 93

F
Fair n-categories, 318
Fat delta, 318
Fundamental
- groupoid, 13, 305
- groupoidal weakly globular n-fold category functor, 303
- Tamsamani n-groupoid, 302
- weakly globular double groupoid of a space, 309

S. Paoli, *Simplicial Methods for Higher Categories*, Algebra and Applications 26,
https://doi.org/10.1007/978-3-030-05674-2

- weakly globular 3-fold groupoid of a space, 311
- weakly globular n-fold groupoid, 305

G
Geometric weak equivalences, 50, 302
Globularity condition, 43, 55
Groupoid, 3
Groupoidal
- Tamsamani n-categories, 300
- weakly globular n-fold categories, 300
- weakly globular Tamsamani n-categories, 298

H
Higher
- cohomology operations, 7
- homotopy operations, 7
- operads, 317
- stacks, 5
Hochschild cohomology, 7
Hom-$(n-1)$-category, 59, 103, 115
Homotopically discrete n-fold category, 93
Homotopy
- category, 24
- coherence, 5
- hypothesis, 11, 302, 303
- type theory, 6
Hypercrossed complexes, 13
Hypercube, 155

I
Induced
- Segal map, 20
- Segal maps condition, 59, 104, 114
Inductive definition, 57
Infinity categories, 7
Informal discussions, 66
Internal
- category, 27
- equivalence relation, 95
- functor, 27
- groupoid, 27
Isofibration, 75
Isomorphism classes of objects functor, 73

K
Kan loop group functor, 14
k-invariants, 316

L
Left adjoint to the n-fold nerve, 305
Local $(n-1)$-equivalence, 128
Localization, 12
Loop space, 5

M
Models
- of $(\infty, 1)$-categories, 6
- of (∞, n)-categories, 7
- of higher categories, 8
- of ∞-operads, 5
Monoid, 3
Multi-diagonal, 105, 304
Multi-nerve functor, 29
Multi-simplicial object, 20

N
n-equivalences, 59, 103, 115
Nerve functor, 29
n-fold
- category, 27
- internal category, 29
- internal groupoid, 29
- simplicial object, 20
Non-abelian cohomology, 5
n-type, 12

O
Operads, 5
Opetopes, 8
Ordinal sum, 304
Orientation, 155

P
Postnikov
- decomposition, 12
- truncation, 307
Pseudo-functor, 78
Pseudo-natural transformations, 82
Pseudo T-algebras, 81
Pursuing Stacks, 13

Q
Quantum computing, 6
Quasicategories, 6
Quillen model structure, 5, 8, 319

R
Relative categories, 6, 296
Repletion under isomorphisms, 27
Rigidification functor, 215

S
Segal
- condition, 51
- map, 18
- maps condition, 30, 167
Segalic pseudo-functors, 166
Segal-type model, 57
Semistrictification hypothesis, 318
Semistrict n-category, 318
Set of (n, t)-hypercubes, 155
Simplicial
- category, 5, 6, 18, 316
- identities, 18
- map, 18
- object, 18
Singular functor, 304
Staircase of (n, t)-hypercubes, 159
Street–Roberts conjecture, 7
Strict
- 3-category, 44
- n-category, 7, 39, 54
- n-groupoid, 8, 13, 40
Strictification
- functor, 78
- of pseudo T-algebras, 83
String of (n, t)-hypercubes, 155
Symmetric group, 20

T
Tamsamani
- 2-categories, 118
- n-categories, 117
TQFT, 5
Transport of structure, 83, 202
Tricategory, 8
Truncatable n-fold categories, 118
Truncated higher categories, 7
Truncation functor, 58
Two-out-of three property, 284

V
Van Kampen theorem, 13

W
Weak
- globularity condition, 53, 58, 114
- n-category, 8
- ω-category, 7
- units conjecture, 318
Weakly globular
- double category, 121
- 3-fold category, 123
- n-fold category, 119
- n-fold groupoids, 304
- Tamsamani 2-category, 116
- Tamsamani n-category, 113
Whitehead products, 13

Z
Zero-types, 105

www.ingramcontent.com/pod-product-compliance
Ingram Content Group UK Ltd.
Pitfield, Milton Keynes, MK11 3LW, UK
UKHW021919270726
14059UKWH00002B/98

* 9 7 8 3 0 3 0 0 5 6 7 3 5 *